DATE DUE

FEB 19 1991			
JUL 29 1991			
APR 24 1993			
JAN 18 1994			

CURRENT ADVANCES IN

LANs, MANs & ISDN

The Artech House Telecommunications Library

Vinton G. Cerf, *Series Editor*

Digital Cellular Radio by George Calhoun

E-Mail by Stephen A. Caswell

Design and Prospects for the ISDN by G. DICENET

Introduction to Satellite Communication by Bruce R. Elbert

Communication Satellites in the Geostationary Orbit by Donald M. Jansky and Michel C. Jeruchim

Telecommunications in the U.S.: Trends and Policies, Leonard Lewin, ed.

Introduction to Telecommunication Electronics by A.Michael Noll

Introduction to Telephones and Telephone Systems by A.Michael Noll

Integrated Services Digital Networks by Anthony M. Rutkowski

Advances in Computer Systems Security: 3 volume set, Rein Turn, ed.

Techniques in Data Communications, by Ralph Glasgal

Measurement of Optical Fibers and Devices by G. Cancellieri and U. Ravaioli

Codes for Error Control and Synchronization by Djimitri Wiggert

Telecommunication Systems by Pierre-Girard Fontolliet

Optical Fiber Transmission Systems by Siegfried Geckeler

Traffic Flow in Switching Systems by G. Hebuterne

Optimization of Digital Transmission Systems by K. Trondle and G. Soder

Quality Measures and the Design of Telecommunications Systems by John H. Fennick

Troposcatter Radio Links by G. Roda

Radiodeterminatin Satellite Services and Standards by Martin Rothblatt

The Manager's Guide to the New Telecommunications Network by Lawrence D. Gasman

Innovations in Internetworking, Craig Partridge, ed.

Principles of Signals and Systems: Deterministic Signals by B. Picinbono

Private Telecommunication Networks by Bruce Elbert

Television Technology: Fundementals and Future Prospects by A. Michael Noll

Minimum Risk Strategy for Aquiring Communications Equipment and Services by Nathan J. Muller

Manager's Guide to CENTREX by John R. Abrahams

Voice Teletraffic System Enigneering by James R. Boucher

Advances in Fiber Optics Communications, Henry F. Taylor, ed.

Broadband LAN Technology by Gary Y. Kim

CURRENT ADVANCES IN

LANs, MANs & ISDN

B. G. KIM

ARTECH HOUSE

Library of Congress Cataloging-in-Publication Data

Current advances in LANs, MANs, and ISDN / B.G. Kim, editor
p. cm.
Includes bibliographical references.
ISBN 0-89006-385-0
1. Local area networks (Computer networks) 2. Fiber optics.
3. Digital communications. I. Kim, B.G., 1953-
TK5105.7.C85 1989 89-17502
004.6--dc20 CIP

ARTECH HOUSE, INC.
685 Canton Street
Norwood, MA 02062

International Standard Book Number: 0-89006-385-0
Library of Congress Catalog Card Number: 89-17502

10 9 8 7 6 5 4 3 2 1

Contents

Preface

The past two decades have witnessed significant advances in various fields of telecommunication. The packet switching technique has been accepted as an efficient means of transporting bursty data traffic. Local area network (LAN) technology has matured. Digital switching techniques are widely used. Numerous field trials of ISDN (integrated services digital network) are reported.

Although we may appreciate these efforts, we are faced with the task of charting the future telecommunication course that will lead us into the next century. On many fronts, we can already observe various activities that will set the standards for the future: ANSI FDDI (fiber distributed data interface) for high-speed LANs; IEEE 802.6 MAN (metropolitan area network); and broadband ISDN for integrated voice-data-video services.

Underlying these various new developments is fiber optic technology as an indispensible centerpiece. This book seeks to unite these developments under a common theme of integrated services in fiber optic networks. In doing so, we attempt not only to provide an overview of recent developments, but also to capture alternative viewpoints.

Optical fiber is an attractive transmission medium because of its noise immunity and wide bandwidth. Based on continuing improvements of fiber optic technology, we focus on applications of the technology in high-speed local, metropolitan, and wide area networks. Of particular interest are those applications capable of providing integrated services of diverse types of traffic. In fact, recent standard-making activities of FDDI, MAN, and broadband ISDN all address the issue of integrated services.

To provide a coherent view of service integration, this book is organized into four chapters.

1. *Lightwave Systems and Voice-Data Integration.* This chapter provides an overview of current fiber optic technology and the evolution of integrated voice-data services. Current standards of ISDN are also described.

2. *Local Area Networks (LANs).* LANs have become a very visible component of data communication equipment. Still, performance of the Ethernet is misunderstood by many people. The performance of existing LANs is reevaluated with experimental measurements. The FDDI protocol and other high-speed LANs are surveyed.

3. *Metropolitan Area Networks (MANs).* A MAN is a new project that attempts to provide fiber optic connectivity to individual homes and offices in a metropolitan area. The recent IEEE 802.6 MAN standard is described. Several alternative approaches to MANs are also explored.

4. *Broadband ISDN.* Broadband ISDN represents an exciting new opportunity by which video, voice, and data are integrated. The presentation is not limited to an overview of current efforts in broadband ISDN, but rather is extended to alternative proposals.

Articles included in this book do not represent an exhaustive list of the literature in relevant topics. The articles are carefully selected to provide a forum of diverse viewpoints. For further details of topics, the reader is advised to consult references cited in the articles as well as additional literature listed in the bibliography.

This book is intended for a broad range of professionals involved in telecommunication.

- *Students and Educators:* This book provides an overview of current issues in telecommunication. So that the reader may be exposed to different perspectives on the issues, alternative viewpoints are included where appropriate.
- *Network Practitioners:* Chapter 2 includes articles that discuss practical issues of LANs and their measured performance. Discrepancies between theoretical predictions and experimental measurements may provide an insight into realistic expectations of future networks.
- *Network Designers:* In addition to keeping the reader up to date on current developments of industrial standards, the book provides a convenient means of exploring diverse viewpoints.
- *Network Managers:* Chapters 3 and 4, in particular, alert the reader to developments of new services and trends in future networks.

Chapter 1
Lightwave Systems and Service Integration

To provide a background for service integration in fiber optic systems, this chapter reviews three topics. The current status of fiber optic research and installations is summarized in the first section. The second section examines various ways of sharing bandwidth between voice and data traffic. Architecture and control structures of the standard ISDN are reviewed in the third section.

1.1 FIBER OPTIC SYSTEMS

The key parameter for measuring the capacity of fiber optic systems is the bandwidth-distance product. This product represents the amount of data which can be supported without repeaters. Current commercial systems achieve 0.4 Gb/s bandwidth with 48 km repeaterless span. The bandwidth-distance product increases annually by a factor of two. If the trend continues, commercial systems with a product of 100 (1 Gb/s in 100 km span or 10 Gb/s in 10 km span) will be feasible by 1990. Laboratory results exceeding a product of 1000 have already been reported.

Improvements in fiber capacity are primarily due to advances in optical sources and fiber technology such as single-mode lasers and single-mode fibers. Improvements in optical receiver sensitivity, however, lag. The next generation of fiber optic systems attempts to enhance receiver sensitivity by developing coherent optical communication systems. Current research efforts in fiber and other components are summarized in the first article by Li. For a more comprehensive treatment of fundamentals of fiber optic technology, the reader is referred to [Taylor, 1988]. The next article by Cohen provides an overview of fiber installation spans: undersea and terrestrial long-haul systems and local loop and distribution systems.

The article by Acampora *et al.* proposed an overall fiber optic system capable of utilizing the full capacity of fibers. The article matches the level of abstraction in which we are interested and proposes a multiplexing scheme whereby channel capacities of hundreds or thousands of Gb/s can be made available. Although not included in this book, optical (photonic) switching techniques are being vigorously researched. Optical switching is intended to eliminate costly conversion between optical signals and electronics and to process optical signals directly [Midwinter and Smith, 1988].

1.2 VOICE-DATA INTEGRATION

Voice traffic is characterized by its real-time delivery requirement (less than 100 ms), whereas data traffic is more concerned about error-free deliveries. When the two types of traffic are combined, voice is typically treated on a priority basis to meet the stringent real-time requirement. Most integration techniques are based on a fixed-length frame structure. They are classified into two frame formats: short-frame and long-frame. In each frame format, several bandwidth allocation strategies exist.

1. *Short-Frame Format:* Characteristics of a frame are inherited from those of the T1 line (DS1 format). A frame lasts for 125 μs and is divided into 24 slots. A user is assigned to a slot and transmits a byte of data in it.
   - *TDM* (*time division multiplexing*). TDM is a widely used technique by which a slot is allocated for the duration of a voice or data call.
   - *Burst* (*Spurt*) *Switching.* A voice call in TDM lasts for several minutes. As is well known, a voice call alternates between talkspurts and pauses (silences) and a voice call utilizes less than 50% of the allocated capacity. Burst switching takes advantage of pauses in a voice call and allocates a slot only during a talkspurt.
2. *Long-Frame Format:* A long frame typically lasts for 10 ms or longer, and is partitioned into voice and data regions. Each region is dedicated to the corresponding traffic type and is divided into fixed-size slots. Depending on the application, sizes of voice and data slots may be different. The size of a slot is sufficiently large to transmit a group of voice samples collected over the frame period or to transmit an entire data packet. Several policies exist which determine the boundary between the regions of contiguous voice and data slots.
   - *Fixed Boundary* (*FB*) *Policy.* The boundary between voice and data regions is predetermined, resulting in fixed bandwidth allocation.
   - *Movable Boundary* (*MB*) *Policy.* A minimal bandwidth is allocated to data traffic. However, data traffic can also employ any unused portion of voice region.
   - *Proportional Boundary* (*PB*) *Policy* [Konheim and Pickholtz, 1985]. The boundary is determined by the ratio of prevailing voice and data loads at the beginning of a frame. The policy allocates more slots to traffic with a heavier load. A variation exists which limits the number of slots allocated to one traffic type at a time.
   - *Variable Frame* (*VF*) *Structure.* The preceding policies are applied to transmission formats which have fixed frame lengths. VF structures, on the other hand, allow for statistical variations of frame lengths.

Traditionally, voice is circuit-switched and data is packet-switched. Two articles are included in this topic. Chen and Messerschmitt's article reviews integrated switching techniques in the context of the two frame formats outlined above. The hybrid switching technique allocates a portion of a frame to circuit-switched voice and the remainder to packet-switched data. Several long-frame formats of the hybrid switching technique are described in the article by Ross and Mowafi.

1.3 ISDN

Because computers and communications penetrate each other's fields, the CCITT (Consultative Committee on International Telephone and Telegraph) adopted standards for interfaces providing digital and integrated services. The ISDN has the potential of being the first step toward a true information society. By bringing digital lines into individual homes and offices, conventional telephone services are digitized and access to remote databases becomes readily available. Current field trials are limited to islands of regional telephone offices in the US, particular European countries, and Japan. When individual ISDN islands are fully connected, users will experience true digital interconnectivity across the US and the continents.

In addition to digital connectivity, the ISDN is specifically designed to accommodate voice and data. The ISDN standard consists of an interface providing two 64 kb/s B channels and a 16 kb/s D channel (so-called $2B + D$ channels).

Control (signaling) information and short data packets are sent over the *D* channel, whereas *B* channels are connected in either a circuit-switched or packet-switched manner for voice, data, or both.

Simultaneous voice and data access is expected to stimulate a variety of applications. Some, such as videotex, which is successfully deployed in European countries, may also be offered in the US. The overall architecture of the ISDN is described in Malek's article, which presents a collection of packet- and circuit-switched subnetworks within the realm of an ISDN network. Roehr's article examines the details of a control structure of the ISDN, namely SS7 (Signaling System Number 7).

ADVANCES IN LIGHTWAVE SYSTEMS RESEARCH

AT&T TECHNICAL JOURNAL

Tingye Li

***Tingye Li** is head of the Lightwave Systems Research Department at AT&T Bell Laboratories in Holmdel, New Jersey. He is responsible for research on various lightwave technologies and systems for long-distance and local-area communications applications. Mr. Li joined the company in 1957 and has a B.Sc. from the University of Witwatersrand, and an M.S. and Ph.D. from Northwestern University, all in electrical engineering.*

Twenty years have passed since the first serious proposal was made to use a glass-fiber waveguide as a telecommunications transmission medium. Since then, vigorous research and development efforts have led to widespread applications of optical-fiber communication throughout the telecommunications industry. Today, lightwave systems research is exploring the vast potential bandwidth of low-loss single-mode fibers for long-haul and local-network applications. Laboratory systems with experimental single-frequency lasers and low-noise avalanche photodiodes have attained multi-gigabit-per-second transmission in single-mode fibers that are well over 100 km long. Coherent technology promises high receiver sensitivity and offers an efficient means for channel selection in a densely packed wavelength-multiplexed system. The first coherent-systems experiments have confirmed theoretical expectations but many challenges lie ahead. Other novel systems for local distribution and networking are also under study. This paper reviews recent research progress in these areas.

Perspective

Historically, telecommunications research has always been motivated to explore new technologies for expanding transmission capacity. Thus, the realization of the first laser in 1960 stirred intense interest in optical communications. The laser was envisaged as the requisite coherent source of an optical carrier on which information could be impressed at an unparalleled rate.

Research began on a broad front. But not until ten years later was a feasible, low-loss, transmission medium demonstrated. It was a doped high-silica optical fiber with a loss of 20 dB/km. Coincidentally, continuous-wave operation at room temperature was achieved in the AlGaAs (aluminum-gallium-arsenide) semiconductor laser. The attainment of these two basic elements of an optical communications system was soon followed by a profusion of research advances that, in less than a decade, made optical-fiber communications a practical reality.[1]

Deployed Fiber Systems. The first-generation fiber systems—introduced commercially in the late 1970s—operated at wavelengths of about 0.82 μm. They used AlGaAs injection lasers and multimode fibers of 3- to 4-dB/km transmission loss. Interoffice trunking was the main arena of application.

As these systems were being tested in the field just before their commercial introduction, two research advances were announced that affected the course of lightwave technology.

- The transmission loss of fiber was lowered dramatically to 0.5 dB/km in wavelength regions around 1.3 and 1.5 μm.
- Continuous-wave, room-temperature operation was achieved in an InGaAsP (indium-gallium-arsenide-phosphide) laser that could emit in the 1.0- to 1.7-μm wavelength range.

The wavelength region near 1.3 μm is especially significant. At this wavelength, the pulse spreading caused by material and waveguide effects (chromatic dispersion) in the fiber is minimal,[2,3] making it possible to transmit over long distances at exceedingly high data rates. Also, because chromatic dispersion is proportional to source spectral width, opening the 1.3-μm transmission window allowed light-emitting diodes (LEDs) to be used in systems with moderate speed and distance requirements. For some applications (such as data links), the simpler LEDs are favored over lasers because they afford superior reliability, lower cost, and much less sensitivity to temperature variations.

Hence, the second-generation of lightwave systems operate at a wavelength of 1.3 μm, using either LEDs or lasers as sources and multimode fibers as the transmission medium. For subscriber-loop applications, both LED- and laser-based systems operate at bit rates of up to 200 Mb/s. Laser-based systems, which can operate at higher speeds over longer distances, are deployed in metropolitan areas for interoffice trunking and in the long-haul plant.

Second-generation fiber systems have now been superseded by 1.3-μm, laser-based, single-mode systems that are designed for intercity and intercontinental service. These third-generation systems operate at several hundred megabits per second with repeater spans that exceed 40 km, and can be upgraded to speeds beyond 1 Gb/s (which is being planned). Many of these third-generation systems are now in commercial operation.

Fiber Systems Research. Current research in lightwave systems is exploring the vast potential bandwidth of low-loss, single-mode fibers. A minimum loss of 0.16 dB/km at $\lambda = 1.57$ μm (which is close to the theoretical limit of Rayleigh scattering in silica) has been demonstrated in the laboratory. Typical high-quality production fibers exhibit minimum losses close to 0.2 dB/km.[2,3]

For the 1.45- to 1.65-μm wavelength region of minimum loss, the potential bandwidth is 25,000 GHz. Imaginative and effective use of even a minute portion of this immense bandwidth can radically affect the telecommunications industry. The challenge to the research community is, therefore, intriguing and complex.

Meanwhile, researchers are trying to approach the limits that fiber loss and dispersion impose on transmission speed and distance. They have done systems experiments at data rates well beyond 1 Gb/s over single-mode fibers that exceed 100 km in length. Experiments that involved direct (intensity) modulation and detection required single-frequency lasers that have low frequency deviations (chirp) during modulation and high-sensitivity, broadband avalanche photodiodes (APDs). Such experiments have obvious implications for near-future long-haul systems.

Coherent lightwave transmission is being investigated for possible application in both long-haul trunking and local networks because:

- It offers higher receiver sensitivity.
- Simple optical amplifiers could be used for boosting the signal.
- It provides an efficient way to select a particular channel from many closely spaced wavelength-multiplexed channels.

Coherent transmission requires narrow-linewidth single-frequency lasers. Initial successes are encouraging, but many challenges lie ahead. Other novel systems for local distribution and networking are also under study.

This paper reviews recent progress in lightwave systems research, emphasizing high-speed transmission and coherent technology. It discusses the limits that single-mode fiber imposes, as well as the requirements of the relevant devices. Applications in the long-haul and local arenas are considered.

Transmission Bandwidth

The maximum rate of data transmission in a single-mode fiber is determined by its chromatic dispersion, which depends on the fiber geometry, refractive-index profile, and material properties.[2,3]

In a conventional single-mode fiber with a step refractive-index profile, the geometry and material effects combine to produce a dispersion minimum at a wavelength λ_0 slightly above 1.3 μm. Thus, the single-channel transmission bandwidth of a fiber that operates at the wavelength λ_0 can be very large, especially if the source has a narrow spectral width.

This is illustrated in Figure 1, which gives the maximum single-channel transmission bandwidth—or data rate—of a conventional single-mode fiber, 100 km long, as a function of wavelength for sources of various spectral widths, $\Delta\lambda$. However, a single-mode fiber with a refractive-index profile that is more complicated than a simple step can have maximum transmission bandwidths that are very different from those shown.[3]

The calculations[4] that Figure 1 represents take into account higher-order waveguide and material effects, as well as performance degradation from intersymbol interference. The case labeled $\Delta\lambda = 0$ is akin to radio transmission in that the spectral width of the optical carrier is much narrower than the signal bandwidth, or bit rate, B. When this condition prevails, the maximum bit rate B_{max} for transmission over a distance L satisfies the relation $(B_{max})^3 L$ = *constant*, for wavelengths near the dispersion minimum λ_0. Away from λ_0, the relation is

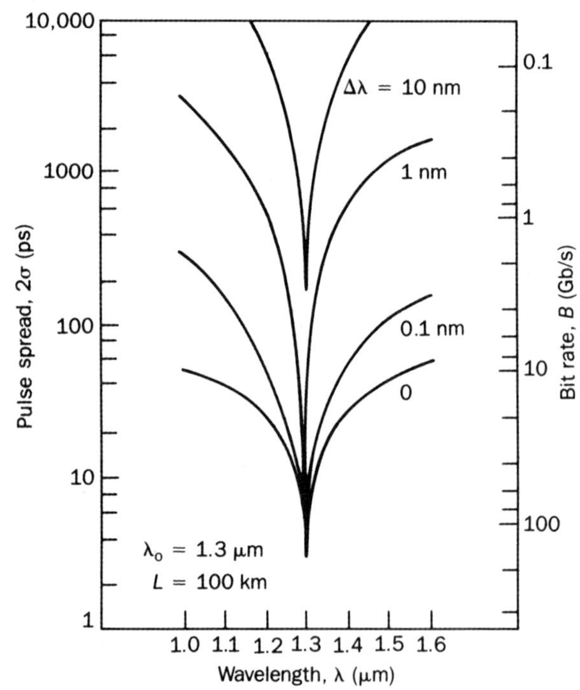

Figure 1. Theoretical maximum transmission bandwidth of a conventional single-mode fiber, 100 km long. The four curves represent sources of different spectral widths $\Delta\lambda$. The wavelength of minimum dispersion λ_0 is 1.3 μm.

$(B_{max})^2 L = constant$.

When the spectral width of the optical carrier is much greater than the bandwidth of the signal, the maximum bit rate is related to the distance by $(B_{max})^2 L = constant$ near λ_0, and $B_{max} L = constant$ away from λ_0. Using these relations and Figure 1, one can scale maximum transmission distances for various data rates.

Figure 1 clearly shows that a source with wide spectral width would severely limit the transmission bandwidth of a conventional single-mode fiber operating at 1.55 μm where fiber attenuation is minimal. The chromatic dispersion there is about 18 ps/km·nm. Therefore, to maximize repeater spacing, one must use lasers with narrow spectral widths or shift the wavelength of minimum dispersion to 1.55 μm. The latter has been achieved but not without the penalty of some increased minimum loss.[2,3]

Single-Frequency Lasers

Conventional InGaAsP semiconductor lasers that operate at 1.3 or 1.5 μm tend to emit in several longitudinal modes (or lines) spread over a spectral width of several nanometers. These multifrequency lasers are not suitable as sources for high-speed, long-distance transmission (Figure 1).

Furthermore, the power in each mode tends to fluctuate, especially under direct modulation, even though the total power remains constant from pulse to pulse. This power fluctuation among modes is converted through chromatic dispersion of the fiber into pulse-shape variations at the receiver, a phenomenon known as *mode partition noise*. Partition noise can impair system performance even for moderate transmission distances.[5] Hence, sources for high-speed systems must operate in a single longitudinal mode.

Single-frequency operation can be obtained in a laser by using a coupled-cavity or distributed-feedback structure. The cleaved-coupled-cavity (C^3) laser, used in early long-distance transmission experiments at λ = 1.5 μm, has two cavity sections that are coupled optically but isolated electrically.[6] Single-frequency operation is obtained by suitably adjusting the drive current through each section. The distributed-feedback (DFB) laser uses a grating to provide a distributed, but frequency-selective, feedback mechanism that confines emission to a single mode.[7] Single-frequency operation is more robust in DFB lasers.

Under direct modulation, the instantaneous frequency of the semiconductor laser will vary with drive current, resulting in pulse spreading at the receiver.[8] This phenomenon, known as *chirp*, arises because the refractive index of the laser depends on carrier density.[9] As the refractive index varies, so does the wavelength (or frequency) of emission.

If one uses a structure to confine the optical field more tightly, nonlinear processes in the laser—when high optical field densities are present—tend to damp the large amplitude and frequency fluctuations, and leave a smaller, steady frequency shift. (A steady frequency shift with negligible transition time will produce negligible pulse spreading.) A low-chirp InGaAsP DFB laser that was amenable to high-speed direct modulation was built[10] and used in long-distance transmission experiments at speeds up to 4 Gb/s.

APDs and High-Sensitivity Receivers

Simple, InGaAs PIN photodiodes have excellent frequency response and perform well in 1.5-μm receivers that operate at multigigabit-per-second data rates. (PIN stands for positive intrinsic negative; some authors may use p-i-n instead.) However, they lack the internal gain needed to enhance receiver sensitivity for high-speed transmission over long distances. Early attempts to make InGaAsP APDs were not successful because of the high dark current from tunneling and the degradation of frequency response from charge accumulation at heterojunction interfaces. Germanium APDs with low dark current and high gain have been developed and are now available commercially, but they contribute more excess noise than InGaAsP devices.

To solve the problems of tunneling dark current

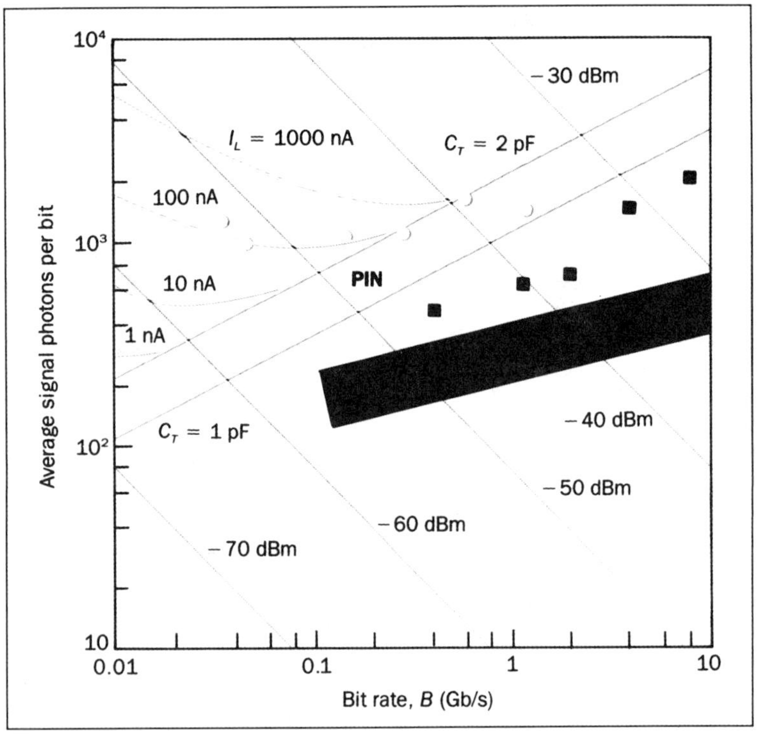

Figure 2. Receiver sensitivities of state-of-the-art optical repeaters, plotted as a function of bit rate. Sensitivities are expressed as the average number of signal photons (per bit) required to achieve an error probability of 10^{-9}. The blue curves are calculated for a PIN photodiode receiver with a GaAs FET preamplifier, while the red band gives the expected performance for an APD receiver. The blue dots and red squares represent the best known-experimental results for PIN and APD receivers, respectively.

and interface charge accumulation, a structure that consists of several layers of InGaAsP in varying compositions was tried and proved successful.[11] Its use of a narrow-bandgap InGaAs layer for light absorption and a wide-bandgap InP (indium-phosphide) layer for carrier multiplication overcame the problem of dark current. A grading layer of InGaAsP added at the heterojunction reduced charge accumulation. The resulting device—known as the separate absorption, grading, multiplication (SAGM) avalanche photodiode—has worked well in experimental high-speed systems. With its gain-bandwidth product as high as 60 GHz, the SAGM APD is the best performing photodetector for high-speed operation at 1.5 μm.

Wideband, low-noise preamplifiers are as crucial as high-performance photodetectors for attaining high sensitivity in optical receivers. (A useful definition of receiver sensitivity in a digital repeater is the minimum input power required to achieve an error probability of 10^{-9}.) The sensitivity of an optical receiver is limited by the quantum (or shot) noise associated with the primary photogenerated signal current in the detector. In practice, excess noise from the photodetector, thermal noise from the input circuit, device noise from the transistor preamplifier, and shot noise from leakage currents all contribute in varying degrees to prevent receiver performance from reaching the quantum-noise limit.

Figure 2 summarizes the performance of experimental optical receivers designed to operate in the spectral

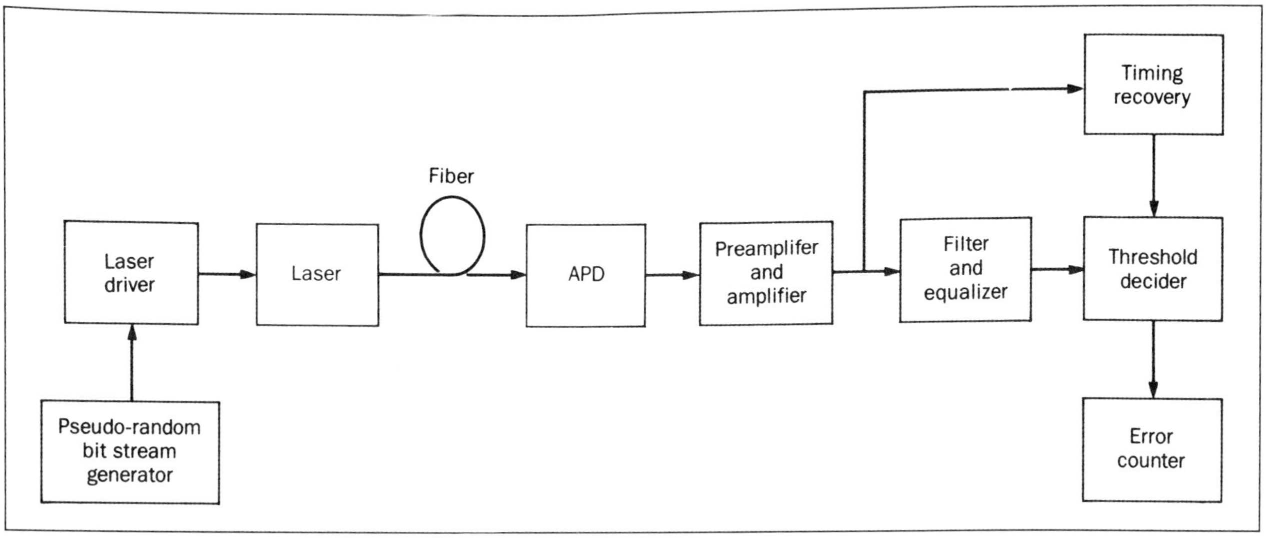

Figure 3. A typical lightwave system experiment.

region of 1.3 to 1.5 μm. The ordinate is the average (minimum) number of photons required to achieve an error probability of 10^{-9} in a time interval equal to the reciprocal of the bit rate, assuming an equal distribution of ones and zeros. This number is proportional to the average energy per bit. The abscissa is the bit rate.

The blue curves are the results of calculations based on parameters of currently available PIN photoreceivers that use GaAs (gallium-arsenide) field-effect transistors (FETs) as preamplifiers. The red shaded band represents the expected performance for APD receivers. The green diagonal grid lines convert the energy per bit into power in decibels referenced to 1 mW (dBm), assuming a 1.5-μm operating wavelength. The horizontal line at the bottom represents the ultimate sensitivity of ten photons per bit, as limited by the fundamental quantum noise. Noises that arise from thermal effects in circuits and devices cause degradations from this limit.

The straight blue lines whose slopes are proportional to the square root of the bit rate represent the theoretical performance for a receiver that consists of an ideal PIN photodiode followed by a GaAs FET preamplifier with a 50-mA/V transconductance, a total capacitance C_T of either 1 or 2 pF, and no leakage current. The blue curves that slope upward to the left represent performance when various leakage currents I_L are present. As can be seen, leakage currents are not important at high bit rates where the determining factor is the FET's channel conductance noise, which is proportional to the square of the total capacitance divided by the transconductance. The blue dots represent the best experimental values for PIN photoreceivers reported to date in the literature, while the red squares represent values for APD receivers.[12]

As Figure 2 shows, the best experimental APD receivers are about 5 dB more sensitive than the best experimental PIN receivers. This sensitivity difference will increase to about 7 dB for production receivers because ultra-high-performance PIN photoreceivers are more difficult to manufacture. Further improvements in device performance may soon enhance the best sensitivities of APD receivers to a level about 15 dB above the quantum-noise limit. Additional improvements will likely require device breakthroughs that will provide considerably less excess noise from the APD.

High-Speed Systems Experiments

Systems experiments are conducted to demonstrate the feasibility of transmission and explore the limits of technology regarding how individual components behave and affect the performance of an entire system. Figure 3 illustrates the basic elements of a system experiment.

To evaluate the behavior of each component in a system and assess the system's performance, one measures the bit error rate (BER) as a function of the received power for various operating conditions. Eye diagrams, which consist of superpositions of traces of the signal in a time slot (of one bit), are used as a diagnostic tool for analyzing component or subsystem behavior.

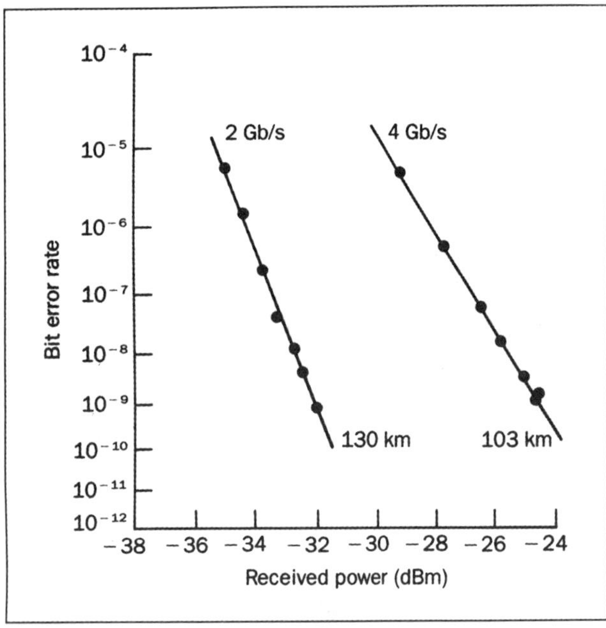

Figure 4. Measured error probability as a function of received power for two high-speed, long-distance system experiments.[13,14] For both, the lasers were biased above threshold to minimize chirp and, therefore, maximize transmission distance.

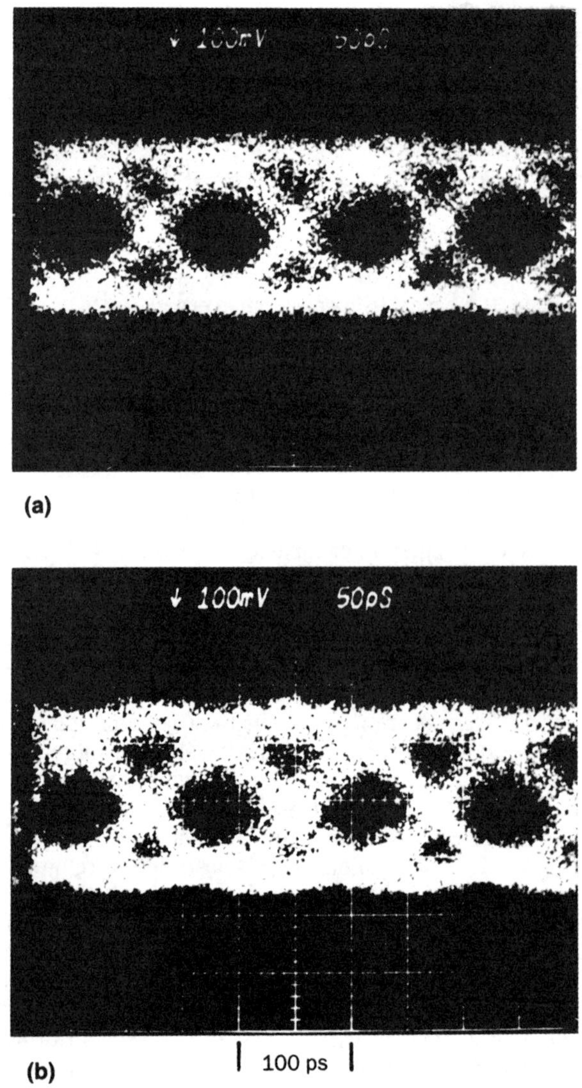

Figure 5. Eye diagrams of (a) transmitted and (b) received signals observed in the 8-Gb/s system experiment.[16] The *open* eye of the received signal after traversing 68 km showed that a bit-error rate of 10^{-9} was obtained.

Figure 4 shows measured BER data taken in two high-speed experiments that operated at 2 and 4 Gb/s and achieved transmission distances of 130 and 103 km, respectively.[13,14] The experimental low-chirp single-frequency DFB laser, high-speed SAGM APD receiver, and very-low-loss single-mode fiber with a core of nearly pure silica (0.21 dB/km) all contributed to the successes of the record-setting experiments.

To avoid laser chirp, one could let the laser operate continuously and gate the optical carrier with an external modulator. Indeed, further record-setting high-speed transmission experiments at 4 and 8 Gb/s attained transmission distances of 117 and 68 km, respectively.[15,16]

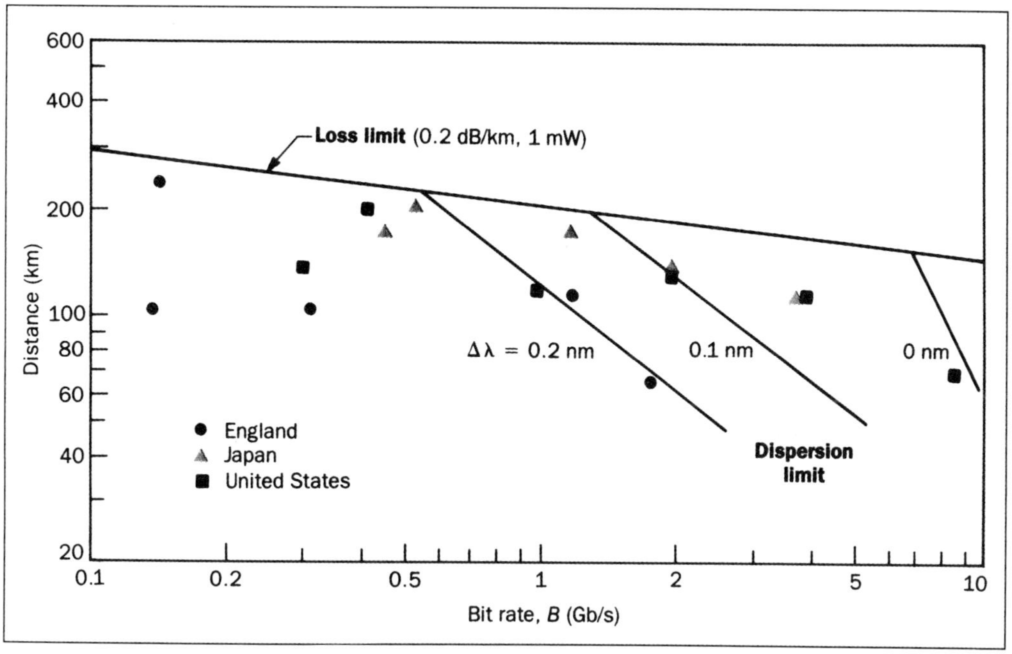

Figure 6. Results of long-distance transmission experiments conducted in laboratories in the United States, Japan, and England in the last few years. The solid lines represent theoretical limits imposed by fiber loss and dispersion, for a system that uses conventional single-mode fiber and state-of-the-art devices, and for various laser spectral widths ($\Delta\lambda$).

High-power C^3 lasers and titanium-diffused lithium-niobate waveguide modulators were used in both experiments. The 8-Gb/s experiment is especially significant because it was the first demonstration that the information bandwidth of the signal was a significant factor in limiting the transmission distance. (See the curve for $\Delta\lambda = 0$ in Figure 1.) Eye diagrams of the 8-Gb/s signal are shown in Figure 5.

High-bit-rate transmission experiments over long distances have been conducted in various laboratories in the last few years. Figure 6 plots the achieved transmission distances versus bit rate. The solid lines represent theoretical limits from fiber loss and dispersion, assuming a conventional single-mode fiber with 0.2-dB loss, 1-mW launched laser power, experimental APD receivers, and laser spectral widths of $\Delta\lambda = 0.2$, 0.1, and 0 nm. Interestingly, even though experimental conditions were diverse, the best observed results were close to the theoretical limits. Clearly, long-distance (>100 km) transmission at ultra-high (>10 Gb/s) data rates will require *dispersion-shifted* fibers, with λ_0 in the 1.5-μm region of minimum loss.

Of course, the transmission capacity of a lightwave transmission system can always be increased by using wavelength-division-multiplexing (WDM). The ability of a WDM system to grow gracefully with traffic demand makes it attractive where traffic growth is unpredictable. The disadvantage is that each multiplexed channel requires its own repeater, resulting in multiplied costs. However, the cost increase may be alleviated somewhat by possible large-scale integration of devices and circuits in a multiple-repeater package.

In a laboratory WDM experiment, light from ten carefully chosen DFB lasers were multiplexed onto a single-mode fiber to simulate a system with a total capacity

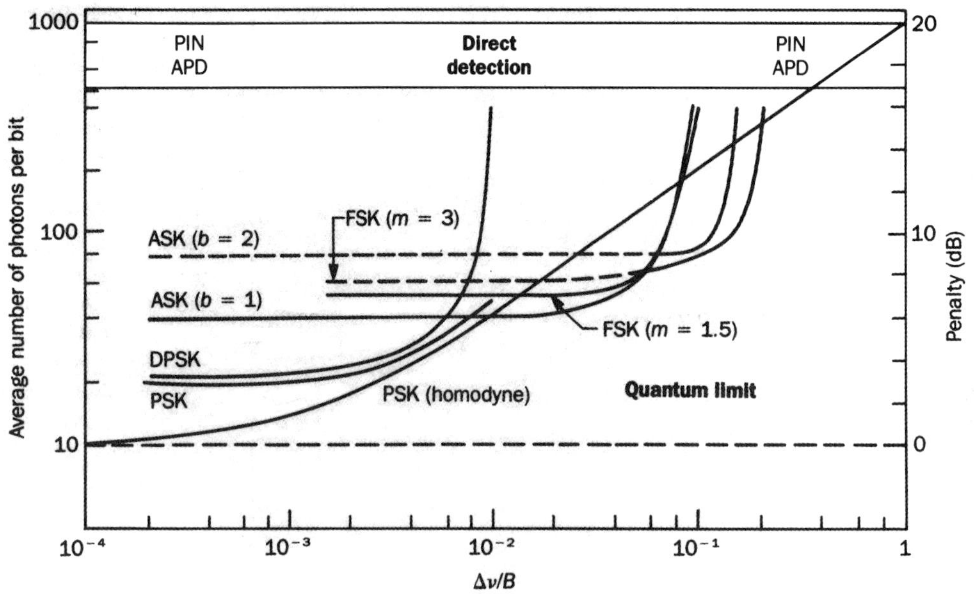

Figure 7. Receiver sensitivities for coherent modulation and detection as functions of the combined transmitter and local-oscillator laser linewidth ($\Delta\nu$) normalized to the bit rate *B*. Receiver sensitivity is expressed as the average number of photons (per bit) required to achieve an error probability of 10^{-9}, or as a power penalty relative to the quantum limit of direct detection. Except for the curve labeled PSK (homodyne), all others are calculated for heterodyne detection. The shaded band shows performance of experimental receivers that use direct detection.

of 20 Gb/s.[17] The grating-type wavelength multiplexer and demultiplexer had a channel spacing of 1.35 nm (17 GHz) centered around 1.545 μm. A transmission distance of 68.5 km was demonstrated.

Coherent Transmission

In coherent lightwave transmission, modulation is applied as amplitude, frequency, or phase variations in the optical carrier. The received signal and the output from a local-oscillator (LO) laser are mixed in the photodetector of a heterodyne receiver to generate an intermediate-frequency (IF) signal that typically falls in the microwave-frequency range. Ordinary microwave techniques can then be used to demodulate the IF signal.

If the LO power is large enough, quantum (or shot) noise in the detector will predominate over thermal and device noise in the preamplifier, thus affording high receiver sensitivity. The following motivates interest in coherent modulation and detection:

- The promise for increased receiver sensitivity
- The ability to select among densely packed WDM channels
- The potential for using wideband optical amplifiers to boost *en masse* a large group of WDM signals.

The theoretical sensitivity of a coherent-detection receiver can approach the *quantum limit* of direct detection of ten photons per bit. For example, the sensitivity limit of an ideal homodyne receiver (in which the intermediate frequency is zero) for phase-shift-keyed (PSK) signals is nine photons per bit. If heterodyne detection is used, the ideal sensitivities are 18 photons per bit for PSK

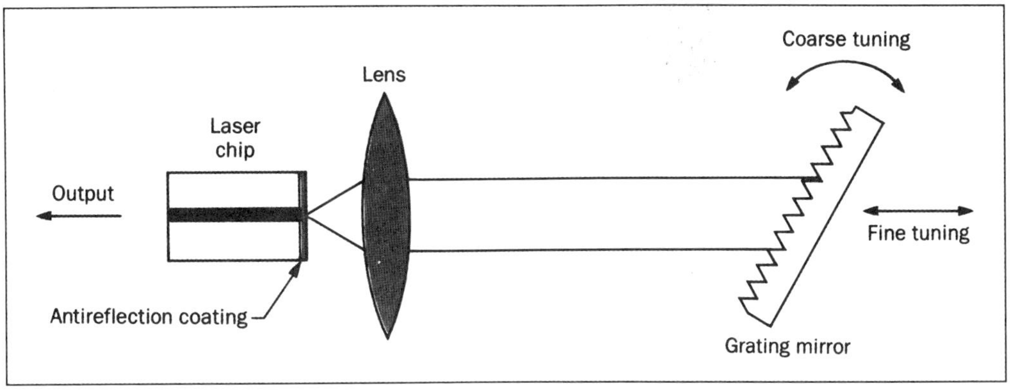

Figure 8. An extended cavity laser used in laboratory system experiments. One facet of the semiconductor laser chip is antireflection coated to extend its cavity into the long, passive region terminated by the grating mirror. Linewidths as narrow as a few kilohertz have been obtained in such lasers.

modulation and 36 photons per bit for both amplitude-shift-keyed (ASK) and frequency-shift-keyed (FSK) modulation.

Many problems must be solved before these ideal sensitivities can be approached in practice. Foremost among them is laser spectral purity. The linewidths of single-frequency semiconductor lasers are very large (compared to microwave sources, for example), typically from several megahertz to a few hundred megahertz. These large spectral widths are caused by phase noise that arises from refractive-index fluctuations (caused by carrier-concentration fluctuations), and from the high level of spontaneous emission in the laser cavity.[18] Phase noise can dramatically degrade receiver performance. The degradation can become so large that a 10^{-9} BER is impossible to attain, no matter how high the received power.

Figure 7 illustrates receiver performance for a variety of modulation and detection conditions. The ordinate is the average (minimum) number of photons per bit required to achieve 10^{-9} error probability. The abscissa is the combined linewidth of the transmitter and LO laser, normalized to the bit rate B. The quantum limit and experimental direct-detection sensitivities are shown as a dashed horizontal line and shaded region, respectively.

Except for the curve labeled PSK (homodyne), all other curves are calculated for heterodyne detection.[19-23] The ASK curves represent two values of b, which is the IF bandwidth normalized to the bit rate B. Those marked FSK are for two values of m, which is the frequency deviation normalized to the bit rate B.

It is possible to narrow the linewidth of a conventional or DFB laser by extending the length of its cavity. The linewidth reduction factor is approximately equal to the square of the ratio of the total cavity length to the active cavity length.[18]

Figure 8 is a schematic of an extended-cavity laser in which a grating mirror is used to achieve single-frequency operation. Coarse tuning of the wavelength of operation is obtained by tilting the grating, while fine tuning or frequency tracking is done by changing the cavity length or the laser drive current. Linewidths around 10 kHz are commonly observed in such extended-cavity lasers designed for laboratory experiments.[24] *Practical* narrow-linewidth *tunable* lasers that are suitable for field applications are not yet available.

Besides laser spectral purity, tunability, and stability, researchers also have been studying other important

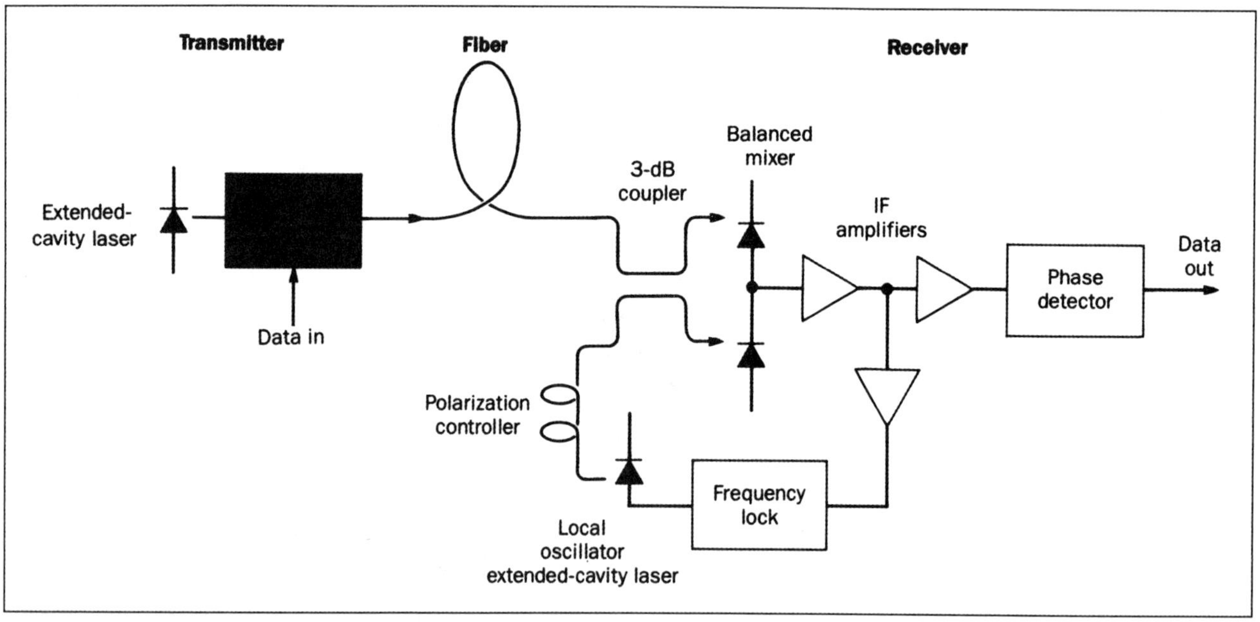

Figure 9. This coherent-transmission system experiment used differential-phase-shift-keyed modulation and heterodyne detection at bit rates of 400 Mb/s and 1 Gb/s.

issues specific to coherent transmission, including:

- Direct frequency modulation of injection lasers
- LO laser power and intensity noise
- Optical phase-locked loops for homodyne detection
- Nonlinear effects in the fiber that impair system performance
- Polarization properties of fibers and components.

Reflected power that is fed back into the laser cavity can cause linewidth variations and must be minimized using isolators.[25]

The issue of polarization is especially important because efficient coherent detection requires that the polarization state of the received signal be matched to that of the LO signal at the photodetector. Because the output polarization of conventional single-mode fibers tends to fluctuate randomly, one must find a way to maintain polarization in the fiber or compensate at the receiver for changes in the polarization. Experimental observations suggest that polarization variations are slow enough to make tracking and compensation possible.

The first coherent-lightwave-transmission experiments, which used *semiconductor* (rather than gas) lasers as transmitters and local oscillators, were reported in 1984.[26,27] The modulation formats were ASK and FSK at bit rates near 100 Mb/s. Later experiments at higher speeds attained receiver sensitivities that were closer to ideal.[28,29]

Figure 9 is a schematic of one of these experiments that operated at 400 Mb/s and 1 Gb/s with differential-phase-shift-keyed (DPSK) modulation.[28] With a balanced mixer to suppress LO intensity noise and use LO power efficiently, the system achieved a record receiver sensitivity of 45 photons per bit at a 400-Mb/s bit rate.

Operation at higher bit rates for coherent transmission is difficult because of the stringent requirements for flat response over a large bandwidth in the receiver front end and the following linear channel. Additionally, higher LO power is required to overcome thermal noise. Nevertheless, a transmission distance of 200 km at 1 Gb/s has been achieved.[28] The longest distance for high-speed coherent transmission reported to date is 290 km, in an experiment at 400 Mb/s using an optimal form of FSK, called continuous-phase (CP) FSK.[29]

Local Distribution and Networking

Until now, lightwave technology has not had a widespread effect in the local arena. Successful introduction of future integrated wideband services will depend on how various photonics technologies will be applied to solve the problems of local distribution and networking.

Therefore, much attention is focused on this area of research.

Coherent transmission is envisioned as a candidate for local distribution of many WDM channels. Channels can be selected by tuning the local oscillator. Recently, a ten-channel WDM experiment was done in a laboratory to show this principle.[30] A tunable heterodyne receiver was used to select a signal from among ten 0.83-μm lasers, each modulated at a 70-Mb/s data rate and separated by a 6-GHz channel spacing. System margins in this experiment permitted a maximum of 64 channels to be distributed.

A novel scheme for channel selection in a densely packed WDM system without resorting to coherent technology is the *phonon star*. It exploits backward Brillouin scattering in a single-mode fiber to provide amplification in a narrow bandwidth.[31] Because the required pump power is around 10 mW, tunable single-frequency semiconductor lasers can be used at the receiver for channel selection. The advantages of this scheme are:

- Polarization is not important.
- Lasers with extremely narrow linewidth are not necessary.
- Simple direct detection receivers with baseband electronics can be used.

A disadvantage is that a requisite length of single-mode fiber (<5 km) is needed to provide the Brillouin gain. In a recent experiment, data at 45 Mb/s were detected error-free in the presence of a signal in an adjacent channel as close as 140 MHz.[31]

A few channels (say about 50 channels of 50 Mb/s each) could be sent using microwave subcarrier multiplexing (SCM).[32,33] In this scheme, the lightwave carrier is intensity-modulated by signals that ride on microwave subcarriers. For example, several 50-Mb/s channels can individually modulate several microwave subcarriers at different microwave frequencies. The combined microwave signal can then intensity-modulate a laser, or be grouped in different frequency bands to modulate different lasers. The lasers need not operate in a single frequency, but their modulation bandwidths must be sufficiently large. A microwave demultiplexer that follows the wideband receiver separates the various subcarriers.

A recent experiment showed SCM with three 44-Mb/s signals multiplexed in a 4-GHz band and modulating a single laser.[32] Intermodulation distortion, which is present when two or more subcarriers modulate the laser, can be a limiting factor in the performance of such systems. However, acceptable transmission quality was obtained in the experiment.

Local networking requires intelligent switching to establish communications channels. A possible scenario for future local networks could be one where high-bit-rate time-division-multiplexed (TDM) lightwave signals carried by fibers are detected, demultiplexed, and then switched by electronic switching systems as lower-bit-rate channels of various services. For example, a 1-Gb/s lightwave channel with multiplexed voice, data, and video would serve each subscriber. Electronic switches that use high-speed silicon or gallium-arsenide integrated circuits would switch the various services that run at bit rates up to 100 or 200 Mb/s. Both circuit and packet switching techniques could be employed. This evolutionary approach can easily be made compatible with today's networks and systems that are under development.

On the other hand, the vast bandwidth of optical fibers challenges the imagination and ingenuity of research workers to innovate revolutionary systems that promise greatly enhanced technical capability and offer important economic advantages. Both WDM and TDM techniques are being explored for local switching and networking. Different network topologies and various protocols with distributed intelligence for accessing a network are being investigated. These system pursuits are exerting strong influences on device research. For example, *tunable* narrow-bandwidth optical filters and *tunable* narrow-linewidth lasers are being pursued for WDM applications. The work on local distribution and networking is just beginning. Which technologies will eventually prevail is difficult to predict.

Conclusions

Current vigorous research in lightwave technology and systems will provide many options for a variety of applications in the long-haul and local-networking arena. High-speed and wavelength-multiplexing techniques are being pushed to explore the vast transmission bandwidth of single-mode fibers.

Progress has been both rapid and abundant, and the fruits of research are being developed into commercial products. Photonics will surely become the dominant technology for information movement and management in the information-dominated society of the future.

References

1. T. Li, "Advances in Optical Fiber Communications: An Historical Perspective," *IEEE Journal on Selected Areas in Communications*, Vol. SAC-1, April 1983, pp. 356-372.
2. S. R. Nagel, J. B. MacChesney, and K. L. Walker, "Modified Chemical Vapor Disposition," *Optical Fiber Communications*, Vol. 1, Fiber Fabrication, edited by T. Li, Academic Press, Orlando, Florida, 1985, pp. 1-64.
3. D. Kalish and L. G. Cohen, "Single-Mode Fiber: From Research and Development to Manufacturing," *AT&T Technical Journal*, Vol. 65, No. 6, November/December 1986, pp. 19-32.
4. D. Marcuse and C. Lin, "Low Dispersion Single-Mode Fiber Transmission—The Question of Practical Versus Theoretical Maximum Transmission Bandwidth," *IEEE Journal on Quantum Electronics*, Vol. QE-17, June 1981, pp. 869-877.
5. R. A. Linke et al., "Mode Power Partition Events in Nearly Single-Frequency Lasers," *IEEE Journal of Lightwave Technology*, Vol. LT-3, June 1985, pp. 706-712.
6. W. T. Tsang, N. A. Olsson, and R. A. Logan, "High-Speed Direct Single-Frequency Modulation with Large Tuning Rate and Frequency Excursion in Cleaved-Coupled-Cavity Semiconductor Lasers," *Applied Physics Letters*, Vol. 42, April 1983, pp. 650-652.
7. H. W. Kogelnik and C. V. Shank, "Stimulated Emission in a Periodic Structure," *Applied Physics Letters*, Vol. 18, February 1971, pp. 152-154.
8. R. A. Linke, "Modulation Induced Transient Chirping in Single-Frequency Lasers," *IEEE Journal of Quantum Electronics*, Vol. QE-21, June 1985, pp. 593-597.
9. T. L. Koch and J. E. Bowers, "Nature of Wavelength Chirping in Directly Modulated Semiconductor Lasers," *Electronics Letters*, Vol. 20, December 1984, pp. 1038-1040.
10. T. L. Koch et al., "1.55-μm InGaAsP Distributed Feedback Vapor Phase Transported Buried Heterostructure Lasers," *Applied Physics Letters*, Vol. 47, July 1985, pp. 12-14.
11. J. C. Campbell et al., "High-Performance Avalanche Photodiode with Separate Absorption Grading and Multiplication Regions," *Electronics Letters*, Vol. 19, September 1983, pp. 818-819.
12. B. L. Kasper, "Sensitivity Limits on Direct Detection Receivers," *Technical Digest*, Conference on Optical Fiber Communication, OSA, Washington, D.C., 1986, paper WJ4.
13. B. L. Kasper et al., "A 130 km Transmission Experiment at 2 Gb/s Using Silica-Core Fiber and a Vapor Phase Transported DFB Laser," *Conference Proceedings* (Post Deadline Papers), 10th European Conference on Optical Communications, VDE-Verlag GmbH, Berlin, 1984, paper PD-6.
14. A. H. Gnauck et al., "4 Gb/s Transmission Over 103 km of Optical Fiber Using a Novel Electronic Multiplexer/Demultiplexer," *Technical Digest* (Post Deadline Papers), Conference on Optical Fiber Communication, OSA, Washington, D.C., 1985, paper PD-2.
15. S. K. Korotky et al., "4 Gb/s Transmission Experiment Over 117 km of Optical Fiber Using a Ti:$LiNbO_3$ External Modulator," *Technical Digest* (Post Deadline Papers), Conference on Optical Fiber Communication, OSA, Washington, D.C., 1985, paper PD-1.
16. A. H. Gnauck et al., "Information-Bandwidth-Limited Transmission at 8 Gb/s Over 68.3 km of Single-Mode Optical Fiber," *Technical Digest* (Post Deadline Papers), Conference on Optical Fiber Communication, OSA, Washington, D.C., 1986, paper PDP-9.
17. N. A. Olsson et al., "68.3 km Transmission with 1.37 Tbit·km/s Capacity Using Wavelength Division Multiplexing of Ten Single-Frequency Lasers at 1.5 μm," *Electronics Letters*, Vol. 21, January 1985, pp. 105-106.
18. C. H. Henry, "Theory of the Linewidth of Semiconductor Lasers," *IEEE Journal of Quantum Electronics*, Vol. QE-18, February 1982, pp. 259-264.
19. J. Salz, "Coherent Lightwave Communications," *AT&T Technical Journal*, Vol. 64, No. 10, December 1985, pp. 2153-2209.
20. B. Glance, "Performance of Homodyne Detection of Binary PSK Optical Signals," *IEEE Journal of Lightwave Technology*, Vol. LT-4, February 1986, pp. 228-235.
21. G. Nicholson, "Probability of Error for Optical Heterodyne DPSK System with Quantum Phase Noise," *Electronics Letters*, Vol. 20, November 1984, pp. 1005-1007.
22. G. Jacobsen and I. Garrett, "Error-Rate Floor in Optical ASK Heterodyne Systems Caused by Nonzero (Semiconductor) Laser Linewidth," *Electronics Letters*, Vol. 21, March 1985, pp. 268-270.
23. I. Garrett and G. Jacobsen, "Influence of (Semiconductor) Laser Linewidth on the Error-Rate Floor in Dual-Filter Optical FSK

Receivers," *Electronics Letters*, Vol. 21, March 1985, pp. 280-282.
24. R. Wyatt and W. J. Devlin, "10 kHz Linewidth 1.5 μm InGaAsP External Cavity Laser with 55 nm Tuning Range," *Electronics Letters*, Vol. 19, February 1983, pp. 110-112.
25. R. W. Tkach and A. R. Chraplyvy, "Regimes of Feedback Effects in 1.5-μm Distributed Feedback Lasers," *IEEE Journal on Lightwave Technology*, Vol. LT-4, November 1986, pp. 1655-1661.
26. M. Shikada et al., "100 Mbit/s ASK Heterodyne Detection Experiment Using 1.3 μm DFB-Laser Diodes," *Electronics Letters*, Vol. 20, February 1984, pp. 164-165.
27. R. Wyatt et al., "140 Mbit/s Optical FSK Fibre Heterodyne Experiment at 1.54 μm," *Electronics Letters*, Vol. 20, October 1984, pp. 912-913.
28. R. A. Linke et al., "Coherent Lightwave Transmission Over 150 km Fiber Lengths at 400 Mbit/s and 1 Gbit/s Data Rates Using Phase Modulation," *Electronics Letters*, Vol. 22, January 1986, pp. 30-31.
29. K. Iwashita et al., "Linewidth Requirement Evaluation and 290 km Transmission Experiment for Optical CPFSK Differential Detection," *Electronics Letters*, Vol. 22, July 1986, pp. 791-792.
30. E.-J. Bachus et al., "Ten-Channel Coherent Optical Fibre Transmission," *Electronics Letters*, Vol. 22, September 1986, pp. 1002-1003.
31. A. R. Chraplyvy and R. W. Tkach, "Narrowband Tunable Optical Filter for Channel Selection in Densely-Packed WDM Systems," *Electronics Letters*, Vol. 22, September 1986, pp. 1084-1085.
32. T. E. Darcie et al., "Lightwave System Using Microwave Subcarrier Multiplexing," *Electronics Letters*, Vol. 22, July 1986, pp. 774-775.
33. J. E. Bowers, "Optical Transmission Using PSK Modulated Subcarriers at Frequencies to 16 GHz," *Electronics Letters*, Vol. 22, October 1986, pp. 1119-1121.

Trends in U.S. Broad-Band Fiber Optic Transmission Systems

LEONARD G. COHEN, FELLOW, IEEE

Abstract —Fiber optic media are rapidly penetrating the telecommunications network. They are used as undersea and terrestial trunk lines, central-office loops, optical data links, and will eventually be included within the distribution plant which connects directly into individual premises. The rapid implementation of lightwave systems is occurring because the high bandwidths and low losses of optical fibers enable broad-band communication services to be provided between widely spaced repeaters and also allows users to upgrade currently installed systems to meet future needs. This paper will describe current applications and show how future trends will depend upon the continuing evolution of lightwave components and optical fiber designs.

Introduction

OPTICAL fiber technology is in the process of revolutionizing the U.S. telecommunications industry. The low losses and gigantic bandwidth available on single-mode fibers far exceeds the requirements for plain old telephone service (POTS). That has allowed telco carriers to embrace the goal of supplying universal information services [1], [2] on a digital network that will enable subscribers to interactively mix voice, data, and video services through terrestial facilities and intercontinental undersea systems.

Backbone long-distance trunk facilities are currently being deployed at a very rapid pace that, in several years, will satisfy long-haul needs. That will be followed by a surge of systems for short-haul local distribution as the wave of optical fiber technology moves ever closer to subscriber premises. This paper will describe how the characteristics of lightwave systems are being tailored for applications encompassing undersea cable routes, central office trunk lines, and local exchange routes. Fig. 1 illustrates a simplified tree-like structure for the telecommunications network. The local exchange loop plant interconnects a central office with subscribers through feeder and distribution cables. A remote interface is used to cross-connect feeder cables, which emanate from a central office, with distribution cables that go to subscribers. Information may be transmitted via a digital loop carrier (DLC) system in the feeder plant, particularly for loops longer than about 4 km. Feeder cable lengths are usually less than 20 km and remote terminal interface sites are selected to make distribution cable lengths shorter than 4 km. Interconnecting trunk lines between central offices

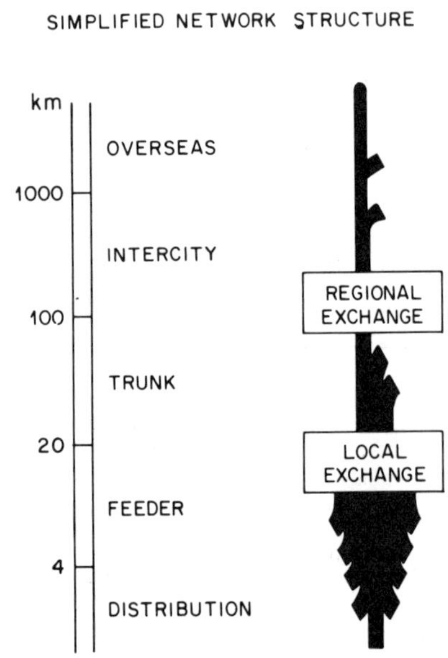

Fig. 1. Simplified tree-like structure for the telecommunications network indicating cable lengths in the local exchange plant (distribution and feeder) and trunk lines (metropolitan areas, intercity, and overseas).

are typically shorter than 20 km in metropolitan areas but can be longer than 200 km within the regional exchange network between cities. Much longer lengths make up undersea routes that can interconnect coastal cities or link continents (> 5000 km). Single-mode fiber technology has overwhelming advantages for all systems applications with the possible exception of short subscriber lines. Current single-mode fibers are being deployed with low losses and nearly zero dispersion at 1.3 μm wavelength. However, second-generation systems will undoubtedly include 1.55 μm wavelengths where silica-based glass fiber losses are minimized.

Transmission Limitations

Fig. 2 illustrates a variety of refractive-index profiles for single-mode fiber media [3]–[12]. Such fibers have to be packaged into cable units that provide protection but do not add significant extrinsic losses. The armored cable unit in Fig. 3(a) is designed for undersea environments that require high strength and very low losses [13]. It has steel-strength members and elastomer filling material that buffers individual fibers from bending-related extrinsic

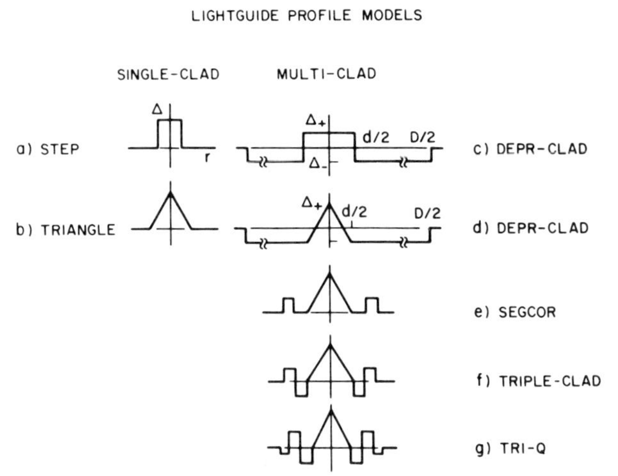

Fig. 2. Refractive-index profiles for single-mode fiber media that can be tailored to control dispersion within the 1.3–1.6 μm spectral region. They include step and graded-index core profiles and claddings with multiple-index wells.

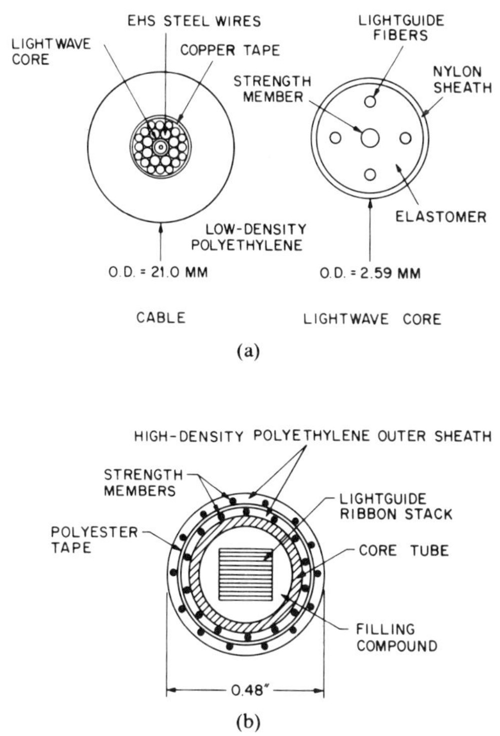

Fig. 3. Examples of cable units for packaging individual fibers. (a) Undersea cable contains steel wires within a copper tape that protects the lightwave cable core unit. The core unit has a central strength member and elastomer fill material which buffers individual fibers to reduce microbending. (b) Ribbon cable designed for containing large numbers of fibers.

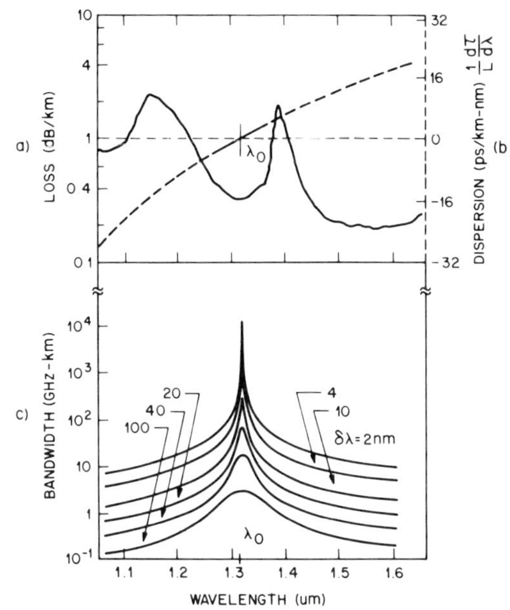

Fig. 4. Single-mode fiber transmission characteristics. (a) Loss spectra (dB/km versus λ in μm) with a 0.2 dB/km minimum value near $\lambda = 1.55$ μm. Loss peaks are due to cutoff of the second propagating mode near $\lambda = 1.25$ μm and OH ion absorption near $\lambda = 1.39$ μm. (b) Dispersion spectra (ps/km-nm versus λ in μm) with $\lambda_0 = 1.3$ μm zero-dispersion cross-over wavelength. (c) Bandwidth-distance spectra (GHz-km versus λ in μm) are drawn for source spectral widths ranging from $\delta\lambda = 2$ nm to $\delta\lambda = 100$ nm. The bandwidths scale with $\delta\lambda^2$ at the peaks and $\delta\lambda$ elsewhere.

forces. The unit in Fig. 3(b) is a ribbon cable designed for terrestial applications that require large numbers of fibers [14]. It can contain a 12×12 matrix of fibers within a 0.5 in sheath diameter.

Single-mode fibers have lower transmission losses than multimode fibers because the lower dopant concentration required in the core leads to lower intrinsic losses due to Rayleigh scattering effects. Fig. 4 shows typical loss and dispersion characteristics plotted versus wavelength, λ. Minimum losses are determined by the combination of Rayleigh scattering (which decreases according to λ^{-4}) and the tail of infrared absorption effects due to multiphonon molecular vibrations. The absorption of water in the form of OH ions causes an excess loss peak of 1 dB/km at $\lambda = 1.39$ μm if the OH concentration is 25 ppb. Conventional fibers have step-index germania-silicate cores that are surrounded by silica [Fig. 2(a)] or fluorosilicate cladding regions [Fig. 2(c)]. Minimum losses (0.16 dB/km near $\lambda = 1.57$ μm) have been achieved in fibers with nearly pure silica cores and deeply depressed fluorosilicate cladding regions (Fig. 2(c) $\Delta_+ \approx 0$) [15].

Dispersion characteristics of fibers can be modified by tailoring the refractive-index profile to make waveguide dispersion cancel material dispersion at desired wavelengths [16], [17]. For example, dispersion-shifted fibers, with linearly graded-index core profiles, have been produced with low losses and zero dispersion at $\lambda_0 \approx 1.55$ μm [Fig. 2(b), (d)–(g)]. Fig. 2(e) illustrates Corning Glass Works' [11] dispersion-shifted SEGCOR design which is already being marketed with losses lower than 0.25 dB/km at 1.55 μm. As the data rate increases in a digital communication system, dispersion effects cause errors which cannot be compensated by increasing the signal power. Temporal broadening in single-mode fibers, σ, occurs because chromatic dispersion effects make different wavelengths (frequencies) of an optical pulse propagate with different velocities such that

$$\sigma = DL\delta\lambda \tag{1}$$

where D is the fiber dispersion, L is the fiber length, and $\delta\lambda$ is the spectral width of the pulse. The ultimate dispersion limitation of a system can be characterized by convolving the dispersion spectrum ($D(\lambda)$ in ps/km-nm) of the fiber [Fig. 4(b)] with the spectrum of an input pulse with linewidth $\delta\lambda$. Minimum-pulse broadening ($\sigma_{\min}$) occurs for a source centered about λ_0 and is proportional to the slope of the dispersion spectrum multiplied by the square of the source linewidth [18].

$$\sigma_{\min} \approx \frac{1}{8}\frac{dD}{d\lambda}L\delta\lambda^2 \tag{2}$$

We assume that the dispersion power penalty is small provided that pulse broadening (σ) is smaller than half a bit period ($B \approx 1.82 B_w < L/2\sigma$). The bit-rate capacity and bandwidth (B_w) spectra of fibers are inversely related to dispersion as shown in Fig. 4(c). The peak values of BL ($B_w L$) increase as $\delta\lambda^2$ and is on the order of 1000 Gbits/s-km at λ_0. The bit-rate (bandwidth)–distance product decreases rapidly with wavelength and is only about 50 Gbits/s-km near $\lambda = 1.55$ μm for conventional fiber and $\delta\lambda = 1$ nm. Single-mode injection laser sources emit power in several longitudinal modes that are separated by about $\Delta\lambda = 1$ nm in wavelength. The maximum capacity of a lightwave system is then limited by a combination of dispersion effects, including intersymbol interference due to pulse broadening [19] and fluctuations in the received signal due to laser-mode-partitioning [20]. The latter effect occurs because the instantaneous power in any mode can fluctuate randomly with time even if the sum of the power in all laser modes is constant. Such random mode fluctuations add noise to the received signal as they become dispersed in time after propagating along a fiber.

The quality of the laser source is parametrized by the rms width, $2\delta\lambda$, of the laser spectrum and the fraction, K, of the fluctuation power not in its dominant mode. System performance limits can then be represented by a bit-rate–distance product (BL) that depends upon the $\delta\lambda$ and K parameters of the laser and the D and $dD/d\lambda$ chromatic dispersion properties of the fiber. Even single-frequency lasers, which emit only one longitudinal mode, can have rapid wavelength variations of that mode with time because of carrier-density changes brought on by high-frequency injection current pulses. Such chirped behavior causes an effective linewidth broadening ($\delta\lambda \approx 0.1$ nm) of a single-longitudinal mode [21]. Equation (3) represents pulse broadening effects on BL, and (4) shows that mode-partition noise can impose even more stringent performance limitations on BL when $K > 0.1$ and λ is away from the wavelength of minimum dispersion. The BL limitation for the system is the lesser of (3) or (4).

$$BL \leqslant \frac{341}{\delta\lambda D}\ \frac{Gb}{s} - \text{km} \tag{3}$$

$$BL \leqslant \frac{130}{\delta\lambda D\sqrt{K}}\ \frac{Gb}{s} - \text{km} \tag{4}$$

Much higher BL products are possible in dispersion-shifted fibers if λ is close to the system operating wavelength. Then $D \approx 0$ and $\sigma \approx (dD/d\lambda)L\delta\lambda^2$ (instead of $DL\delta\lambda$) which lead to dispersion constraints given by the lesser of (5) or (6).

$$BL \leqslant \frac{11207}{\delta\lambda^2 \frac{dD}{d\lambda}}\ \frac{Gb}{s} - \text{km} \tag{5}$$

$$BL \leqslant \frac{1173}{\delta\lambda^2 \frac{dD}{d\lambda}\sqrt{K}}\ \frac{Gb}{s} - \text{km} \tag{6}$$

Finally, a different set of performance limits apply to single-mode lasers with very narrow linewidths characterized by $\delta\lambda \leqslant (\lambda^2/c)B$. Then, pulsewidths become fixed by the modulation speed of the source. Within that transform-limited regime, $\delta\lambda$ should be replaced by $(\lambda^2/c)B$ in (3) and (5).

The maximum separation between repeaters in a lightwave system can be calculated by comparing cable losses (including splices, connectors, etc.) to the differences between the power levels available from optical sources (lasers) and the power levels (including system margin allowances) required for digital reception at various data rates. Fig. 5 illustrates receiver sensitivity limits (in decibels below 1 mW) for achieving 10^{-11} bit-error rates [22]. Quantum photonic noise determines intrinsic limits shown in the lower band. Present technological limits, for direct detection, are achieved with InGaAs avalanche photodiodes [23]. They are approximately 12 dB higher than the quantum noise limit. Coherent heterodyne detection schemes have been used to increase receiver sensitivities by as much as 6 dB [24]. Fig. 6 plots the achievable repeater spacing (for laboratory experiments at $\lambda = 1.55$ μm without margin allowances) versus bit rate assuming 0 dBm average power is launched into the fiber from an injection laser and the receiver can detect -43 dBm power (when $B = 420$ Mbits/s) at the fiber output. Solid lines indicate performance limitations due to fiber attenuation (0.2 dB/km and 0.3 dB/km). The diagonal lines indicate capacity limitations due to mode-partition noise ($K = 0.15$, $\delta\lambda = 4$ nm) and pulse-broadening effects ($K = 0$, $\delta\lambda = 4$ nm and 0.2 nm). The dispersion-shifted performance line ($K = 0$, $\delta\lambda = 4$ nm) assumes that the system wavelength is within 20 nm of the average zero dispersion wavelength along the fiber ($\lambda_0 = 1.55 \pm 0.02$ μm).

Fig. 7 illustrates the chronological progress of laboratory demonstrations of lightwave system performance. It shows that the BL product has increased by approximately a factor of two per year over the past decade. Initial experiments used multimode fiber and operated near $\lambda \approx 0.8$ μm (○-data). Better results are obtained with single-mode systems operating near $\lambda = 1.3$ μm (■-data) and $\lambda = 1.55$ μm (●-data). Recent experiments have demonstrated approximately 500 Gbits-km/s for a single channel [25] and 2000 Gbits-km/s for 10 WDM channels (▲-data) between $\lambda = 1.5$ μm and 1.6 μm [26].

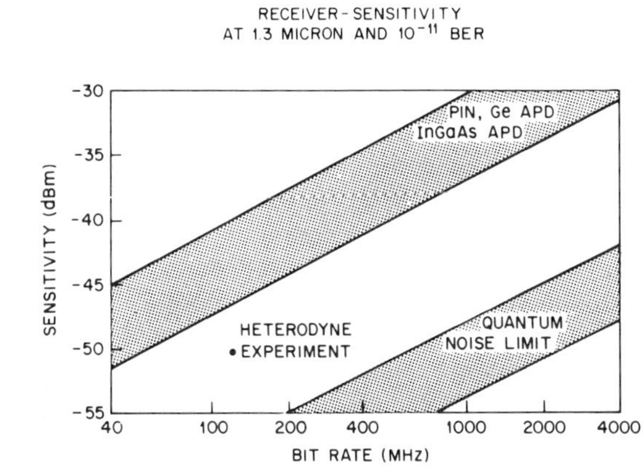

Fig. 5. Receiver sensitivity limits for achieving 10^{-11} bit-error rates. Present technological limits, for direct detection, are approximately 12 dB lower than quantum noise limits. Coherent detection has been used to increase receiver sensitivities by approximately 3 dB.

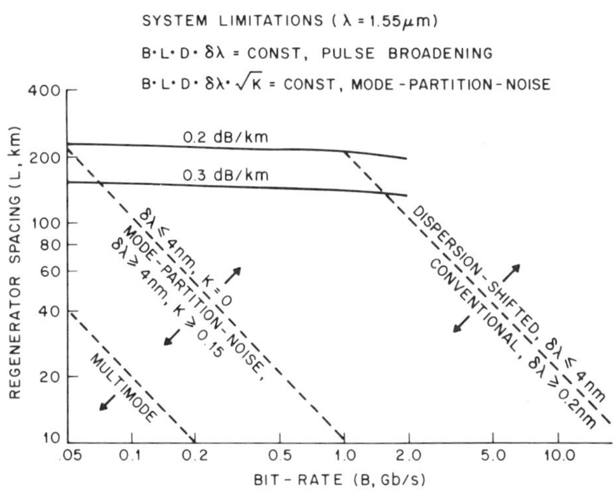

Fig. 6. System limitations near $\lambda = 1.55$ μm are characterized by plotting regenerator spacing (in km) versus bit rate (in Gbits/s). Horizontal lines indicate loss limitations. Diagonal lines indicate dispersion limitations due to mode-partition noise and pulse broadening with multimode and quasi-single-mode lasers. For comparison, intermodal dispersion limits are also shown for high-bandwidth multimode fibers.

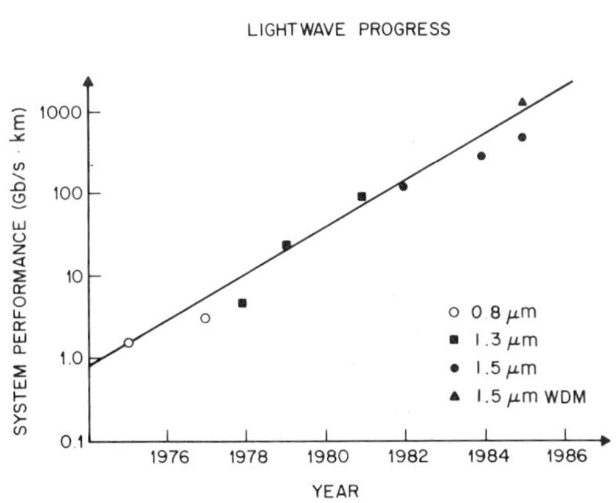

Fig. 7. Chronology of progress in lightwave system performance as characterized by bit-rate x length product (BL in Gbits/s-km). Recent experiments have demonstrated approximately 500 Gbits-km/s for a single channel near $\lambda = 1.55$ μm, and 2000 Gbits-km/s for 10 WDM channels between $\lambda = 1.5$ μm and 1.6 μm.

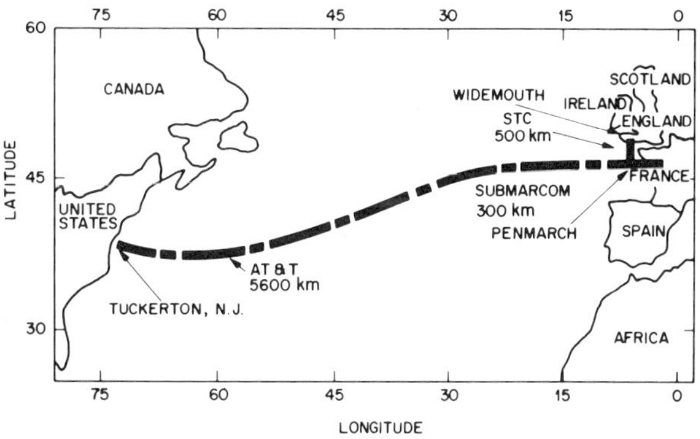

Fig. 8. Undersea cable route for the TAT-8 lightwave system linking the United States and Europe.

Undersea Long-Distance Systems

In order to meet increasing traffic needs to link the U.S. with Europe, undersea systems in the North Atlantic have been installed approximately every five years over the past thirty years. By 1988, after the installation of the first transatlantic lightwave system, TAT-8 [13], the total capacity will be about 20000 voice trunks without the use of TASI or digital compression techniques. Ocean cable links, between coastal cities in the U.S. and with Pacific and Carribean islands, are also in the planning stages. If the growth rate continues to increase at 25 percent per year then the total necessary capacity should triple by 1992. That would require another new system with twice the capacity of TAT-8.

Major goals for new systems are to provide high reliability with a lower cost per circuit mile. The primary driving force for cutting costs is to reduce the number of regenerators needed [27]. Fig. 8 shows the lightwave cable link between the United States and Europe for the TAT-8 system. Construction contracts were awarded to AT&T, Standard Telephone and Cables, Ltd., and Submarcom. The cable is 6650 km long and has a branching point at the European continental shelf to allow for transmission to England and France. The system will operate at $\lambda \approx 1.3$ μm with 296 Mbit/s transmission rates and approximately 50 km repeater spacings (130 regenerators) through six single-mode fiber links including two spares. It is obvious that 1.55 μm technology can be used to further increase information capacity and repeater spacings. Appropriate laser, fiber, cable, and receiver technology have already been demonstrated in systems experiments [28], [29].

One demonstration used a quasi-single-mode DFB laser and a silica-core fiber (Fig. 2(c), $\Delta_+ \approx 0$) with $\lambda_0 = 1.29$ μm [5], [28]. The link loss was 0.204 dB/km including 26 splices. Approximately 0.2 dBm laser power was coupled into the fiber and the receiver sensitivity was -41.4 dBm at 420 Mbits/s. The resultant dispersion penalty was negligible.

Another experiment used multilongitudinal-mode lasers ($2\delta\lambda = 2.8$ nm) with dispersion-shifted fiber [8], [29]. The fiber had a triangular core profile and a depressed-cladding lightguide structure [Fig. 2(d)] that was tailored to have minimum dispersion at $\lambda_0 \approx 1.55$ μm. A laser sparing strategy was demonstrated by using a passive polarizing beam-splitter to couple two transmitters into the long-transmission fiber. The span loss, of a 150 km link including splices, was 0.267 dB/km at $\lambda = 1.526$ μm and 0.258 dB/km at $\lambda = 1.542$ μm. Since the dispersion penalty was negligible, it is evident that high-bit-rate system spans can be longer than 100 km even with multilongitudinal-mode lasers. In order to further demonstrate feasibility, a continuous 29.6 km undersea cable core was manufactured to show that low fiber losses can be maintained in cables that are suitable for deep-sea applications [30]. Fig. 3(a) shows a cable cross-section in which three silica-core and three dispersion-shifted fibers are embedded in an elastomer cable core that cushions individual fibers in order to reduce cabling-induced bending losses. The elastomer core is surrounded by a thin nylon layer, a series of steel strands for strength, a continuously welded copper cylinder for powering, and finally a layer of polyethylene insulation. Fig. 4 shows spectral loss curves for the fiber cable. Minimum loss values were 0.194 dB/km for the silica-core fibers and 0.246 dB/km for the dispersion-shifted fibers. The losses did not change significantly during the cable manufacturing process or during environmental testing when the entire cable core was temperature cycled between 0°C and 40°C.

Terrestial Trunking Systems

The primary requirement for trunk transmission systems is to provide high information capacity between central offices along routes that can be in congested metropolitan areas, suburban locales, or between cities. In many cases, the quality of such systems can be most effectively upgraded by increasing bit rates rather than by increasing span lengths. The first test within a commercial U.S. environment was done eight years ago when the Illinois Bell Telephone Operating Company and Bell Laboratories collaborated to test a working fiber optic system within the downtown area of Chicago.

Extensive single-mode fiber network expansions are being planned by long-haul telecommunications carriers including AT&T, Electra, Lasernet, LDX Net, Lightnet, Litel, and United Telecom. Fig. 9 illustrates intercity long-haul fiber optic links in place and being planned through 1987 [31]. If most of these plans come to fruition, then the long-haul fiber optic cable network will contain more than one million fiber-km. AT&T has already shipped 1000 fiber megameters for deployment with the Bell Operating Companies. In addition, AT&T and the Bell Operating Companies have announced plans for deploying an additional 16 000 cable kilometers to form nationwide supertrunks that will link an east coast system (stretching from Boston to Miami) with a west coast system (running almost the entire length of California). Transmission rates currently range up to 417 Mbits/s and will later be increased to 1.7 Gbits/s. The highest acceptable error rate is 10^{-11} in one repeater span, but fewer errors are achieved in working systems. That implies that fewer than 10^7 errors can occur for a total data throughput of 10^{18} bits during a 20-year lifetime of a system operating at 400 Mbits/s. Typical unrepeatered span lengths range up to 48 km. Current system wavelengths are near 1.3 μm. However, the 1.5 μm wavelength region may become advantageous in order to increase the system margin at higher bit rates, to increase repeater spacings within new installations, or to increase information capacity by wavelength multiplexing.

So far, this paper has described the peak performance of selected components under laboratory conditions. However, field systems use long lengths of spliced fiber and large volumes of lightwave components which have a statistical spread in performance. Fig. 10 illustrates this by plotting the percent probability for achieving specific span lengths at 400 Mbits/s [22]. For example, at $\lambda = 1.31$ μm, the spread in span length ranges from less than 30 km in many applications to more than 200 km in laboratory experiments. Therefore, system objectives involve economical choices that depend on the selection of higher grade components in the interest of achieving better performance. There is a tradeoff between using short span lengths and more repeaters as opposed to maximizing the span length in order to use fewer repeaters. The spread in the distribution at $\lambda = 1.55$ μm has been assumed to be the same as at $\lambda = 1.31$ μm and the shift to longer span lengths is equal to the factor of two lower losses expected for spliced and cabled fibers at $\lambda = 1.55$ μm. It is interesting to note the substantial overlap of the two distributions which makes the choice between the two wavelengths depend on the cost premium at $\lambda = 1.55$ μm.

As stated earlier, the trend for intercity lightwave systems is to provide high information capacity over moderately long repeater span lengths. Fig. 11 illustrates two strategies that depend on regenerator costs [22]. If n regenerators operating at a bit-rate B are less expensive than one unit at rate nB (where n is an integer greater than one), then the clear choice would be to employ wavelength-division multiplexing on separate regenerators rather than time-division multiplexing at a higher bit rate to minimize the number of fibers. The converse is true if it is less expensive to use one regenerator at a higher bit rate as in the sublinear region on the left. Single-mode fibers, lasers, and detectors have already been designed for gigabit performance and milestone experiments have already demonstrated 4 Gbit/s transmission over more than 100 km [25]. Therefore, the boundary line between the time and wavelength multiplexing regimes is primarily determined by electronic circuit capability. Silicon IC's are now readily available for operation below 600 Mbits/s and have the potential for use at gigabit rates. There is also a strong effort to use GaAs technology which features much higher electronic mobility than silicon. If reliable GaAs integrated circuits become readily available, then 2 Gbit regenerators could become as readily available as current 417 Mbit/s regenerators.

LONG-HAUL FIBEROPTIC LINKS PLANNED AND IN PLACE

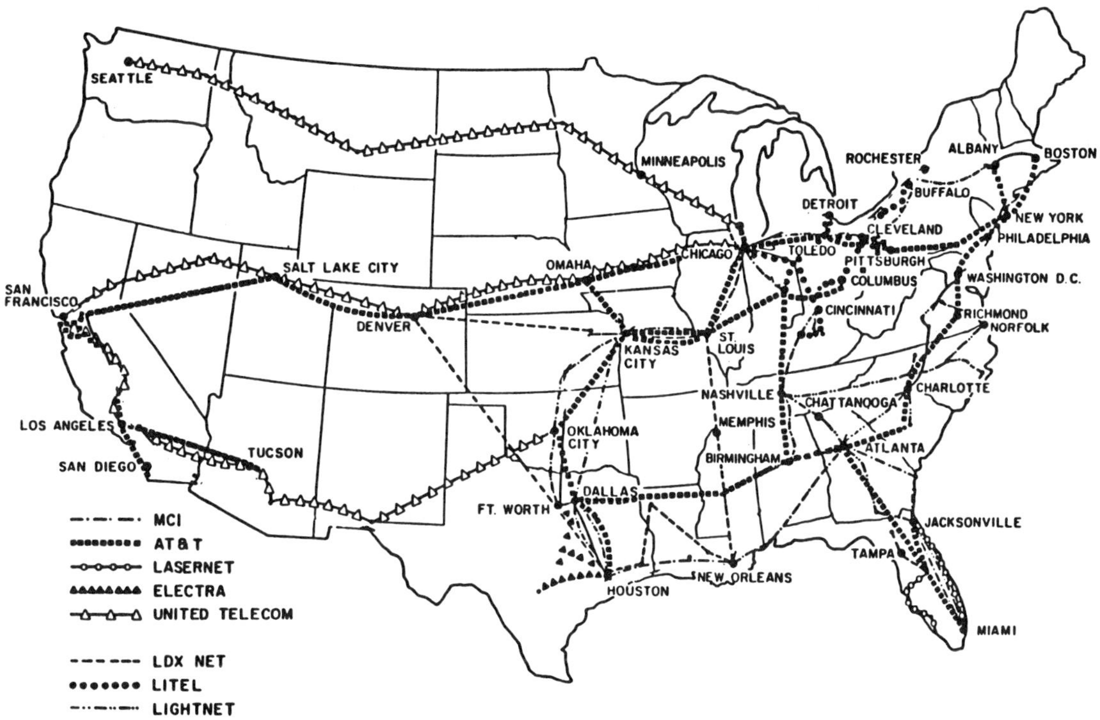

Fig. 9. Existing and projected long-haul intercity fiber optic digital network within the United States. Trunks are delineated for individual telecommunication carriers [31].

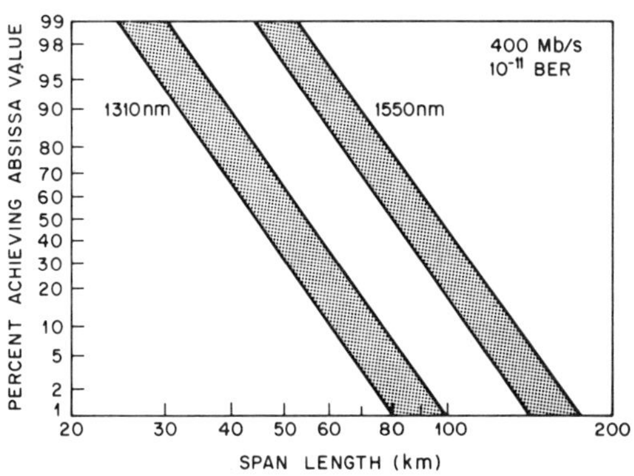

Fig. 10. Percent probability for achieving specific span lengths at 400 Mbits/s. Fiber losses are assumed to be a factor of two lower at $\lambda = 1.55\ \mu m$ than at $\lambda = 1.3\ \mu m$. The bands for each wavelength are due to the statistical spread in performance of fibers and lightwave components.

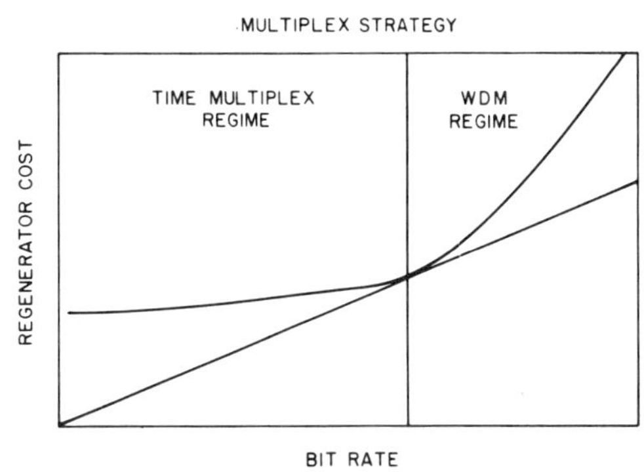

Fig. 11. Strategies for multiplexing information depend on the regenerator cost at various bit rates. Costs for WDM systems are linear with bit rate because each λ-channel requires a separate regenerator. Costs for time-division multiplexing are constant until high bit rates make high-frequency electronic circuits and lightwave components prohibitively expensive.

AT&T, Nippon Electric Corporation, Ericsson, Fujitsu, Northern Telecom, Plessey/Stromberg Carlson, Rockwell/Collins, and Telco Systems are among the suppliers of high bit-rate systems above 400 Mbits/s. The AT&T FT Series G [32] is an example of a modern and versatile system which can provide point-to-point links for applications (metropolitan or intercity) involving repeater spacings up to 40 km. It can operate at transmission rates of 417 Mbits/s and 1.7 Gbits/s. At 417 Mbits/s the system provides 6048 two-way 64 kbit/s channels per fiber pair. This configuration can be upgraded to 1.7 Gbits/s to provide 24 192 channels per fiber pair using the same fibers and repeater spacings. Essentially all the digital processing functions are done at 45 Mbit/s rates. A line interface interleaves the synchronized bit streams and converts them to optical signals at higher bit rates up to 1.7 Gbits/s. Preliminary regenerator models, at 1.7 Gbits/s, have been assembled using APD's, GaAs amplifier stages, a SAW retiming filter, silicon decision circuits, and a buried-heterostructure multimode laser which can couple -2 dBm into a single-mode fiber with an 8 μm core diameter. Binary transmission at 1.7 Gbits/s over 52 km of fiber occurs with 10^{-11} BER with a -34.5 dBm received power level. No dispersion or mode-partition noise occurs if laser operating wavelengths are within 5 nm of the zero dispersion wavelength, $\lambda_0 = 1305$ nm.

Loop Transmission and Distribution to Subscribers

A major goal of public network exchange carriers is to implement switched integrated services digital networks (ISDN) which can provide cost-effective end-to-end digital connectivity for a wide range of services including voice, video, telemetry, and data [1], [2], [33], [34]. The seven regional Bell Operating Companies (RBOC's) and the independent telephone companies reach into almost every home and office in the U.S. Forecasters have predicted that there will be millions of "smart" terminals and well over 100 000 digital PBX's and local area networks in the U.S. by the 1990's. This will provide a huge market for optical fibers to serve as digital pathways that link the enormous public user community.

Present networks contain a loop feeder segment which links a central office with a remote interface that interconnects subscribers through a loop distribution segment. Digital loop carrier systems may be used to convey information in the feeder plant, particularly for transmission paths that are longer than about 4 km. The SLC[R] 96 system was introduced by AT&T to link 96 customers with two-way voice and data services. It is accomplished by multiplexing 96 64 kbit/s bit streams into four 1.544 Mbit/s bit streams and transmitting them along four T1 copper lines [34]. In 1982, the system was made compatible with wide-band fiber optics media in the feeder plant. This fiber SLC system uses one fiber to transmit the combined 6.443 Mbit/s bit stream in each direction between a central office and a remote terminal. Next-generation systems will allow systems users (RBOC's, CATV) to improve system performance and increase services simply by adding terminal electronics, without cable plant rearrangements or additions [35]. An early example of this concept is being provided by AT&T's VIVID™ system [36] for videoconferencing over a telephone network. The system uses optical fibers to connect a digital video terminal, on the customer's premises, with a wide-band (45 Mbit/s) video switch in a telephone company central office. That enables customers to configure and control private conference hookups over a public telephone network.

However, there are still three technical challenges that need to be addressed before publicly accessable fiber optics media can reach large numbers of individual subscribers. First, large quantities of cabled fiber media must be deployed at costs approaching 5 cents/cabled mile. Second, some architectures require a remote wide-band switch to provide video services without the need for supplying multigigabit information capacity onto every customer's premises [33]. Present broad-band lightwave switching is done on baseband electrical signals and therefore requires optical–electrical–optical conversion. This costly requirement could be eliminated by using integrated-optical switching within electrooptically active $LiNbO_3$ waveguide circuits [37] that could be shared by subscribers through a remote terminal. The third challenge is to reduce the manufacturing costs of reliable lightwave transmitters and receivers that can be deployed on subscriber premises.

Customer-premises fiber optic facilities can consist of self-contained data communications links within a building, or a group of buildings within a relatively small geographic area. Such fiber optic local area networks (FLAN's) are usually arranged in a bus, ring, or star topology (Fig. 12) and standardization efforts are proceeding to develop common protocols for sharing resources through the common carrier network. The number of stations which can share the media is limited since optical fiber taps have relatively large insertion losses (1 dB is typical). Therefore, point-to-point transmission through active nodes is desirable. Although the need for low-cost lightwave components and interconnection hardware has pushed current systems towards using LED's and multimode fiber, there is a pervasive push towards using single-mode technology in the future. The first high-speed LAN was Ethernet developed by the Xerox corporation [38], [39]. Thus far, fiber optic LAN's have been very slow to come to market. However, there are currently more than fifty vendors in the U.S. (including AT&T, HP, Fibronics, IBM, Siecor FiberLAN, Wang, and Xerox) who are offering or planning to offer FLAN's [40]. For example, the AT&T ODL 200 and ODL 50 series are point-to-point data links which can carry 200 Mbit/s or 50 Mbit/s duplex channels on (up to 2 km) fiber linking computer terminals and peripherals [41]. Fiber data links, operating at rates up to 8 Mbits/s, are also used to interconnect the nodes of AT&T DATAKIT™-VCS (virtual circuit switch systems) for multiplexed transmission of a multiplicity of FLAN channels [42].

Fig. 13(a) illustrates an all fiber local exchange network that uses a broad-band double star switching architecture with one star switch at the serving central office and the other at a remote terminal [34]. It is capable of switching voice, data, and a number of video channels simultaneously to many subscribers. Bandwidth requirements are 45 Mbits/s for each color TV channel while multiple services including voice and data occupy up to 64 kbits/s per channel. The disadvantage of the star architecture is that it uses more fiber than necessary per subscriber. Other strategies are to use ring [43] or tree structures [Fig. 12(b), (c)]. For example, the former has a trunk emanating from the remote terminal and short fiber drop cable branches allocated to a customer. However, that requires an efficient and cost-effective means for coupling light between the trunk and its branches. Alternatively, the lightwave transmitter and receiver components could be located at nodes between the fiber trunk and coaxial cable drop lines allocated to a customer. This kind of topology is well suited for ribbon-cable units [Fig. 3(b)] that can package up to 144 fibers within a 12×12 fiber matrix that fits into a 0.5

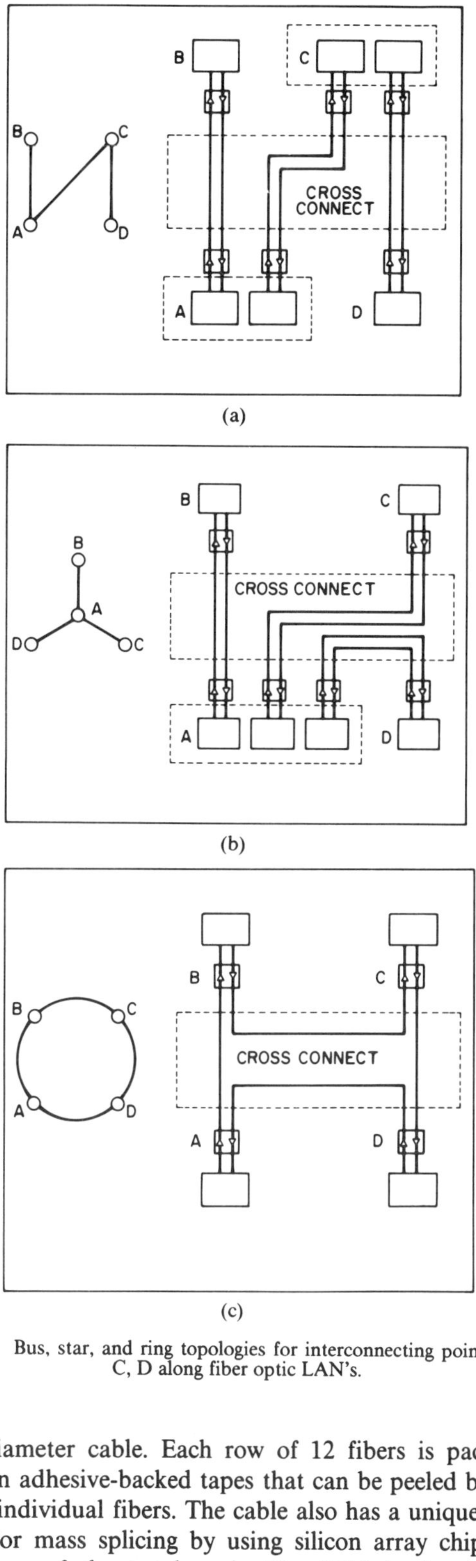

Fig. 12. Bus, star, and ring topologies for interconnecting points A, B, C, D along fiber optic LAN's.

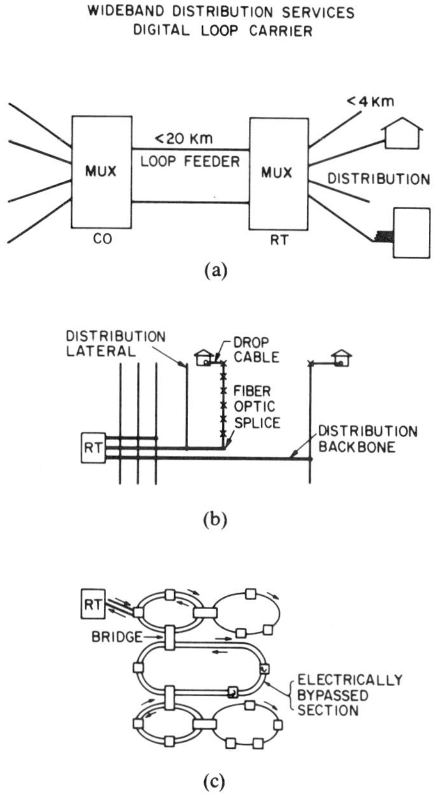

Fig. 13. Proposed architectures for local distribution of wide-band services on digital loop carrier systems. Channel banks would be used to multiplex data lines onto optical fibers. A remote termial (RT) interconnects loop feeder cable from central offices (CO) with distribution cable to individual subscribers. The interconnection from the remote terminal can be through (a) star, (b) tree, or (c) ring distribution networks.

inch diameter cable. Each row of 12 fibers is packaged between adhesive-backed tapes that can be peeled back to access individual fibers. The cable also has a unique capability for mass splicing by using silicon array chips. An advantage of the topology in Fig. 12(c) is that bridges couple ring networks that are self-contained links which do not require back and forth communication with the remote terminal.

Several regional operating companies have indicated that they will begin experimenting with optical fiber subscriber lines which will evolve into an integrated services digital network (ISDN) [2]. The Southern Bell Operating Company will initiate demonstrations to provide four simultaneous video channels into a residence, from a total of 108 video channels [33]. The implication is that multigigabit/s information rates will be available at remote terminals capable of switching 200 Mbits/s into individual premises. Typical cable lengths are 10 km in a feeder segment and several km, in the distribution plant, from a remote terminal into a customer's premises. Therefore single-mode fibers will be required for the feeder (50 Gbits-km/s), but there is still considerable discussion about whether single-mode or multimode technology will be sufficient and most cost-effective for the distribution cable into a customer's premises (0.5 Gbits-km/s). Whatever the outcome, there is a potential for millions of fiber-kilometers which need not have the very lowest losses, but must be cheap to manufacture and interconnect.

Oversized multimode fibers (62.5 μm core dia, 125 μm cladding dia; 85 μm core dia, 125 μm cladding dia; or 100 μm core dia, 140 μm cladding dia) can be used to efficiently collect light from edge-emitting LED's. Their large cores also relax geometric tolerances and could result in cheaper interconnection hardware. On the other hand, multimode fibers are not compatible with electrooptic switching technology that may be useful in the remote terminal. In addition, wide-band subscriber applications will require moderately high-quality graded-index profiles that are relatively expensive to manufacture. By comparison, single-mode fibers with relatively simple profile structures, supply enormous bandwidths and low losses. Disadvantages of single-mode fibers relate to their small mode-field sizes which make splicing and interconnection requirements more stringent. However, local distribution applications have short span lengths that allow a loss margin for imperfect connectors; and are within the dynamic range for coupling light between LED's and single-mode fibers [44], [45]. At present, LED's couple about -13 dBm into multimode fibers (versus 0 dBm for lasers). When coupled to a single-mode fiber, the power available from an LED is reduced by the ratio of $a^2\Delta$ if the active area of the LED is less than the collection area of the fiber. Experiments in that direction have produced approximately -27 dBm power coupling from edge-emitting LED's into single-mode fibers. The best published experimental results have demonstrated error-free transmission of a 140 Mbit/s signal through 22.5 km of single-mode fiber [45]. Some additional improvement may become possible by developing ultrasmall LED's. In any case, there should be 6–12 dB usable loss margin which is sufficient to transmit data at several hundred Mbits/s over several kilometers.

Conclusion

The information capacity of single-mode optical fibers can allow higher than 10 Gbit/s transmission rates along 100 km span lengths for every nanometer of spectral width within the 1.3–1.6 μm wavelength region. Ultimately, the exploitation of this enormous bandwidth will depend upon the development of high-speed integrated electrical and optical components to utilize the capacity of the media. There are several technical options that can be used even if the system wavelength is displaced from the minimum dispersion region of the fiber. They include using quasi-single-frequency laser transmitters that can be directly modulated above 10 Gbits/s; cw lasers that are followed by external electrooptic waveguide modulators; or multi-longitudinal-mode lasers that can be used with dispersion-shifted fibers. The receiver options include using sensitive APD intensity detectors or optical heterodyne down conversion to detect the phase of a broad-band modulated light signal.

Different telecommunications applications shift the need for increased bandwidth–distance products in different directions. Transoceanic systems demand high performance and long repeater spans because the length of cable is fixed by the choice of landing sites and the cost of regenerators is very high. That pushes the technology choice towards the 1.55 μm spectral region. Laboratory experiments have already demonstrated 420 Mbit/s transmission rates over 203 km in low-loss silica-core fiber, and 296 Mbit/s transmission rates over 150 km of dispersion-shifted fiber.

Terrestial systems are generally deployed along congested routes which constrain most repeater lengths to be shorter than 50 km. There is more pressure to increase system capacity by going to high bit rates and later employing wavelength-division multiplexing if necessary. Once the long-haul fiber cable is in place, a network for broad-band digital services will emerge. It will then be necessary to provide voice, video, and data traffic along relatively short loop feeder cables to and from the subscriber. Perhaps the most challenging and largest growth area for fiber optics telecommunications will be in this short-haul network. Since enormous quantities of fiber cables and light-wave components will be required, there is great pressure to reduce costs. A proposed cost-effective solution is to share resources at a remote terminal which would be able to switch information on demand to individual subscribers. However, there is still considerable debate about how to accomplish it.

Acknowledgment

Tha author gratefully appreciates valuable input from D. Gloge, I. Jacobs, P. J. Lauriello, R. A. Linke, S. R. Nagel, and R. E. Wagner.

References

[1] P. M. Villiere, J. S. Mayo, and W. Weeks, "Universal information services: A marketplace perspective; a technological evolution; AT&T network systems commitment," presented at the USTA Southern Telecommun. Showcase, New Orleans, LA, Feb. 1985.

[2] I. Dorros, "Shaping the ISDN," *Telephone Engineer & Management*, pp. 107–110, Dec. 1983.

[3] P. D. Lazay and A. D. Pearson, "Developments in single-mode fiber design, materials, and performance at Bell Laboratories," *IEEE J. Quantum Electron.*, vol. QE-18, pp. 504–510, Apr. 1982.

[4] W. M. Flegal, E. A. Haney, R. S. Elliott, J. T. Kamino, and D. N. Ernst, "Mass production of depressed clad single-mode fiber," in *Conf. Opt. Fiber Commun. (OFC'84) Tech. Dig.*, Jan. 23–25, 1984, New Orleans, LA, pp. 56–57.

[5] K. Ciemiecki Nelson *et al.*, "Fabrication and performance of long lengths of silica core fiber," in *OFC'85 Tech. Dig.*, pp. 98–99.

[6] B. J. Ainslie, K. J. Beales, D. M. Cooper, D. R. Day and J. D. Rush, "Monomode Fibre with Ultra-Low Loss and Minimum Dispersion at 1.55 μm," *Electron. Lett.*, vol. 18, pp. 842–843, 1982.

[7] A. D. Pearson *et al.*, "Optical transmission in dispersion-shifted single-mode spliced fibers and cables," *J. Lightwave Technol.*, vol. LT-2, pp. 346–349, Aug. 1984.

[8] H. T. Shang, T. A. Lenahan, P. F. Glodis, and D. Kalish, "Dispersion-shifted depressed-clad triangular-profile single-mode fiber," in *OFC'85 Tech. Dig.*, Feb. 1985, pp. 92–93.

[9] L. G. Cohen, W. L. Mammel, and S. J. Jang, "Low-loss quadruple-clad single-mode lightguides with dispersion below 2ps/km-nm over the 1.28 μm–1.65 μm wavelength Range," *Electron. Lett.*, vol. 18, pp. 1023–1024, Nov. 1982.

[10] S. J. Jang, J. Sanchez, K. D. Pohl, and L. D. L'Esperance, "Graded-Index single-mode fibers with multiple claddings," in *Int. Conf. Integrated Opt. Fiber Commun. (IOOC'83) Tech. Dig.*, June 27–30, 1983, Tokyo, Japan, pp. 396–397.

[11] T. D. Croft, J. E. Ritter, and V. A. Bhagavatula, "Low-loss dispersion-shifted single-mode fiber manufactured by the OVD process," in *OFC'85 Tech. Dig.*, Feb. 1985, pp. 94–95.

[12] D. M. Cooper, S. P. Craig, C. R. Day, and B. J. Ainslie, "Multiple-index structures for dispersion-shifted single-mode fibers," in *OFC'85 Tech. Dig.*, Feb. 1985, pp. 96–97.

[13] P. K. Runge and P. R. Trischitta, "The SL Undersea Lightguide System," *IEEE J. Select. Areas Commun.*, vol. SAC-1, pp. 459–466, Apr. 1983.

[14] C. H. Gartside and J. L. Baden, "Single-mode ribbon cable and array splicing," in *OFC'85 Tech. Dig.*, pp. 106–107.

[15] R. Csencsits, P. J. Lemaire, W. A. Reed, D. S. Shenk, and K. L. Walker, in *OFC'84 Tech. Dig.*, Jan. 1984, New Orleans, LA, pp. 54–55.

[16] L. G. Cohen, C. Lin, and W. G. French, "Tailoring zero chromatic dispersion into the 1.5–1.6 μm low-loss spectral region of single-mode fibers," *Electron. Lett.*, vol. 15, pp. 334–335, June 1979.

[17] L. G. Cohen, W. L. Mammel, and S. Lumish, "Tailoring the shapes of dispersion spectra to control bandwidths in single-mode fibers," *Opt. Lett.*, vol. 7, pp. 183–185, Apr. 1982.

[18] F. P. Kapron, "Maximum Information Capacity of Fiber-Optic Waveguides," *Elecron. Lett.*, vol. 13, pp. 96–97, Feb. 1977.

[19] L. G. Cohen, W. L. Mammel, J. Stone, and A. D. Pearson, "Transmission studies of a long single-mode fiber—Measurements and considerations for bandwidth optimization," *Bell Syst. Tech. J.*, vol. 60, pp. 1713–1725, Oct. 1981.

[20] K. Ogawa, "Considerations for single-mode fiber systems," *Bell Syst. Tech. J.*, vol. 61, pp. 1919–1931, Oct. 1982.

[21] R. A. Linke, "Transient chirping in single-frequency lasers: Lightwave systems considerations," *Electron. Lett.*, vol. 20, pp. 472–474, May 1984.

[22] D. C. Gloge and K. Ogawa, "Technology for intercity lightwave systems," in *OFC'85 Tech. Dig.*, Feb. 1985, pp. 84–85.

[23] J. C. Campbell, A. G. Dentai, W. S. Holden, and B. L. Kasper, "High-performance avalanche photodiode with separate absorption, grading, and multiplication regions," *Electron. Lett.*, vol. 19, pp. 818–820, 1983.

[24] M. Shikada *et al.*, "100 Mbit/s ASK heterodyne detection experiment using 1.3 μm DFB-laser diodes," *Electron. Lett.*, vol. 20, pp. 164–165, Feb. 1984.

[25] S. K. Korotky *et al*, "4 Gb/s transmission experiment over 117 km of optical fiber using a Ti:$LiNbO_3$ external modulator," in *OFC'85*, post deadline paper PDO*1*, *Feb.* 1985.

[26] N. A. Olsson *et al.*, "68.3 km transmission with 1.37 Tbit-km/s capacity using wavelength-division multiplexing of ten single-frequency lasers at 1.5 μm," *Electron. Lett.*, vol. 1, pp. 105–106, Jan. 1985.

[27] R. E. Wagner, "Future 1.55-μmm undersea lightwave system," *J. Lightwave Technol.*, vol. LT-2, pp. 1007–1015, Dec. 1984.

[28] V. J. Mazurczyk *et al.*, "420 Mbs transmission through 203 km using very low loss fiber and a DFB laser," in *10th European Conf. Opt. Commun.* (*ECOC*), post-deadline proc., Stuttgart, Germany 1984.

[29] N. S. Bergano, R. E. Wagner, H. T. Shang, and P. F. Glodis, "150-km 296-Mbit/sec dispersion-shifted fiber system experiment," in *OFC'85 Tech. Dig.*, Feb. 1985, pp. 50–51.

[30] R. E. Wagner, L. G. Cohen, D. Kalish, and D. A. Meade, "Performance of a 29.6-km cable core for 1.55-mum undersea systems," in *OFC'85 Tech. Dig.*, Feb. 1985, pp. 106–107.

[31] "The long-haul fiberoptic carriers," Kessler Marketing Intelligence Rep., 22 Farewell St., Newport, RI, Oct. 1984.

[32] "AT&T FT Series G Lightwave System," AT&T Network Systems Marketing Communications, Sept. 1984.

[33] R. K. Snelling, K. Wilson Kaplan, and W. B. Holmes, "Bringing fiber to the home: What will it take?," in *OFC'85 Tech. Dig.*, Feb. 1985, pp. 6–7.

[34] "The SLC 96 subscriber loop carrier system," Special Issue of the *AT&T Bell Lab. Tech. J.*, vol. 63, Dec. 1984.

[35] P. P. Bohn, C. A. Brackett, M. J. Buckler, T. N. Rao, and R. H. Saul, "The fiber SLC carrier system," ibid., pp. 2389–2416.

[36] P. E. Rubin and M. E. Schweig, "VIVID™ video teleconferencing goes crosstown with lightwave," *AT&T Bell Lab. Rec.*, pp. 20–25, Sep. 1985.

[37] R. C. Alferness, "Guided-wave devices for optical communication," *IEEE J. Quantum Electron.*, vol. QE-17, pp. 946–959, 1981.

[38] R. M. Metcalfe and D. R. Boggs, "Ethernet: Distributed packet switching for local computer networks," *Commun. Ass. Comput. Mach.*, vol. 19, no. 7, pp. 395–404, July 1976.

[39] E. G. Rawson and R. K. Schmidt, "Fibernet II: An ethernet-compatable fiber optic local area network," in *Proc. LAN'82*, Sept. 1982, pp. 86–90.

[40] D. C. Hansen, "Fiber Optics in Local Area Network Applications," *Advances in Electronics and Electron Physics*, vol. 57, New York: Academic, 1981.

[41] "AT&T unveils lightwave links for data communications," AT&T News Release, Sept. 20, 1984.

[42] A. G. Fraser, "DATAKIT—A modular network for synchronous and asynchronous traffic," in *Proc. ICC*, June 1979, pp. 20.1.1–20.1.3.

[43] D. Hanson, "OFC'85 Minitutorial: Fiber-Optic Local Area Networks," Feb. 1985.

[44] G. Arnold and O. K. Krumpholz, "Coupling of monomode fibers to edge-emitting diodes," in *OFC'85 Tech. Dig.*, Feb. 1985, pp. 48–50.

[45] P. W. Shumate, J. L. Gimlett, M. Stern, M. B. Romeiser, and N. K. Cheung, "Transmission of 140 Mb/s signals over single-mode fiber using surface-and-edge emitting 1.3 μm LED's," in *OFC'85*, post deadline paper PDO3, Feb. 1985.

Leonard G. Cohen (M'68–SM'77–F'85) received the B.E.E. degree from City College, City University of New York, NY, and the Sc.M. and Ph.D. degrees from Brown University, Providence, RI, in 1962, 1964, and 1968, respectively.

In 1968, he joined the Guided Wave Research Laboratory, AT&T Bell Laboratories, Holmdel, NJ, where he worked on optical fiber transmission studies. At present, he is Supervisor of the Lightguide Materials, Diagnostics, and Applications Group in the Materials Research Laboratory, Murray Hill, NJ.

Dr. Cohen received the 1980 IEEE Electronics Letters Premium Award for a paper relating to optical fibers; the 1981 Bell System Technical Journal Best Paper Award for Physical Sciences and Devices; and the 1982 Bell System Technical Journal Honorable Mention Award for a paper relating to physical sciences and devices. He is a member of Sigma Xi, Tau Beta Pi, Eta Kappa Nu, and the Optical Society of America.

TERABIT LIGHTWAVE NETWORKS: THE MULTIHOP APPROACH

Anthony S. Acampora, Mark J. Karol, and Michael G. Hluchyj

AT&T TECHNICAL JOURNAL

Anthony S. Acampora *is director of the Transmission Technology Laboratory and* ***Mark J. Karol*** *is a member of technical staff in the Network Systems Research Department. Both are at AT&T Bell Laboratories in Holmdel, New Jersey, and are involved with basic and applied research on local communications networks.* ***Michael G. Hluchyj*** *is a former member of technical staff in the Network Systems Research Department. Mr. Acampora joined the company in 1968 and has a B.S.E.E., M.S.E.E., and Ph.D. in electrical engineering from the Polytechnic Institute of Brooklyn. Mr. Karol joined the company in 1985 and has a B.S.E.E. and a B.S. in mathematics from Case Western Reserve University, and an M.S.E.E., M.A., and Ph.D. from Princeton University.*

For multiuser communications networks, lightwave technology offers the potential to supply a pool of capacity—to be shared among users—far above that provided by any alternative technology. However, the bandwidth limitation of the electro-optic converters needed to attach each user to the optical medium prevents any one user from accessing more than a tiny fraction of the overall capacity. This paper discusses the problems with conventional approaches for tapping the capacity contained within the optical communications band. It then proposes a new network architecture that permits tapping lightwave's vast capacity potential without requiring a technological breakthrough. With this approach, it becomes possible to create networks that offer hundreds of thousands of gigabits-per-second total capacity, to be shared among users, each limited to a peak rate of 1 Gb/s.

Background

In long-distance point-to-point communications, lightwave has clearly emerged as the technology of choice.[1,2] For this application, the figure of merit for comparing alternative approaches and technologies traditionally has been the product of the achievable bit rate and the unrepeatered distance. It is in this regard that lightwave technology excels. What finally limits the figure of merit is the loss and dispersion in the optical transmission medium, with the following combining to increase the achievable figure-of-merit:

- Improvements in fiber fabrication processes
- The transitions from multimode to single-mode fiber and from multimode to single-mode lasers
- Operation near the zero-dispersion wavelength of 1.3 μm
- Availability of narrow-linewidth lasers
- The introduction of coherent demodulation in receivers with performance that approaches the quantum limit.

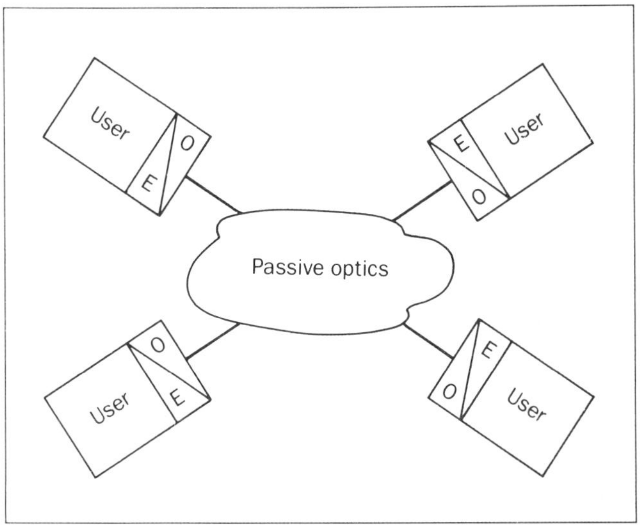

Figure 1. Ideal lightwave network. E/O is an electro-optic interface.

These impressive accomplishments have raised the obvious follow-up question: Are the advantages of lightwave technology limited in utility to long-distance point-to-point communications? In particular, what opportunities, advantages, and problems arise, if any, when lightwave technology is applied to the domain of multiuser communication networks? Perhaps the most dramatic appeal of lightwave technology stems from the opportunity to share an unprecedented amount of capacity among all network users.

If we limit the radius of the network to 20 km, then dispersion and transmission loss become much less of an issue, and the network could still interconnect users located within a good-sized city. Conservatively, if we estimate that the optical band lies in the 200-nm window centered at the zero-dispersion wavelength of 1.3 μm, we find that the medium possesses the enormous bandwidth of 35 terahertz (THz). In contrast, a "broadband" coaxial-cable-based system has a bandwidth of about 300 MHz (five orders of magnitude lower than fiber), and ordinary twisted-pair copper wiring has a bandwidth of only several megahertz (seven orders of magnitude lower). The appeal of lightwave is obvious.

But although the optical medium possesses enormous bandwidth, any one user may tap into—at most—a bandwidth of "only" several gigahertz because each user is limited by the bandwidth constraint that electro-optic access establishes. In a fundamental sense, the bandwidth associated with modulation of photons by electrons and electronic signals generated by photodetection will always be orders of magnitude lower than the medium's inherent bandwidth. Thus, perhaps the most basic issue to arise in multiuser lightwave networks is: How can we tap the optical medium's vast bandwidth potential and effectively share it among many users, each constrained by the far lower bandwidth of electro-optic conversion?

Figure 1, which shows a simplified version of an ideal lightwave network, depicts this situation. The network consists of three parts:

- The *passive optical media*. This is the "glass" portion of the network that contains all the bandwidth.
- The *generic "users"* themselves, electronic in nature. Each user may be a PBX (private branch exchange) trunk, LAN (local-area network) gateway, host computer, graphics workstation, high-resolution video camera or monitor, fast document scanner, high-speed printer, digital telephone, etc.
- The *electro-optic conversion* that changes electronic signals to optical signals, and vice versa.

The peak data rate that any user generates or receives can be no greater than that allowed by the electro-optic bandwidth constraint. (We envision a user interface in which traffic is converted to a packet format. User data—generated in bursts or streams—is broken into packets, and each packet's transmission rate is the maximum that the electro-optic bandwidth constraint permits.) We see that each user can access (transmit or receive) only a tiny portion of the optical bandwidth.

To tap the vast bandwidth potential, it is essential

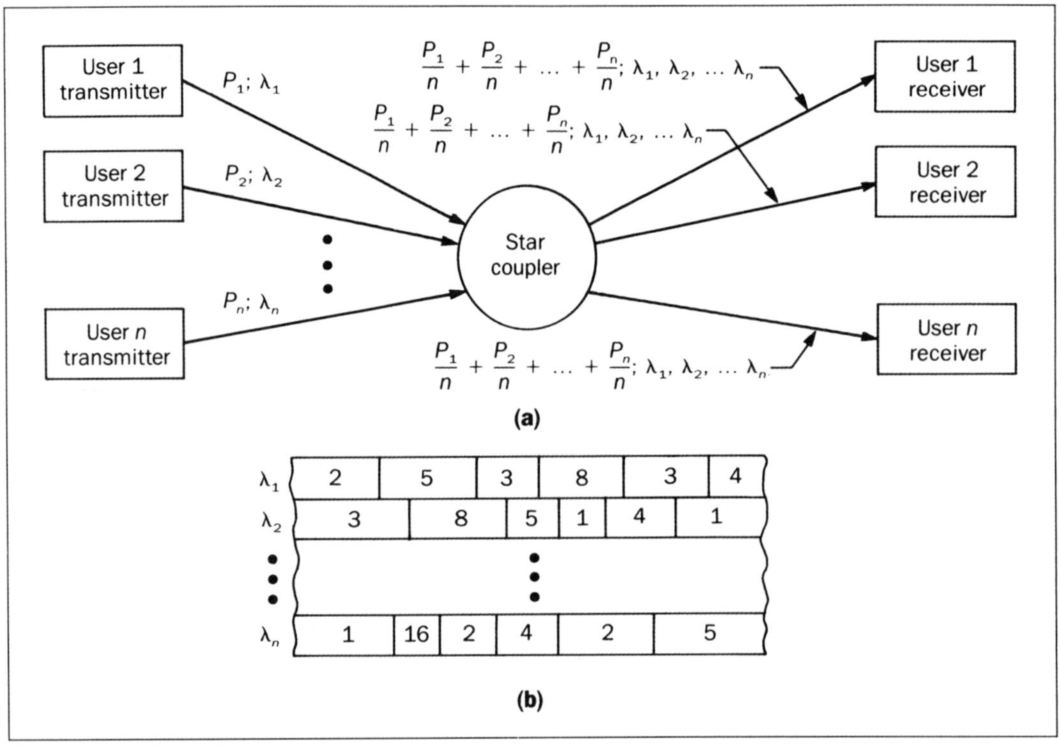

Figure 2. Conventional approach to achieving concurrency. (a) A wavelength-division multiplexed system; (b) interconnection pattern. *P* = power; λ = wavelength.

that a multitude of messages among the various users concurrently reside on the optical medium. Thus, the great challenge is to devise a networking arrangement that achieves a high level of concurrency. Simply put, we want to build a network that offers a total capacity, or throughput, measured in terabits per second from electronic components that operate in the gigabit-per-second range.

At first glance, it might appear that concurrency is easy to achieve with a straightforward, conventional approach. Simply assign to each user a unique wavelength on which each packet may be transmitted in the network. Figure 2a illustrates the approach.

Here, user j has been assigned to transmit on wavelength λ_j, $j = 1, 2, \ldots, n$. All signals are linearly combined in a centrally located passive-star coupler, and the superimposed signals are made available to all receivers. Notice that splitting each signal n ways at a star coupler causes a power loss of $1/n$ on each signal toward each receiver. The star topology best minimizes the loss for this broadcast arrangement, while providing acceptable detectability.

Each receiver, then, simply tunes to the wavelength of interest, much like a listener tunes a radio receiver to pick up the desired station from the sea of stations being broadcasted. Figure 2b illustrates a typical interconnection pattern among the n users, showing the time sequence of the channels to which each receiver must tune. Two problems are apparent:

- Each receiver must be instructed about which channel must be tuned for each interval to receive the appropri-

Figure 3. Multihop lightwave network. NIU = network interface unit; λ = wavelength.

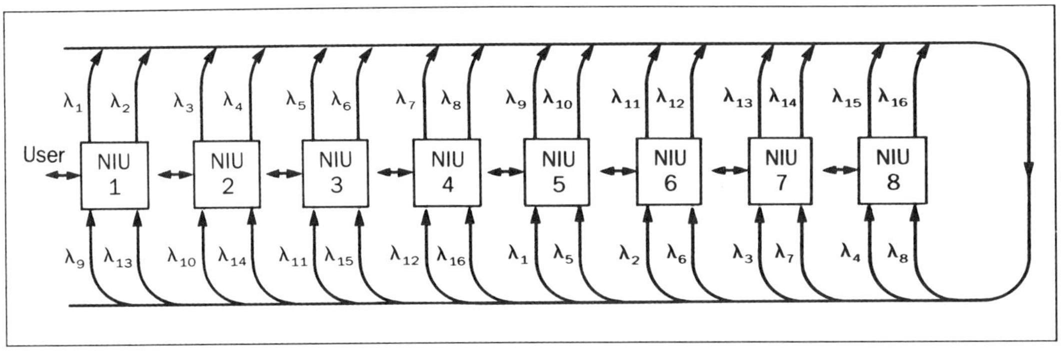

ate messages. Such pretransmission coordination requires communication among the users and is generally done using a separate signaling or reservation channel that is common to all users.

- By far the larger problem, each receiver must tune rapidly over the entire optical band in the prescribed interconnection sequence. Optical receivers that can be tuned over such a broad range on a time scale comparable with the packet transmission rate (i.e., tune in several tens of nanoseconds, and then listen for several microseconds) are well beyond the current state of the art and may require a profound technological breakthrough.

Thus, we have a dilemma: Lightwave networks appear to offer unique possibilities by virtue of the vast bandwidth potential, but suffer unique constraints by virtue of limitations of the technology. The challenge, then, is to devise a new architectural approach to resolve this dilemma.

The multihop approach to lightwave networking, described in this paper, resolves this dilemma without requiring a technological breakthrough. In fact, after illustrating the approach with an embodiment that exploits today's technology, we conclude with an embodiment that can be realized with yesterday's technology.

The multihop approach is based on the observation that optical bandwidth is so plentiful that—for the first time—it is acceptable to "squander" some bandwidth if this permits tapping the rest. Like the approach of Figure 2, multihop relies on wavelength-division multiplexing (WDM), except that all transmit and receive wavelengths are permanently assigned to each user; that is, wavelength agility or tunability is not needed.

The wavelength assignments are made such that at least one user can receive each transmission. But that user, on determining that a received message is intended for a different user, simply retransmits the message onto one of its transmit wavelengths where either the destination user receives it or yet another user relays it. The wavelengths are assigned in such a way that a path exists between all user pairs and, for a given number of transmitters and receivers per user, the average number of hops a packet must take to reach its destination is close to the minimum. It is worth noting that multihop networks are common in packet communications systems—such as packet radio networks and local, metropolitan, and wide-area networks.[3-8]

Typical capacity or throughput results are as follows. Let the network contain about 900 users that each can transmit on two wavelengths and receive on two wavelengths. Let each transmitter and receiver operate at a peak rate of 1 Gb/s. Then, after accounting for the effects of multiple hops, the network will provide a total capacity of 200 Gb/s to be shared among all users. Moreover, each user enjoys a peak transmission rate of 1 Gb/s and can present a sustained, effective average-traffic load of about 220 Mb/s. The throughput of a multihop lightwave network scales linearly with the user transmission rate, so each

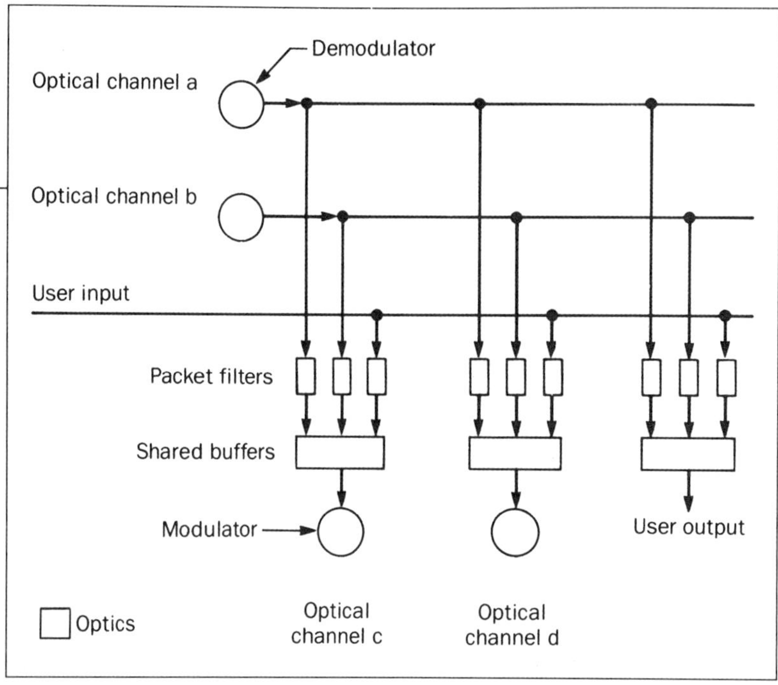

Figure 4. Block diagram of two-receiver, two-transmitter network interface unit.

user can present an average load of about 22 Mb/s if the transmission rate is 100 Mb/s.

Each user needs equipment similar to that for a simple network in which all users share a single 1-Gb/s channel. However, for such a single-channel network, a total capacity of only 1 Gb/s is to be shared among all users, and each of 900 users connected to the network could present an average traffic load of only 1.1 Mb/s. Thus, by doing little more than permanently assigning different users to different wavelengths, both network capacity and sustainable throughput per user are tremendously increased. Moreover, as more users attach to the network, each brings its own channels, or resources, increasing the total capacity.

We will illustrate the multihop concept first for the case where each network port is equipped with two fixed-wavelength transmitters and two fixed-wavelength receivers. This then will be generalized to an arbitrary number p of transmitters and receivers. The benefits of such a generalization include:

1. The ability to concentrate the traffic of several users through each port, which increases the sustainable throughput per user
2. The ability to grow the network modularly without rearranging existing channel assignments
3. Improved reliability.

Finally, we shall show how to implement the multihop concept without wavelength-division multiplexing, thereby permitting the use of yesterday's optical technology. The same approach, coincidentally, also provides a way to realize a large, centrally located space-division packet switch as well as a distributed-lightwave network.

The Multihop Approach

In the multihop approach to lightwave networks, wavelength-division multiplexing is the key to tapping the bandwidth potential of the optical medium, allowing simultaneous transmissions among multiple, electronic speed-limited users. But to avoid the problems of pretransmission coordination and wavelength agility common in conventional approaches, a multihop network uses only fixed-wavelength transmitters and receivers. Transmitting a packet from one user to another may require routing the packet through intermediate users, each repeating the packet on a new wavelength, until the packet is finally transmitted on a wavelength that the destination user receives. In other words, a packet may have to take multiple hops to reach its destination.

Although the physical topology for a multihop network can take a variety of forms (e.g., star, bus, or tree), for simplicity, Figure 3 illustrates an eight-user multihop network in a bus topology. The users interface to the unidirectional optical bus through network interface units (NIUs) distributed along the bus. Each NIU has two fixed-wavelength transmitters and two fixed-wavelength receivers. The example requires sixteen WDM channels. All information transmitted onto the bus passively loops back past every NIU, but each NIU hears only information transmitted on channels for which it has a receiver.

The scheme works as follows. For illustration, suppose user 1 wants to send a message to user 5. NIU 1 receives the message from user 1, appropriately addressed with a header that identifies user 5 as the destination, and transmits the message on wavelength λ_1. NIU 5 receives the message, recognizes (by examining the destination address) that it is intended for "local reception," and passes it to user 5 through the user port. Because NIU 5 is (permanently) assigned to receive on one of the two channels on which NIU 1 is assigned to transmit, namely λ_1, messages pass from NIU 1 to NIU 5 with just a single hop through the network.

But if user 1 wants to send a message to user 2, we recognize that NIU 2 does not receive any of the transmit channels assigned to NIU 1. However, if we use NIU 5 as an intermediate-relay node, we can establish a two-hop path between NIU 1 and NIU 2. Specifically, when NIU 1 receives a message from user 1 that is destined for user 2, it transmits the message on wavelength λ_1. When NIU 5 receives the message, it recognizes that the message is not intended for local reception and, with user 2 as the destination address, retransmits the message over wavelength λ_{10}. NIU 2 is assigned to receive wavelength λ_{10}, so it receives the message, and then passes it on to user 2 through the user port.

As a final example, consider messages that user 1 sends to user 7. These messages will require three passes through the network, using first NIU 5 and then NIU 2 as intermediate-relay nodes. NIU 1 transmits to NIU 5 using wavelength λ_1, which demodulates the message and remodulates it onto λ_{10}. NIU 2 receives the message over λ_{10} and relays it to the destination NIU 7 using wavelength λ_3.

Channels are permanently assigned to users such that after one or more passes through the network, a message is assured reception by the proper user's receiver. In the example just presented, three hops is the maximum

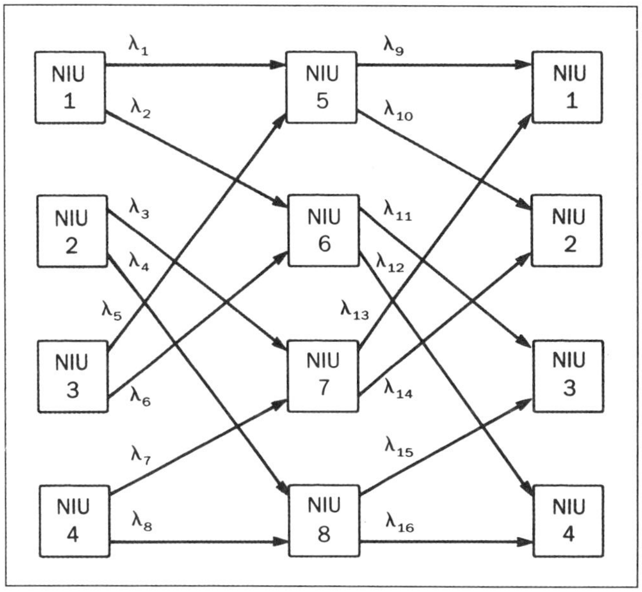

Figure 5. Connectivity of network interface units (NIUs), for the example in Figure 3. λ = wavelength.

required of any source-destination pair.

With a single user attached to each NIU, Figure 4 shows a functional block diagram of a two-transmitter, two-receiver NIU. The NIU functions as a three-by-three nonblocking packet switch: two inputs and two outputs for the optical channels, and an electronic input and output for the user. An additional input and output can be included for control information.

The incoming optical signals for the two assigned receive wavelengths are demodulated to baseband electrical signals that, together with the generated user data, enter the three-by-three nonblocking packet switch on separate inputs. The destination address in the header of each arriving packet determines to which of the three outputs the packet is to be routed. Buffers are provided at the outputs, because multiple packets may arrive on different inputs destined for the same output. Two outputs drive optical modulators for the two

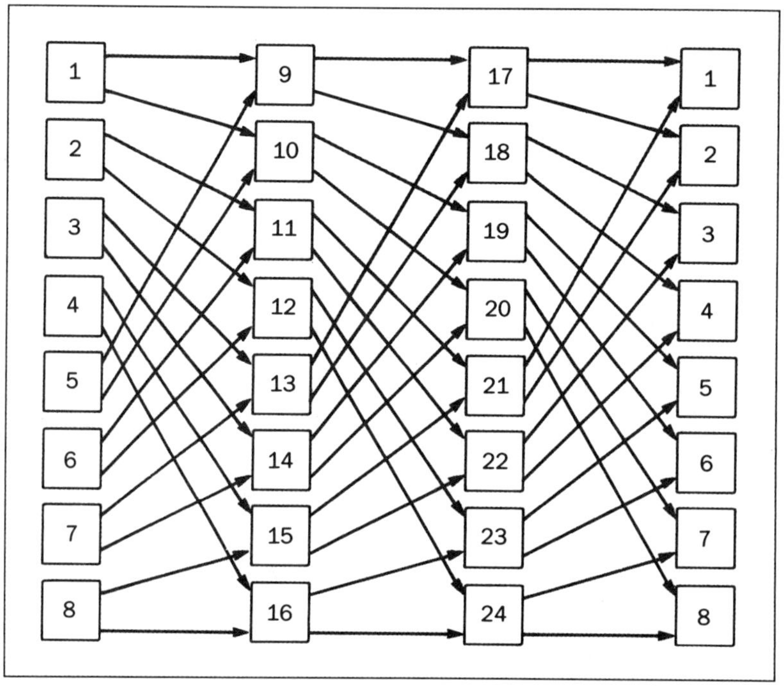

Figure 6. Multihop connectivity of 24 network interface units.

assigned transmit wavelengths at the NIU. The third output goes directly to the user.

Figure 5 shows the connectivity graph associated with the example of Figure 3. The connectivity graph indicates all (ordered) pairs of NIUs that are separated by a single hop. Specifically, for each NIU in the network, we draw a directed arc from an NIU that transmits on a particular wavelength to all NIUs that receive the wavelength. In this connectivity graph (Figure 5), the eight NIUs are placed in two columns of four NIUs each. The assignment of transmit and receive wavelengths is such that the interconnection pattern between columns is a perfect shuffle,[9] with the right side of the second column connected to the left side of the first column as if the entire graph were wrapped around a cylinder.

Figure 6 shows the connectivity graph for a 24-NIU multihop network. Here, using 48 WDM channels, 24 NIUs are arranged in three columns of eight NIUs each. In general, for two transmitters and two receivers per NIU, $n = k2^k$ NIUs ($k = 2, 3, \ldots$) are arranged in k columns of 2^k NIUs each, making use of $W = k2^{k+1}$ WDM channels.[10]

The recirculating perfect-shuffle connectivity graph can be further generalized for the case of p transmitters and p receivers ($p = 1, 2, \ldots$) per NIU.[11] Figure 7 shows an 18-NIU connectivity graph with $p = 3$ and $k = 2$. The resulting pattern of arcs between adjacent columns is called a p-shuffle,[12] being a generalization of the ($p = 2$) perfect shuffle.[9] A (p,k) recirculating perfect-shuffle connectivity graph has $n = kp^k$ NIUs ($k = 2, 3, \ldots; p = 1, 2, \ldots$) in k columns of p^k NIUs each, using $W = kp^{k+1}$ WDM channels.

Although the transmit and receive wavelengths for the NIUs in a multihop network could be assigned using other connectivity graphs, the class of recirculating perfect-shuffle graphs has several desirable features

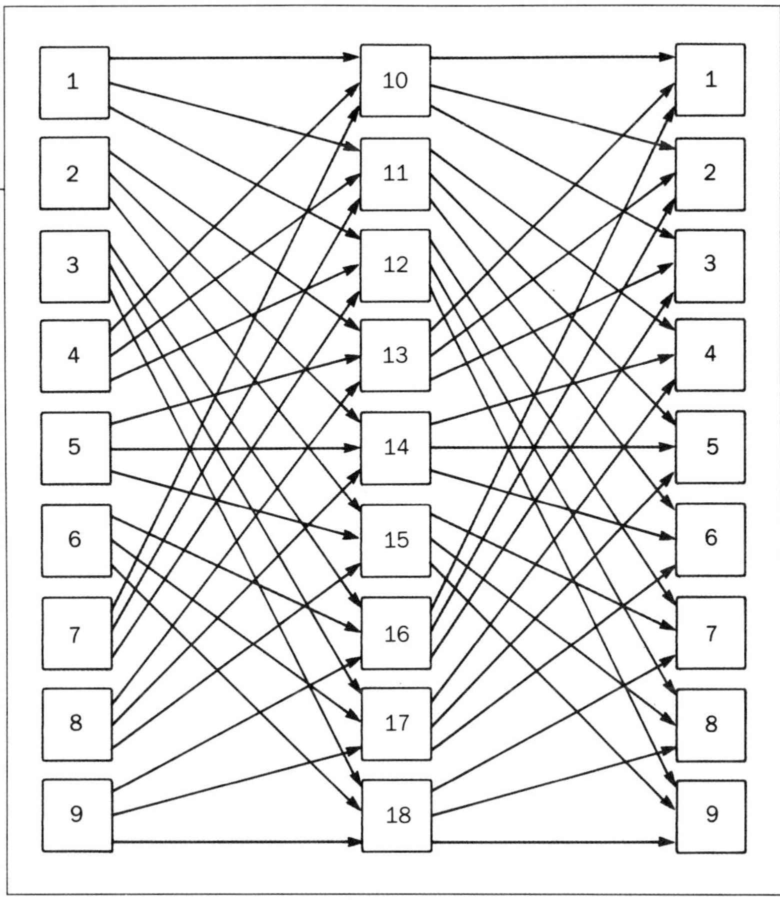

Figure 7. Multihop connectivity of 18 network interface units ($p = 3, k = 2$).

related to the network's performance, modularity, and reliability. We will now focus on performance and then cover modularity and reliability in the next section.

Performance. For a recirculating perfect-shuffle connectivity graph, a path that has at most $2k - 1$ hops exists from each NIU to all others. Figure 8 illustrates this for the ($p = 2$, $k = 3$) 24-NIU connectivity graph, where we show a spanning tree for routing packets from NIU 1 to all other NIUs. Because of the symmetric nature of the graph, we can draw a topologically equivalent spanning tree for all other NIUs. Notice from Figure 8 that the depth of the spanning tree grows as $2k - 1$, which is about proportional to the base p logarithm of the number of NIUs. More specifically, we have

$$\text{Number of NIUs } h \text{ hops from source} = \begin{cases} p^h & h = 1, 2, \ldots, k-1 \\ p^k - p^{h-k} & h = k, k+1, \ldots, 2k-1 \end{cases}$$

so that the expected number of hops between two randomly selected users is given by

$$E[\textit{number of hops}] = \frac{kp^k(p-1)(3k-1) - 2k(p^k-1)}{2(p-1)(kp^k-1)} \quad (1)$$

If the network is uniformly loaded (i.e., each user generates an equal amount of traffic destined for all other users) and all WDM channels are loaded equally, the channel efficiency η for the network is given by

$$\eta = \frac{1}{E[\textit{number of hops}]} = \frac{2(p-1)(kp^k-1)}{kp^k(p-1)(3k-1) - 2k(p^k-1)} \quad (2)$$

Figure 9 shows the channel efficiency η as a function of the connectivity graph degree p for various numbers

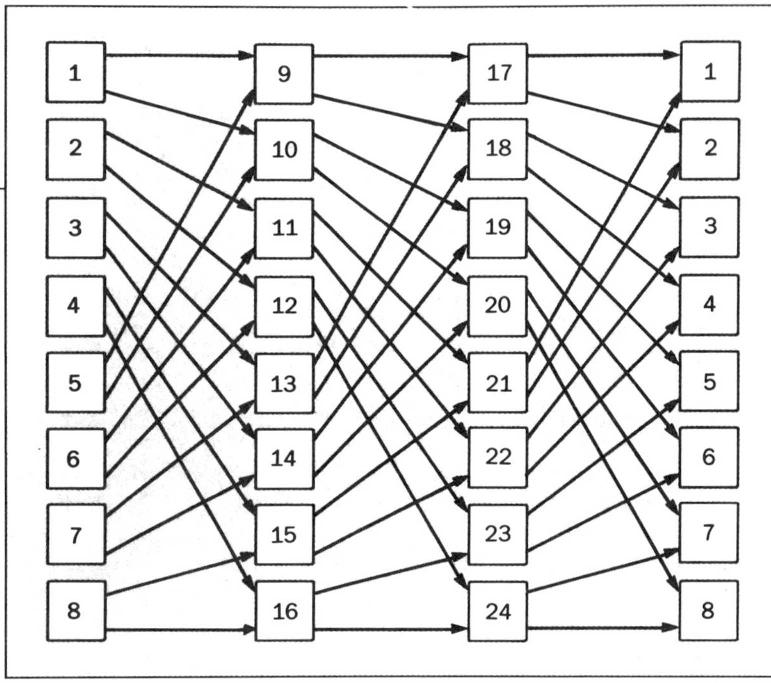

Figure 8. Spanning tree (paths in color) for multihop connectivity graph of 24 network interface units ($p = 2$, $k = 3$).

of columns k. Also plotted in Figure 9, for each k, is an upper bound on the channel efficiency for any regular directed graph of degree p. To obtain the upper bound, we assume that an ideal (but not necessarily realizable) connectivity graph exists with an associated routing such that each NIU has a perfect p-ary spanning tree for routing packets and all channels are load balanced.[11] Notice how close the performance of the recirculating, perfect-shuffle connectivity graph is to that of the ideal graph. For large p, Equation (2) becomes

$$\eta = \frac{2}{3k - 1} \quad \text{for } p = \infty \tag{3}$$

which shows clearly how the channel efficiency decreases as the inverse of the number of columns k.

Our main concern, however, is not with the channel efficiency, but with achieving concurrency in the network as measured by the network throughput. The total network throughput and the throughput per NIU are simply $\eta W = \eta k p^{k+1}$ and ηp, respectively, normalized to the channel transmission rate. Table I shows, for various p and k, some of the possible network configurations in terms of the number of NIUs and total network throughput with all kp^{k+1} WDM channels operating at 1 Gb/s. (All throughput values given in this paper scale linearly with the user transmission rate.) Observe from Table I that total throughputs in the hundreds and even thousands of gigabits per second are achievable even though each user is limited to 1 Gb/s.

Figure 10 shows the throughput per NIU as a function of the number of NIUs in the network for various values of p, the number of transmitters and receivers per NIU. Each curve is parameterized in k, with the top left side corresponding to $k = 2$. For comparison, we also show the throughput per user of a single-channel network that operates at 100-percent media-access efficiency. The throughput per user for the single-channel network decreases as $1/n$ versus the nearly $1/\log_p n$ decrease for the multihop network. In particular, with 896 NIUs, the throughput per user of a single-channel network is 1.1 Mb/s versus 222 Mb/s (i.e., a factor of 200 greater) for the multihop network.

Propagation delay is related to the geographical extent of the network, but multiuser lightwave networks are generally characterized by a large ratio of propagation delay to packet transmission time. For example, with a transmission rate of 1 Gb/s over a 10-km length of fiber, the propagation delay is 50 times the transmission time of a 1000-bit packet (assuming the speed of light in fiber is $2 \cdot 10^8$ m/s). Reference 10 presents an approximate expression for the mean queueing delay and buffer overflow probability.

Although the focus of this paper is on the dedicated-

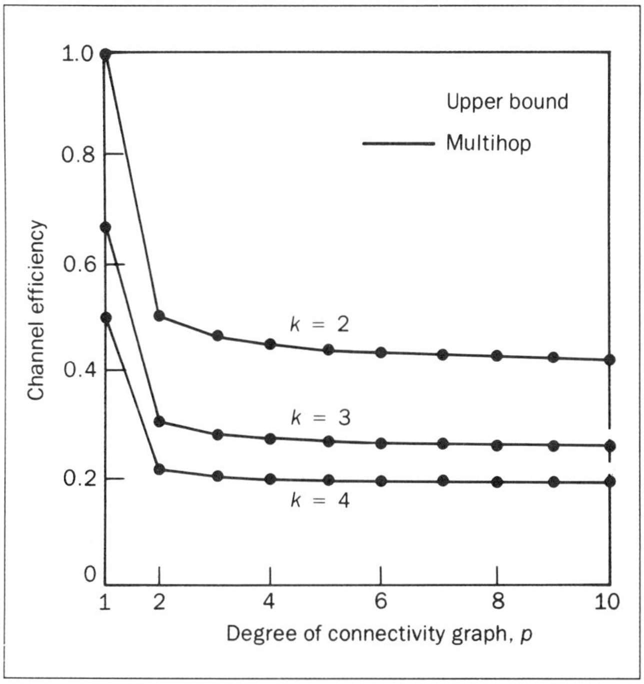

Figure 9. Multihop channel efficiency. k = number of columns in connectivity graph.

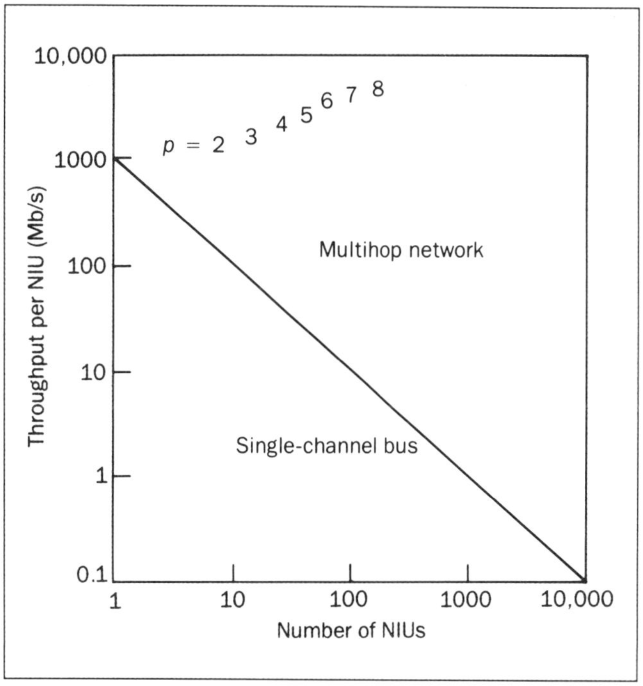

Figure 10. Multihop throughput per network interface unit (NIU) for values of p (number of transmitters and receivers per NIU).

channel multihop network (i.e., each arc in the connectivity graph represents a separate WDM channel), it is worth noting that the same connectivity graphs apply to the use of shared channels among the NIUs. Figure 11 shows the same connectivity graph as Figure 5, but here only four WDM channels are shared among the eight NIUs. Each NIU has only one fixed-wavelength transmitter and one fixed-wavelength receiver, with NIU pairs (1,3), (2,4), (5,7), and (6,8) each sharing a common channel. In general, with a connectivity graph of degree p, each NIU still requires only one transmitter and one receiver, but now p NIUs share each of the kp^{k-1} channels. Because multiple NIUs are transmitting on a common channel, we have to contend with a multiple access problem and the possible inefficiencies that result. In the next section, we show how—with concentration—one can have the equivalent of one transmitter and one receiver per user with dedicated channels.

Concentration, Modularity, and Reliability

A characteristic of fully distributed networks, such as found today in many local-area networks, is network interface circuitry that is located close to each user. Although this may minimize network interface equipment, one must also consider issues such as media-access efficiency, network capacity, security, maintenance, and administration. Networks with desirable characteristics along these lines typically have more of a centralized or, at least, hierarchical structure (see, for example, reference 4).

Figure 12 illustrates how a multihop network

Table I. Multihop Total Throughput, 1-Gb/s User Transmission Rate

(k,p)	Number of NIUs	Total throughput (Gb/s)
(2,2)	8	8.000
(2,3)	18	24.81
(3,2)	24	14.72
(2,4)	32	56.69
(2,5)	50	108.4
(4,2)	64	27.62
(2,6)	72	184.8
(2,7)	98	290.6
(3,3)	81	68.21
(5,2)	160	52.73
(3,4)	192	208.1
(4,3)	324	193.6
(3,5)	375	498.4
(6,2)	384	101.9
(3,6)	648	1,021
(7,2)	896	198.7
(4,4)	1,024	791.8
(3,7)	1,029	1,877
(5,3)	1,215	560.1
(4,5)	2,500	2,380
(6,3)	4,374	1,640
(5,4)	5,120	3,071
(4,6)	5,184	5,867
(4,7)	9,604	12,604
(7,3)	15,309	4,834
(5,5)	15,625	11,573
(6,4)	24,576	12,037
(5,6)	38,880	34,305

NOTE: k = number of columns in connectivity graph; p = number of transmitters (receivers) per NIU; NIU = network interface unit.

might be physically configured to support a large number of users with a smaller number of network interface units. Each NIU attaches to users on one side via separate access lines (possibly fiber) and to the other NIUs on the other side via dedicated WDM channels that are combined at a star coupler. Each user receives only that information destined for it; the NIUs forward packets that require multiple hops in the network. The NIUs can be geographically distributed—housed in satellite closets in a building or "remote terminals" in a city—or centrally located in a common equipment frame, such as in a central office.

Besides improved network security, monitoring, and maintenance, this approach can achieve the equivalent of a single transmitter and single receiver per user, while increasing the network capacity and eliminating the media access associated with shared channels. Specifically, suppose each NIU serves p users and contains p transmitters and p receivers (i.e., one pair per user) to form a p-ary multihop network with the other $kp^k - 1$ NIUs in the network. (The degree of concentration can be selected to yield a desired "equivalent number" of transmitters and receivers per user.) The throughput per user is then simply the channel efficiency η, as given in Equation (2), times the channel transmission rate.

For a 1-Gb/s channel transmission rate, Figure 13 shows the throughput per user as a function of the number of users for various NIU sizes. For comparison, we see that the throughput per user of a single-channel network—if we assume 100-percent media-access efficiency—decreases linearly with the number of users. For multihop, the throughput per user decreases (nearly) proportional to the logarithm of the number of users. Figure 13 also illustrates the performance advantages associated with increasing p (i.e., the NIU size) to grow the network while keeping k small.

Within the context of the concentrator, we also have a convenient way of modularly growing a multihop network. Figure 14 shows that the connectivity graph for a (p = 2, k = 2) 8-NIU multihop network is a subgraph of the (p = 3, k = 2) 18-NIU connectivity graph. This holds true for any given k, as p is increased. Hence, one

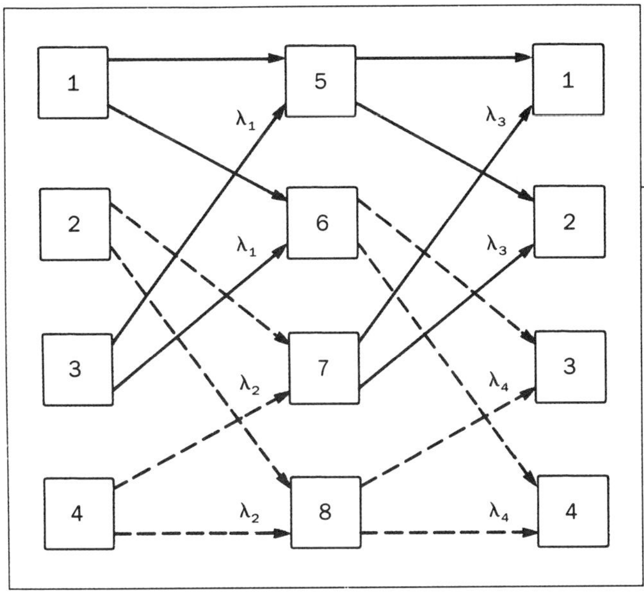

Figure 11. Multihop connectivity graph for eight users with four shared channels (p = 2, k = 2). λ = wavelength.

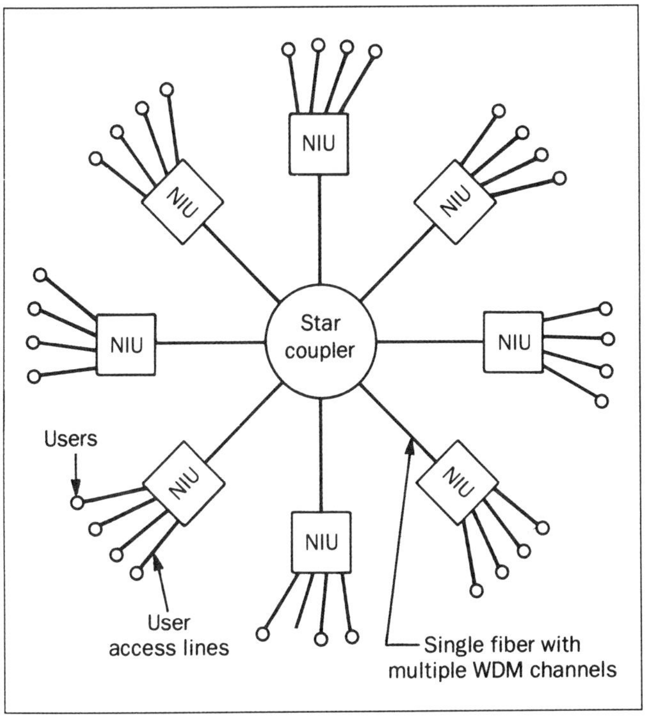

Figure 12. A multihop physical topology with network interface units (NIUs) as concentrators. WDM = wavelength-division multiplexing.

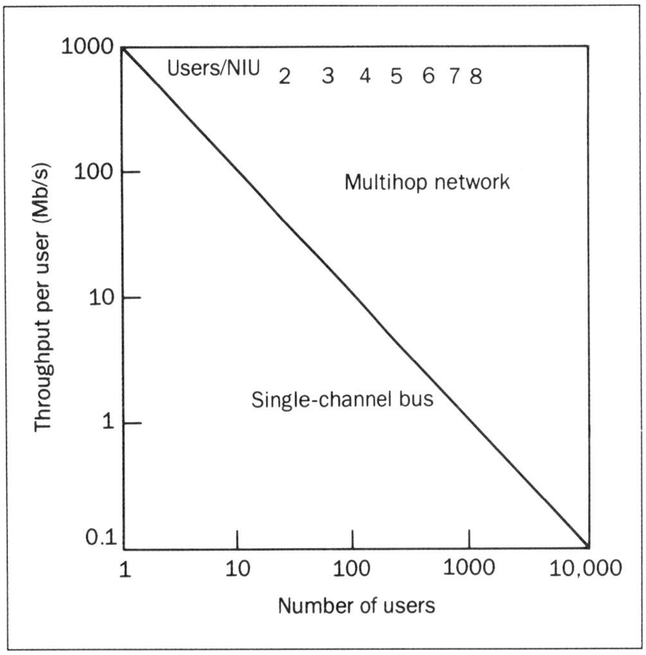

Figure 13. User throughput with concentrators; one transmitter and one receiver per user. NIU = network interface unit.

can grow from small to large multihop networks by adding new NIUs and adding (or enabling) additional transmitters and receivers to existing NIUs. As long as the number of columns k in the recirculating perfect-shuffle connectivity graph remains fixed, growing the network by increasing p requires no changes in the assignment of transmit and receive wavelengths to existing NIUs. To ensure that a path exists between all NIU pairs, while growing a multihop network from one value of p to the next, requires adding at most k new NIUs to the network (one per column in the connectivity graph) at any one time. Thus, the growth in the network can take place in small increments.

Finally, the cylindrical nature of the connectivity graph permits us a degree of fault tolerance. If, for example, NIU 5 in Figure 5 were to fail, the normal path from

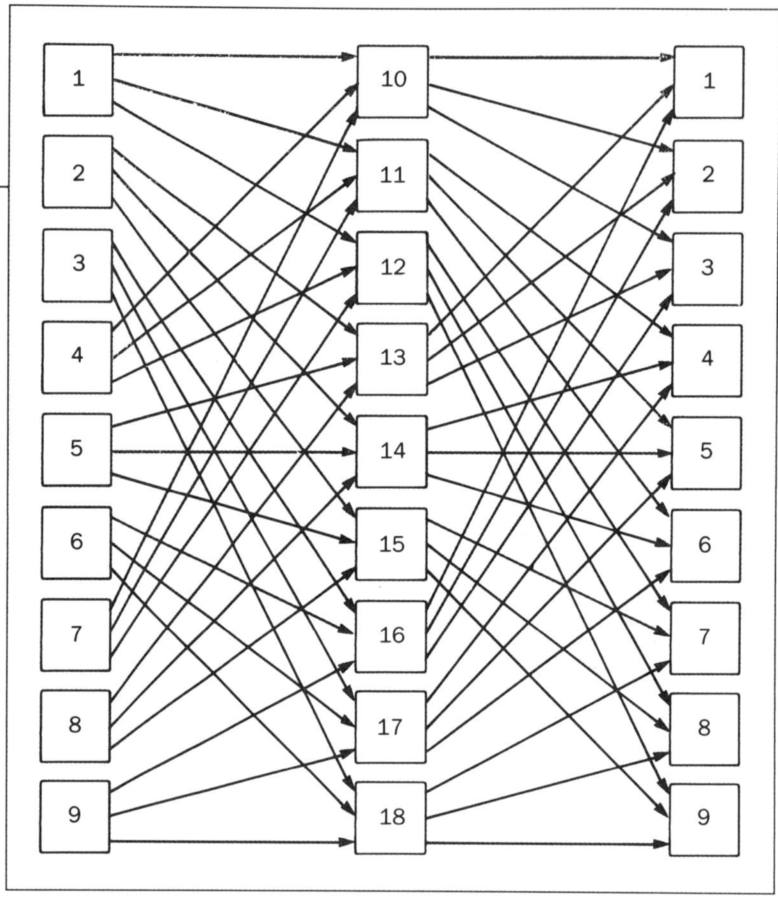

Figure 14. Modular growth of multihop networks. The paths in color represent a connectivity graph for a multihop network with eight network interface units.

NIU 1 to NIU 2 would be disrupted. However, the packet could still be routed to NIU 2 via the path 1-6-4-7-2. When NIU 5 fails, users attached to NIU 5 are cut off from the network, but users attached to all other NIUs feel the effects of the failure only through (perhaps) longer source-destination routing paths. At least p NIUs must fail, and in certain combinations, before users attached to nonfailed NIUs can be cut off from the network. For example, in Figure 5, if NIU 5 and NIU 6 both fail, NIU 1 and NIU 3 are cut off.

Alternative routing paths may also be useful even when NIUs have not failed. The existence of multiple paths (often of the same length) from a source to a destination provides the opportunity to route packets around NIUs that are temporarily congested with a large backlog of packets.

A Non-WDM Approach to Multihop Networks

Thus far, we have considered multihop networks for which the channels required by the connectivity graph were created by multiplexing different carriers onto a common fiber. However, as Figure 15 illustrates, the "WDM channels" referred to could instead be separate fibers, with a fiber-patch panel replacing the star coupler at the center (see, for example, Figure 12). The patch panel is used to make the appropriate interconnections among the incoming and outgoing NIU fibers according to the recirculating, perfect-shuffle connectivity graph. Because all connections in the network are strictly point-to-point, we can make use of yesterday's technology in building a high-capacity lightwave network.

Conclusions

We have proposed a multihop approach for achieving concurrency in distributed-lightwave networks. Multihop networks avoid two serious drawbacks of standard multichannel approaches:

- The requirement of wavelength-agile transmitters or receivers
- Pretransmission coordination between two users that want to communicate.

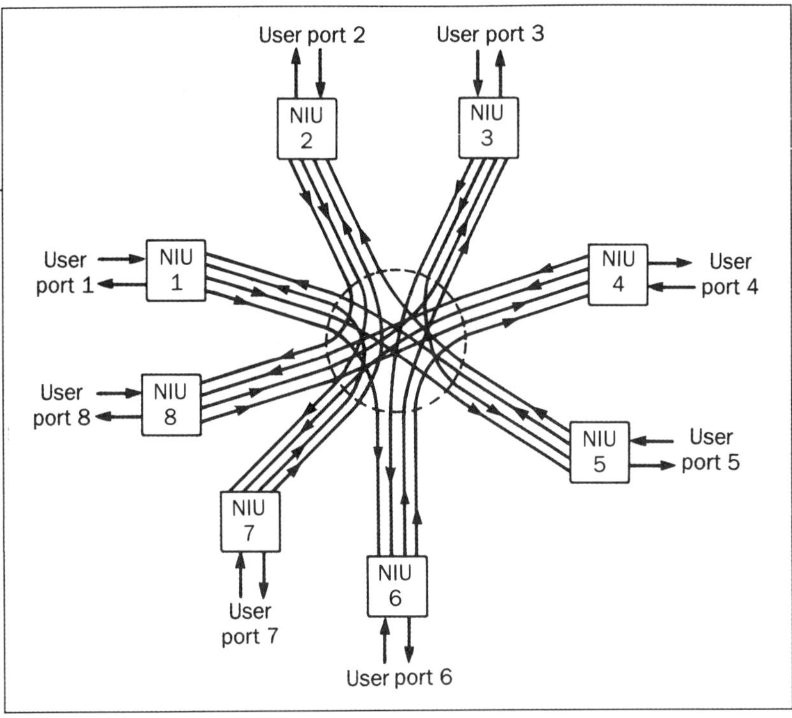

Figure 15. An approach to multihop networks without using wavelength-division multiplexing. The circled area (center) is a fiber-patch panel. NIU = network interface unit.

By creating dedicated channels out of separate fibers, multiple WDM channels on a single fiber, or a combination of both, one can build highly concurrent lightwave networks with hundreds or thousands of gigabits-per-second throughput, even while the users are limited to rates of 1 Gb/s or lower. Although transmitting a packet from one user to another may require routing the packet through intermediate NIUs, the network connectivity is specifically designed to:

- Achieve efficient use of the channel bandwidth.
- Allow modular growth of the network from small to large configurations.
- Provide a degree of network reliability.

References

1. P. S. Henry, "Introduction to Lightwave Transmission," *IEEE Communications Magazine*, Vol. 23, No. 5, May 1985, pp. 12-16.
2. T. Li, "Advances in Optical Fiber Communications: An Historical Perspective," *IEEE Journal on Selected Areas in Communications*, Vol. SAC-1, April 1983, pp. 356-372.
3. R. E. Kahn et al., "Advances in Packet Radio Networks," *Proceedings of the IEEE*, Vol. 66, No. 11, November 1978, pp. 1468-1496.
4. A. S. Acampora and M. G. Hluchyj, "A New Local Area Network Architecture Using a Centralizd Bus," *IEEE Communications Magazine*, Vol. 22, No. 8, August 1984, pp. 12-21.
5. A. G. Fraser, "Toward a Universal Data Transport System," *IEEE Journal on Selected Areas in Communications*, Vol. SAC-1, No. 5, November 1983, pp. 803-816.
6. N. F. Maxemchuk, "Regular Mesh Topologies in Local and Metropolitan Area Networks," *AT&T Technical Journal*, Vol. 64, No. 7, September 1985, pp. 1659-1685.
7. A. Tanenbaum, *Computer Networks*, Prentice-Hall, Englewood Cliffs, N. J., 1981.
8. I. Chlamtac and A. Ganz, "Toward Alternative High Speed Networks: The Swift Architecture," *INFOCOM '87*, San Francisco, California, March 1987.
9. H. S. Stone, "Parallel Processing with the Perfect Shuffle," *IEEE Transactions on Computers*, Vol. C-20, No. 2, February 1971, pp. 153-161.
10. A. S. Acampora, "A Multichannel Multihop Local Lightwave Network," *GLOBECOM '87*, Tokyo, Japan, November 1987.
11. M. G. Hluchyj and M. J. Karol, "ShuffleNet: An Application of Generalized Perfect Shuffles to Multihop Lightwave Networks," to be presented at *INFOCOM '88*, New Orleans, Louisiana, March 1988.
12. J. H. Patel, "Performance of Processor-Memory Interconnections for Multipıocessors," *IEEE Transactions on Computers*, Vol. C-30, No. 10, October 1981, pp. 771-780.

Integrated Voice/ Data Switching

Thomas M. Chen
David G. Messerschmitt

Introduction

Historically, voice and data communications have been handled by different communication networks. For example, the primary carrier for voice has been the public switched telephone network whereas data has been handled for the most part by specialized data networks. One reason has been the traditional separation between voice and data applications. Consider a person speaking on the telephone and then sending an electronic message from a computer terminal. The two actions are handled by separate instruments and are perceived as serving separate purposes. Telephony permits immediate, personal, and interactive contact; data messages allow for more pre-meditated, formal, and non-interactive communication. A tremendous need for a unified treatment of voice and data has not existed, so separate special-purpose communication networks have been largely satisfactory.

Another reason for different networks is the fundamentally different characteristics of voice and data signals. Voice is inherently a real-time, analog signal generated by human speakers. Voice signal characteristics, such as spectral density and average activity, are well-known and consistent between different speakers. On the other hand, most data is machine-generated and digital. Data characteristics, such as bit rate and message length, vary widely depending on the particular application.

Interest in "integrating" voice and data communications has been stimulated recently by deregulation of the U.S. telephone industry and international activities in planning the standards for the Integrated Services Digital Network, or ISDN [1-3]. ISDN will be a worldwide digital network offering a wide range of voice and data services based on 64 kbits/s channels. Although ISDN will most likely be comprised of logically separate networks as shown in Figure 1, it will provide subscribers with the functionality of a single, integrated network by offering a standardized, integrated user access to services [4].

Offering voice and data services in a single network promises several benefits. Users achieve convenience, flexibility, and economy. An integrated user interface allows different terminal equipment to be moved and plugged into any interface in the same way that different electrical appliances can use any standard electrical power outlet. Furthermore, services can be customized to individual needs without having to be concerned with the compatibility of different special-purpose networks. For network providers, integration promises benefits in efficiency and economy. Sharing facilities not only increases efficiency, but should also simplify network operations and maintenance, items which will quickly become very complex in a non-integrated network with a proliferation of new services. Reduced network costs should result in lower service prices to users.

This research was supported by a grant from Pacific Bell, with a matching grant from the University of California MICRO Program.

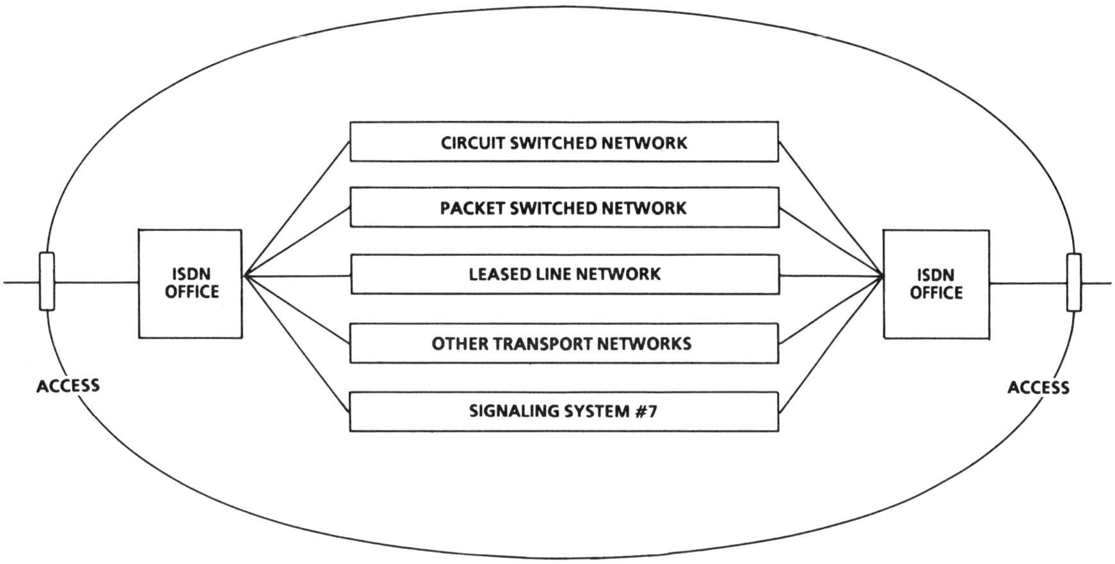

Fig. 1. Probable ISDN architecture.

Integration of voice and data communications is also motivated by advances in computer and communications technology. The telephone system is gradually converting from analog to an entirely digital network because of decreasing costs, increasing data transmission, and increased capability for integration of services. In addition, fiber optics make large bandwidths available at modest cost. This abundant bandwidth combined with increased network capabilities are driving forces in expanding traditional telephone services.

Finally, there is also an expectation that technology will result in the merging of voice and data applications, which have been traditionally separate. For example, artificial intelligence might eventually allow computers to accept and interpret voice commands in person-to-machine conversations. In the futuristic office, messages containing voice and video as well as text might be edited, stored, and transmitted like electronic mail today. Terminal equipment will become more functional and integrate the separate functions of the telephone, terminal, and printer. There will be a need to process and transmit voice and data, and in fact all types of information, in a unified manner.

The purpose of this paper is to provide a basic understanding of the technical problem in integrating voice and data. Although integration is discussed specifically in terms of voice and data, the problem can be generalized to all types of real-time and non-real-time services. First, the different types of traffic found in communication systems are examined. Integration is investigated at different levels. In particular, this paper focuses on the integration of voice and data at the switching level. Different switching approaches are compared and some current integrated switching systems are described.

Types of Traffic

We might think of voice and data information as types of "traffic" to be transported through the communication network. Designing a communication network specifically for either voice or data is relatively uncomplicated; the difficulty in designing a network for both voice and data lies in the different requirements of voice and data traffic. We will describe these requirements by first examining the characteristics of voice and data signals. There are three general classes of traffic in existing communication networks, although more classes might arise in future networks [5].

Voice and video are representatives of the inherently real-time Class I traffic (for our purposes, video can be considered similar to voice except at a higher bandwidth). Voice signals are generated in real-time, person-to-person calls. Due to the conversational nature of speech, only one direction is usually active at any time. Voice traffic can tolerate a certain amount of degradation (e.g., noise, clipping, compression) and occasional blocking (i.e., the connection of a call is refused) without becoming objectionable. However, large transmission delays accentuate subjective problems with echos (unless echo cancelers are used) and themselves disrupt a conversation. Although the exact amount of subjectively acceptable delay is subject to debate, it seems generally agreed that the maximum allowable delay is in the approximate range between 100 and 500 ms [6].

Classes II and III traffic are collectively referred to as "data". Class II traffic consists of person-to-machine (or possibly machine-to-machine) "interactive data", such as videotex. Although not strictly real-time, this traffic has certain delay limitations; a subscriber can wait a fraction of a second, but not minutes, for a computer to respond to a command. Communication is characteristically "bursty" and asymmetric; that is, this type of traffic takes the form of intermittent bursts of information separated by intervals of silence at unequal rates in the two directions. Class II messages can tolerate short transmission delays but not errors. Class III traffic consists of machine-to-machine "bulk data". Messages are typically unidirectional and relatively long. Not being real-time in nature, messages may be delayed substantially longer than Class II messages and arrive in any random sequential order, but they must arrive without errors.

We can illustrate the differences between traffic classes by placing them in a coordinate system representing their characteristics. For the purpose of illustration, the three coordinates shown in Figure 2 were chosen to represent the traffic's degree of tolerance to: network delay, blocking, or degradation (e.g., source rate reduction, transmission errors, and loss of messages). Class I traffic is relatively tolerant of degradation and blocking but rather intolerant of delay. Class II traffic is more tolerant of delay, less tolerant of blocking, and not tolerant of degradation. Class III traffic is much more tolerant of delay and not tolerant of blocking.

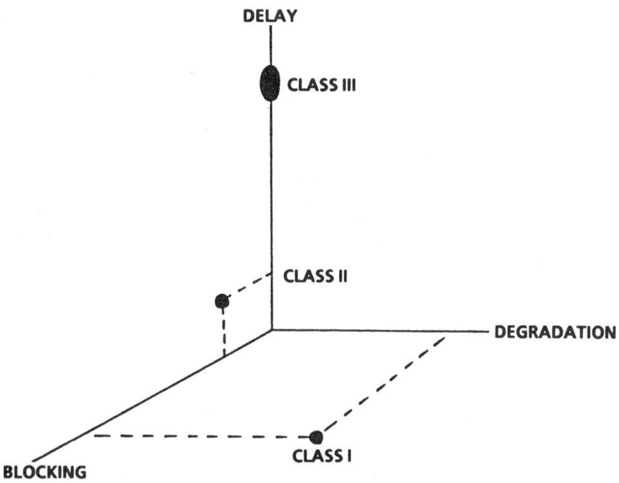

Fig. 2. Traffic tolerances.

These three characteristics were chosen to illustrate the relationship between traffic requirements and switching characteristics. The characteristics of a switching system can be placed in the same coordinate system. An example shown in Figure 3 is a telephone system which has blocking and does not correct for transmission errors. Another example is ARPANET which imposes a variable end-to-end delay but no blocking, and corrects for transmission errors through an elaborate protocol involving error-checking and acknowledgments. If we superimpose the two coordinate systems, it becomes clear that the telephone system most closely meets Class I traffic requirements and ARPANET matches Class II and III traffic requirements. Of course, there are other relevant parameters, such as bandwidth and peak-to-average traffic ratio, that have not been included in this simplified example. Clearly, a switching system that does not impose blocking, degradation, or delay would be ideal for all classes of traffic, but practical systems generally must resort to blocking, delay, or some type of degradation (e.g., discarding messages) to handle the problems of congestion and excessive traffic conditions.

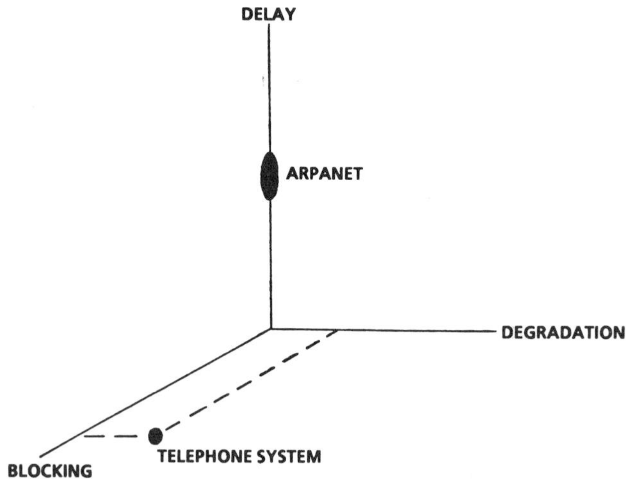

Fig. 3. Switching system characteristics.

Levels of Integration

In a broad sense, integration means that voice and data can be handled by the same communications network. Specifically, voice and data traffic can be integrated at three recognized levels [7], as illustrated in Figure 4. ISDN can be viewed as an example of integration at the first level: integrated access. Voice and data services are accessible through a single user access interface, allowing voice and data terminal equipment to share a common network interface. Voice and data are carried on the same transmission link between the subscriber and network switch called the local loop. However, traffic within the network is routed to different transport networks which are separately optimized for a specific type of service. As planning for ISDN progresses, it is becoming more evident that the real issue at this level is agreement on the standards for the user access interface.

At the second level, called integrated transmission, these transport networks share common transmission facilities between switches but maintain separate switching facilities. Voice and data traffic would share the same transmission link, for example, by Time-Division Multiplexing (TDM) where messages are interleaved in time, or by Frequency-Division Multiplexing (FDM) where messages are sent at separate frequencies. TDM is analogous to the automobile traffic situation where automobiles and pedestrians are allowed use of a road at non-overlapping time intervals. In the FDM case, automobiles and pedestrians are allowed on the same road but in separate lanes. TDM is already used extensively for multiplexing voice signals on long-distance trunks in the telephone network. In the future, transmission links carrying voice and data

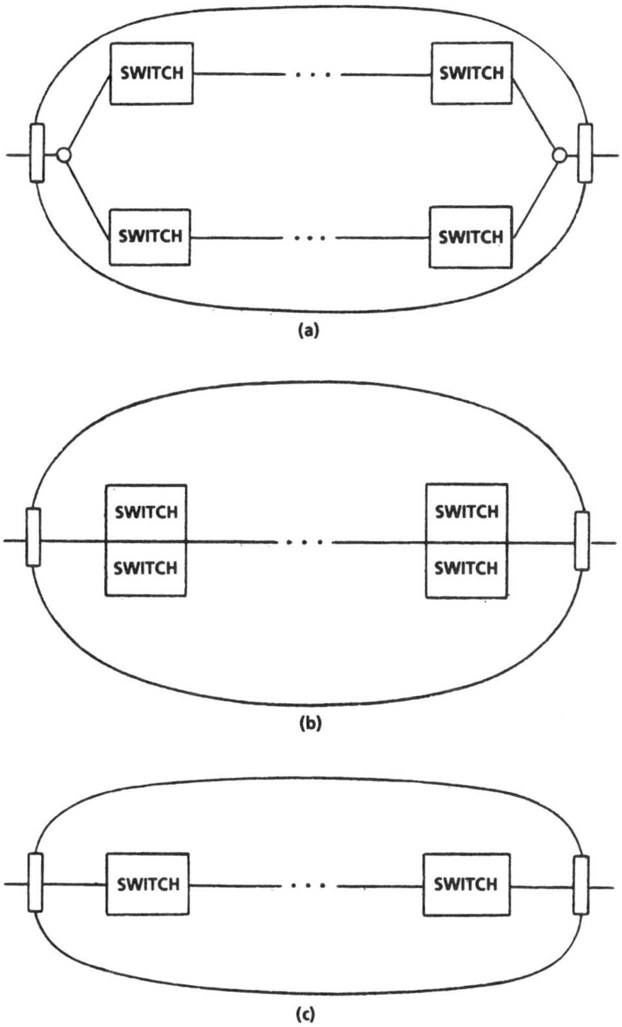

Fig. 4. Levels of integration: (a) Integrated access. (b) Integrated transmission. (c) Integrated switching.

will be predominantly optical fibers and will provide tremendous bandwidth, analogous to 10,000-lane superhighways.

At the third level, called integrated switching, switching facilities are shared as well as transmission links and network access. A network integrated at this level might be considered to be completely integrated since all types of traffic are handled entirely by the same facilities. This level is also the most technically challenging due to the different switching requirements for voice and data in terms of delay, degradation, blocking, and other parameters. Initial approaches have attempted to adapt conventional circuit switching to handle data or conventional packet switching to handle voice. Other approaches have attempted to develop new schemes specifically for both voice and data such as burst switching and hybrid switching. Most recently, the feasibility of an advanced version of packet switching, called wideband or "fast" packet switching, has been demonstrated. It is currently gaining favor as the most promising approach in the telecommunications industry.

Circuit Switching

Conventional switching approaches are circuit switching, packet switching, and message switching. Circuit switching is probably the most familiar since the public telephone system is the primary example of this approach [8]. Circuit switching was developed during the early days of telephony when calls were connected by an operator at a manual switchboard. Although the switching operation is now performed electronically, the principles of circuit switching are still much the same.

The distinguishing feature of circuit switching is the exclusive dedication of a channel of fixed bandwidth between two users for the duration of a call. A circuit or direct electrical path is established during a call set-up procedure, as illustrated in Figure 5. A signaling message indicating a call request is passed through the network to find an available circuit. If a circuit is found and the call is accepted, a signaling message indicating call acceptance is returned. The circuit is held exclusively until the call is disconnected, even if it is not actually utilized for transmission of information.

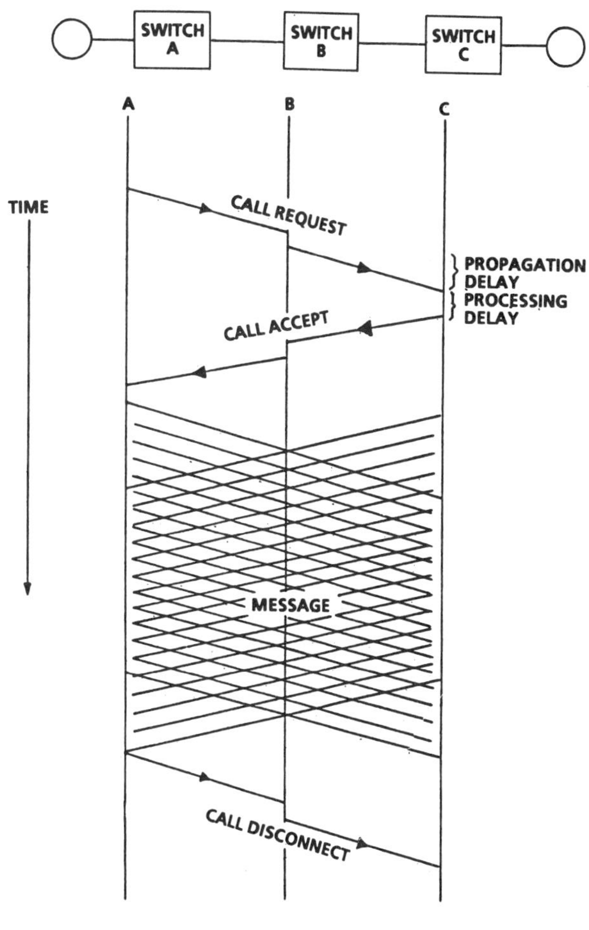

Fig. 5. Timing diagram for circuit switching.

The basic circuit switch, shown in Figure 6, consists of three functional blocks: terminal interface, switching network, and controller. The terminal interface handles signaling functions such as call request and ringing and ensures that the incoming signals are compatible with the electrical characteristics of the switch facilities. The switching network provides the physical transmission paths for the signals. The controller contains the intelligence responsible for the proper operation of the switch hardware.

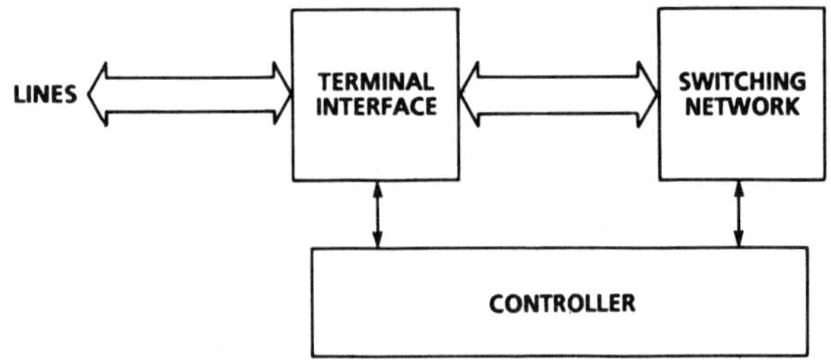

Fig. 6. Basic circuit switch architecture.

Circuit switching has a number of important features. First, traffic is handled on a blocking basis; the network handles excessive traffic conditions by refusing to connect a new call while continuing to hold those calls already connected. The blocked calls may be either delayed for a later time or cleared (e.g., the telephone system clears blocked calls by returning a fast busy signal). Second, this approach is efficient only if the set-up and disconnect times are small compared to the duration or holding time of the call. Third, bandwidth is utilized efficiently only if the circuit is active fairly constantly since the channel is dedicated for the entire call. Fourth, a dedicated channel permits communication with minimal transmission delay, a particularly important consideration for real-time traffic. Finally, traffic is carried by the network without regard to content; that is, information (except for signaling information) is not processed or altered in transit by the network. As a consequence, the network is unable to correct for transmission errors; the responsibility for error detection and correction is relegated to the users.

These properties make conventional circuit switching most suitable for real-time services like voice. Typical conversations last about 300 seconds, much longer than the set-up times, which are approximately a second in the telephone system. During a call, voices are active about 40 to 50 percent of the time, and the network imposes virtually no delay other than propagation time (approximately 20 ms on terrestrial links) from source to destination.

Circuit switching has several disadvantages for most data applications, and in particular for Class II [9]. First, bandwidth is utilized much less efficiently because most data is usually active only about 5 to 15 percent of the time. This means that the channel is nearly always idle. Second, data bursts are often short and require only a brief connection. It is inefficient to circuit-switch messages when the holding time becomes comparable to the call set-up time. Third, circuit switching works well for voice because a 4-kHz bandwidth is characteristic of all speech signals. However, a dynamic allocation of a bandwidth is needed for data communications due to the diversity of bit rates required for various data applications. Finally, circuit switching does not have the capability of node-to-node error-checking since the network is insensitive to the content of the transmitted information. In contrast, if messages are processed within the network, they can be checked for errors before delivery to the destination. This is a desirable feature for error-sensitive data traffic.

Modifications to circuit switching have been proposed to overcome these difficulties. Bandwidth can be utilized more efficiently with statistical multiplexing techniques such as Time Assignment Speech Interpolation (TASI) [10] or Digital Speech Interpolation (DSI) [11]. Several sources can share a fewer number of channels by dynamically selecting only active sources. The number of channels is selected such that the probability that a greater number of sources will be active simultaneously is low. In these infrequent cases, signals would be "clipped" (partially cut off) or compressed (quantized more roughly).

The inefficiency in call connections for very brief messages can be improved by reducing call set-up and disconnect times. The idea of fast circuit switching is to develop switches which perform signaling and call set-up so quickly that it becomes efficient to switch very brief data bursts. Set-up times less than 140 ms are commonly presumed, although such fast digital switches are not presently commercially available [9].

Packet Switching

The concept of packet switching originated as a distributed switching system for survivable military communications [12,13]. The first important application was the development of the ARPANET to link together time-shared research computers in a nationwide network. Unlike circuit switching, packet switching was designed particularly for data communications rather than voice communications. After the success of ARPANET, many other packet-switched networks were started for public data services.

Packet switching is based on the idea of message switching, or store-and-forward switching, shown in Figure 7. Message switching resembles the method of mail delivery in the postal system. A message is formed by concatenating the data information with a header and an end-of-message flag, which is similar to putting a letter into an envelope. The header contains all the information necessary for routing the message through the network information such as source, destination, identity number, and checksum for error. The message is stored in a buffer at each switch which decodes the message header and determines the next node in the route. When the appropriate link becomes available, the message is forwarded to the buffer in the next switch. Usually, an acknowledgment is returned by the receiving node if the message is received without error; otherwise, the sender retransmits the message after some time.

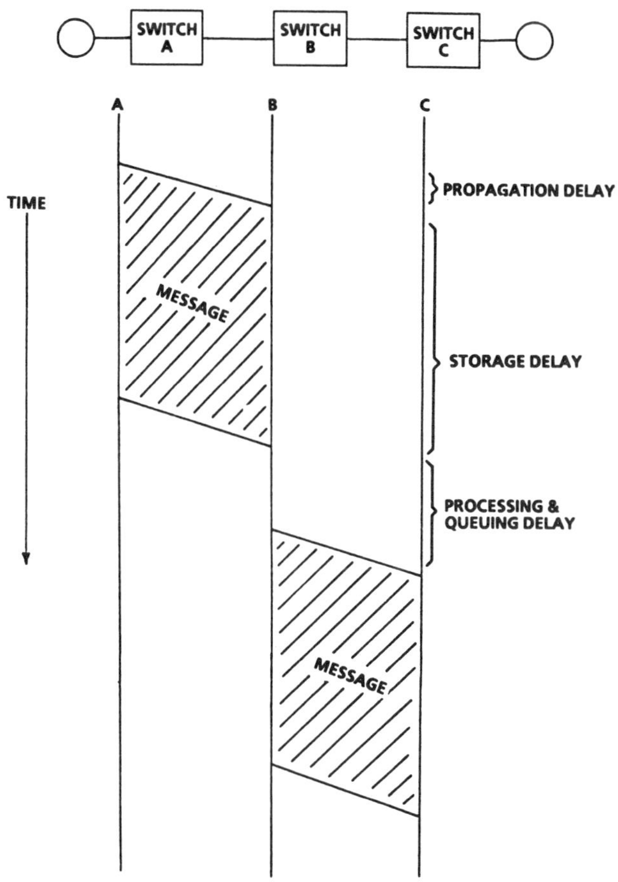

Fig. 7. Timing diagram for message switching.

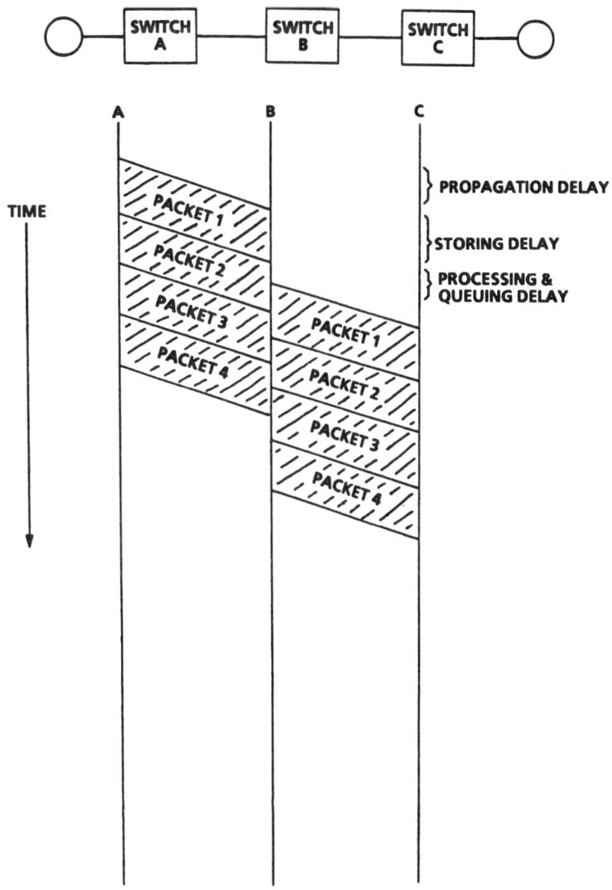

Fig. 8. Timing diagram for packet switching (datagram mode).

Packet switching [14,15] is the same as message switching with one difference; messages are divided into smaller segments of limited length, called packets, each with its own packet header. The shorter packets need less storage time at each switch, as seen in Figure 8. The end-to-end delay for packets is much less than for messages, particularly on routes involving many hops. The exact length of the packet is a trade-off between delay and overhead. Shorter packet lengths decrease the queueing delay at each switch but increase the percentage of the packet taken up by the header bits.

The basic packet switch, shown in Figure 9, consists of four functional blocks: input buffers, output buffers, switch fabric, and controller. Incoming packets are stored in the input buffers, and their headers are decoded. When the appropriate route is determined, they are placed in the proper output buffers through the switch fabric. The switch fabric might be a simple bus or a multi-stage interconnection network. All routing, processing, and control functions are performed by the controller.

There are two general methods for routing packets through networks: datagrams and virtual circuits. In datagram mode, packets are routed independently. Packets travel any available route to the destination and arrive in a random sequential order. In virtual circuit mode, a "VC request" packet sets up a logical connection between the source and destination. Every subsequent packet between this source and destination will travel this same virtual circuit identified by a number in the header. The virtual circuit is cleared eventually by a "VC disconnect" packet. A virtual circuit is not the same as a dedicated channel in circuit switching

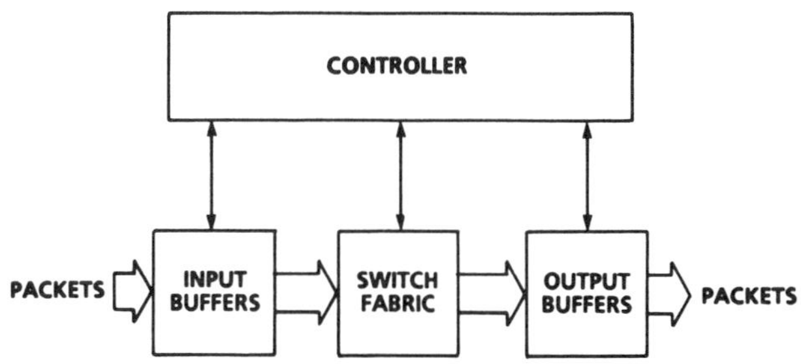

Fig. 9. Basic packet switch architecture.

because several virtual circuits can share the same physical circuit. A virtual circuit means only that the routing decision does not need to be performed for every packet. Hence, processing is simpler for multi-packet messages, but every packet still needs to be queued and processed at each switch. Routing by virtual circuit is more vulnerable to node failures and less adaptive to changing traffic conditions than datagrams.

Packet switching has important differences from circuit switching. First, traffic is handled on a delay basis, i.e., the network will accept new packets when traffic is heavy but might impose extremely long queueing delays. The delay at each switch consists of time to store the entire packet, process the header, and wait for an available link. The delay times depend on packet length, header complexity, and network traffic conditions, respectively. Second, there is no signaling involved in call set-up (in datagram mode), but there is overhead associated with the packet headers. These header bits require a portion of the channel capacity and processing at each switch. Third, packet switching allocates bandwidth dynamically instead of pre-allocating bandwidth like in circuit switching. Channel capacity is used only when there is information to be sent. Bandwidth is utilized more efficiently than circuit switching at the expense of increased processing and queueing delays. Finally, and perhaps most importantly, packets are processed as they are being transmitted. This makes it possible to error-check, copy, and even alter packets within the network. Thus, packet switching permits much more direct control over user information than circuit switching.

For its efficiency with "bursty" traffic, packet switching has been used increasingly in commercial data networks. It has not been used for real-time services like voice except in the context of numerous experiments. A basic problem in packet switching voice is the reconstruction of a continuous stream of speech from packets experiencing random and possibly excessive transit delays [16]. It is necessary to impose an additional delay at the speech decoder so that the voice packets are "played-out" with relatively uniform delays. Increasing the delay at the decoder reduces the fraction of "lost" packets (i.e., packets that arrive later than their target playout times), but the delay becomes subjectively annoying beyond a certain limit.

However, for several reasons, packet switching is still attractive for voice. First, packet switching is extremely flexible. For example, a packet can be copied and broadcast to a group of stations as easily as sent point-to-point. Second, packet switching is particularly efficient with bursty traffic, and voice exhibits some "burstiness". Speech actually consists of short, discrete bursts called talkspursts, lasting on the order of milliseconds to seconds. Packetizing these talkspurst and ignoring silence intervals performs a statistical multiplexing function by using channel capacity only when voices are active. Finally, packet switching has the capability of processing and altering speech information within the network, unlike circuit switching, which transports information without processing. This feature is useful, for example, for judiciously abridging or discarding voice packets when traffic becomes congested.

Several modifications have been proposed to reduce the transit delays for voice packets [17-23]. Packet storage delays can be reduced by shorter packet lengths or by "virtual cut-through" switching [24] which allows a packet to be forwarded before it is stored entirely. Packet headers can be simplified by using virtual circuits and not requiring acknowledgments or retransmissions since the reconstructed speech can tolerate a certain amount of degradation. Voice packets can be given higher priority than data packets to reduce queueing delays. This priority can be pre-emptive so that arriving voice packets can interrupt a data packet in progress, or non-preemptive, where arriving voice packets must wait for the packet in progress to finish. When packet delays become excessive, packets can be judiciously abridged or discarded within the network, or the speech source can be instructed to reduce its encoding rate by means of "feedback" control packets. Finally, fast packet switching, an important approach favored by many in the telecommunications industry, is currently being researched and developed. It is discussed in a later section.

Another approach is to minimize the detrimental effect of late voice packets by encoding the speech in a certain manner. In embedded coding [18,19], the speech is encoded at a number of different rates. The encoded information is placed or "embedded" in packets of different priority such that lower priority packets can be discarded without affecting the continuity of the reconstructed speech. The loss of the lower priority packets causes a graceful degradation in the speech quality. For example, a speech encoder operating at the embedded rates of 8, 16, 32, and 64 kbits/s will generate voice packets with four different priority levels. The loss of the lowest priority packets will result in an effective bit rate of 32 kbits/s.

Burst Switching

Burst switching is a form of message switching that combines different features of fast circuit switching [25-27]. The idea is to take advantage of the bursty nature of voice and data traffic. A "burst" is a variable-length message consisting of a 4-byte header followed by an information field and an end-of-burst flag. In a data burst, the information is the data message; in a voice burst, the information is a talkspurt. A third type, a "command" burst, is used to carry network information between burst switches. Like message switching, burst switching uses headers for routing and queueing for bandwidth contention. However, a burst can begin to be forwarded before it is buffered completely because a burst is always transmitted at the same rate, unlike a store-and-forward message that must be buffered and then transmitted at the full rate of the output link. In this way, burst switching resembles fast circuit switching.

Instead of the traditional central control located at the switching center, control in burst switching is partially distributed to numerous small "link switches". Each link switch has limited processing capability. Its own program is capable of exchanging messages with other link switches to execute a service. Link switches handle network access and local switching. Some processor functions remain in centralized, high-capacity "hub switches" used at points of high concentration. With this distributed architecture, burst switching accommodates network expansion and is less sensitive to node failures and overloads.

Routing information is based on the location of the destination in the network. Local destinations can be handled by the appropriate link switch. Longer distance routes involve accessing a Translation & Routing (T&R) processor for the routing information. The T&R processor returns the appropriate routing information in response to a routing request message. After the routing information is placed in the header, a virtual circuit is made by exchanging a series of messages. The routing information in the header instructs each switch how to route the burst. The virtual circuit is held until it is broken down by a disconnect message at the end of a call.

Although bursts are handled by the same switching facilities, different types of burst are switched differently. Command bursts have highest priority, voice second priority, and data lowest priority. When bursts contend for a link, the queued burst with highest priority is sent first. The protocol for command and data bursts involves aborting and retransmitting the bursts in case of incorrect delivery. Voice bursts are not retransmitted because speech information is extremely time-sensitive. In addition, when 2 ms of speech samples accumulate at a burst switch before an output link becomes available, the information is discarded by the switch, resulting in clipping of the reconstructed speech.

Hybrid Switching

Hybrid switching attempts to provide both circuit and packet switching features [5,28,29]. This is accomplished by time-multiplexing voice and data as shown in Figure 10. A master time frame is established on each link consisting of a synchronizing Start-Of-Frame (SOF) marker and a number of time slots. Voice traffic is allotted a certain number of slots while data is allowed on the remaining slots. The voice slots are circuit switched while the data slots are packet switched.

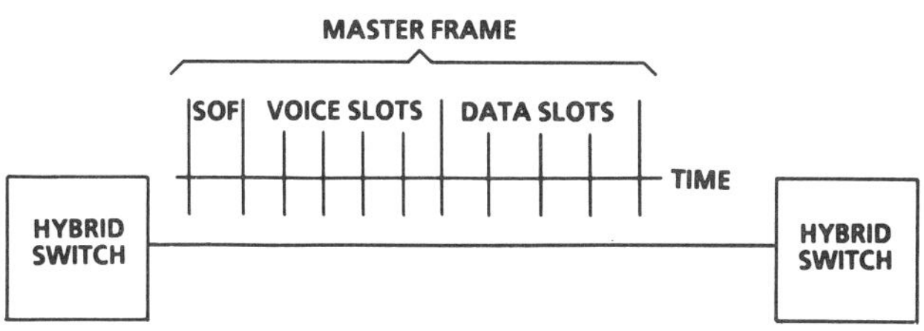

Fig. 10. Hybrid switching.

In the basic scheme, the frame length and voice/data boundary are fixed, and time slots are equal lengths. For example, a master frame of 10 ms on 1.544 Mbits/s links would provide frames of 15,440 bits at a rate of 100 frames/s. Voice sources at 16 kbits/s would need 160-bit time slots. However, a fixed voice/data boundary is inefficient due to the statistical fluctuations in voice and data traffic. Temporarily unused voice slots cannot be used for data nor unused data slots for voice. It is difficult to place the voice/data boundary for maximum efficiency under changing traffic patterns. For more efficiency, the voice/data boundary can be "movable". Voice calls can be allotted up to a maximum number of time slots; data is allowed to use temporarily silent voice slots as well as slots allocated just to data. Furthermore, the frame length and time slots can be dynamic for more adaptability. As might be expected, increased efficiency is achieved at the cost of complicated analysis and switch operations.

The idea of hybrid switching is to combine the features of circuit and packet switching in order to handle each type of traffic in the conventional manner, i.e., voice is handled on a blocking basis and data on a delay basis. Although voice and data traffic must be separated to different fabrics within the switch, the hybrid switch presents the functionality of a single, integrated switch at each link. With both circuit and packet switching capabilities, hybrid switching will effectively accommodate any mix of traffic. Furthermore, hybrid switching is more compatible with the present telephone system than packet or burst switching because existing circuit switches can be updated to handle the data switching requirement. Thus, hybrid switching allows for a more graceful evolution of the telephone network.

Fast Packet Switching

Fast packet switching, also called Asynchronous Time Division (ATD) or Asynchronous Transfer Mode (ATM), is an advanced version of packet switching based on fiber optic links and simplified protocols [30-37]. A protocol is the set of rules that implement the various functions involved in the transfer of data between two users (e.g., routing, error control, flow control, etc.). Due to their high degree of complexity, conventional protocols are structured hierarchically as a set of layers referred to as the network architecture [15,38,39]. Each layer performs specific functions of the protocol and provides certain functions or "services" to higher layers while relying on the services provided by lower layers.

The most well-known network architecture is the Open Systems Interconnection (OSI) reference model developed by the International Standards Organization (ISO) as a framework for standardization of protocols [40]. Its seven layers are shown in Figure 11. When an "application", (the highest level entity), communicates with another "application", its data message is passed down the layers with each layer appending its own header and containing information to be used in the same layer at the destination. At the destination, the appropriate headers are removed as the data is passed up the layers. Each layer communicates with its "peer" in the same layer.

The lowest layer, called the physical layer, covers all the physical aspects of transmitting a bitstream between adjacent nodes. Above the physical layer, the data link layer is primarily responsible for making the

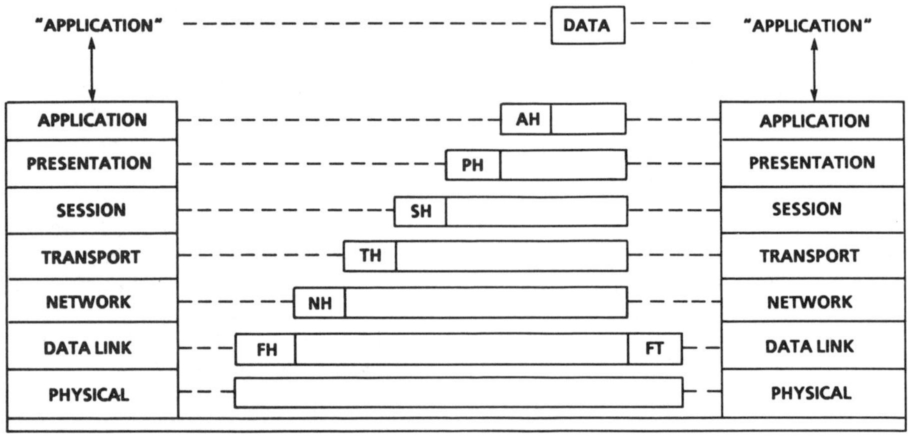

Fig. 11. OSI protocol reference model.

physical link reliable. This is accomplished through a usually complicated procedure involving error detection, acknowledgments, and retransmissions. The next higher layer is the network layer, which encompasses the functions of routing and congestion control. The four remaining layers, above the network layer, involve end-to-end protocols that are less relevant to the switching network.

The key aspect of fast packet switching is simplification of the lower layer protocols. Specifically, the functions of error correction and flow control can be removed from the lower layers because of the abundant bandwidth and low error rates possible with fiber optic links. Error correction and flow control are provided on an end-to-end basis by higher layer protocols as needed (i.e., for data but not for voice). The philosophy is to handle all types of traffic in a common manner in the lower layers and to overlay more traffic-specific protocols in the higher layers. In addition, routing is simplified. The simplified protocols and routing result in a concise header format, similar to the example shown in Figure 12, which enables the processing of packets to be performed entirely in hardware. The combination of high-speed links and hardware processing makes possible fast packet switches with low delay

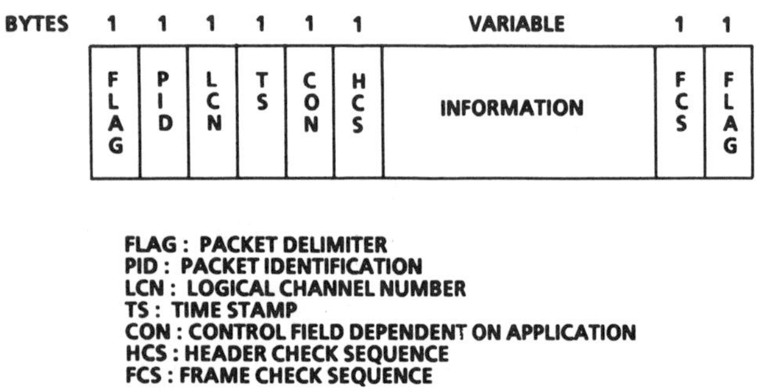

Fig. 12. A fast packet format.

and high throughput sufficient for voice and possibly video services [31,41-44].

A Comparative Discussion of Switching Approaches

Comparisons of switching approaches have focused on circuit and packet switching for many reasons. These technologies are the predominant ones today and will continue to dominate because of the enormous investment in existing telephone and data networks. At least, any future switching techniques will still be based on a circuit or packet switched backbone network. It is more practical to adapt a proven technology than to implement a new technology. Another reason for their dominance is their proven efficiency and effectiveness for voice or data. The final reason is that circuit and packet switching will soon be implemented in the worldwide ISDN. If successful, ISDN will evolve into the broad-band-ISDN [45,46]. The broad-band-ISDN will also be circuit and packet-switched and will provide the user with much higher bit rates at the interfaces to the network.

Harrington [9] analyzed circuit switching techniques for voice and data, and concluded that circuit switching is not the preferable approach for data services. Coviello [47] compared circuit and packet switching for voice and concluded that conventional packet switching was inadequate without major modifications. The general consensus in comparisons of circuit and packet switching [5,48,49] is that each technology has a different area of usefulness depending on parameters such as message length, network topology, traffic patterns, etc. Thus, both circuit and packet switching are commonly expected to be useful in communication networks carrying a mix of different types of traffic, for at least the near future, which is encouraging to the proponents of the hybrid switching approach.

Performance evaluations of burst switching have been limited. An analytical comparison between burst switching and fast packet switching [50] indicated that both techniques performed roughly equivalently for both voice and data. However, burst switching distinguishes between voice and data at the switching level whereas fast packet switching does not. Given that packet switching is a proven technology and will be implemented more widely in the near future, burst switching seems to be disadvantageous in terms of implementation without offering significant performance advantages.

Hybrid switching seems to be a practical short-term approach because present network facilities are based on circuit and packet technologies and hybrid switching is capable of effectively handling a wide mix or different types of traffic. However, hybrid switching is essentially a packaging of two switching fabrics into a single functional switch, and it is not clear that the potential benefits derived from integration can be achieved with hybrid switching. It has the disadvantages of complicated network design and analysis [29] and switch architectures [51,52].

On the other hand, the fast packet switching approach is leading to switches capable of extremely low delays and high throughput. The concept overcomes the drawbacks usually presented in arguments against packet switching while retaining the inherent advantages of flexibility and efficiency. Also, voice and data are truly transported in an integrated manner at the switching level. Consequently, fast packet switching is rapidly emerging in the telecommunications industry as the favored long-term approach.

Conclusions

We have presented a discussion of the technical problem of integrating voice and data, while focusing on the switching level. In particular, we have considered modifications of circuit switching and packet switching, which are conventionally used separately for voice and data, plus burst and hybrid switching.

As might be expected, conventional circuit and packet switching have different regions or usefulness, and neither is suitable without significant modifications. Burst switching suffers from the disadvantages of incompatibility with present networks, without appearing to offer any major performance advantages. Hybrid switching is promising as a short-term approach in terms of network evolution and performance, but presents some significant technical difficulties. On the other hand, fast packet switching is a promising solution to the integration problem, but more research and development is needed.

An example of a problem that lacks thorough understanding is congestion control. Although flow control is exercised end-to-end (e.g., sliding window for data and call blocking for voice), congestion control might still be necessary on a node-to-node basis due to the random nature of traffic flows in packet networks. We speculate that congestion control can be implemented by simply discarding packets, in which case a mechanism for choosing which packets to discard might be needed. A possible mechanism could be a priority system based on an implicit expiration time associated with each packet. However, the implications of this conjectured scheme need to be studied further.

Acknowledgment

The authors are grateful to Dr. Patrick White of Bell Communications Research for his helpful comments on an earlier version of this manuscript.

References

[1] I. Dorros, "ISDN," *IEEE Commun. Mag.*, vol. 19, pp. 16-19, Mar. 1981.

[2] I. Dorros, "Telephone nets go digital," *IEEE Spectrum*, vol. 20, pp. 45-53, Apr. 1983.

[3] W. Tang, "ISDN- new vistas in information processing," *IEEE Commun. Mag.*, vol. 24, pp. 11-16, Nov. 1986.

[4] W. Gifford, "ISDN user-network interfaces," *IEEE J. on Sel. Areas in Commun.*, vol. SAC-4, pp. 343-348, May 1986.

[5] M. Ross, "Circuit versus packet switching," in *Fundamentals of Digital Switching*, (J. McDonald, ed.). NY: Plenum, 1983.

[6] J. Turner, "Design of an integrated services packet network," *IEEE J. on Sel. Areas in Commun.*, vol. SAC-4, pp. 1373-1380, Nov. 1986.

[7] M. Gerla, R. Pazos-Rangel, "Bandwidth allocation and routing in ISDN's," *IEEE Commun. Mag.*, vol. 22, pp. 16-26, Feb. 1984.

[8] A. Joel, Jr., "Circuit-switching fundamentals," in *Fundamentals of Digital Switching*, (J. McDonald, ed.). NY: Plenum, 1983.

[9] E. Harrington, "Voice/data integration using circuit-switched networks," *IEEE Trans. on Commun.*, vol. COM-28, pp. 781-793, June 1980.

[10] K. Bullington, J. Fraser, "Engineering aspects of TASI," *Bell Sys. Tech. J.*, vol. 38, pp. 353-364, Mar. 1959.

[11] S. Campanella, "Digital speech interpolation," *COMSAT Tech. Rev.*, vol. 6, pp. 127-158, Spring 1976.

[12] L. Roberts, "The evolution of packet switching," *Proc IEEE*, vol. 66, pp. 1307-1313, Nov. 1978.

[13] P. Green, Jr., "Computer communications: milestones and prophecies," *IEEE Commun. Mag.*, vol. 22, pp. 49-63, May 1984.

[14] H. Heggestad, "An overview of packet-switching communications," *IEEE Commun. Mag.*, vol. 22, pp. 24-31, Apr. 1984.

[15] A. Tanenbaum, *Computer Networks*. NJ: Prentice-Hall, 1981.

[16] W. Montgomery, "Techniques for packet voice synchronization," *IEEE J. on Sel. Areas in Commun.*, vol. SAC-1, pp. 1022-1028, Dec. 1983.

[17] B. Gold, "Digital speech networks," *Proc. IEEE*, vol. 65, pp. 1636-1658, Dec. 1977.

[18] T. Bially, *et al.*, "Voice communications in integrated digital voice and data networks," *IEEE Trans. on Commun.*, vol. COM-28, pp. 1478-1490, Sept. 1980.

[19] T. Bially, *et al.*, "A technique for adaptive voice flow control in integrated packet networks," *IEEE Trans. on Commun.*, vol. COM-28, pp. 325-333, March 1980.

[20] G. Barberis, *et al.*, "Coded speech in packet-switched networks: models and experiments," *IEEE J. on Sel. Areas in Commun.*, vol. SAC-1, pp. 1028-1038, Dec. 1983.

[21] C. Weinstein, J. Forgie, "Experience with speech communications in packet networks," *IEEE J. on Sel. Areas in Commun.*, vol. SAC-1, pp. 963-980, Dec. 1980.

[22] J. Forgie, A Nemeth, "An efficient packetized voice/data network using statistical flow control," *ICC*, pp. 38.2.44-48, June 1977.

[23] M. Listanti, F. Villani, "An X.25-compatible protocol for packet voice communications," *Computer Commun.*, vol. 6, pp. 23-31, Feb. 1983.

[24] P. Kermani, L. Kleinrock, "Virtual cut-through: a new computer communication switching technique," *Computer Networks,* vol. 3, pp. 267-286, 1979.

[25] S. Amstutz, "Burst switching — an introduction," *IEEE Commun. Mag.*, vol. 21, pp. 36-42, Nov. 1983.

[26] E. Haselton, "A PCM frame switching concept leading to burst switching network architecture," *IEEE Commun. Mag.*, pp. 13-19, Sept. 1983.

[27] J. Haughney, "Application of burst-switching technology to the defense communication systems," *IEEE Commun. Mag.*, vol. 22, pp. 15-21, Oct. 1984.

[28] G. Coviello, P. Vena, "Integration of circuit/packet switching in a SENET (slotted envelope network) concept," *NTC '75,* pp. 42.12-42.17.

[29] I. Gitman, *et al.*, "Analysis and design of hybrid switching networks," *IEEE Trans. on Commun.*, vol. COM-29, no. 9, pp. 1290-1300, 1981.

[30] J. Turner, "New directions in communications (or which way to the information age?)," *IEEE Commun. Mag.*, vol. 24, pp. 8-15, Oct. 1986.

[31] R. Muise, *et al.*, "Experiments in wideband packet technology," *1986 Zurich Sem. on Dig. Commun.*, pp. D4.1-5.

[32] M. Dieudonne, M. Quinquis, "Switching techniques for asynchronous time division multiplexing (or fast packet switching)," *ISS '87,* pp. B5.1.1-6.

[33] P. Gerke, "Fast packet switching- a principle for future system generations?," *ISS '87,* pp. B5.2.1-7.

[34] P. Kirton, *et al.*, "Fast packet switching for integrated network evolution," *ISS '87,* pp. B6.2.1-7.

[35] G. Luderer, *et al.*, "Wideband packet technology for switching systems," *ISS '87,* pp. B6.1.1-7.

[36] W. Standish, S. Sistla, "Network switching in the 1990's," *ISS '87,* pp. C5.1.1-9.

[37] K. Takami, T. Tatenaka, "Architectural and functional aspects of a multi-media packet switched network," *ISS '87,* pp. B6.4.1-5.

[38] W. Stallings, *Data and Computer Communications.* NY: MacMillan, 1985.

[39] M. Schwartz, *Telecommunication Networks: Protocols, Modeling and Analysis.* MA: Addison-Wesley, 1987.

[40] H. Zimmermann, "OSI reference model- the ISO model of architecture for Open System Interconnection," *IEEE Trans. on Commun.*, vol. COM-28, pp. 425-432, Apr. 1980.

[41] J. Hui, E. Arthurs, "A broadband packet switch for integrated transport," *IEEE J. on Sel. Areas in Commun.*, vol. SAC-5, pp. 1264-1272, Oct. 1987.

[42] H. Ichikawa, *et al.*, "High-speed packet switching systems for multimedia communications," *IEEE J. on Sel. Areas in Commun.*, vol. SAC-5, pp. 1336-1345, Oct. 1987.

[43] S. Nojima, *et al.*, "Integrated services packet network using bus matrix switch," *IEEE J. on Sel. Areas in Commun.*, vol. SAC-5, pp. 1284-1291, Oct. 1987.

[44] Y. Yeh, *et al.*, "The Knockout switch: a simple, modular architecture for high-performance packet switching," *IEEE J. on Sel. Areas in Commun.*, vol. SAC-5, pp. 1274-1283, Oct. 1987.

[45] C. Hoppitt, "ISDN evolution: from copper to fiber in easy stages," *IEEE Commun. Mag.*, vol. 24, pp. 17-22, Nov. 1986.

[46] B. Schaffer, "Switching in the broad-band ISDN," *IEEE J. on Sel. Areas in Commun.*, vol. SAC-4, pp. 536-541, July 1986.

[47] G. Coviello, "Comparative discussion of circuit vs packet switched voice," *IEEE Trans. on Commun.*, vol. COM-27, no. 8, pp. 1153-1159, 1979.

[48] P. Kermani, L. Kleinrock, "A trade-off study of switching systems," *ICC '79,* pp. 20.4.1-20.4.8.

[49] K. Kummerle, H. Rudin, "Packet and circuit switching: cost and performance boundaries," *Computer Networks,* vol. 2, pp. 3-17, 1978.

[50] P. O'Reilly, "Burst and fast-packet switching: performance comparisons," *Infocom '86,* pp. 653-666.

[51] H. Rudin, "Studies on the integration of circuit and packet switching," *ICC '78,* pp. 20.2.1-20.2.7.

[52] N. Keyes, M. Gerla, "Report on experience in developing a hybrid packet and circuit switched network," *ICC '78,* pp. 20.6.1-20.6.6.

Thomas M. Chen (S '81) received his B.S. and M.S. degrees in 1984 from Massachusetts Institute of Technology, Cambridge.

He has worked on image compression at the IBM San Jose Research Laboratory and on ISDN at Pacific Bell. He is currently pursuing his Ph.D. in electrical engineering at the University of California, Berkeley.

David G. Messerschmitt (IEEE Fellow) received a B.S. degree from the University of Colorado in 1967, and an M.S. and Ph.D. from the University of Michigan in 1968 and 1971, respectively. From 1968 to 1977 he was a member of the technical staff and later, Supervisor at Bell Laboratories, Holmdel N.J., where he did systems engineering, development, and research on digital transmission and digital signal processing (particularly relating to speech processing). He is currently a Professor of Electrical Engineering and Computer Sciences at the University of California at Berkeley. Current research interests include applications of digital signal processing, adaptive filtering, digital communications (on the subscriber loop and fiber optics), architecture and software approaches to programmable and dedicated hardware digital signal processing, communication network design and protocols, and computer-aided design of communications and signal processing systems. He has published over 100 papers and has ten patents issued or pending in these fields. Since 1977, he has also served as a consultant to a number of companies.

Dr. Messerschmitt is a member of Eta Kappa Nu, Tau Beta Pi, Sigma Xi, and has several best paper awards. He has served as a Senior Editor of the *Communications Magazine* and as Editor for Transmission of the *Transactions on Communications.* Also, he was a member of the Board of Governors of the Communications Society. He has organized and participated in a number of short courses and seminars devoted to continuing engineering education.

Performance Analysis of Hybrid Switching Concepts for Integrated Voice/Data Communications

MYRON J. ROSS, SENIOR MEMBER, IEEE, AND OSAMA A. MOWAFI, MEMBER, IEEE

Abstract—**This paper describes and evaluates variations of a generic hybrid switching concept for integrated voice and data communications. Concepts examined include the Slotted ENvelope NETwork approach, the enhanced hybrid (SENET with time assignment speech interpolation/time assignment data interpolation), the flexible hybrid, and the variable length frame/packet approach. Primary criteria used to evaluate the concepts are transmission throughput efficiency and delay versus information throughput. Other criteria, such as voice continuity and routing/flow control, were also applied. It is seen that several of the hybrid approaches perform very well in comparison with circuit and packet switching techniques, in terms of their ability to service different traffic mixtures both efficiently and effectively.**

I. INTRODUCTION

BECAUSE of the increasing emphasis on digitized voice and the great strides being made in computer technology, there is general agreement in industry and government that switching and transmission of voice, data, image, and text through common facilities is feasible and will greatly improve the efficiency and decrease the cost of telecommunications. Although no true integrated switching facilities are now in production for commercial or military users, many schemes are in the stages of study and experiment [1]-[5]. These range all the way from pure circuit switching, through hybrid switches that circuit-switch voice and packet-switch data, to all packet techniques for both voice and data. Each scheme for integration of voice and data has its advantages, but at this time there is no clear indication of the direction in which development should proceed.

There are several reasons why we consider a hybrid arrangement. In [6], circuit switching was shown to be ill suited for satisfying all classes of traffic associated with future networks. In [7], the question of whether packet switching could be used to handle all classes was addressed. The results were encouraging; however, further effort is needed to evaluate the approaches to packetization and their effects on speech quality, and the increased processor loading as compared to circuit-switching techniques. The transmission savings due to TASI-type concepts also need to be compared between circuit- and packet-switching approaches.

Integration of diverse traffic types is justified by a more efficient use of the transmission bandwidth, which results from a match between the data being serviced and the switching concept to which they are best suited. Various studies comparing circuit switching and packet switching have concluded that circuit switching is more cost-effective for traffic characterized by long contiguous messages, and packet switching is more cost-effective for traffic with short messages [8]-[13]. The exact message size at which this crossover occurs, however, is a function of the various models used by each investigator, particularly those for calculating switching and transmission costs.

In this paper, we first provide a general description of a hybrid concept and delineate the variations of this generic approach. A comparative evaluation of the various alternatives is made with respect to transmission throughput efficiency, delay versus information throughput, and other factors such as voice continuity and flow control. The performance characteristics of hybrid and packet switching are compared in terms of their ability to service different mixtures of voice, interactive, and bulk data traffic. It is seen that several of the hybrid approaches perform very well in comparison with circuit and packet switching techniques, in terms of their ability to service different traffic mixtures both efficiently and effectively.

II. DESCRIPTION OF HYBRID SWITCHING CONCEPTS

In the hybrid switching concept, switching and transmission facilities of the network are shared between traffic, with a single facility providing both circuit- and packet-switched modes of operation. The transmission capacity of the link between two nodes is used to carry circuit-switched traffic at one instant and packet-switched traffic at another instant. There are several variations of this approach. In one variation, voice and other real-time traffic are circuit-switched while interactive data and other nonreal-time traffic are packet-switched. Another alternative is to packet-switch the voice traffic and circuit-switch bulk data and facsimile, with any necessary error control for the circuit-switched data provided by packets or a small transmission capacity in the reverse direction. At the switch, the programs and functions needed to perform either circuit or packet switching are present. The switch processing capacity is dynamically shared by these functions; the load at any instant is dependent on the traffic mix, routing, etc., at that time.

Fig. 1(a) illustrates the approach for the case where the voice is circuit-switched. Wide-band digital trunk capacity is divided into contiguous, constant-period master frames in the time domain. The period, e.g., 10 or 20 ms, once selected,

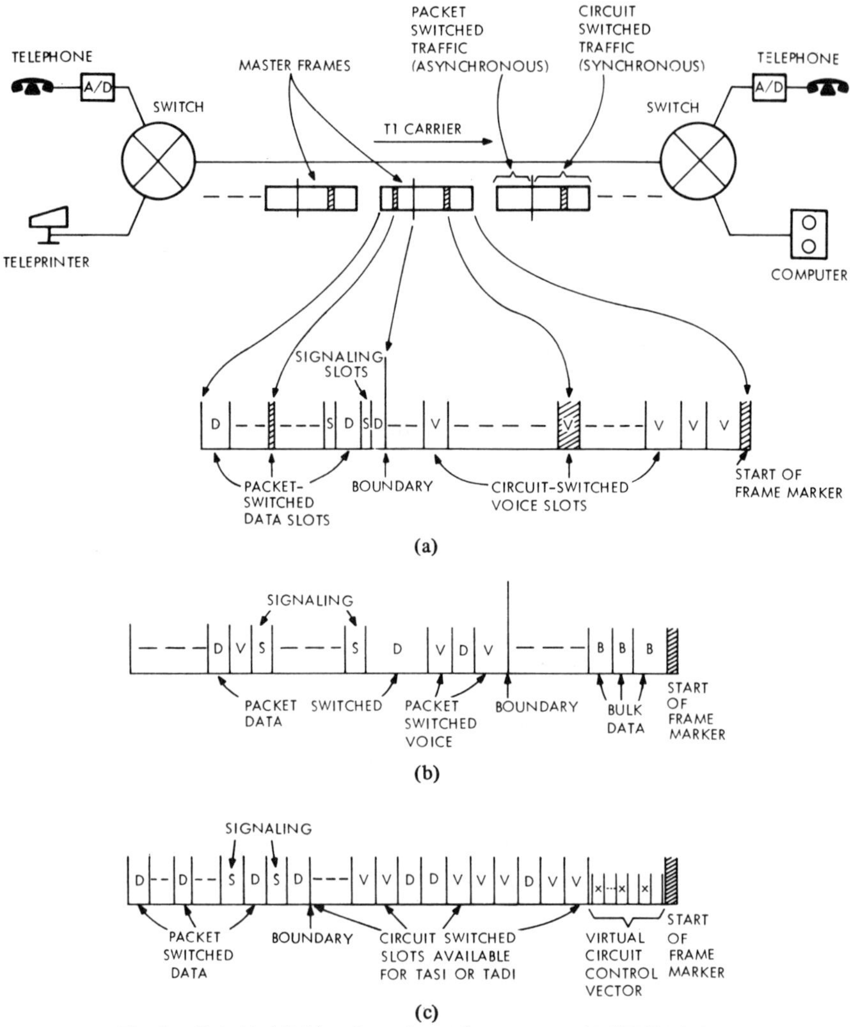

Fig. 1. Hybrid switching channel capacity structure. (a) SENET hybrid switching channel capacity structure. (b) Flexible hybrid switching channel capacity structure. (c) Channel capacity structure using TASI/TADI techniques.

would be constant through the network. Within a frame, portions are utilized as 1) a marker to indicate the start of a frame; 2) a region that multiplexes those synchronous types of traffic normally associated with circuit-switched traffic; and 3) a region that incorporates packet-switched or message-switched traffic, and is handled as packet-switched data. Here, voice, facsimile, and video would be circuit-switched; interactive data and signaling messages for circuit setup and disconnection would be packet-switched; and bulk data would be either circuit- or packet-switched, depending on the application. Within the frame, traffic of varying bandwidth can be accommodated simultaneously by allocating time shares as needed, until the total capacity of the wide-band trunk has been allocated.

In Fig. 1(a), the period of the master frame is 10 ms and the transmission rate is 1.544 Mbits/s. Thus, 15 440 bits are provided within each master frame, with frames transmitted at a rate of 100 per second. Digitized voice at 16 kbits/s would require an allotment of 160 bits/frame for the duration of the conversation. For example, if a call lasts 5 min, then 160 bits must be reserved for each of 30 000 master frames (100 frames/s times 300 s). Interactive data between a terminal and computer are transmitted in the form of packets. Incorporated into the packet format will be information necessary for routing, security, identity, and precedence.

There are various approaches to the sharing of transmission capacity in a hybrid system. The frame boundary is software-controlled, and the partitioning of frame capacity to circuit-switched or packet-switched regions need not be fixed or identical on different network links. In the following sections, variations of this basic generic approach will be investigated. As we shall see, the various hybrid approaches are not necessarily discrete and exclusive, but rather are entries on a continuum of techniques in which performance and control blend from one technique to another.

A. SENET (Slotted Envelope Network) Approach

The communication channel linking two nodes in a hybrid system utilizes a TDM master-frame format. Fig. 1(a) illustrated this concept for the SENET approach [14]. The frame partitions link capacity into two regions: a circuit-switched synchronous traffic region (Class I) and a packet-switched asynchronous traffic region (Class II). Circuit-switched operation is similar to that described in [6]. A complete end-to-end circuit is established for each pair of voice and data users and dedicated for the full duration of use. Signaling messages are used to convey the information needed in order for the network to interconnect one subscriber with another. Connections are allocated and maintained in accordance with link maps at the end of each link. Changes in the link maps are implemented by signaling messages, which are error-controlled and link-coordinated to prevent misconnection. These signaling messages are handled as packet-switched data.

In packet-switched operation, messages are divided into fixed-bit format segments for transmission through the network. Each segment, or packet, has additional bits added for address and administrative purposes (e.g., the origination and destination addresses, sequence number, priority). Ideally, packets are treated individually, each forwarded along the best available path at any given instant. Data packets are error-checked along the way, e.g., each time another wide-band link is traversed. At the destination node, another switch reassembles packets into complete messages, which are then presented to the destination terminal.

Input and output processing of the traffic stream at each link is performed by peripheral processors using direct memory access (DMA). Mapping of the incoming slots on input frames and outgoing slots on output frames is accomplished by the switch CPU. During call setup for circuit switching, the incoming signaling message is used to determine the outgoing link and frame slot associated with that call. A circuit-switched slot is assigned and the signaling message is forwarded to the next switch. The frame structure for the outgoing link is also communicated to the next switch. Once an end-to-end connection is made through the network, the maps at each node specify the incoming link, frame, and slot of each call with the appropriate outgoing link, frame, and slot. Since error detection and correction are not applied to circuit-switched calls, no additional processing by the switch is necessary once the connection is established.

The outgoing link for an incoming packet is determined from processing the header, and is a function of the destination switch, priority, routing strategy, etc. (Datagram mode of operation). Received packets are queued at the outgoing link for transmission on a subsequent frame. For some asynchronous data transactions, e.g., interactive data applications, signaling messages would set up the transaction in analogous fashion to that of voice calls, i.e., the availability of a receiver for a transmission would be assured prior to accepting data from the originator (virtual circuit mode of operation).

There are various methods of assigning transmission capacity to the different regions. In the fixed-boundary arrangement fixed portions of the transmission capacity are reserved for circuit-switched traffic (e.g., voice) and packet-switched traffic (e.g., interactive data). Each form of traffic coexists with the other on the same transmission link but cannot make use of capacity that has not been preassigned to it. The chief disadvantage of this approach is an inefficient use of transmission capacity, since one type of traffic cannot utilize capacity reserved for the other even though unused capacity may be available.

The movable-boundary arrangment, while somewhat more difficult to implement, allows both types of traffic to coexist on the same link according to a flexible-boundary rule in which transmission capacity is dynamically shared between the two regions. Although a boundary is normally assigned between packet- and circuit-switched capacity, one type of traffic can utilize idle channel capacity normally assigned to the other. This leads to a system that requires less overall capacity than the fixed-boundary arrangement, at the expense of processing complexity and the need for precise coordination. The SENET approach is an example of a flexible-boundary, mixed-traffic arrangement [15]-[17].

Other ways to allocate the frame capacity include designing two movable boundaries or no boundary at all between switching regions. In the former case, each switching category has a guaranteed allocation for dedicated use; however, they may each use a shared common capacity on either a demand or priority basis. In the latter case, the entire frame capacity is shared between circuit-switched and packet-switched traffic. However, care must be taken to ensure that one type of traffic, e.g., circuit-switched, maintains a certain level of performance in the presence of overload on the other. This could be accomplished through the use of priority assignment or by storing call-origination requests.

Fig. 2 shows (for illustrative purposes) the channel map allocations for a typical SENET frame. The Class I region of each frame is specified by a start-of-frame (SOF) marker and by allocation maps maintained at both ends of each link. These allocation maps are specific as to call identity, starting points of the allocation in the master frame, magnitude of allocation per virtual connection, and precedence of the virtually connected call. Master-frame allocation sequences are maintained in the form of lists, one list per outgoing trunk and one list per received trunk. Connection of a received virtual channel to a transmit virtual channel is by means of list processing.

Virtual channel allocations of each of 15 different rates of transmission of ten representative kinds of real- and near-real-time communications are shown. Allocations to the Class I region are always made following the last previous allocation (i.e., at the Class II end of the Class I region), thus reducing common channel interswitch signaling (CCIS) to that concerning the allocated call. As calls terminate, all subsequent allocations are incremented toward the SOF marker by the magnitude of the terminated allocation, based on the terminating CCIS message; that is, a CCIS message directly canceling a single call will result in reallocation of all calls occurring farther away from the SOF marker than did the terminated call, without the necessity for individual CCIS messages for each of these reallocated calls. As a result of these call terminations and dynamic reallocations, the position of the Class I/II boundary will vary dynamically, but will always vary in such a direction as to require the minimum size for the Class I calls currently being serviced, and will thereby provide the maximum amount of master-frame capacity for other than Class I transmissions.

The Class II region of a master frame is essentially self-

TYPICAL MASTER FRAME CLASS II ALLOCATIONS

ALLOC. START	ALLOC. WIDTH	CALL IDENTITY AND PRECEDENCE	ALLOCATION PURPOSE [TYPICAL]
9896	8	CLASS I/II BOUND	8-BIT ADCCP FLAG SEQUENCE
9904	112	CCIS-1:Y	CALL ANSWERBACK, RELEASE, PREEMPT RELEASE, OPERATOR RECALL
10016	104	CCIS-2:Z	ACKNOWLEDGE, GLARE, OUT-OF-SERVICE SYNC, CALLED PARTY UNAVAILABLE, INVALID NUMBER, INCOMPATIBLE CONN.
10120	752	PACKET-1:Z	TYPICAL TRANSMITTED INTERACTIVE, QUERY, DATA BASE UPDATE - SINGLE PACKET MESSAGES
10872	712	PACKET-2:O	TYPICAL TRANSMITTED/RECEIVED: NARRATIVE/RECORD LINE BLOCKS [AVERAGE: 11.6 PER MESSAGE]
11584	104	CCIS-3:O	ALL TRUNKS BUSY, EQUIPMENT BUSY, INVALID ROUTE
11688	1080	PACKET-3:P	TYPICAL RECEIVED: INTERACTIVE OR RESPONSE - ONE OF MULTIPLE PACKETS PER TYPICAL INTERCHANGE
12768	224	CCIS-4:P	CALL INITIATE
12992	160	CCIS-5:R	CALL COMPLETE, ACKNOWLEDGE CALL INITIATE
13152	2080	PACKET-4:R	BULK-1 DATA, NON-SENSOR BULK-2 DATA, ARQ-FACSIMILE
15232	208	NONE	IDLE CAPACITY

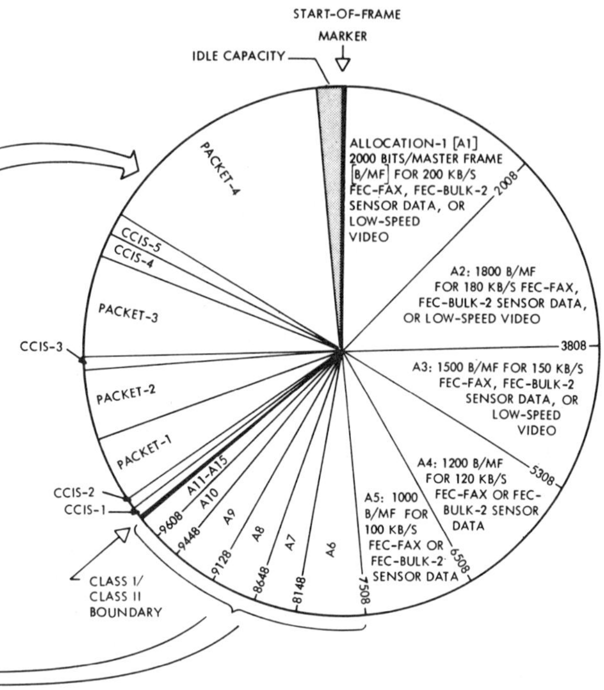

DIGITIZED VOICE

ADDITIONAL CLASS I ALLOCATIONS

CHAN. ALLOC.	START'G POINT	CHAN. WIDTH	FEC-FAX	FAX	PCM	DPCM	CVSD	APC	LPC	VOCODER
6	7508	640	X		X					
7	8148	500	X							
8	8648	480	X		X	X				
9	9128	320	X			X	X			
10	9448	160	X				X	X		
11	9608	96		X				X		
12	9704	80						X	X	
13	9784	48		X						X
14	9832	40							X	
15	9872	24		X						X

Fig. 2. Typical master-frame map allocations for the SENET concept.

identifying in that its contents are individually recognizable packets enveloped in accordance with ADCCP procedure. Consequently, at the first occurrence of an ADCCP flag sequence following the SOF marker, it may be assumed that the Class II region has been entered. This may be cross-checked by integration of the total allocation for Class I with respect to the individual width per allocation. Allocations in the Class II region are made according to precedence as long as Class II space is available. No distinction is made between interactive, query/response, database update, narrative/record data packets, and Bulk-1 data transfers and nonsensor Bulk-2 data transmissions.[1] Bulk-2 sensor data are handled as a Class I virtual connection; Bulk-1 and Bulk-2 nonsensor bulk data are handled as Class II data transmission (i.e., divided into, and transmitted and handled, as packet-switched transmissions).

There is no boundary marker to signify end of frame. The SOF of each successive master frame serves as the end of frame of the preceding master frame. Should the contents of a master frame not occupy all of the available capacity in that frame, the remaining space following the end of the last complete packet in Class II will be filled with synchronous characters (e.g., ADCCP flag sequences).

B. Flexible-Hybrid Approach

One of the disadvantages of the SENET approach is that transmission efficiency suffers as a direct consequence of the inefficiency in transmitting circuit-switched voice. Time assignment speech interpolation (TASI) is one method of improving the efficiency but at the price of adding a considerable amount of complexity. Here we describe a variation of the basic SENET approach that would improve the efficiency of the hybrid system considerably. The approach, which we have termed the flexible-hybrid concept, handles bulk data in a circuit-switched mode while voice and interactive data are handled in a packet-switched mode. Several investigators have suggested this scheme as a potentially cost-effective strategy for integrating voice and data in a common system [11], [18].

The flexible-hybrid scheme is illustrated in Fig. 1(b). Voice is transmitted as packet-switched data without error control by a fixed routing algorithm. Interactive and narrative/record data that require error control are transmitted as error-protected packet-switched data by adaptive routing. Bulk, facsimile, and burst data information are transmitted in simplex form as circuit-switched data, with any necessary error control or acknowledgments provided by packets or a small transmission capacity allocated in the reverse direction. This method has several advantages. The bulk data class can be transmitted more efficiently and with less delay using circuit switching since, in general, long messages favor circuit-switched operation and short messages favor packet-switched operation [13]. This requires modification of the SENET approach to allow the bulk data to be transmitted over a two-wire path in simplex form, similar to the way bulk data are transmitted via fixed routing in a packet-switching scheme. The price to be paid is asymmetrical link maps and the neces-

[1] Bulk-data is identified as the transmission of entire files, programs or processing results; Bulk-2 data is identified as the transfer of extremely lengthy information such as an entire database or sensor data.

sity to provide error control in the reverse direction. The SENET scheme specifies symmetrical link maps for allocation of data regardless of direction of transmission (e.g., see Fig. 2). The simplex transmission of bulk-like data would necessitate separate frame map pairs at the ends of each link—one pair of maps for each direction. As stated previously, error control could be provided either through circuit-switched allocation of a small control channel in the reverse direction or by packets sent in the Class II region.

The packet switching of voice provides a TASI-like efficiency advantage identical to that found in packet-switching systems. During speech silences, transmission ceases; hence, the capacity is available for servicing other users. Packet-switched traffic or message-switched traffic requiring error protection is handled conveniently within the packet-switched mode of operation.

Several modifications to the approach could be used to improve operational performance. The circuit-switched region could be used to carry synchronous encrypted voice or data. In addition, high-priority voice or data that require low delay (e.g., voice over satellite links) or delay that is independent of processor load, or voice traffic requiring significant interaction between subscribers to ensure that the information transmitted is correct, could also be circuit-switched.

C. Enhanced-Hybrid Approach

Other variations of the SENET concept can provide enhancements by leading to an improvement in channel utilization. These include the use of time assignment speech interpolation (TASI) and time assignment data interpolation TADI). One advantage of the SENET approach of transmitting synchronous voice in Region 1 is that it lends itself to the use of TASI and TADI techniques by taking advantage of the inherent properties of speech. During the speech-silence intervals of a conversation, transmission can cease, and the available capacity can be used to transmit other data. The original transmitted speech is said to be interpolated. Either data packets or voice segments from other voice sources can be transmitted during the silent periods.

The use of TASI and TADI would serve to bring the transmission efficiency of the SENET concept closer to that of packet switching. Fig. 1(c) illustrates one approach to implementing TASI/TADI in a hybrid system. A control vector in the circuit-switched portion of every frame indicates which user slots are in talk spurt and which are in silence. These are identified by x's in the diagram. Virtual circuits in which the voice source is in an idle state can be used to carry data or additional speech from a second voice source. The former is designated time-assigned data interpolation (TADI) and the latter is time-assigned speech interpolation (TASI). Voice sources that leave the silent state and resume speech generation may be "frozen out," i.e., they may not find an available slot in which to transmit. A sufficient number of slots must be provided to ensure that the percentage of speech "frozen out" is within an acceptable percentage (e.g., 0.5 percent) [19], [20].

Several approaches to the implementation of tandem TASI have been studied; however, all appear to have significant problems associated with them [5]. One problem in applying TASI to a SENET-like scheme concerns the maintenance of slot allocations for voice calls by maps located in the switches at the end of each link (see Fig. 2). Meticulous control of the alteration of these maps is necessary. Transmitting and receiving switches must coordinate precisely any changes in these maps, e.g., how much of the frame a new allocation requires, where it will be located, the exact timing of a change in map allocation. Changes in slot utilization, either from errors or from late or out-of-sequence messages, could disrupt every active user in the system. A multibandwidth requirement to service voice subscribers in a SENET system at various transmission rates adds further complication. Thus, the resulting coordination complexity will seriously affect the practicality of a scheme to interpolate speech into the silent periods of other speech.

Interpolation of packet-switched data into voice-silent periods using TADI appears to be a practical alternative. In this instance, "silence" slots in the frame would be treated as a dispersed string of sequential bit spaces, followed contiguously by the formal Class II region. The switch would start with the first packet of the oldest message of the highest precedence, and would allocate the bits of this packet sequentially into as many of the silent allocations in the Class I region of that frame as are necessary. When a complete packet had been mapped into the region of silences, the next queued packet in order would be similarly and contiguously mapped into the region of silences. If the bits available in the totality of silence allocations in a frame were not enough to contain all the packets queued and ready for transmission during that frame, the Class II region would be used as a contiguous extension of the silent allocation. This technique would avoid many of the coordination problems associated with the use of tandem TASI, particularly the effects of servicing multibandwidth voice.

It is also possible to eliminate the separate regions altogether, allowing the circuit-switched data to fill the entire master frame except for the SOF marker and a frame-silence map. Packet-switched data would be switched into the silence periods (as indicated by the frame-silence map) using TADI techniques.

D. Variable-Length Frame/Packet

Miyahara and Hasegawa have proposed an integrated multiplex scheme for the hybrid system which will help to eliminate unused frame capacity [21], [22]. In an integrated circuit- and packet-switched multiplex structure, such as the SENET, the communication link is dedicated to circuit switching at one instant and packet switching at another. The frame structure is divided into a circuit- and a packet-switched region. Frame intervals are kept constant for the synchronous assignment of circuit-switched time slots. Packets are fit into that part of the frame following the circuit-switched region. It is desirable under this scheme to transmit as many packets as possible, and thus minimize the idle capacity at the end of the frame. But, because of the constant frame format, either packets are too long for transmission in that frame, in which case capacity is wasted, or the packet length is shortened, which increases the overhead of that packet relative to its information field. However, if the frame interval is allowed to

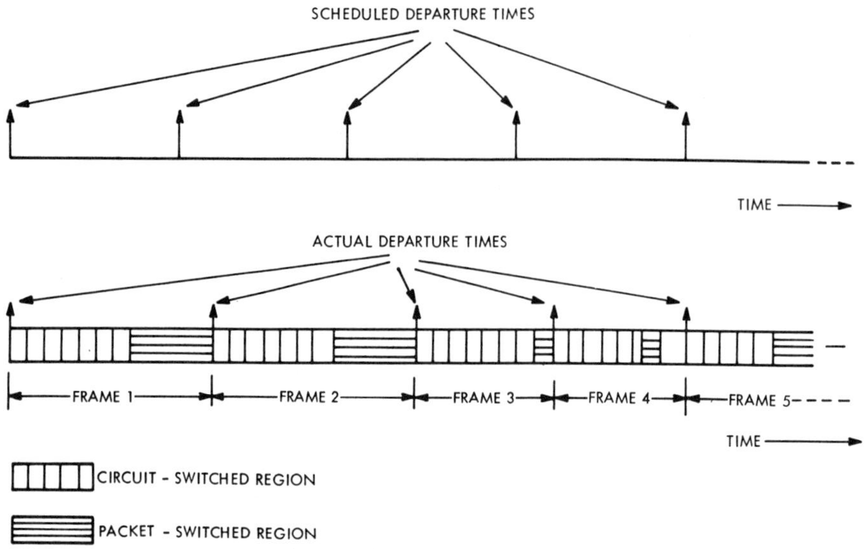

Fig. 3. Variable-length frame/packet departure times at a node.

have variable length, a packet that exceeds the residual capacity of a frame can be transmitted over two consecutive frame intervals. This will result in an increased transmission efficiency, an increased flexibility in handling varying traffic demands, and an improvement in average packet transmission delay.

One problem with this method is that if the packet is being transmitted over two consecutive frame intervals, the circuit-switched region cannot interrupt at that time to begin its scheduled transmission. Under these conditions, the transmission delay of the circuit-switched portion is no longer constant. The interarrival time of successive circuit-switched bit groups will no longer be equal to the constant frame interval (10 ms); thus, the transparency to delay dispersion, which is a natural consequence of circuit switching, would not be preserved. However, if the dispersion of frame interarrival times can be upper-bounded to a maximum value, and that value is sufficient for regaining synchronization at the destination node, and in addition, there is little or no impact on voice quality and intelligibility, then the transparency of circuit switching can still be offered.

Fig. 3 illustrates this process. Frames are scheduled for departure at fixed time intervals. However, the new frame will not be transmitted if a packet is in transmission at that time. Note that, if the traffic on the link decreases sufficiently, the delay dispersion can readjust itself to the nominal value. This is the case for frames 4 and 5, where the actual and scheduled departure times for frame 5 coincide.

Thus, if the outgoing link is available when a packet arrives, the packet can be transmitted in the capacity coming just after its arrival, without waiting for the next frame cycle. This causes a decrease in the expected delay for packet-switched data. Although the delay for circuit-switched traffic will vary, it can be bounded to a maximum value acceptable for the transmission of digitized voice traffic [21].

III. TRANSMISSION EFFICIENCY

In evaluating the various concepts we considered two types of transmission efficiency: overhead efficiency and throughput efficiency [5]. Overhead efficiency provides a measure of the number of information bits that can be transmitted on the line relative to the number of overhead bits required. This represents a maximum achievable efficiency, i.e., the efficiency obtainable with a solidly packed line. The throughput efficiency takes into account the stochastic arrivals of packets, their lengths, and the effects of loading at the queues. It provides a more conservative estimate of efficiency, but is a value closer to that which would be obtained in a realistic environment.

Throughput efficiency is defined as the ratio of maximum achievable information bit/second (bit/s) throughput rate to the total available bandwidth of the transmission system. For a given hybrid-switched link, the throughput efficiency can be expressed by

$$E = (R/C) * 100 \tag{1}$$

where E is the link efficiency expressed as a percentage, R is the maximum information throughput, both voice and data, that the link can support (bits/s), and C is the link bandwidth (bits/s).

All the hybrid schemes discussed in this paper are based on the time-division multiplexing of voice and data over the hybrid frame. The synchronous or circuit-switched portion of the frame services its traffic in a synchronized fashion and uses blocking as part of its operational mechanism. The nonsynchronous portion of the frame services its traffic in a totally asynchronous fashion and uses buffering as part of its operational mechahism.

Letting E_1 and E_2 be the throughput efficiency corresponding to the synchronous and nonsynchronous portions of a hybrid system, respectively, then

$$E = \sum_{i=1}^{2} \alpha_i E_i \tag{2}$$

where α_i is the fraction of the time during which the integrated link is available for traffic type i. The throughput effi-

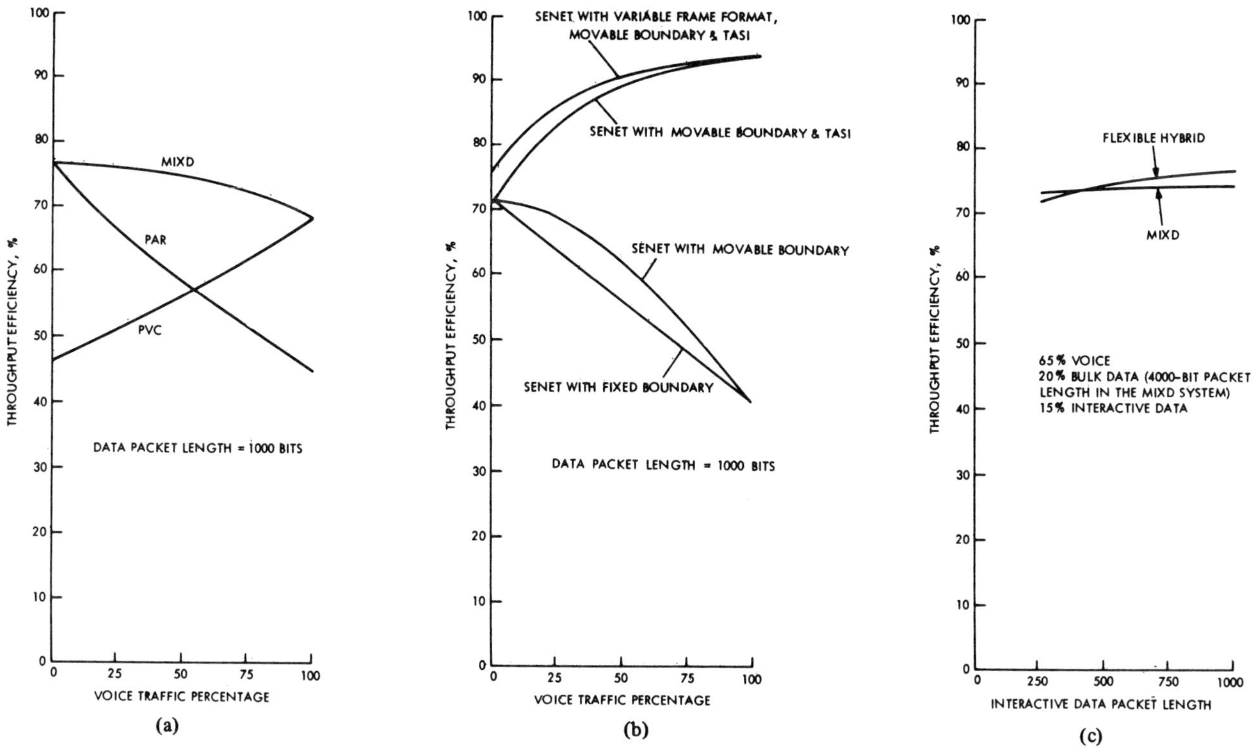

Fig. 4. Comparison of throughput efficiency for packet and hybrid switching concepts. (a) For packet-switching concepts as a function of percentage voice content. (b) For hybrid switching concepts as a function of percentage voice content. (c) Versus interactive data packet length for the flexible hybrid and MIXD concepts.

ciency of the synchronous portion is given by

$$E_1 = (1 - P_B) * A * T * 100 \tag{3}$$

where A is the voice activity factor (the ratio of time that speech is present to the total elapsed time), T is the TASI advantage, and P_B is the blocking probability for circuit-switched traffic. This can be approximated by the Erlang B formula:

$$P_B = \frac{a^2/s!}{\sum_{j=0}^{s} a^j/j!} \tag{4}$$

where a is offered circuit-switched load in erlangs and s is the number of circuit-switched slots per frame. E_2, the throughput efficiency for the nonsynchronous part (operating on a queuing basis), is given by

$$E_2 = (R_2/C_2) * 100 \tag{5}$$

where R_2 is the maximum information throughput rate supported by the packet-switched portion and C_2 is the portion of the channel allocated to the nonsynchronous traffic (referred to as the "packet switched subchannel"). R_2 is the information throughput rate corresponding to the maximum tolerable data-packet delay.

Observe that E_2 is computed for a given prescribed maximum tolerable average data-packet delay, while E_1 is computed for a maximum tolerable average blocking probability. Therefore, the throughput efficiency is computed for a prescribed average voice blocking and mean data delay. Pure packet-switching concepts, on the other hand, operate on a delay basis; therefore, their throughput efficiencies are evaluated only for a given maximum tolerable delay. This poses a conceptual difficulty when we try to compare the hybrid and the packet-switching concepts.

In order to find R_2, the delay-throughput relationship of the packet-switched subchannel is obtained using a single-server queuing model. The derivation and all the assumptions associated with the model are given in Section IV.

Fig. 4(a) and (b) shows the throughput efficiency for packet and hybrid switching concepts, except for the flexible hybrid, as a function of percentage voice traffic. Specific assumptions for voice and data information parcel and overhead lengths for the hybrid schemes are given in Table I. Corresponding parameters for the packet switching concepts are also given in Table I.

There are three basic packet switching concepts to integrate voice and data [23]. These are essentially based on fundamental variations in packet transport protocols. In the PVC, or **packetized virtual circuit** approach, both voice and data packets are routed along fixed paths, while in the PAR, or **packet adaptive routing approach**, voice and data packets are independently routed to their destination. In the MIXD system, voice packets are routed over a fixed path while data packets are independently routed to their destination. Under PVC, throughput efficiency increases with increasing voice

TABLE I
INFORMATION PARCEL AND OVERHEAD LENGTHS FOR PACKET-SWITCHED VOICE AND DATA TRAFFIC

	SENET Packet Switched Region	Flexible Hybrid Packet Switched Region	PVC Concept	PAR Concept	MIXD Concept
Voice Packets					
Link control header length (except CRC)*	0 bits	32 bits	32 bits	32 bits	32 bits
Packet header	0 bits	32 bits	32 bits	150 bits	32 bits
Information bits	0 bits	160 bits	160 bits	160 bits	160 bits
Link control CRC	0 bits	0 bits	0 bits	0 bits	0 bits
Setup overhead	0 bits	928 bits	928 bits	0 bits	928 bits
Data Packets					
Link control header length (except CRC)	32 bits	32 bits	32 bits	32 bits	32 bits
Packet header	150 bits	150 bits	32 bits	150 bits	150 bits
Information bits (interactive)	250-1000 bits	250-1000 bits	250-1000 bits	250-1000 bits	250-1000 bits
Information bits (bulk)	4000 bits	0 bits	4000 bits	4000 bits	4000 bits
Link control CRC	32 bits	32 bits	32 bits	32 bits	32 bits
Setup overhead	0 bits	0 bits	928 bits	0 bits	0 bits
Voice Packet Acknowledgments					
Link control header length (except CRC)	0 bits	0 bits	0 bits	0 bits	0 bits
Link control CRC	0 bits	0 bits	0 bits	0 bits	0 bits
Frequency	0 packets/sec	0 packets/sec	0 packets/sec	0 packets/sec	0 packets/sec
Data Packet Acknowledgments					
Link control header (except CRC)	32 bits	32 bits	32 bits	32 bits	32 bits
Link control CRC	32 bits	32 bits	32 bits	32 bits	32 bits
Frequency	20% of data packet rate**	20% of data packet rate**	20% of data packet rate**	20% of data packet rate**	20% of data packet rate**
Other Overhead					
Routing packet length	750 bits	750 bits	0 bits	750 bits	750 bits
Routing update frequency	2 packets/sec	2 packets/sec	0 packets/sec	2 packets/sec	2 packets/sec

* Cyclic Redundancy Check
** This means that 80 percent of the link level ACKS are piggybacked over data packet transmissions in the reverse direction.

voice traffic percentage because voice packets do not carry error control overhead nor are they acknowledged. Under PAR, as the percentage voice content is increased, the efficiency decreases. This is because voice packets are short by design and the PAR overhead bits represent a large percentage of the total voice packet budget.

For systems with a low percentage of voice traffic (0–25 percent) several concepts exhibit good performance. These include the MIXD and PAR packet-switching schemes, the SENET with movable boundary, and the hybrid schemes with TASI. However, TASI is not impressive at the low voice/data ratios, showing only about a 7 percent advantage in throughput efficiency over the MIXD system. In the medium ranges (25-75 percent), the MIXD system and the hybrid schemes with TASI give the best performance, with the MIXD efficiency about 15 percent below that of the hybrid schemes. At high-percentage voice, performance is best for the hybrid schemes with TASI, with efficiencies 25-30 percent above that of the MIXD system. Over the whole range of voice percentages, the MIXD system is superior to all packet- and hybrid-switched systems except those that utilize TASI.[2] However, improvements obtained through the use of TASI are paid for by added system complexity.

Fig. 4(c) shows the throughput efficiency of the flexible-hybrid and MIXD packet-switching systems as a function of interactive data packet length. The results are given for a distribution of 65 percent voice, 20 percent bulk data, and 15 percent interactive data. Both systems show comparable performance, i.e., throughput efficiency in the mid-to-high 70's over a moderate range of interactive data packet lengths. Above 400 bits, the flexible hybrid shows slightly better performance. Thus, the flexible-hybrid system, like the MIXD, yields good throughput efficiency performance.

[2] This plot assumed no bulk data; therefore, the flexible-hybrid would exhibit the same performance as the MIXD systems.

IV. DELAY-THROUGHPUT ANALYSIS

The delay characteristics of traditional circuit switching following call setup are excellent, being on the order of milliseconds. Delay is limited by the propagation through the interconnecting links of the transmission medium. No queueing of data takes place at nodes.

Delays for packet-switching networks are more of an issue than was the case for circuit-switching networks. Although data delays are on the order of tens of milliseconds, end-to-end delays for voice are on the order of 75 ms [7]. However, any voice delay that is less than 250 ms will be acceptable for transmissions over a terrestrial network. Because of the assumed use of T1 links (terrestrial) for transmission, propagation delay is not an important component of overall end-to-end delay. Processing delay becomes the important consideration.

In this section we will carry out the analysis for an "integrated link" model, with the purpose of evaluating the delay-throughput tradeoff curves, which describe the behavior of the packet-switched subchannel in a hybrid-switching system. Since the main objective is to provide a comparative evaluation of different switching techniques, a single-link model is considered here to avoid the corruption of our conclusions by the specific topology or the particular routing procedure of any network.

With the above discussion, we proceed to describe the different delay components encountered by a data packet as it hops from one node to another. Over an integrated link, a data packet encounters the following delay components:

1) *Propagation delay due to the finite speed of the electromagnetic waves*—Defined to be the delay for the first bit of the packet. This component depends on the length of the communication links.

2) *Transmission delay due to the finite transmission rate of the links*—The time for the bits of a packet to be clocked out on the link. A function of the packet length, the transmission bandwidth, and the number of hops between the source and destination nodes.

3) *Nodal processing delay*—The time it takes the node to process one packet. The nodal processing delay has two components, a fixed component that corresponds to the processing time of a single packet and a random component that corresponds to the internal queueing of packets in front of the processor. This component depends on the processing capacity of the node.

4) *Queueing delay*—The time that the packet must wait in front of the output trunk for the transmission of all the packets that precede it, including the residual service time of the packet being transmitted. This delay depends on the link utilization, the priority structure, and the number of links in the source–destination path.

In order to decouple our delay model from the specifics of node processing capacity (i.e., type of computer at the switching model, or the physical length of the trunks), we will consider only the transmission and waiting delay components.

Under the fixed-boundary frame management policy, the partition of link capacity between the synchronous and nonsynchronous traffic is fixed. Under the movable-boundary-frame management policy, however, the packet-switched subchannel can dynamically utilize the idle circuit-switched slots. Fig. 5 displays a single-server queuing model for the integrated voice/data link. Voice frames arrive in a totally synchronized fashion. The data stream, however, can be assumed to arrive at the node according to a Poisson point process with rate λ.

The server services the data queue during a certain period of time, then goes on to service the incoming synchronous voice traffic during another period of time. Therefore, as far as interactive data packets are concerned, the server acts as a single-server queue with rest period (during which time it services the voice traffic). Such a queueing system has been studied [24], and its behavior shows that under heavy traffic the throughput-delay performance is quite similar to an equivalent single-server queueing system without a rest period, but with a reduced service capacity. Since our delay models will be essentially used to find the throughput efficiency, the heavy traffic edge of the delay-throughput relationship is the one of interest; therefore, the above approximation will be used. The capacity reduction factor will be given by α_2, where α_2 is the fraction of the time during which the integrated link is available for traffic type 2 (nonsynchronous traffic).

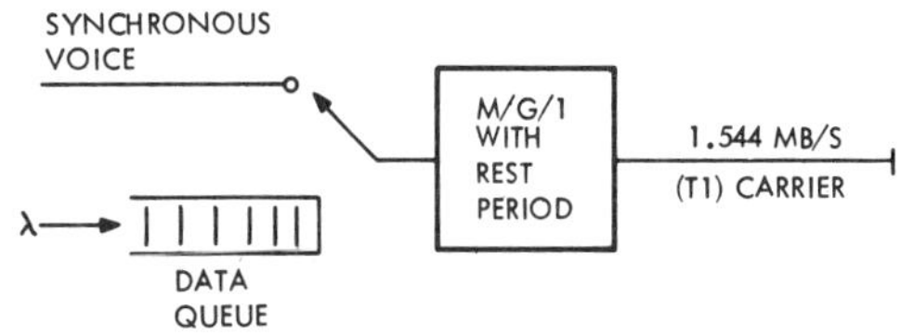

Fig. 5. Single-server queueing model for an integrated voice/data link.

In order to make our analysis general enough to cover the flexible hybrid, we consider a system that carries more than one traffic type over the nonsynchronous portion of the frame. The service time per packet and packet length are related by the linear relationship

$$X_i = \frac{L_i}{\alpha_2 \beta C} \tag{6}$$

where L_i is the length of packets with priority i. This length includes an information parcel (voice or data) along with a packet header and a line protocol overhead. C is the link capacity, and β is a scaling factor (less than one), used here to account for the loss of part of the channel bandwidth in retransmitting packets received in error and in transmitting other control packets such as the acknowledgments, routing updates, and connection setup and takedown signaling messages.

Therefore

$$\beta = 1 - \sum_k \gamma_k \tag{7}$$

where γ_k is the fractional link bandwidth lost due to the kth control overhead.

Clearly, if $i = 1$, we have the case of a SENET frame. If $i = 2$, we obtain a model that would be suitable for analyzing the flexible-hybrid-switching scheme.

The following control overheads are considered.

1) γ_1 is the fractional bandwidth lost due to the fixed-frame format.

2) γ_2 is the fractional bandwidth lost in packet retransmission.

3) γ_3 is the fractional bandwidth lost in transmitting routing update packets. This factor is proportional to both the length and the frequency of the update packets.

4) γ_4 is the fractional bandwidth lost in transmitting acknowledgments.

5) γ_5 is the fractional bandwidth lost in the connection setup and takedown for both voice and data traffic.

In order to evaluate the queueing behavior over the packet-switched subchannel, the moments of the service times need to be evaluated. Since the moment is itself a linear operation, the kth moment of the service time is then given by

$$\overline{X_i^k} = \frac{\overline{L_i^k}}{(\alpha_2 \beta C)^k}. \tag{8}$$

For the above queueing model with nonpreemptive head-of-the-line priority discipline, the first two moments of the queueing time are given by

$$\overline{W}_j = \frac{\lambda\overline{X^2}}{2\left[1-\sum_{i=1}^{j-1}\lambda_i\overline{x_i}\right]\left[1-\sum_{i=1}^{j}\lambda_i\overline{x_i}\right]} \tag{9}$$

and

$$\overline{W_j^2} = \frac{\lambda\overline{X^3}}{3\left[1-\sum_{i=1}^{j-1}\lambda_i\overline{X_i}\right]^2\left[1-\sum_{i=1}^{j}\lambda_i\overline{X_i}\right]} + \frac{\lambda\overline{X^2}\sum_{i=1}^{j}\lambda_i\overline{X_i^2}}{2\left[1-\sum_{i=1}^{j-1}\lambda_i\overline{X_i}\right]^2\left[1-\sum_{i=1}^{j}\lambda_i\overline{X_i}\right]^2} + \frac{\lambda\overline{X^2}\sum_{i=1}^{j}\lambda_i\overline{X_i^2}}{2\left[1-\sum_{i=1}^{j-1}\lambda_i\overline{X_i}\right]^3\left[1-\sum_{i=1}^{j}\lambda_i\overline{X_i}\right]} \tag{10}$$

where

$\overline{W}_i$ = expected value of the waiting time for messages with priority i
$\overline{W_i^2}$ = second moment of the waiting time of messages with priority i
λ_i = arrival rate of messages with priority i
λ = arrival rate of messages of all priorities
$\overline{X_i}$ = expected value of service time of priority i
$\overline{X_i^2}$ = second moment of service time of priority i
$\overline{X^2}$ = second moment of service time, all priorities
$\overline{X^3}$ = third moment of service time, all priorities.

The above equations for the waiting-time moments simplify considerably for a flexible-hybrid system using both voice and data frames of constant length, and a two-priority service discipline, with priority 1 given to the voice and priority 2 to the data. The above equations then take the form

$$\overline{W_1} = \frac{\lambda_1(\overline{X}_1)^2+\lambda_2(\overline{X}_2)^2}{2(1-\lambda_1\overline{X}_1)} \tag{11}$$

$$\overline{W_2} = \frac{\lambda_1(\overline{X}_1)^2+\lambda_2(\overline{X}_2)^2}{2(1-\lambda_1\overline{X}_1)(1-\lambda_1\overline{X}_1-\lambda_2\overline{X}_2)} \tag{12}$$

$$\overline{W_1^2} = \frac{\lambda_i(X_1)^3+\lambda_2(X_2)^3}{3(1-\lambda_1\overline{X}_1)} + \frac{\lambda_1(X_1)^2[\lambda_1(X_1)^2+\lambda_2(X_2)^2]}{2(1-\lambda_1\overline{X}_1)^2} \tag{13}$$

$$\overline{W_2^2} = \frac{\lambda_1(\overline{X}_1)^3+\lambda_2(\overline{X}_2)^3}{3(1-\lambda_1\overline{X}_1)^2(1-\lambda_1\overline{X}_1-\lambda_2\overline{X}_2)} + \frac{[\lambda_1(\overline{X}_1)^2+\lambda_2(\overline{X}_2)^2]^2}{2(1-\lambda_1\overline{X}_1)^2(1-\lambda_1\overline{X}_1-\lambda_2 X_2)^2} + \frac{\lambda_1(\overline{X}_1)^2[\lambda_1(\overline{X}_1)^2+\lambda_2(\overline{X}_2)^2]}{2(1-\lambda_1\overline{X}_1)^3(1-\lambda_1\overline{X}_1-\lambda_2\overline{X}_2)}. \tag{14}$$

For the case of SENET, where we have pure interactive data over the packet-switched subchannel, the above equations simplify to

$$\overline{W} = \frac{\lambda\overline{X^2}}{2(1-\lambda\overline{X})} \tag{15}$$

and

$$\overline{W^2} = \frac{\lambda\overline{X^3}}{3(1-\lambda\overline{X})} \tag{16}$$

where λ is the data packet arrival rate (packets/s) and $\overline{X^k}$ is the kth moment of data packet service time. The variance of the waiting time is given by

$$\sigma_{W_j}{}^2 = \overline{W_j^2} - (\overline{W}_j)^2. \tag{17}$$

Assuming independence between the service- and waiting-time delays, the mean and variance of the total delay time per packet are given by

$$\overline{T_j} = \overline{X_j} + \overline{W_j} \tag{18}$$

and

$$\sigma_{T_j}{}^2 = \sigma_{W_j}{}^2 \tag{19}$$

where j = 1, 2 correspond to voice and data response times, respectively.

For a typical packet (voice or data), response times are given by the weighted sum

$$T = a\overline{T_1} + (1-a)\overline{T_2} \tag{20}$$

and

$$\sigma_T{}^2 = a\sigma_{T_1}{}^2 + (1-a)\sigma_{T_2}{}^2 \tag{21}$$

where

$$a = \frac{b}{b+(1-b)\left(\frac{L_1}{L_2}\right)}$$

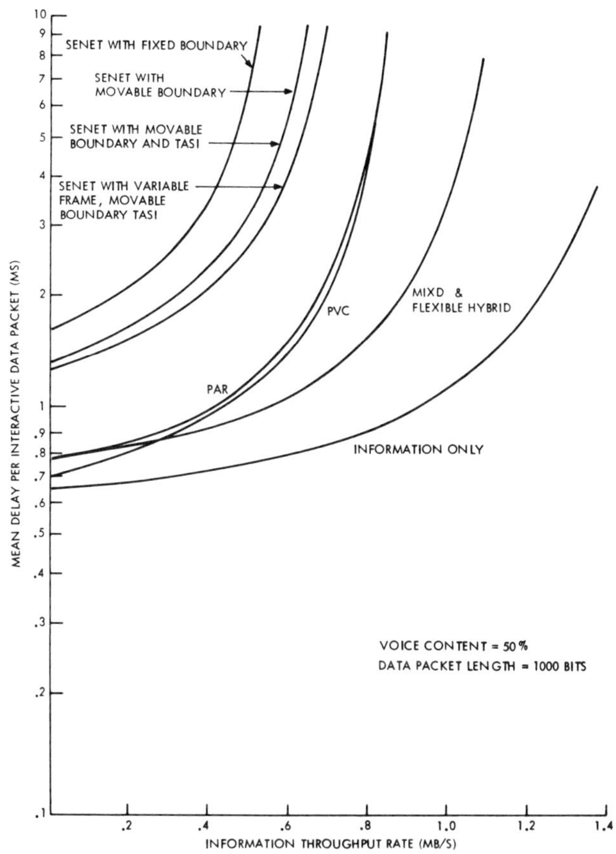

Fig. 6. Mean delay per interactive data packet versus throughput for various packet and hybrid concepts.

and

$$b = \frac{\text{voice information rate}}{\text{total information rate}}$$

L_1 = length of a voice information parcel

L_2 = length of a data information parcel.

The specific assumptions are found in Table I.

Data delay in the hybrid schemes is not equal to that in the packet-switching concepts. Fig. 6 shows the mean delay per interactive data packet versus throughput for various hybrid- and packet-switching schemes. This is for the case of 50 percent voice content and an interactive data packet length of 1000 bits. The closer a delay curve is to the "information only" curve, the better is the performance. For a given throughput, the packet-switching schemes exhibit the best delay characteristics. The flexible hybrid and MIXD systems display a superior performance over all other concepts presented.

The voice delay in hybrid systems is somewhat greater than in circuit switching, once the connection has been established. This is chiefly a result of the 10 ms frame interval and the placement of the voice bits in the output frame. The hybrid systems exhibit less variation in voice delay with changes in throughput than do the packet-switching systems. Thus, in packet-switching systems, delays vary about a mean value, with extremes becoming more prevalent at the higher throughput rates. The voice delays for the hybrid system are more constant with increase in load.

V. OTHER PERFORMANCE FACTORS

Concepts were also examined as a function of additional evaluation criteria. These included routing and flow control, voice continuity, precedence/preemption, synchronization, service features, and classmarking flexibility.

In a circuit-switching approach, and in the circuit-switched portion of the hybrid approach, a synchronous path is provided between communicating parties. Thus, an excellent medium is provided for maintaining voice continuity over the network. The price for this, however, is transmission inefficiency, since net speech-channel activity is on the order of 40 percent. Packet switching provides efficient transmission but does not present as good a medium for voice continuity, since the voice waveform is broken up into packets and transmitted asynchronously. Individual packets can be lost, creating gaps in the voice bit stream. On the other hand, when the addition of TASI to circuit or hybrid switching is considered, circuit switching loses its advantage in voice continuity performance. Voice is handled in talk spurts, which may be clipped or cut out entirely.

Both hybrid and packet-switching concepts are capable of incorporating elaborate and powerful protocol, error control, flow control, and acknowledgment schemes. Circuit-switching network protocols are very limited in comparison with some specific problems, including an inability to handle incompatible subscribers and no provision for link-by-link error control. Part of this derives from the fact that circuit-switched flow control operates on a loss or blocking probability strategy, whereas hybrid or packet flow control introduces delay management of traffic resources as well. However, flow control and routing are generally regarded as research areas in packet (and hybrid) systems that require much more work. This is further complicated by the introduction of packetized voice for the packet and flexible-hybrid concepts. Hence, although packet-switching techniques appear to have more capability than circuit switching with regard to flow control and routing, there is still unproven ground to tread, and one should not assume that existing solutions, e.g., the ARPANET flow control mechanism, will apply. Protocol, flow control, routing, and error control complexity in the hybrid system are expected to be greater than those of packet switching, since two different types of data must be sent—synchronous as well as asynchronous.

Investigation of precedence/preemption led to the conclusion that the preemption of voice and data connections is a complex problem in any concept. Networks using packet or hybrid switching should exhibit a better precedence and preemption performance than circuit switching. This derives from the use of a delay approach as well as a loss approach to manage the precedence/preemption strategy and tailor it to a particular voice or data traffic distribution. The reconciliation of voice and data precedence structure, and the relationships of control message precedence to voice/data content precedence, present a problem for all concepts, and must be studied carefully for any candidate concept. Ultimately, the relationship between voice and data precedence structures will probably be solved by military requirements, however arrived at.

Routing strategies can be very elaborate for all of the generic concepts. For circuit switching, the objective might be to minimize the time interval between capacity allocation and its use by a subscriber, since network resources are wasted during this interval. In packet switching, the objective may be to minimize delivery delay by routing traffic around congested links or switches.

If the network is lightly loaded, the routing scheme does not matter, and if it is very heavily loaded, the routing scheme does not matter either, because all routes are saturated. It is only in a moderately loaded network that the proper selection of a routing scheme can cause the network to make the best use of its resources. Furthermore, whether the concept used is packet or hybrid, we must be careful to distinguish between the processes of route selection (for a call) and routing table update. It was also noted that per-packet dynamic routing is more complex than a virtual circuit approach, but the virtual circuit approach is not necessarily better. Further investigation must compare both approaches from the standpoint of processing time, buffer requirements, and software complexity.

All approaches need bit synchronous techniques to maintain timing relationships within the network. Synchronous concepts (e.g., circuit switching) require framing hardware, whereas asynchronous concepts (e.g., packet switching) require packet buffers. The hybrid concept requires a combination of the hardware/memory utilized in packet and circuit switching.

Service features, in general, are concept-independent; that is, they do not have a significant efficiency or delay effect within any particular concept. Voice conferencing is a feature that is seen as a problem for any of the concepts considered and, in fact, for any digital system.

Classmarking flexibility does not appear to be a strength for some concepts and a weakness for others (although it is more limited in a circuit-switched network than in a hybrid- or packet-switched network). Classmarking appears to be flexible for all concepts, just as it does for a wide variety of stored-program, controlled-circuit, message, and packet systems being produced today.

VI. DELAY VERSUS THROUGHPUT AS A FUNCTION OF TRAFFIC TYPE

Three single-traffic conditions (pure voice, pure interactive data, and pure bulk data) were examined in order to determine the region(s) of traffic mix over which a particular integrated concept is best suited. Fig. 7 shows the delay-throughput relationship for various concepts as a function of single-traffic type.

Fig. 7(a) shows the delay-throughput relationship for a T1 carrier transmitting pure voice. Both the MIXD and flexible hybrid techniques reduce to the PVC approach, i.e., voice packets sent via the virtual circuit (or fixed routing) mode without error control. The PAR technique incorporates additional bits for supervision and routing. Since voice packets are short by design, this overhead represents a rather high percentage as compared to the information being transmitted. Thus the MIXD, flexible hybrid, and PVC concepts exhibit superior delay performance over that of PAR over the whole range of throughput.

All of the SENET-like schemes reduce to time division multiplexing (TDM). In this case, it is observed that at the SENET start-of-frame instant, synchronous trunk transmission starts immediately with data extracted from input memory. The total time taken by a voice parcel between the start of frame and the completion of transmission is a random variable that has its values in the range (τ, T) where τ is the SENET slot duration and T is the SENET frame period. This time takes the lower value if the voice call is assigned to the first slot, and it takes the higher one if the call is assigned to the last slot in the frame. For a voice conversation picked at random, the expected value of the total delay in front of the transmission trunk would, therefore, be given by $(\tau + T)/2$. Frames arrive at the output trunk synchronously at a rate of one each T seconds. Each frame has at most one voice parcel per active voice call. Therefore, each TDM slot can be modeled as a $D/D/1$ server, with a maximum arrival rate of one voice parcel each T seconds. The actual rate, however, depends on whether or not TASI is applied. A voice activity factor of 0.4 is assumed here, and the trunk delay performances for both SENET and enhanced SENET are shown.

Fig. 7(b) shows the delay-throughput performance for a T1 carrier transmitting pure interactive data. For this case, the SENET with variable frame, MIXD, and flexible hybrid schemes reduce to the PAR scheme, i.e., error-protected interactive data packets sent via the Datagram (or adaptive) routing mode. The PVC technique for interactive data transmission requires additional bits for data link control, as well as the overhead necessary to set up and break down virtual path connections. The result is that, except at very low throughput rates, the PAR-reduced systems show superior performance over both the SENET scheme with fixed boundary and the PVC technique.

Finally, Fig. 7(c) provides the delay performance for bulk data under each of the different integrated voice/data schemes. Notice that the difference between these curves and the "information only" curve is extremely small for this case, indicating that the overhead efficiency is high due to the relatively long packet length used for bulk data (4000 bits). For the low and mid-throughput range, PVC shows lower delay for bulk data than any of the other schemes. At the high throughput range, the flexible hybrid provides the smallest mean delay for bulk data.

The above discussion focuses on the performance of the different integration techniques for a single-traffic condition. The total traffic in an integrated network, however, will generally consist of a combination of all three types of traffic. Fig. 8(a) shows the volume that bounds the different mixture ratios for all three traffic types. Fig. 8(b) and (c) displays the recommended integration technique for the different informa-

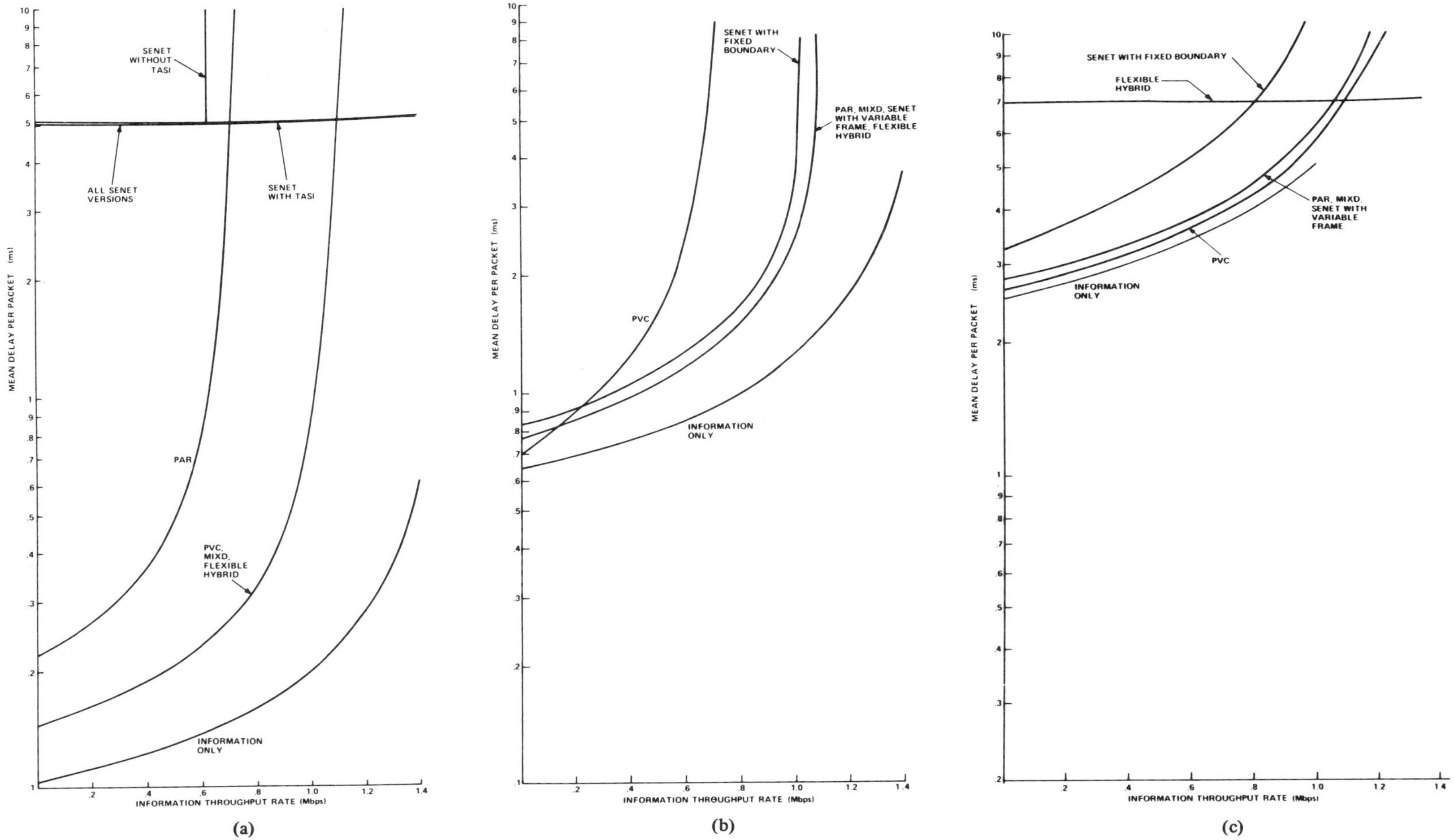

Fig. 7. Delay–throughput relationship for a T1 carrier transmitting different traffic mixtures. (a) Pure voice. (b) Pure interactive data. (c) Pure bulk data.

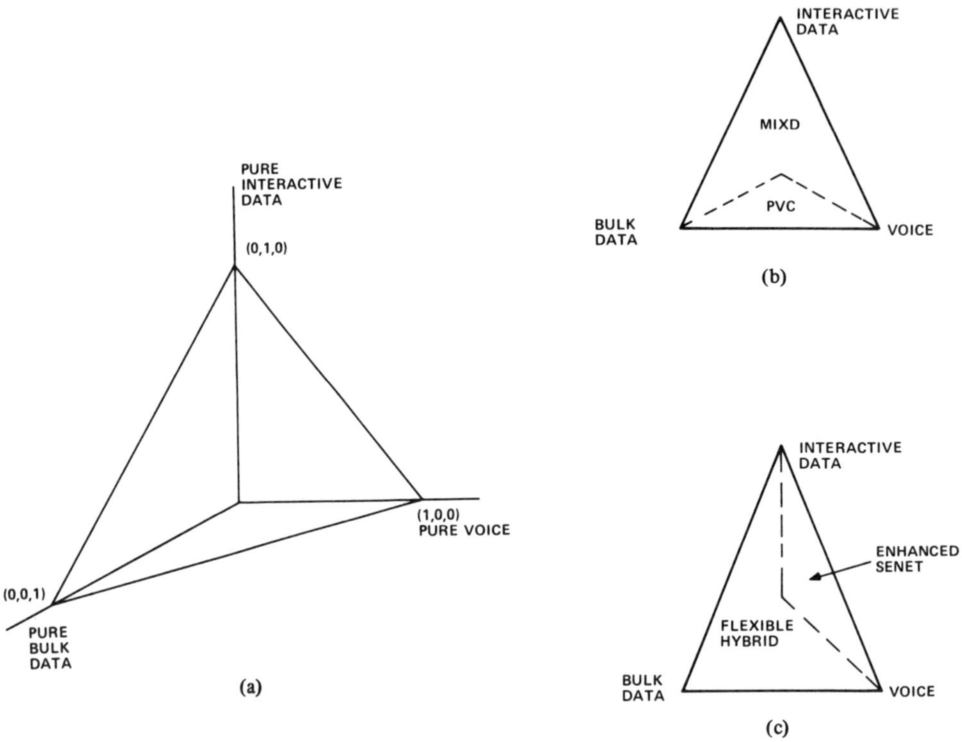

Fig. 8. Guidelines for recommended integration techniques. (a) Traffic mixture volume. (b) Low information throughput rates. (c) High information throughput rates.

tion mixing ratios within the traffic volume. Based on the above results, the following guidelines are recommended. Under low information throughput, either the PVC or the MIXD scheme should be implemented in the switches. Under high information throughput conditions, enhanced SENET and flexible hybrid are the best concepts from the viewpoint of delay versus throughput.

VII. CONCLUSIONS

Evaluation of the hybrid switching concepts has led to the following conclusions.

Two of the SENET variations examined—enhanced hybrid and variable frame and packet—showed superior transmission throughput efficiency (70–95 percent) when they were supplemented with TASI techniques. Varying the frame boundary provided only about 5 percent improvement. The classic SENET approach, without the use of TASI, showed only a fair transmission efficiency (40–70 percent) except at the low voice percentages. The flexible hybrid concept showed a transmission efficiency comparable to that of the MIXD packet-switching system (i.e., in the mid-to-upper 70's).

Data throughput performance, for all hybrid approaches other than the flexible hybrid concept, appears to be poorer than that of packet-switching systems. The flexible hybrid exhibits a delay performance similar to that of the MIXD packet-switching system.

The complexity of hybrid concepts is expected to be higher than that of the other two generic families, for two reasons: the need to incorporate both packet and circuit switching techniques in the same system, and the difficulty of implementing TASI in the circuit-switching portion of some of the hybrid approaches, where frame maps showing speech slot assignments must be maintained at the ends of each link.

The problem of achieving tandem TASI/TADI in the hybrid system is a difficult one. Several approaches have been considered, however, more work is required before the implementation of tandem TASI/TADI can be put on a firm basis. In addition, tandem TASI/TADI techniques imply a large amount of control overhead, far in excess of that needed for the link-by-link transfer of information. The results presented here on efficiency are for the single-link case only. Improvements as a result of using TASI/TADI techniques may be optimistic when applied to a multilink network, because of the additional control required.

The performance of the hybrid system is highly dependent on the mix of voice and data traffic. Thus, a multibandwidth requirement to service voice subscribers at various transmission rates could lead to problems in hybrid systems where voice is circuit-switched. For example, how does one reconcile the servicing of low-bandwidth subscribers versus that of large-bandwidth subscribers? Is it justified to preempt, e.g., 12 voice subscribers of low bandwidth by one voice subscriber of high bandwidth who has a precedence level slightly higher than the low bandwidth subscribers? Although this has little impact on transmission efficiency, it can have the effect of introducing a poorer performance from the point of view of the individual subscriber.

In those techniques that employ the packet switching of voice, if a constant sampling interval at the switch is used in the formation of voice packets, the packet size will vary depending upon the voice rate. Thus, the time needed to accumulate enough digitized speech at the destination for output to a listener will vary. The effect will be greater for voice at lower bit rates, e.g., 2400 bits/s, and, because of delay in the network, could lead to a problem in delivering the reconstructed speech to a listener in a continuous stream. This can be avoided by slightly modifying the flexible-hybrid system to use packet switching for high-bit-rate voice transmission and circuit switching for low-bit-rate voice transmission.

In summary, the flexible hybrid emerges as a viable candidate for integrated voice/data communication. In addition, the SENET with TASI also yields good performance, but its implementation will require further investigation.

REFERENCES

[1] B. Occhiogrosso, I. Gitman, W. Hsieh, and H. Frank, "Performance analysis of integrated switching communications systems," presented at the Nat. Telecommun. Conf., 1977.

[2] "Economic analysis of DoD voice and data networks," Network Analysis Corp., Sept. 1978.

[3] M. J. Ross and C. M. Sidlo, "Approaches to the integration of voice and data telecommunications," presented at the Nat. Telecommun. Conf., Nov. 1979.

[4] O. A. Mowafi and M. J. Ross, "Switching techniques for integrated voice/data networks—Comparative evaluation," presented at the Networks Int. Conf., London, England, Mar. 1980.

[5] "Experimentation and evaluation of advanced integrated system concepts," GTE Products Corp., Final Rep., DCA Contr. 100-79-C-0024, Sept. 1980.

[6] E. A. Harrington, "Voice/data integration using circuit switched networks," *IEEE Trans. Commun.*, vol. COM-28, pp. 781–793, June 1980.

[7] G. Coviello, "Comparative discussion of circuit- vs. packet-switched voice," *IEEE Trans. Commun.*, vol. COM-27, Aug. 1979.

[8] R. Esterling and P. Hahn, "A comparison of digital data network switching alternatives," presented at the Nat. Telecommun. Conf., Dec. 1975.

[9] H. Miyahara *et al.*, "A comparative evaluation of switching methods in computer communication networks," presented at the Int. Conf. Commun., San Francisco, CA, June 1975.

[10] J. Blackman, "Integrated switching of voice and non-voice traffic," presented at the Int. Conf. Commun., Philadelphia, PA, June 1976.

[11] R. Rosner and B. Springer, "Circuit and packet switching. A cost and performance tradeoff study," *Comput. Networks*, 1976.

[12] G. Cicchetti and Luborsky, "Hybrid integrated digital network," presented at the World Telecommun. Forum, Geneva, Switzerland, 1975.

[13] H. Rudin, "Studies on the integration of circuit and packet switching," presented at the Int. Conf. Commun., Toronto, Ont. Canada, June 1978.

[14] G. Coviello and P. Vena, "Integration of circuit/packet switching by a SENET (slotted envelope network) concept," presented at the Nat. Telecommun. Conf., New Orleans, LA, 1975.

[15] M. Ross, A. Tabbot, and J. Waite, "Design approaches and performance criteria for integrated voice and data switching," *Proc. IEEE*, vol. 65, Sept. 1977.

[16] "The application of multiple processor computer systems to digital communications networks," Carnegie-Mellon Univ., Pittsburgh, PA, DCA Contr. DCA-100-75-C-0071, June 1976.

[17] "SENET—DAX study," GTE Sylvania, Final Rep., DCA Cont. 100-75-0071, June 1976.

[18] I. Gitman and H. Frank, "Economic analysis of integrated voice and data networks: A case study," *Proc. IEEE*, vol. 66, Nov. 1978.

[19] K. Bullington and J. Fraser, "Engineering aspects of TASI," *Bell Syst. Tech. J.*, vol. 38, Mar. 1959.

[20] C. Weinstein, "Fractional speech loss and talker activity model for TASI and packet-switched speech," *IEEE Trans. Commun.*, vol. COM-26, Aug. 1978.

[21] H. Miyahara and T. Hasegawa, "Performance evaluation of modified multiplexing technique with two types of packets for circuit and packet switched traffic," presented at the Int. Conf. Commun., Boston, MA, 1979.

[22] ——, "Integrated switching with variable frame and packet," presented at the Int. Conf. Commun., Toronto, Ont., Canada, June 1978.

[23] O. A. Mowafi and W. J. Kelley, "Integrated voice/data packet switching techniques for future military networks," *IEEE Trans. Commun.*, vol. COM-28, Sept. 1980.

[24] S. S. Lam, "Delay analysis of a time division multiple access (TDMA) channel," *IEEE Trans. Commun.*, vol. COM-25, Dec. 1977.

★

Myron J. Ross (S'62–M'66–SM'80) was born in Winthrop, MA, on June 10, 1942. He received the B.S., M.S., and Ph.D. degrees in electrical engineering from Northeastern University, Boston, MA, in 1964, 1966, and 1970, respectively.

Associated with GTE Sylvania Systems Group since 1972, he has since been engaged in the design and development of telecommunication switching systems, and in studies of predictive coding and secure voice processing systems. Some of his recent activities include technical supervision of a group performing research and exploratory development of integrated voice/data systems, investigating applications of digital switch models to tactical military communications, and serving as Principal Investigator for a message and packet switching network studies program. Prior to 1972, he worked at Aberdeen Proving Ground, MD, as a Test Director, at Draper Laboratory, Massachusetts Institute of Technology, Cambridge, on the Apollo Project, and taught in the Graduate Schools of George Washington University, Washington, DC, and Northeastern University.

Dr. Ross is a member of Phi Kappa Phi, Tau Beta Pi, Sigma Xi, and Eta Kappa Nu.

★

Osama A. Mowafi (S'73–M'78) was born in Cairo, Egypt, on November 27, 1944. He received the B.S. degree in 1965 from Cairo University, Cairo, the M.S. degree in electrical engineering in 1970 from the Brazilian Space Institute (INPE), Campos, Brazil, and the Ph.D. degree in 1978 from the University of Maryland, College Park.

From 1965 to 1968, he was a Research Engineer with the National Institute of Standards, Cairo, where he worked on frequency and time measurement techniques. From 1968 to 1969 he was with Telefunken Telecommunication Development Labs in Sao Paulo, Brazil. From 1969 to 1973 he was with the Brazilian Space Institute working on the analysis of satellite communication systems. He is currently the Manager of the Networks Architecture Department of Computer Sciences Corporation) (CSC), Falls Church, VA, where he is responsible for the design and analysis of various computer communication networks. His current interests include queueing theory, packet broadcast communications, integrated voice/data networks, and network adaptive routing techniques.

Dr. Mowafi is a member of the Association for Computing Machinery.

Integrated Voice and Data Communications Overview

Manu Malek

Introduction

Until the early 1950's, the Public Switched Telecommunication Network (PSTN) had evolved to support voice communication. At the same time, data transmission was limited to teletypewriter information exchange at ten characters/s corresponding to 50 bits/s. With the introduction of the digital computer during the 1950's and with the arrival of time sharing, remote terminals, and computer networking in the 1960's, data communication gained more importance. Initially, some methods were devised to use the PSTN for transporting data. There were, however, some major limitations in using a network designed and optimized for voice communication to transport data. To alleviate these limitations, separate dedicated networks were created to transport data.

It is, however, inefficient to use disjoint networks for voice and data applications. This is especially true considering that data traffic has been steadily growing since the 1960's and that the rate of growth is also increasing. Table 1 shows the anticipated relative change in the volume of voice and data communications by the year 1991. Within five years the volume of data communications will rise by about 35% per year while the volume of voice communications will only rise by about 4% [1]. Thus, it is important to devise networks that can efficiently support both voice and data transport and to realize the economic benefits from their synergy.

This paper provides an overview of issues and techniques relating to integrated voice and data communications. It is mainly concerned with switched networks. However, some of the concepts discussed are applicable to both switched networks and special services networks. The Voice Communication section discusses the voice communication network and the evolution of transmission and switching technologies. The Data Communication section presents data communication concepts, switching techniques, and network architecture. Voice digitization and its benefits are presented in the Voice Digitization section. Transmission and switching issues in integrated voice and data communications are discussed in the Integrated Voice and Data Transport Issues section. The Integrated Services Digital Network section introduces the Integrated Services Digital Network (ISDN) concept, and discusses ISDN architecture, standards and its evolution. The Integrated Network Operations section discusses integrated network operations. A final section summarizes the paper.

Voice Communication

A communication network may be viewed as having two major parts: access and backbone (or subnet) transport. Access refers to the communication facilities used to connect customers' Terminal Equipment (TE) to the subnet. In voice networks, the TE's are telephone sets connected to the subnet by transport paths referred to as loops. The subnet provides connectivity between the TE's; it recognizes the address specified by a requesting TE and sets up a connection to the requested TE. The

TABLE I
ANTICIPATED RELATIVE CHANGE IN VOICE AND DATA TRAFFIC VOLUMES [1]

Year	% Voice	% Data
1987	87	13
1988	83	17
1989	76	24
1990	68	32
1991	57	43

nodes of the subnet are switches, interconnected by transmission lines, referred to as trunks. A trunk is a communications path that is not dedicated to a particular user. A group of trunks on the same physical path is referred to as a trunk group. Switches directly serving customers are located in End Offices (EO's), sometimes also referred to as Central Offices (CO's). We may use the term EO loosely to indicate the switching equipment residing there.

Access Architecture

The access part of the voice network has a tree architecture. Each telephone set has an electrical path (the loop) to its serving EO. Telephone sets may be interconnected locally by Private Branch Exchanges (PBX's). Each PBX is connected to the subnet by a trunk group (see Figure 1). The loop may be a copper wire pair (metallic) or a channel in an analog or a digital carrier, generally referred to as a loop carrier.

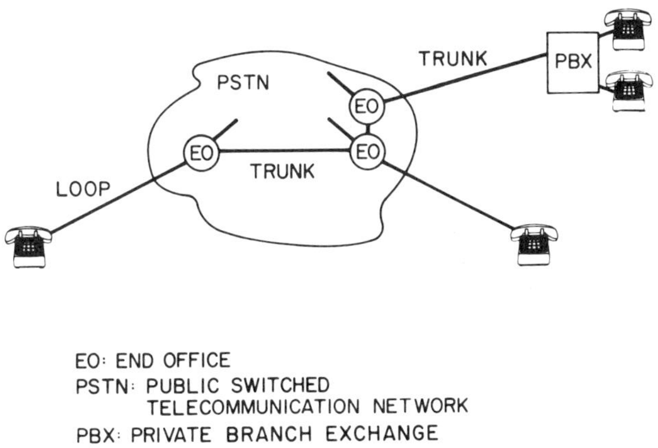

Fig. 1. Voice communication environment.

If a loop carrier is used, the TE's in certain geographical proximity are connected to the Remote Terminal (RT) of the carrier. The RT performs the required multiplexing/demultiplexing functions. At the EO, the so-called Central Office Terminal (COT) of the carrier performs demultiplexing/multiplexing functions, and it provides TE "appearances" to the switch. If the loop carrier and the switch are both digital, the loop carrier is often integrated into the switch at the EO without the need for a COT.

Exchange Network Architecture

By an "exchange network" we mean the subnet interconnecting EO's in a relatively small geographical area such as a Local Access and Transport Area (LATA), which roughly corresponds to the area served by one Numbering Plan Area Code. Within such a subnet, at least one path must be provided from each EO to every other EO. Complete direct interconnection of n EO's, requiring $\frac{n(n-1)}{2}$ trunk groups, is of course neither practical nor economical. An efficient method of interconnecting EO's in a local network is to use another level of switches, called tandem switches, residing in Tandem Offices (TO's). These switches do not initiate or terminate traffic; their function is merely to route the traffic between EO's. Each EO is connected to a TO by a so-called final trunk group. The EO is then said to "home" on that TO. This arrangement creates a network with two levels of hierarchy, as shown in Figure 2, and represents the architecture of most operating telephone company networks today.

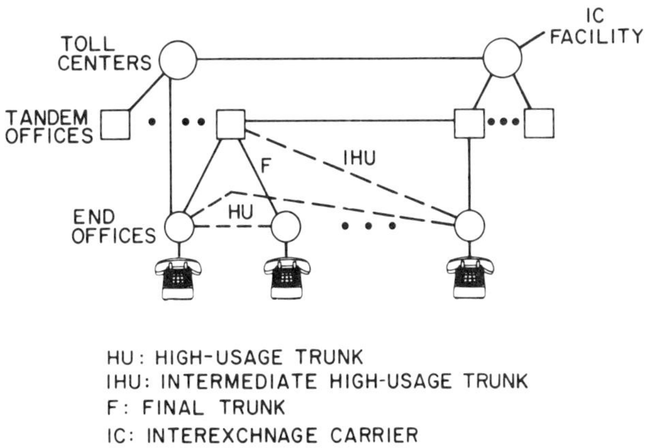

Fig. 2. Local network architecture.

Tandem office locations and switch capacities must be determined based on location and traffic patterns of EO's to minimize the total subnet cost. If there is heavy traffic between a pair of EO's, it may be more cost effective to connect them directly by a High Usage (HU) trunk group. Similarly, a TO and an EO may be directly connected by an Intermediate High Usage (IHU) trunk group (see Figure 2). The traffic between two EO's is first routed over HU trunks or an IHU trunk, if they exist. If all these trunks are busy, the traffic will be routed over a final trunk. The non-coincidence of peak periods for traffic between pairs of EO's may be used advantageously to realize economic efficiencies [2,3].

Interexchange Network Architecture

Communication between EO's in an exchange network and among exchange networks takes place via toll switches located in Toll Centers (TC's), also referred to as Class 4 offices. (In most exchange networks dialing a "1" directs the call to a TC.) Several EO's or TO's may be connected via final trunks to a TC as depicted in Figure 2. The function of the subnet in the interexchange (or inter-LATA) network is to provide connectivity between toll switches. Higher level switches may be used here based on the geographical distribution of the TC's and their traffic patterns. These higher level switches are located in Primary Centers (PC's), also referred to as Class 3 offices. This concept can be extended to even higher level switches. The North American long-distance network (serving the U.S. and Canada) adopted a five-level plan in the late 1940's. The corresponding architecture, shown in Figure 3, has been in place since the early 1950's. The number of switches in each level is, of course, less than that in the lower level. For example, there are about 20,000 end offices and 1,340 toll centers in the North American network. Note that HU and IHU trunk groups are used in this network wherever traffic economically warrants them. Reference [2] describes a method of locating offices in each level and determining the optimal homing configuration.

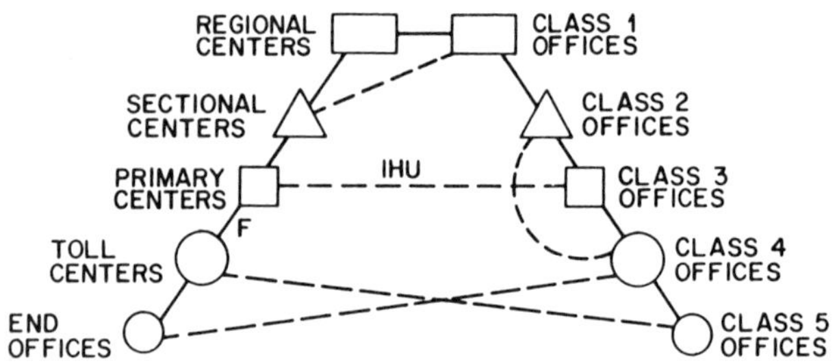

Fig. 3. Long distance network architecture.

The network architecture shown in Figure 3 represents the total network view that existed prior to the divestiture of AT&T on January 1, 1984. In today's post-divestiture network, the toll centers are the boundary between the exchange network and the interexchange network. That is, the exchange network architecture of Figure 2 subtends from the toll centers in the interexchange network architecture of Figure 3.

Dynamic Nonhierarchical Routing

The routing strategy in the long-distance network is as follows. For routing between two centers, the first choice is an HU trunk, if an idle one can be found. If all trunks in an HU trunk group are busy, alternate routing takes place to an IHU trunk (if an idle one exists), or to a final trunk, in that order. The routing decisions are made by individual switches in each level of the hierarchy. Because of its rigid nature and low blocking requirement on final trunks, this routing strategy (with the associated network architecture) does not allow efficient usage of network transmission and switching resources. A more efficient operation would be possible if route patterns could change based on varying traffic patterns (e.g. owing to business/residential traffic, different time zones, and seasonal variations).

Dynamic Nonhierarchical Routing (DNHR) provides that flexibility. With DNHR, the interexchange network will evolve to an essentially two-level structure. A separate high-speed signaling system allows a path to be selected between two nodes in the network before a connection is attempted on the path. If one leg of a selected path is blocked (i.e., if all links on that leg are busy), the call attempts another path. The number and the choice of paths vary with the day of week and the time of day [3]. In general, there may be up to 15 choices for a two-link connection. Figure 4 shows an example of possible routes between toll centers in Wichita and Nashville in the AT&T network.

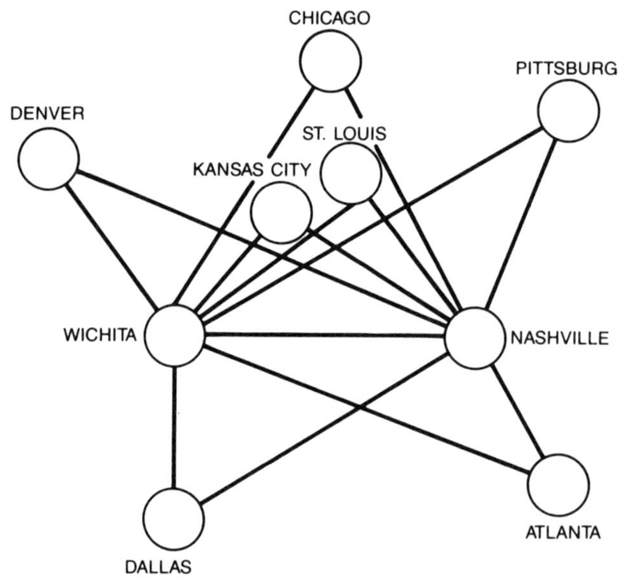

Fig. 4. Dynamic nonhierarchical routing.

Studies show that nonhierarchical routing provides up to 15% cost reduction compared to hierarchical routing [3]. AT&T implemented the first phase of DNHR on 25 No. 4 ESS™ switches in April 1986.

Introduction of Digital Transmission and Switching Technologies

Until the early 1960's the telecommunications network consisted of electromechanical switches interconnected by analog transmission systems. The AT&T T1 carrier system, using Pulse Code Modulation (PCM) and Time Division Multiplexing (TDM), was introduced in 1962. It was first used to derive more channels over the metallic cable facilities to relieve congestion in

---

ESS™ is a trademark of AT&T.

metropolitan areas and then to replace analog carrier systems in intercity and short-haul interstate routes. In the same time frame, analog electronic switches that implemented Stored Program Control (SPC) capabilities were introduced. Examples of these switches are the AT&T No. 1 ESS™ switch, the Siemens' ESK™ switch, and the GTE No. 1 EAX™ switch. The latter two switches are quasi-electronic in that they use electronic controls to operate mechanical cross points.

Up to the 1970's customer access had been via copper-based physical connections. In the 1970's the analog loop carrier was introduced to relieve congestion in the loop plant. The Digital Loop Carrier (DLC) was introduced later in the same decade. During the 1970's the digital toll switch was also introduced. The AT&T No. 4 ESS was introduced into the long-distance network in 1976. Northern Telecom DMS200™ is another digital toll switch which was introduced in early 1980.

The 1980's witnessed the introduction of digital local switches in the public telecommunication network. The AT&T No. 5 ESS switch, Northern Telecom DMS100, the GTE GTD-5™, and the Siemens' EWSD are examples of these switches. We have also witnessed the integration of the digital loop carrier with digital local switches and the rapid penetration of optical fiber-based transmission systems both in the access and in the subnet.

Data Communication

In digital data communications the TE generates a bit stream for transport across the subnet to the other TE's. The TE could be a host computer, a terminal, a graphic workstation, etc. Concentrators may also be used to concentrate the data from several TE's to realize efficiencies in transport (see Figure 5). They are used in situations where each TE has traffic only a fraction of the time. The output line of a concentrator has less capacity than the sum of the input line capacities.

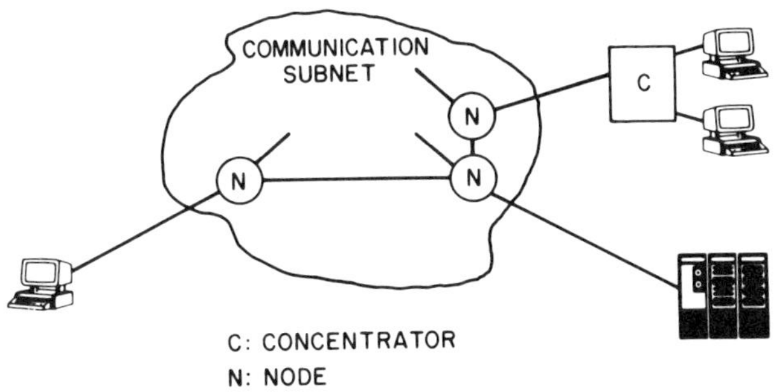

Fig. 5. Data communication environment.

ESS™ is a trademark of AT&T.
ESK™ is a trademark of Siemens AG.
EAX™ is a trademark of GTE.
DMS™ is a trademark of Northern Telecom, Inc.
GTD™ is a trademark of GTE.

Access to the subnet is provided via links commonly supporting data rates of 2.4 kbits/s to 1.5 Mbits/s (sometimes higher). The nodes in the subnet are high-speed processors (computers) capable of storing, processing, and routing digital data. They are interconnected by transmission lines capable of supporting data rates of 56 kbits/s to 45 Mbits/s and higher.

Switching Techniques

The two basic switching schemes in digital communications are: circuit switching and store-and-forward switching. In circuit switching an electrical circuit is assigned and dedicated to each call and remains busy during the call. This situation is similar to today's public voice communication. To establish a circuit between two users in a circuit-switched network, a call set-up procedure must be followed. Because transmission resources remain available after call set-up, no storage of data takes place at the subnet nodes.

In store-and-forward switching the message is transported one hop at a time toward its destination. That is, the message is transported from one node in the subnet to an adjacent node, and it is stored there pending the availability of an outgoing link, while the link on which it was transported is released. This process repeats until the message reaches its destination. A characteristic of store-and-forward switching is that delay becomes a significant performance criterion because of storage delay at subnet nodes (in addition to propagation delay over the transmission facilities). However, this method allows for the efficient use of transmission facilities bandwidth.

There are two variants of store-and-forward switching: message switching and packet switching. In message switching the whole message is transmitted through the network as one block of data. In packet switching, however, the message is broken up into smaller, often uniform-sized packets before being offered to the subnet. The packets are then transported through the subnet as independent blocks. They are recombined at the destination to form the original message. Message switching is usually inadequate for interactive traffic owing to its delay performance.

The Open System Interconnection (OSI) reference model, proposed by the International Standards Organization (ISO), provides a framework for allowing communication between processes in host computers built by different manufacturers or with different operating systems [4].

Digital Data Communications Over the Voice Network

The PSTN has been designed and optimized primarily to transport voice signals (i.e., continuous signals in the frequency range of 300 Hz to 3,300 Hz). To transport digital data over the PSTN, the data must be converted to "voice-like" signals at the interface to the network. This method is referred to as voice-band data communication. Modems (modulator/demodulators) using Frequency Shift Keying (FSK), Phase Shift Keying (PSK), or their variants convert digital signals into voice-band analog signals [5]. Because the PSTN is accessible from almost anywhere, this method allows the extension of digital data communication services to many customers. Modems have been in use for about 20 years, and their performance has been steadily improving. The maximum achievable transmission rate today is 4.8 kbits/s on the public telephone network and 19.2 kbits/s on conditioned private lines using Quadrature-Amplitude Modulation (QAM) - a combination of amplitude and phase shift keying. From the network provider's point of view, the use of modems may be objectionable because a customer may engage the resources of the network for an extended time period on a flat-rate call without providing sufficient revenue to the provider. Another problem with voice-band data communication is the susceptibility of analog signals to transmission impairments, which often have more severe effects on data transmission than on voice transmission. Despite these limitations, however, voice-band data communication is a popular and efficient method of data communication owing to the ubiquity of the PSTN. Private lines are often used effectively for some applications requiring high data rates, but they are expensive and require a long lead time to be provisioned.

Voice Digitization

One method to integrate voice and data communication is to digitize voice at the source and use a digital transport fabric. Thus, voice is presented to the network over the access media as digital data (i.e., a bit stream). For network congestion control purposes, however, the digital data resulting from voice is treated differently from genuine digital data. Indeed, voice digitization has been in use since the early 1960's when the T1 carrier was introduced in metropolitan areas to reduce cable congestion. Here, Pulse Code Modulation (PCM) converts speech into a 64 kbits/s stream (8,000 samples/s, 8 bits/sample), referred to as the DS0 signal. The T1 carrier multiplexes digitized voice signals from 24 channels into a 1.544 Mbits/s stream, referred to as the DS1 signal. A "frame" in the T1 carrier consists of 24 x 8 = 192 bits plus one extra bit (the F bit) for frame synchronization. With a sampling rate of 8000/s, this results in a gross data rate of 8000 x 193 bits/s or 1.544 Mbits/s.

With more sophisticated coding schemes this 64 kbits/s rate may be reduced. For example, in the differential PCM method instead of encoding the whole sample, only the difference between the magnitudes of the current and the predicted future sample is encoded. In Adaptive Differential PCM (ADPCM), the quantizer step size is adaptively changed to be roughly proportional to the amplitude of the difference. A data rate of 32 kbits/s for encoding toll quality speech has been achieved using ADPCM [6].

Benefits of Digitization

Integrated voice and data communications using digitized voice allows end-to-end digital connectivity. Extending digital transport to the customer's premises and transporting signals over a digital network improves transmission quality, increases transmission efficiency, and simplifies signal processing. Specifically:

- In analog transmission, impairments (such as noise) accumulate with circuit length. In digital transmission, however, impairments are not, in general, dependent on length owing to the periodic regeneration of digital signals along the transmission route. Also, voice can be integrated with data.
- Digital signals allow efficient storage and lend themselves easily to the application of various digital signal processing techniques such as data compression, echo cancellation, firmware control of switching, and encryption of data for security purposes.
- Recent advances in Large-Scale Integration (LSI) technology have reduced the size, complexity, and cost of supporting circuitry for digital networks [7].
- Fiber optics technology has made the introduction of high-capacity digital transmission lines (e.g., 45 Mbits/s and higher) in the interoffice and customer access environments economically feasible.

As mentioned earlier, a major advantage of speech digitization is that it allows efficient transport of signals generated by analog or digital sources. This, in turn, allows for the provision of integrated services over the same access line to customers. We will discuss this concept further in the section on *ISDN*.

Integrated Voice and Data Transport Issues

The communications technology has thus far been generally application-specific. That is, today's various communication networks are each tailored to support a specific service. Examples are the PSTN for voice, the cable television networks for TV signal transmissions, and various packet-switched networks for data transmission. A customer needs to have separate access lines to each of these networks to use their services. This environment is a result of the evolving nature of the applications, and because application-specific networks often cannot efficiently accommodate new services. The long-term solution is to develop an integrated network that can support a variety of applications. Such a network should be capable of providing a set of basic transport capabilities (i.e., a platform, such as offering a wide range of performance to the users). Specific applications could then be built on this platform. We will discuss the elements of such an integrated network in the following subsections. We will start with a discussion of the characteristics of voice and data traffic, and then we will discuss integrated switching and integrated access. We will also present an implementation method for integrated voice and data communications.

Traffic Characteristics

To be able to effectively integrate voice and data transport, the characteristics of these two types of traffic and their transport requirements must be understood. Voice (or voiceband data) is sensitive to absolute delay. In voice communication the delay must be small and uniform; otherwise, voice will be unintelligible. Subjective tests have shown that generally there is no adverse reaction by customers to one-way absolute delays in the range of 300 to 600 ms [8]. The network objective, however, is much less than this (about 150 ms). On the other hand, for digital data the absolute delay is not always critical, and it may be negotiated between the customer and the network provider.

Voice signals tolerate relatively high error rates. In fact, in the T1 carrier, the least significant bit of each voice channel is substituted in the 6th and 12th frames with signaling information — a process referred to as robbed-bit signaling. In digital data transport, however, the error rate must be negligible. Error detection or correction techniques are often used to ensure virtual error-free communication. From the viewpoint of call duration, voice communication is considered to have relatively short duration (three minutes average holding time), but bulk data communication (e.g., file transfer) has moderate (longer than voice) duration. Interactive data communication (as in terminal-to-host) is bursty and is, thus, amenable to packet switching.

Integrated Switching

In a fully integrated network, the same switching fabric is used for handling all varieties of traffic. Choices for such a switching fabric have been discussed in the literature [9-12]. One method is to packetize voice and data traffic and deploy a packet switched network. In fact, the evolving packet switching technology is capable of supporting a variety of communication services with differing requirements [13]. An example of a technique for integrating voice and data in a packet-switched network is the so-called virtual cut-through switching [9]. In this method the packet header is examined for destination address as the packets arrive at a subnet node. Without waiting for its complete arrival, the packet is sent through over an appropriate outgoing line, provided that it is idle. If the line is not idle, the packet is stored and forwarded as usual. Thus, under light traffic conditions, most packets pass through the nodes without suffering queuing delays. This method may be used for services such as voice communication, which require minimal delay.

Given today's application-specific communications network environment, some intermediate steps need to be taken before achieving a fully integrated switching fabric. In this context, integrated switching connotes using the same switching mechanism in the central office to direct the customer call to the proper network. Figure 6 shows an example of integrated switching when two types of transport networks (i.e., a circuit-switched network and a packet-switched network) are involved. The integrated switch should have the "intelligence" to recognize circuit-switched and packet-switched traffic and route the traffic to the proper network. Also, at the receiving end, an integrated switch should route the call to the proper terminal equipment.

Integrated Access

As previously mentioned, customers currently have separate access mechanisms to the different networks serving them. By integrated access we mean an access method that can be used by the customer for all requested services (now provided by several different networks). Figure 7 shows voice and data terminal equipment connected via a multiplexer/demultiplexer (MULDEM) to an access line. At the central office, the access line is connected to an integrated switch that includes a corresponding MULDEM to separate traffic types. Similarly, the traffic from the networks to the terminal equipment is sent over the multiplexed access line. The multiplexing/demultiplexing function may be digital or analog. In the former, the voice signals from the voice terminal must be digitized. This is then multiplexed, with the digital data from the digital terminal equipment, and sent over a digital carrier to the central office. In the analog case, the digital data may be converted to a voice-like signal by a modem and multiplexed with voice signals over an analog access line. In the Data Over Voice (DOV) technique, the digital signal is appropriately modulated onto an analog line at a frequency band above 4 kHz, while voice occupies the frequency spectrum below 4 kHz [14].

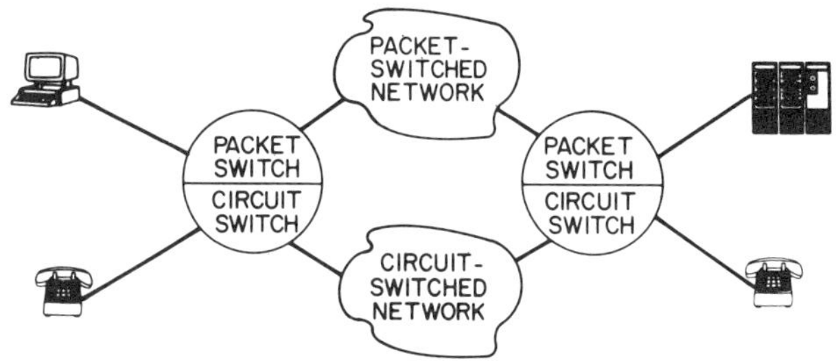

Fig. 6. Integrated switching.

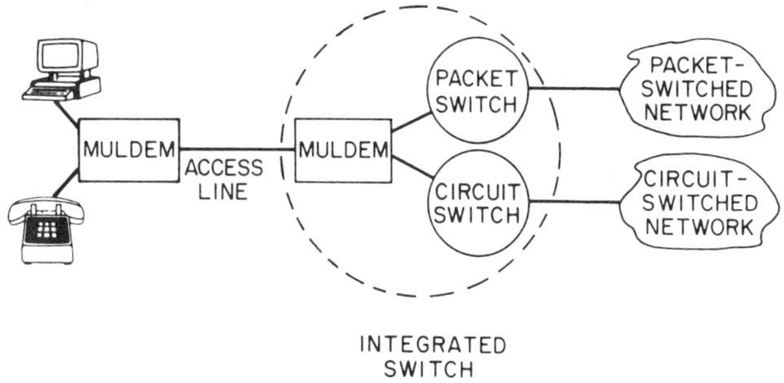

Fig. 7. Integrated access.

Implementation Examples

Integrated Voice and Data Terminal (IVDT) is an example of a device that integrates voice and data communication. An IVDT includes a telephone and modem to allow processing of both voice and data. Separate telephone lines may be used for simultaneous voice and data communications. In that case, an IVDT is essentially similar to having a separate telephone set and a terminal. Recently, there have been some products that allow using an IVDT with the combined transmission of voice and data on a single channel. This amounts to integrating the PBX and the local area network (LAN), which considerably reduces the wiring requirement at the user premises. Examples of this technology are the AT&T System 85, the Northern Telecom Meridian DV-1, and the IBM/Rolm CBX II.

Figure 8 shows an implementation of integrated access/switching. To route the digital data traffic a Datakit™ Virtual Circuit Switch (VCS) is used in the Central Office (CO) in conjunction with the voice switch. Datakit VCS is a digital virtual circuit switch that uses an internal packet switch protocol capable of switching 44,000 packets/s. It is used in local or wide area network applications. The arrangement shown in Figure 8 is sometimes referred to as the Datakit VCS CO/LAN. In the method shown in Figure 8, an Integrated Voice/Data Multiplexer (IVDM) is used at the customer's premises to multiplex the data from a synchronous or asynchronous terminal onto the analog line. Another IVDM unit at the central office separates voice and data traffic and routes them properly. Data rates of up to 19.2 kbits/s can be supported over metallic non- loaded loops. The Datakit VCS also supports multiplexed interfaces for host computers. The voice switch is, of course, a node of the PSTN. The Datakit switches are also networked together for connectivity. Datakit VCS CO/LAN allows the use of existing nonloaded loops to support integrated voice and data communications without the need for special wiring in the CO. Because of the quick installation and low start-up cost, it provides a graceful evolution to an integrated network.

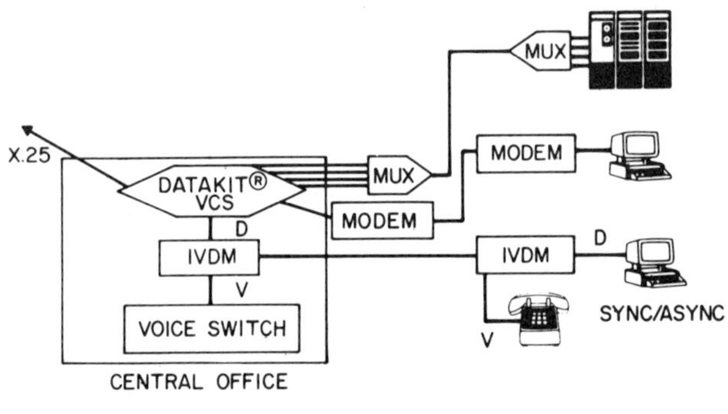

Fig. 8. Datakit VCS central office solution.

Integrated Services Digital Network

During the last few years there has been an accelerating rate of activity in ISDN research and development. ISDN provides flexibility, user control and responsiveness, thus benefiting network providers as well as customers. The CCITT I Series Preamble [15] refers to ISDN as a network "that provides end-to-end digital connectivity to support a wide range of services by a limited set of standard multipurpose user-network interfaces."

To realize the ISDN principles, the present network must evolve. In this section the key network features needed to realize ISDN are discussed along with ISDN architecture and channel structures. A major obstacle in realizing end-to-end digital connectivity is customer access to the network, which is largely analog today. This issue will also be discussed in this section, as well as ISDN standards and ISDN evolution and its status.

Key Network Features

One of the objectives of ISDN is end-to-end digital connectivity. This implies the use of digital customer access and a digital subnet. Of course, it is impractical to replace the existing communications infra-structure with a fully digital network. It is possible, however, to provide end-to-end digital connectivity from the user's point of view. To do this, digital access must be provided to the existing PSTN and other special-purpose networks (see Figure 9). Over this integrated access the user can be provided with a variety of services including voice and various data services such as facsimile, teletex, and videotex. Furthermore, the customer can have many optional features such as teleconferencing, private virtual networking, and local area networking; and these options can be exercised from their terminal equipment.

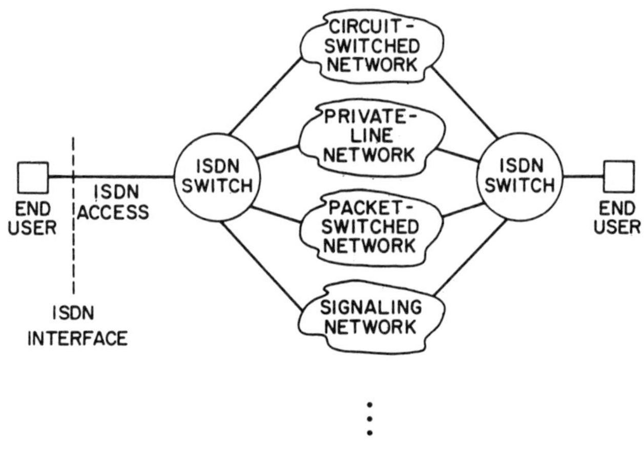

Fig. 9. ISDN architecture.

The key to providing these capabilities is out-of-band message-oriented signaling. With in-band signaling, switches communicate with each other over the same channels that transport customer's information signals. This method of signaling has several drawbacks. It is facility-dependent and inflexible. Furthermore, the call set-up process generally results in inefficient use of network resources. In out-of-band signaling, a separate network is used for signaling, thus allowing for more flexibility. The out-of-band signaling network allows users to access new services by using

DATAKIT™ is a trademark of AT&T.

functional message signaling. AT&T first introduced out-of-band signaling by the Common Channel Interoffice Signaling (CCIS) network with the stored-program controlled No. 4 ESS switch.

Another key objective of ISDN is Clear Channel Capability (CCC) (i.e., the ability of a 64 kbits/s channel to support unrestricted bit patterns in the information stream). The existing digital transmission facilities have restrictions on the minimum one's density and maximum number of consecutive zeros in the bit stream for timing recovery. For example, the T1 carrier requires a 12.5% minimum one's density and no more than 15 consecutive zeros. To allow global end-to-end service compatibility in ISDN, these restrictions on the bit stream must be removed. Compatible line coding must be used between adjacent central offices to accomodate these restrictions. This issue is further discussed later in this section.

ISDN Architecture and Standards

Figure 9 permits viewing of the ISDN architecture as having two parts: the customer access and the network. The ISDN switch is the critical point separating these two architectures. The architecture on the access side is shown in Figure 10. As illustrated, there are four reference points: R, S, T, and U. Network Termination 1 (NT1) performs OSI layer one functions (i.e. functions associated with the physical and electrical terminations of access lines such as maintenance, performance monitoring, and timing functions). NT1 isolates the customer from the loop transmission technology[1]. NT2 perform OSI layers two and three functions such as switching and concentration. NT2 may be a digital PBX, a LAN, a terminal controller, or a multiplexer. Terminal Equipment 1 (TE1) is an ISDN terminal (i.e. a device that complies with ISDN S interface recommendations, such as a digital telephone). TE2 is a non-ISDN terminal that requires a Terminal Adapter (TA) to connect to the ISDN interface reference point S or T. Examples of TE2 are ordinary voice telephones and devices with RS-232 interfaces.

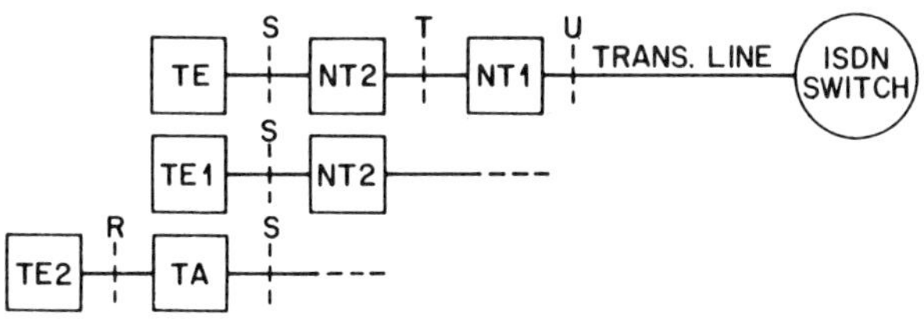

Fig. 10. ISDN interfaces.

[1] The issue of NT1 ownership by customer or network provider is under discussion in the U.S., and ISDN standards groups are working on developing a standard for the U interface. In Europe, however, NT1 is considered to be a part of the network.

The physical access connecting customers to an ISDN node is a digital transmission facility that can support a number of communications channels. The following channel types have been defined for ISDN [15]:

B (for Bearer) channel:	64 kbits/s
D (for Data) channel:	16 or 64 kbits/s
H0 channel:	384 kbits/s
H11 channel:	1536 kbits/s
H12 channel:	1920 kbits/s
H4 channel:	135.168 Mbits/s

The B, H0, H11, and H12 channels support clear transmission. The D channel supports signaling, but it can also carry user information. The above channels form the building blocks for various interfaces provided to the customers. The two major interfaces are:

- The Basic-Rate Interface (BRI) consisting of two B channels and a 16 kbits/s D channel (2B + D), at an overall data rate of 144 kbits/s available to the user.
- The Primary-Rate Interface (PRI) consisting of n B channels and one 64 kbits/s D channel (nB + D), where $n = 23$ for North America and Japan (corresponding to 1.544 Mbits/s) and $n = 30$ for Europe (corresponding to 2.048 Mbits/s)[2]. PRI may include H rather than B channels: 4H0 or 3H0 + D at 1.544 Mbits/s and 5H0 + D at 2.048 Mbits/s.

The above interfaces support narrow-band applications including voice, voice-band data, and digital data up to 2.048 Mbits/s. Other interfaces, including an H4 channel and an assortment of H0 and H1 with B and D channels, have also been either proposed or standardized and deal with wideband applications [15].

On the network side of the ISDN interface, the switch must separate the channels and route them appropriately (see Figure 11). The B channel is routed to a circuit-switched network, packet-switched network, or private-line network. Note that the circuit-switched network does not need to be wholly digital; however, it must be able to support clear digital transmission. Because it may contain both signaling and data, the D channel is split accordingly. The signaling information is diverted to the signaling network, and the data part is routed to the packet-switched network.

The ISDN protocol reference model is based on the OSI model and applies to information transfer as well as signaling and management functions. For signaling between the user and the network, the D channel uses LAPD data link layer protocol, specified in CCITT recommendation Q.921 (or I.441). Layer 3 of ISDN access signaling protocol, specified in CCITT recommendation Q.931 (or I.451), provides user-network interface specifications and procedures for call establishment, maintenance, and termination [16].

[2] This difference is due to the embedded digital carriers based on different digital hierarchies in these regions.

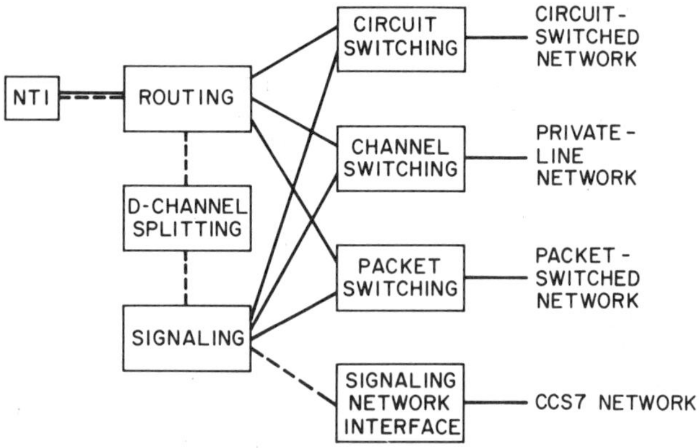

Fig. 11. ISDN architecture - network side.

ISDN Evolution

The present telecommunications network has been planned, designed, and optimized primarily for voice communication. It must evolve to support the ISDN concept. This evolution involves three important aspects: access, interoffice transmission, and switching. The following subsections discuss these aspects.

Access

In the present telecommunications network, access from customer equipment to the serving CO is generally via two-wire copper loops planned for voice transmission with a nominal bandwidth of three kHz. Using voice-band modems, this loop can support digital data rates up to 4.8 kbits/s. The two-wire loop has a limited range of approximately a few miles. Longer loops use inductive loading, which makes them unsuitable for digital data transmission through modems.

To support the ISDN basic rate, the loop must be able to transport digital data at a 144 kbits/s (2B+D) rate. One possible approach is to use four-wire loops. This, however, is expensive and is feasible only for supporting the primary rate (e.g., using a T1 system). Two technologies are now available to support the basic rate on two-wire loops: Time Compression Multiplexing (TCM) and Echo Canceler Hybrid (ECH). Also, optical fiber technology is penetrating the access network, allowing basic and primary rate support [17].

In TCM, transmission from each direction is "bursted" in alternate disjoint intervals to simulate full-duplex communication [18]. The loop data rate is slightly more than 2 x 144 kbits/s. The TCM circuitry restores the data transmission by reading out data at the original rate. In ECH two hybrid circuits, one at the subscriber terminal and one at the network terminal, convert the two-wire loop to four-wire. The resulting reflection of the transmitted signal is removed by two echo cancelers, one at each end. The loop rate here is the same as the transmission rate (i.e. 144 kbits/s). ECH is more expensive than TCM, but it provides a longer range than TCM.

Interoffice Transmission

As stated earlier, the two key network features in support of ISDN are out-of-band signaling and Clear Channel Capability (CCC). The major impediments of the present interoffice network are that it generally uses in-band signaling and that the widely deployed T1 system imposes restrictions on the bit stream. As for signaling, the Common Channel Signaling (CCS) system has been implemented to a limited extent in the exchange carrier and the interexchange carrier networks. It must be expanded to support all switches in these networks. And as for the CCC, two proposals are now being considered by various networks for deployment: Bipolar with 8 Zero Substitution (B8ZS) and Zero Byte Time Slot Interchange (ZBTSI). B8ZS is based on CCITT's standardized line code. It ensures that no more than 7 consecutive zeros occur in the bit stream. ZBTSI acts on the entire DS1 frame, which consists of 24 DS0 channels, and suppresses all zero codes at the originating end, thus permitting recovery at the terminating end.

Switching

The initial deployment of ISDN will include both circuit and packet-switched fabrics. However, as ISDN matures, the packet-switched fabric may predominate [10]. In a mature ISDN, voice will be digitized, may be packetized, and will be treated like digital data with the specific attributes of low delay and some tolerance to error. Note that ISDN services will be offered initially in both circuit and packet modes. In the circuit mode a channel is wholly dedicated to applications. It is used for applications such as voice and video which require a guaranteed bandwidth and a small delay. On the other hand, packet mode is used for such applications as interactive data, which are bursty, delay-tolerant, and require minimal error rates.

ISDN Trials

A number of ISDN trials have taken place, are underway, or have been scheduled in the U.S. and internationally. These activities reflect the global interest and commitment to implement ISDN. Internationally, ISDN trials have been conducted in Belgium, Canada, England, France, Italy, Japan, Sweden, and West Germany, and some pilot services were planned in 1987 [19]. In Europe and the Far East the governments having control over telecommunications are driving ISDN forward and will offer ISDN services to primarily business customers within the decade. In North America, however, ISDN will be offered as its economic benefits are demonstrated to the market (particularly the business market). In the U.S. exchange carrier area, several Local Exchange Carriers (LEC's) conducted ISDN trials in 1986 and are planning to offer limited ISDN services. Illinois Bell successfully conducted an ISDN trial with McDonald's Corp. in 1986 using the AT&T No. 5 ESS switch [20]. This represented the first major U.S. application of the

ISDN concept. Another ISDN trial was performed by Mountain Bell, in late 1986, in Phoenix using Northern Telecom's DMS100 switch. Other trials are planned by Bell South Corp., Pacific Bell, Southwestern Bell Telephone Co., and U.S. West during 1988. Furthermore, some big U.S. corporations, including GM Corp. and Bank of America, have planned their own integrated voice and data networks. The U.S. government is also planning a multibillion-dollar network called the Federal Telecommunications System 2000, to provide voice, data, video, and packet switching services for many government agencies [21].

Beyond ISDN, where multiples of 64 kbits/s channels are provided to the customer, the concept of Universal Information Services (UIS) has been introduced. The goal of UIS is to provide customers access to any kind of voice, data, or image service in any combination, in any place, and at any time with maximum convenience and economy [22]. Key to realizing the UIS concept is the capability of providing customers with bandwidth on demand (i.e. supporting any bandwidth up to the maximum data transfer rate on the user-network interface).

Integrated Network Operations

Operation of a telecommunications network refers to the activities that ensure effective utilization and maintenance of the network. The activities include provisioning, maintenance, and administration. Provisioning, which accounts for a considerable portion of the cost of operating the network, refers to the activities related to providing necessary resources to the network as well as providing services supported by the network to customers. Resource provisioning consists of planning for, ordering, and installing the necessary Network Elements (NE's) and transmission facilities. It involves forecasting traffic demands, translating forecasts to equipment and facility needs, design, installation, inventory update, and acceptance testing. Service provisioning is the process of providing specific services supported by the network to the customers. It includes all the activities between the arrival of a service request by a customer and service activation.

Maintenance is the process that ensures the network is functioning properly. Maintenance functions deal with the detection of problems at the NE's or the transmission facilities connected to them, as well as performing the necessary tasks to alleviate the problems and restore normal operation. Network administration (or network management) involves real-time surveillance and control of the network. Congestion control in the network is an important function here. Billing (i.e., the preparation of customer bills for rendered services) is also considered a network administration function.

Operations System (OS's) are used by telecommunication companies to perform operations functions. However, these OS's are typically specific to the types of service or the switching or transmission technology used in the network. The reasons for this are due to limitations in the capabilities of the NE's, and sometimes because of expediency, separate networks have been traditionally used to provide different types of services. This has given rise to separate OS's to support these networks. For example, now there are separate OS's to support voice and data services or switched and special services. Also, differing philosophies in separate operations have evolved over time to support operations in different disciplines such as switching, interoffice transmission, loop, customer premises equipment, etc. Yet, separate OS's may be in use to support various operations areas such as maintenance, provisioning, and billing. Further, the OS's are typically implemented on processors of various sizes with sometimes incompatible operating systems. In this environment, supporting new functions are cumbersome, often requiring many modifications in the existing OS's, and sometimes even requiring new OS's.

Rapid introduction of the new transmission and switching technologies is paving the way toward the physical integration of networks supporting various services. At the same time, telecommunication services are proliferating due to the increased sophistication of customers and the increased capability of networks to carry various services. To support the new services and technologies, the network operations capabilities must also evolve. The physical network integration provides opportunities to consolidate network operations functions, resulting in more economical and more responsive OS's. For example, a consequence of the new network technology has been to drive some operations functions into the network itself. This has been possible by the introduction of Intelligent Network Elements (INE's). INE's are processor-controlled equipment with their associated memory capabilities, which can be locally or remotely configured and controlled. Examples of INE's are add-drop multiplexers and digital switching systems. With INE's, some portions of the operations functions may be integrated with the element, rather than using external systems, to reduce cost.

How can the environment of traditional service-and technology-dependent operations be changed to an integrated operations environment? This change of environment requires a new approach to partition operations functions, develop methods and algorithms, and the corresponding data structures. New operations architectures must be developed without regard to traditional administrative and organizational boundaries. The architecture should provide a unified framework for all operations areas. It should identify operations primitive functions and group them (e.g., in a hierarchical or layered form) so that various operations activities could be identified as a combination of these functions. An essential part of the operations architecture is the specification of protocols for NE to NE, OS to NE, and OS to OS communications. Recently, some operations architectures have been developed to meet these challenges [23,24].

Summary and Conclusions

We have reviewed architectures and traffic characteristics for voice and data communications, and addressed important issues in integrated voice and data communications. We discussed some possible methods of integrating voice and data, and presented an example of implementing integrated voice and data communications. We also discussed ISDN as a vehicle for supporting integrated services including voice and data as well as image and video.

The forces driving voice and data integration are the increased penetration of digital technology into the network and the strong demand by the customers for a common interface for various services. To achieve integrated voice and data communications, the issues of integrated switching and integrated access must be tackled. A major challenge, in addition to the technical aspects, is the development of standards. Furthermore, while the integration of voice and data provides advantages in economy and flexibility, it also poses new challenges in the area of network operations. This issue also needs to be addressed by planning and developing integrated operations systems. Finally, although the paper has concentrated on the integration of voice and data, the discussion could be generalized to other services such as facsimile, graphics, and video services.

Acknowledgment

The author wishes to thank Shervin Erfani, Shahriar Shokoohi, Curtis Siller, David Tsay, John Williams and the anonymous referees for reviewing this paper and providing helpful comments.

References

[1] H. Anderson, "How to Build an Intelligent Network," *Telecommunication Products and Technology,* pp. 19-27, Aug. 1986.

[2] M. Malek, "Computer-Aided Design of Telephone Systems," *Computers and Electrical Engineering,* vol. 4, pp. 37-51, 1977.

[3] G. R. Ash, R. H. Cardwell and R. P. Murray, "Design and Optimization of Networks with Dynamic Routing," *Bell System Technical Journal,* vol. 60, pp. 1787-1820, Oct. 1981.

[4] J. Voelcker, "Helping Computers Communicate," *IEEE Spectrum,* pp. 61-70, Mar. 1986.

[5] R. W. Lucky, J. Salz and E. J. Weldon, Jr., *Principles of Data Communications,* McGraw-Hill, 1968.

[6] N. S. Jayant, "Coding Speech at Low Bit Rates," *IEEE Spectrum,* vol. 23, pp. 58-63, Aug. 1986.

[7] N. Weste and K. Eshraghian, *Principles of CMOS VLSI Design - A Systems Perspective,* Addison Wesley, 1985.

[8] *Transmission Systems for Communications,* Fifth Edition, AT&T Bell Laboratories, 1982.

[9] P. Kermani and L. Kleinrock, "Virtual Cut-Through: A New Computer Communication Switching Technique," *Computer Networks,* vol. 3, pp. 267-286, 1979.

[10] J. S. Turner, "Design of an Integrated Services Packet Network," *IEEE J. Selected Areas in Communications,* vol. SAC-4, pp. 1373-80, Nov. 1986.

[11] S. R. Amstutz, "Burst Switching - An Introduction," *IEEE Communications Magazine,* pp. 36-42, Nov. 1983.

[12] J. J. Kulzer and W. A. Montgomery, "Statistical Switching Architectures for Future Services," *Proc. Int. Switching Symposium,* May 1984.

[13] S. Erfani and M. Malek, "Capacity Allocation for Multiservice Integrated Networks," *Proc. IEEE Montech '87,* pp. 142-45, Montreal, Nov. 9-11.

[14] J. C. Bellamy, *Digital Telephony,* John Wiley & Sons, 1982.

[15] CCITT Study Group XVIII I Series Recommendations, Oct. 1984.

[16] H. Aldermeshian, "ISDN Standards Evolutions," *AT&T Technical Journal,* vol. 65, pp. 19-25, Jan./Feb., 1986.

[17] C. Hoppitt, "ISDN Evolution: From Copper to Fiber in Easy Stages," *IEEE Communications Magazine,* vol. 24, pp. 17-22, Nov. 1986.

[18] B. S. Bosik and S. V. Kartalopoulos, "A Loop Access System Transmission for Digital Subscriber Loops," *Proc. International Symposium on Subscriber Loops and Systems (ISSLS),* pp. 26-30, 1980.

[19] R. Kenedi and C. L. Wong, "Architectures for Implementation," Special Issue of *IEEE Communications Magazine* on ISDN, vol. 24, Mar. 1986.

[20] B. R. DeMaeyer, W. Heinmiller, F. C. Iffland, K. E. Martersteck and P. Krause, "Implementing the Illinois Bell/McDonald's ISDN Trial: A Technical Overview," *Proceedings ISS '87,* Phoenix, Arizona.

[21] S. N. Pandhi, "The Universal Data Connection," *IEEE Spectrum,* vol. 24, pp. 31-37, July 1987.

[22] P. M. Villiere, "Universal Information Services - A Marketplace Perspective," *USTA Southern Telecom. Showcase,* Feb. 12, 1985.

[23] L. Bernstein, "AT&T Universal Operations Systems (UOS)," *Proc. National Communications Forum,* Chicago, Oct. 1986.

[24] S. E. Aidarous, P. A. Birkwood, and R. M. K. Tam, "An Architectural View for Integrated Network Operations," *IEEE Comm. Magazine,* vol. 25, pp. 31-43, Sept. 1987.

Manu Malek (Senior Member since '79) received his Ph.D. degree in EE/CS from the University of California, Berkeley, in 1970.

Since 1970 he has held various academic and industrial positions including Professor and Chairman of the Electrical Engineering Department at Shiraz University in Iran, Visiting Professor of Electrical and Computer Engineering at the University of New Mexico, Senior Engineer with Hughes Aircraft Company, District Manager of the Special Services Standard Designs Engineering Group at Bell Communications Research, and Supervisor of Special Studies at AT&T Bell Laboratories. He is a distinguished member of the technical staff at AT&T Bell Laboratories, where he is now involved in the area of forward-looking service and facility designs in the Network Management Laboratory. He is also an Adjunct Professor of Electrical Engineering and Computer Science at Stevens Institute of Technology. He is the author or coauthor of three books and over 50 technical papers in the areas of control systems, optimization, and communications.

Dr. Malek is a member of Sigma Xi and Eta Kappa Nu. He was the Chairman of the Computer Society Chapter of the IEEE New Jersey Coast Section in 1982-1984. ■

Inside SS No. 7: A detailed look at ISDN's signaling system plan

Here is the integrated network's central nervous system. But few users know about it or understand what it provides.

Signaling System No. 7, in the final stages of specification by the International Telegraph and Telephone Consultative Committee (CCITT) and formulated in basic accord with the Open Systems Interconnection (OSI) reference model recommended by the International Organization for Standardization (ISO), differs greatly from most other data communications signaling techniques. The differences are neither accidental nor capricious. Rather, they are due to the real-time functions it provides. And these differences account for the power of SS No. 7 to flexibly interconnect clear 64-kbit/s channels—the essence of the Integrated Services Digital Network. Without the new signaling system, ISDN could probably offer none of its features and services. Yet, for all its power and importance at the heart of ISDN, SS No. 7 remains unknown to most users.

SS No. 7 came into being because the telephone industry planned a future array of enhanced features and services that users could reach rapidly and without disruption. To provide the access, the telephone industry needed something—in this case, SS No. 7—analogous to the central nervous system of a living organism, something to coordinate the functions while remaining completely separate from the organism's other parts and not actually performing their function. Without the nervous system's coordination, the organism—in this case, ISDN—could not perform.

SS No. 7 has much to coordinate for ISDN. Among its many features and services, ISDN will offer:

■ User access to a full 64-kbit/s channel with no bits "stolen" for signaling or synchronization.
■ Rapid establishment or modification of the destination and characteristics of the connection.
■ Transmission of user data without disturbance to the existing connection.

Such features and services require the "central nervous system" function provided by a "common channel signaling system." On ISDN access facilities, this duty is performed by the D channel, part of the ISDN 2B + D interface.

On the backbone network, between telephone company switches, SS No. 7 will take on the common channel signaling task. A common channel signaling link is basically just a specialized point-to-point data link. There are two reasons for making it special:

1. It has a set of special tasks to perform.
2. It must operate as a network, rather than as a collection of independent links, so that it can perform as part of the total signaling complex.

In a modern telephone network, where switches are computer-controlled devices, use of a data link to interconnect switches appears trivially reasonable—why do it any other way? However, it is important to realize that telephone switches predate computers by many years. And it is only recently that computers and communications protocols have become sufficiently reliable to put entire trunk groups under their real-time control. Beyond that, sufficiently large trunks, and the signaling protocols to handle these trunks, must come into existence before common channel signaling becomes economical. Dedicating voice-channel-size capacity to a signaling link requires that the common channel handle one or hundreds of channels before it is justified.

As a backdrop to SS No. 7, it would be helpful to review ISDN. ISDN's fundamental concept is subscriber access, via standardized interfaces, to the digital transmission capability of the public telephone network. After more than 10 years of debate and discussion, the CCITT has decided that the standard bearer channel for this access will be an unrestricted 64-kbit/s path,

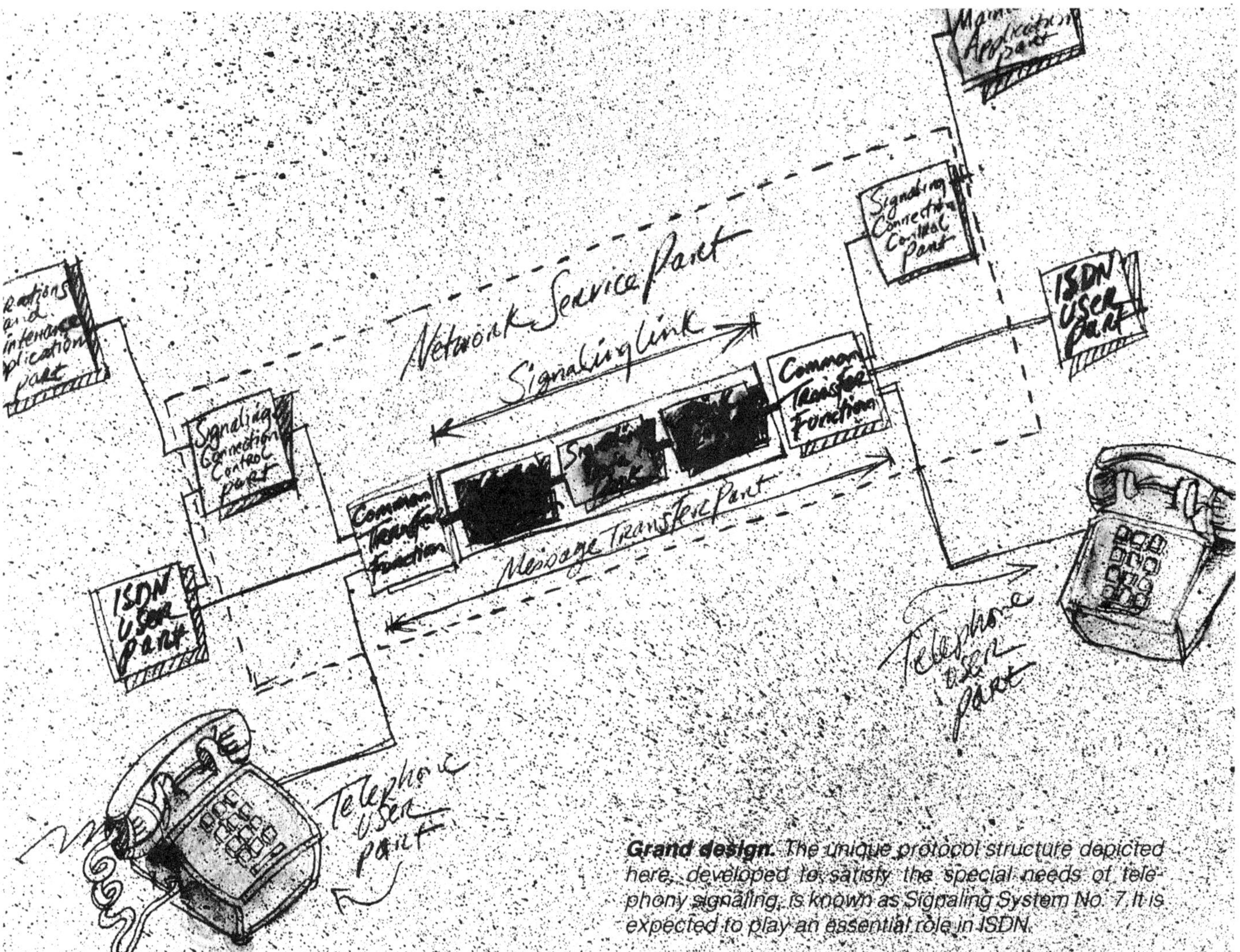

Grand design. *The unique protocol structure depicted here, developed to satisfy the special needs of telephony signaling, is known as Signaling System No. 7. It is expected to play an essential role in ISDN.*

referred to as "B channel." All bit patterns are acceptable user data and none of the bits are to be "stolen" for signaling. Any digital source—such as a digitally encoded voice signal, a facsimile signal, a graphics file, or a data file—can be transported without prior arrangement if it fits in 64 kbit/s.

The flexibility of the B channel leads to problems in providing signaling capability. Since the 64-kbit/s channel can be used for a variety of purposes, there is the desire to allow rapid reconnection of the channel (for example, "Put this connection to the word processing facility on hold and connect me to the database so I can retrieve some facts for this report"). Furthermore, because any bit pattern can be included in the subscriber's data, there cannot be special "escape" or "break" sequences that indicate to the network that the associated information is for the network (for example, "I want to hang up. Please disconnect and stop charging me"). Provision must be made for a signaling channel outside the subscriber's channel. Both of these requirements are satisfied by the common channel signaling capabilities built into ISDN.

Essentially, common channel signaling provides a separate transmission capability for signaling. On ISDN access links, this signaling capability is known as the D channel and is multiplexed with a number of B channels—two B channels for a "Basic Interface" yielding the 2B + D capability and 23 or 30 B channels for a "Primary Rate Interface." (Signaling via a stolen eighth bit every sixth frame is the current North American standard. These stolen bits, and the sensitivity of 1.544-Mbit/s T-carrier repeaters to long strings of zeros, account for the limitation of digital data links in North America generally to a maximum rate of 56 kbit/s—8 kbit/s of the nominal 64 kbit/s provide the signaling and timing capability.) On the backbone trunks between telephone company switches, much larger groupings of B channels must be handled and there are more functions to be supported (for example, billing, routing, credit card checks, 800 service, and maintenance). A different, more efficient signaling technique was needed, and so SS No. 7 came into being.

With the emergence of its fundamental requirements, it became apparent that the signaling system must support, in real time, call establishment and disconnection, billing and financial administration, and supervision of the connection.

To do that, the signaling system must possess the following characteristics:

Speed. Routes must be found and established during the post-dialing interval while the caller waits on the line. During this interval significant resources are dedicated to an attempting call, but billing cannot begin until the connection is established. Similarly, once a caller disconnects (hangs up), facilities must be released and charging stopped.

Accuracy. Errors will cause, at best, lost or misconnected calls. More serious consequences can include billing errors or unnecessarily removing facilities from service.

Reliability. If the signaling system fails, the entire communications facility is useless. Each component of the signaling system must be highly reliable, be thoroughly tested before it is put in service, and incorporate self-checking capabilities. The signaling architecture (software, hardware, and operating philosophy) must accommodate redundancy so that a single failed component does not halt operation of the entire telephone system.

Transaction orientation. Each phase of every call is treated as a separate transaction. Establishing and holding a separate signaling connection (physical or virtual) for every call would not be practical; the time delay would be intolerable and literally thousands of simultaneous connections would be needed. Therefore, a connectionless mode must be used.

The next major development in telephony signaling was the use of registers that receive the customer's dialed digits, store them, and output a switch operating signal. The register's output need not be identical to the input signal; the register is an electromechanical implementation of a lookup table. Direct control of switches by the customer's dial pulses does not allow for alternate routing, uniform numbers for Direct Distance Dialing, or flexible assignment of switch ports, but registers provide for all these capabilities. Simultaneously, the register disassociates the representation used to transfer the information (the signaling) and the operating signal used by the switching machine. Signaling by tones, which can be much faster and more reliable, is possible.

By the 1960s, computer controlled switches changed the role of the register. Instead of many registers working in parallel to directly control a relatively slowly responding electromechanical device, the registers now operated as data buffers—accepting the slow signaling messages from the lines and rapidly inputting these messages into the central processor. For coordination between backbone switches a much simpler solution becomes obvious—interconnect switch processors with high-speed data links and completely avoid the need for registers. This high-speed data link arrangement is known as a common channel signaling system.

Since end subscribers were still using simple analog telephones without the capability of participating in a data link dialog with the access switch, there were no plans for common channel signaling on access lines. (ISDN, with its digital terminals will change this situation. Terminal devices also will have the capability of participating in a high-speed data link exchange with the access switch, and thus be capable of using a form of common channel signaling.)

The CCITT has developed two common channel signaling system recommendations: SS No. 6 and SS No. 7. No. 6 was designed in the late 1960s and early 1970s to operate over analog trunks, using 2.4-kbit/s modems. The low data rate made efficiency a prime consideration. The data link protocols were very carefully optimized to carry trunk identifiers and address digits: Signaling messages are carried in "signaling units" composed of 20 information bits and eight check bits.

One SS No. 6 frame is composed of 11 of these information signal units, along with a twelfth signal unit (of identical 20 + 8 format) used for control. Clearly this is not a protocol designed by computer scientists (no binary value numbers); it is designed by signaling gurus. The protocol is not layered—it is one monolithic specification from the link control to the application code. There are two reasons for this monolithic structure: (1) in the 1960s layered protocols were virtually unknown, and (2) there was an overriding demand for the efficiencies that monolithic protocols can deliver.

"A signaling system for digital [pulse-code modulation] trunks" was the mid-1970s CCITT charter given to the SS No. 7 working group. Some of the group's initial considerations centered on minimum modifications to the No. 6 system, but it soon became apparent that the higher data rates (56 or 64 kbit/s rather than 2.4 or 4.8 kbit/s) and the lower error rates (typically one error in 10^9 rather than one error in 10^5) demanded a dramatically different protocol architecture. By the time that the 1980 CCITT recommendations were published (CCITT operates on a four-year study/publish cycle) the value of layering was well appreciated, bit-oriented protocols (HDLC, SDLC, ADCCP) were widely accepted, and at least the lower layers of SS No. 7 resembled standard computer protocols.

While SS No. 7 is layered, and it uses many of the devices of other bit oriented protocols (for example, the eight-bit flag sequence of 01111110 to open and close frames) it still is a unique protocol suite tailored to the special needs of telephony signaling.

Historical development

Revisions and realignments of SS No. 7 are still being made. Much of the current work is bringing No. 7 into greater conformance with the ISO's seven layer OSI reference model (CCITT Recommendation X.200). This continuing evolution is more understandable when the context in which No. 7 developed is appreciated. In the late 1970s, when work started on No. 7, open systems and a universal hierarchy of protocols were still academic concepts. Furthermore, the designers of No. 7 considered themselves to be primarily telephone signaling people rather than computer data link people.

While ISDN was under study (Special Study Group D had been given this mission during CCITT's 1972-1976 study period), it became apparent that No. 7 was likely to be the signaling system for the integrated network. But there certainly was no consensus as to what an ISDN should look like or what services it should provide. And it must be remembered that SS No. 7 had a destiny of its own, independent of ISDN. No. 7 is the signaling system for all circuit-switched digital net-

1. The layers compared. *SS No. 7 does not match the OSI reference model layer for layer. For one thing, No. 7 was developed before the wide acceptance of OSI and was kept "closed." For another, No. 7 was not intended to be a general-purpose network model. Still, recent changes to No. 7 have been made in line with OSI.*

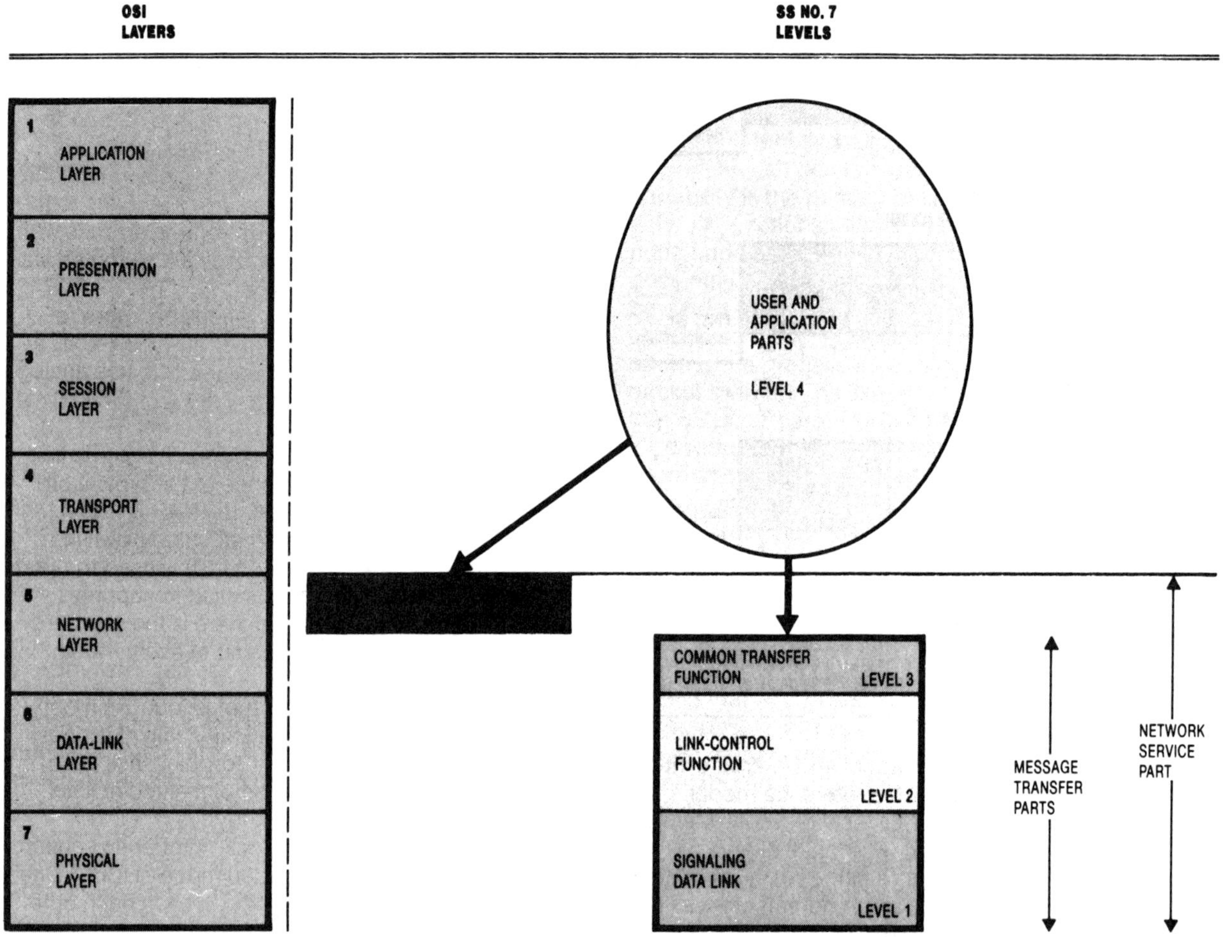

works. It so happens that ISDN is the most discussed network at this time, but No. 7 has provisions for plain pulse-code modulation telephone networks and other provisions for circuit-switched data networks.

The signaling system was defined as all of these environmental factors evolved. It is probable that No. 7 would look somewhat different if it were first conceived today—there undoubtedly would be a greater use of OSI layering principles, standard lower-layer protocols that are now available in large-scale integrated hardware, and more consideration of ISDN. However, the signaling system has been evolving (mainly through the addition of new modules rather than the elimination of existing layers) and further conformance with the OSI model is to be expected.

This ability to evolve gracefully is one of the reasons that there is virtually no talk of starting work on a replacement for SS No. 7—everyone is confident that No. 7 will be able to be extended to accommodate their needs. In addition, there is widespread appreciation of the tremendous investment of time that would be necessary to develop a replacement. The current emphasis within the switching community is focused on getting on with the work of practical implementation and application of No. 7.

The layers

The original SS No. 7 design focused on efficient transport of circuit-switching control messages between switch processors—a closed system application. The objective for the cross-office delay at a relay point (last bit received to last bit transmitted) is 25 to 90 milliseconds—there is little room for unnecessary protocol features. Thus, though No. 7 is layered, the layers are not identical to the OSI reference model layering scheme. Figure 1 compares the layering of the OSI model to the layers used in SS No. 7. In some large part the differences are due to the quest for efficiency in the signaling system.

The value of a reference model is the structure it provides for defining information exchange procedures (protocols and interfaces). A new user function, that has been independently developed, can readily interface to an existing network if both the network and the

2. Link control. *While the frame format looks unique, SS No. 7's link control function shows the influence of typical bit-oriented link protocols.*

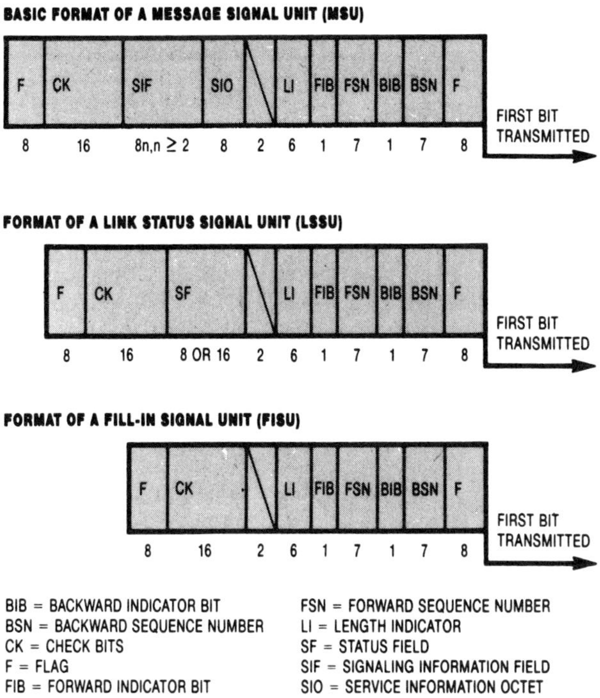

application have built their protocol and interface interchanges in accord with the same reference model.

Since SS No. 7 was being built as a closed system, there was little concern with how additional applications would interface. This was a signaling system, not a general-purpose network. Furthermore, there was not a good (that is, widely accepted) reference model during the early definition stages. The growth in the use of data networking within the telephone industry (for administration and control of the network) and the current widespread acceptance of the OSI reference model has changed the situation. Many of the recent additions to No. 7 have either been in accord with the reference model or have increased the compliance of the signaling system.

In the following descriptions of each of the layers the degree of OSI compliance and recent shifts will be specifically noted. In addition, those aspects of the CCITT recommendations that have been changed by ANSI's T1X1 committee, for use in North America, will also be noted:

The message transfer part (the lower three levels of SS No. 7). Within the recommendation documents, the three lower levels of SS No. 7—collectively called the "message transfer part"—are universally referred to as the MTP. Each of the levels are never referred to by their full names but rather as "level 1, 2, or 3 of the MTP." (It should be noted that the No. 7 recommendations use the terminology "level" rather than OSI's "layer." This may seem trivial, but semantics are a vital issue when working in an international organization that publishes its recommendations in French, Spanish, and English.)

The MTP provides a service similar (but not quite equivalent) to an X.25 network. It accepts packets of data and reliably delivers them to their destination.

Signaling data link (OSI's physical layer). No. 7's signaling data link is in full compliance with the OSI definition of a physical layer. The signaling system is conceived for use on digital (for example, pulse-code modulation) links and has such things as timers and block lengths chosen for use on 64-kbit/s connections. CCITT allows for the use of switched connections to the data link, lower speeds, and the use of analog links with modems. The link is bidirectional and can be routed via a satellite.

The North American implementations are more restrictive, demanding the use of either 56- or 64-kbit/s links. In those cases where it is necessary to use analog facilities to support a No. 7 link, a wideband (56 kbit/s) modem must be employed.

Link control function (OSI's data link layer). The link control function shows the influence of the typical bit-oriented link protocols (HDLC, SDLC, LAP-B) but it definitely has a unique frame format, as is shown in Figure 2. The standard flag (01111110) is used to open and close the frame, bit stuffing is used to suppress strings of more than five 1's, and there is the standard CCITT CRC-16 checksum at the end of every frame. That much looks familiar.

The single-octet address field that is found in most bit-oriented link protocols is gone. It usually does not serve much purpose on the point-to-point links used with packet switching and the use of multipoint links with No. 7 is a topic for further study.

The forward and backward block-numbering schema is pretty much standard "extended mode" HDLC. One octet is allocated for each direction, a bit in each octet is used for control (ACK/NAK or new/retransmit), and the remaining seven bits allow modulo 128 block numbering. (Blocks are numbered from zero through 127. The block following block 127 is numbered zero.) Only signal units used for message transfer are uniquely numbered; overhead signal units use the number of the last message signal unit.

Two retransmission strategies are available. The "basic error correction method" is a typical go-back-N technique; upon receipt of a NAK the transmitter rolls back to the block received in error and retransmits everything from that point. When there is no data to be transmitted fill-in signal units are transmitted. The second strategy, "error correction by preventive cyclic retransmission," is designed for satellite and other high propagation delay links.

The basic design insights are recognition that (1) the link is frequently idle, (2) most of the messages are short, and (3) the signaling system operates in a connectionless mode wherein each signal unit is independent and delivery order is not critical. In that environment it is advisable to not wait for a NAK before making a retransmission. Whenever the link is idle cyclic retransmission of all unacknowledged signal units is started. Only positive acknowledgements are

3. Signaling work. *SS No. 7's signaling network functions cover routing, discrimination, and distribution, as well as traffic, link, and route management.*

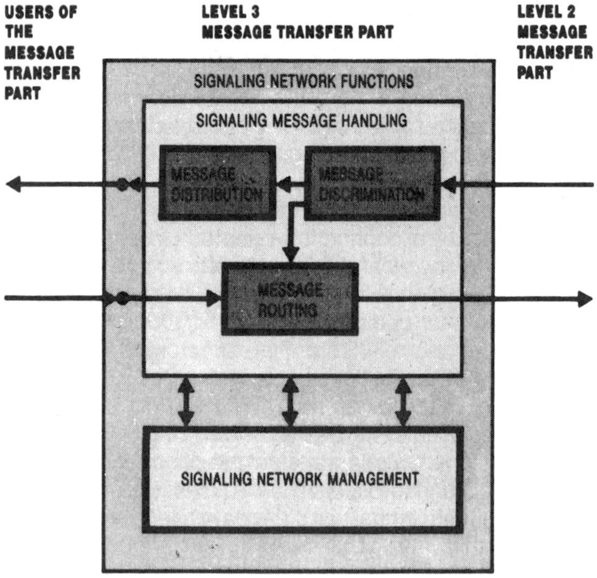

sent by the receiver and provision is made to retransmit unacknowledged signal units on a priority basis whenever the transmitter runs out of sequence numbers.

The next field in the signaling link control frame is a length indicator. In addition to allowing a cross-check on the closing flag and pre-allocation of buffer space (normal functions of a length indicator), the length indicator provides the signal unit type. Message signal units must have data portions larger than two octets. Status signal units have a data field of one or two octets, and fill-in signal units have a zero length indicator. All messages are an integral number of octets long. Flow control is provided through use of a status signal unit sent back by the receiver whenever congestion occurs. Timeouts on the flow condition alerts the level 3 entity if the "stop transmitting" condition lasts longer than 200 milliseconds without contact (including repeated "stop transmitting" status signal units) or 10 seconds with contact.

Up-down counts are used to confirm initial alignment and detect high error-rate environments. Prescaling of "good" counts by a factor of 256 allows detection of noisy physical links. A sustained block error rate of 1 error in 255 blocks will result in an alarm and taking the link out of service.

The signaling network functions (bottom half of OSI's network layer). The signaling network functions are divided into three categories:

1. Routing—determines where a message is to be sent.
2. Discrimination—determines if the message is at its destination or needs relaying.
3. Distribution—determines the user part to which a message should be delivered.

The relationship between these functions is shown in Figure 3. In addition, there are signaling system management functions (management of the traffic, link, and routes) that must be handled by this level.

A 32-bit label is used for level 3 addressing. It is composed of 14-bit origination and destination addresses (called signal point codes) and a four-bit signaling link selection field that is used to distribute the traffic among alternative routes. The North American version of No. 7 uses 21-bit signal point codes to allow utilization of the signaling system in a larger network.

From an OSI point of view, this limited addressing capability is the primary deficiency of the signaling network functions. Signal point codes are basically node addresses; level 3 of the message service part does not provide for addressing (per se) anything beyond a node. (Roughly the equivalent of being able to address a packet switch and not being able to address the processors or processes attached to that packet switch.) The fixed-format user-part messages that SS No. 7 carries provide the information necessary for distribution. Message signal units contain a service information octet that designates the user part, or equivalent, and the network national or international.

This "peeking inside the envelope" is in direct violation of the spirit of the OSI philosophy, wherein each layer is responsible for providing all of the information necessary for its proper functioning. If each layer accepts this responsibility, the upper layers can be changed without changing the lower layers. The corner cutting that No. 7 uses makes it efficient for its signaling purpose. However, applications, such as signaling system management, have arisen that find the rigid format requirements of SS No. 7's third level onerous. Therefore, an additional level that provides full OSI addressing, called the signaling connection control part, has been established.

The signaling link selection field is made part of the address (presented by the user part) so that conflicting demands can be satisfied. From a user part point of view, it is useful if messages that must be associated upon receipt (such as all the messages associated with a single call) are directed along the same route and therefore generally arrive in order. Although the signaling system's connectionless mode allows for proper functioning despite out-of-sequence delivery, sequenced delivery reduces the user part's work load.

From the message transfer part's viewpoint traffic should be distributed uniformly over all available alternate routes. Both levels can be satisfied if the user part puts a value in the selection field that appears to be selected at random and the message transfer part uses a function of this field to determine the route to be taken. CCITT recommends that the least significant bits of the circuit number or a similar entity be used for the signaling link selection field. Such a number will be the same for the entire call, but it will vary uniformly from call to call.

The bulk of the signaling network functions specification is concerned with overcoming link degradations (failures or congestion). Procedures are detailed for rapidly determining that a link has degraded, removing it from service, rerouting traffic, and later bringing the link back into service. The overriding concern for signaling network reliability is evident in these procedures. The goal is no more than 10 minutes of unavailability per year for any route (available 99.998 percent of the time). This goal is achieved through redundancy and rerouting rather than emphasis on super reliability of any given link. Quasi-associated signaling, wherein the signals take a path different than the call to reach a given end point, is a fundamental part of this scheme.

The North American versions of the signaling system accept the CCITT recommendations and then go several steps further in specifying the redundancy that is to be available and the alternate routing techniques that are to be used. There is heavy reliance on a configuration called a quad, shown in Figure 4, which consists of signal transfer points deployed in pairs. The two members of the pair are interconnected by a link (shown as a vertical link in the diagram) that is used solely for restoration purposes. A source or sink node (typically a circuit switch) would be connected to both members of the pair. Backbone interconnections are made to other paired signaling transfer points by means of four signaling links.

Signaling connection control part (upper half of OSI's network layer). To provide full OSI addressing capability the message transfer part is supplemented with the signaling connection control part (SCCP). The combined capability is called the "network service part." In addition, it is planned to have the signaling connect control part eventually provide connections, while the message transfer part will only provide connectionless service.

The message transfer part addresses in the international version of SS No. 7 are limited to 14-bit signaling point codes. This would be equivalent to only allowing addressing packet switches in an X.25 network. For signaling applications this is an adequate solution. All signaling messages have a format that allows extraction of the additional information required for message distribution.

However, this can lead to additional complexity when nonsignaling applications are supported by the signaling network. Each of these nonsignaling applications could be designed as user parts, and provide their own routing and distribution functions. This would result in a number of applications implementing the same function, and clearly would not be in the spirit of the OSI model.

To solve this problem, the designers of SS No. 7 have chosen to add another layer, the SCCP, that looks to the message transfer part to be another user part. This added layer is responsible for services the OSI model assigns to the network layer. This not only allows the message transfer part to retain its efficient character, but also permits SS No. 7 to offer full OSI network layer service. In North America, there is greater emphasis on the SCCP and more tendency to use OSI layering.

The SCCP can accept signaling point codes (as does the message transfer part) or global titles (such as dialed numbers) or subsystem numbers as addresses. The SCCP has access to routing information that allows associating any of these addresses with a signaling point code, which the message transfer part understands. At the destination signaling point, the message is passed to the local SCCP, which takes responsibility for delivery to the user.

There are five classes of service that are defined in the SCCP:

Connectionless

0. Basic (unsequenced) connectionless class.

1. Sequenced (fixed signaling link selection number) connectionless class.

Connection-oriented

2. Basic connection-oriented class.

3. Flow control connection-oriented class.

4. Error recovery and flow control connection-oriented class.

SS No. 7 user parts (upper layers of OSI reference model). The original SS No. 7 architecture had monolithic user protocols. These protocols directly interfaced with the message transfer part. The well-defined services and closed system environment of a classical signaling system made transport, session, and presentation levels unnecessary. The two oldest user parts—the telephone user part and the data user part—were designed this way.

Each user part is complete and independent. The user parts can generally connect directly to the message transfer part, thus providing end-to-end signaling service with only four levels of protocol. Several different user parts could operate in parallel (multiplexed) over a single message transfer part facility.

As other uses for the pervasive signaling network were found, the need for OSI type services increased. Thus, the SCCP and the newer user parts are taking on more of an OSI appearance.

The telephone user part and data user part recommendations jump immediately into long lists of signaling message groups—forward address messages, forward set-up messages, and backward set-up messages. Within each group there are a number of messages—initial address messages, and subsequent address messages. The hierarchical structure that is typical of modern layered protocols is not in evidence. A hierarchical structure allows the addition of newer user features with little fear of conflict with the existing messages.

Detailed instructions are given as to the content of each field of every message, which fields are mandatory, optional, fixed length, and variable length. These rigidly defined formats allow rapid parsing of message content and efficiency in setting up circuit-switched connections.

The telephone user part is still being actively refined for use outside North America. The data user part has largely been supplanted by the ISDN user part. And several more interesting user or application protocols are starting to emerge. The ISDN user part is one of them. The rigid formats of the older user parts are difficult to use with the more fluid parts of the ISDN. Therefore, while the routine signaling functions of the ISDN user part are similar to the old telephone user part, the more innovative features, such as end-to-end signaling (used to modify the nature of a call that has already been set up) and user-to-user data are likely to require the services of the SCCP. (This particular topic is still a matter for discussion within the standards community.) Since the ISDN user part can be used for non-ISDN telephony or data networks, it can completely supplant those user parts. In North America, only the ISDN user part will be supported.

Even more OSI-like in nature is the operations and maintenance application part. It is structured to support the SS No. 7 network, providing functions such as routing data management, route verification, long-term measurement collection, on-occurrence measurement reporting, delay measurements, clock initialization, real-time control, operations, and testing. Many of these services are less well-defined in nature. (Addressing a trouble reporting process running in a number of different switch control units is an example of this type of service.) To support these unstructured tasks there is heavy reliance on the OSI-like services and a complete dependence upon the SCCP.

In North America, an interesting set of application protocols have been developed to allow transport of SS No. 6 messages over the SS No. 7 network. These protocols do not allow a No. 6 end point to interact with a No. 7 end point, but do allow continued economical transport for the well-established No. 6 facilities.

As SS No. 7 matures, users can expect the new application protocols to look like the operations and maintenance application part. A key to identifying these newer protocols is likely to be the term "application," rather than "user," in their title. Within North America, this trend is already well underway. ANSI's T1X1 committee has recently introduced a complete OSI layered structure entirely within the SCCP for the purpose of maintaining the SCCP's routing tables.

Carrying No. 6 traffic over the No. 7 network

Within North America, there is extensive deployment of CCIS6 (an analog common channel system based on CCITT Signaling System No. 6). Several internetworking protocols that provide SS No. 7 support for these CCIS6 installations have been defined. These include:

- AT&T's Embedded CCIS6 (ECIS6) for transmission of CCIS6 messages over a single No. 7 link. At each No. 7 signaling point traversed, there must be a CCIS6 process to accept the message and reroute it, if necessary.
- AT&T's Destination CCIS6 (DCIS6) that uses the routing capabilities of the SCCP to allow routing to distant CCIS6 nodes.
- Telecom Canada's CCIS6 (TCIS6) that adds enough processing in the "interworking unit" (protocol convertor) to provide valid No. 7 envelopes (including destination point codes) for CCIS6 messages.

The problem is how to route CCIS6 messages consistently through a No. 7 network. ECIS6 avoids the problem by falling back to a simple "hop at a time" routing to CCIS6 entities. DCIS6 uses the full OSI routing capability of the SCCP in No. 7. TCIS6 puts the routing burden on the CCIS6 side of the interface.

Note that none of these internetworking techniques allows transfers between CCIS6 and No. 7 end points—they are just a means of allowing CCIS6 end points to communicate via a No. 7 network.

4. North American variations. *One twist to SS No. 7 concerns redundancy, with heavy reliance on the quad configuration—signal transfer points deployed in pairs.*

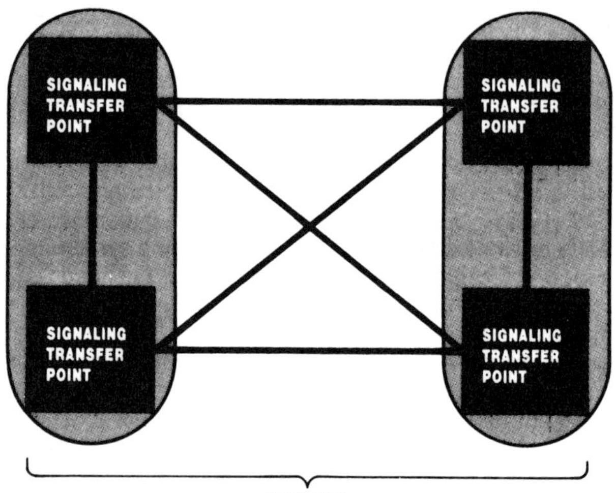

When and where

SS No. 7 is just now reaching the point where it will be operationally deployed in North America. Overseas, there have been a number of installations, mostly employing versions of the telephone users part. In the United States, the main driving forces are the installation of ISDN and the desire of the regional Bell operating companies to take on more sophisticated control functions, particularly In-WATS (800 number) and credit card validation services.

In the switch market, those driving forces have resulted in several organizations (ITT, Digital Switch Corp., Integrated Technology Inc., and AT&T) announcing signal transfer points (STPs). An STP is a processor hub joined to a number (typically over a hundred) of switches via No. 7 paths. The STP can be just a relay point, acting as a packet switch, but, generally, additional database capabilities are installed at the same location. These databases contain 800 routing and billing information and credit-card validity lists. ITT has sold the central processor units of its System 12 switches (without the switching matrix) to Southern Bell and Pacific Bell for use as STPs. ITT has announced that it will enter into an agreement with a computer vendor to provide the database function. Digital Switch Corp. has announced an STP wherein the database is kept on a Stratus fault-tolerant computer. Integrated Technology Inc. has an STP that it has jointly developed with Tandem Computers Inc., the fault tolerant computer pioneer. AT&T has developed what it calls a 2STP that will support both SS No. 7 and SS No. 6. The database in an AT&T installation will reside on an AT&T 3B computer.

An STP is a major purchase, typically selling for close

to $1 million. However, it is expected that the total market is relatively modest, totaling only $500 million through 1990. Vendors see the market's main value to be the opening an STP sale provides for winning future telephone switch sales.

It is worthwhile to underscore the fact that SS No. 7 will be the trunk signaling system for ISDN. A common channel signaling system is essential for providing the services of ISDN. The only common channel signaling system being developed for digital channels is SS No. 7. SS No. 7 will come to fruition independent of the fate of ISDN, but ISDN could not succeed if there were not a responsive common channel signaling system to support it. Therefore, all major manufacturers of central-office switches have, or soon will have, No. 7 appliques for their equipment. In general, these appliques are just new software programs for the switch, but some manufacturers are adding hardware modules to accommodate the additional processing load.

While the prime emphasis in the development of SS No. 7 has been interexchange (trunk) signaling, there is likely to be increased opportunity for users to avail themselves of SS No. 7 once ISDN access lines are available. Interfaces to the ISDN will consist of several 64-kbit/s B channels (used for voice, circuit-switched data, or wideband packet-switched services) and a D channel (used for signaling, telemetry, or low-speed packet-switched services). The D channel (and it's deviant brother, the E channel) are our primary focus.

The link layer protocol to be used on the D channel is LAP-D. It is a protocol that is similar to LAP-B (the balanced HDLC version used on X.25 connections) except for its two-octet address field that is used for statistically multiplexing the various services provided on the D channel. The signaling formats to be used above LAP-D, and when LAP-D should be replaced with the message transfer part of SS No. 7 (thus becoming the deviant E channel), require discussion.

The E channel is a leading candidate for PBX access. Using the SS No. 7 message transfer part on PBX links would allow the PBX to fully exploit the ISDN opportunities. The PBX would become more nearly an exchange switch. Signaling would be rapid and more features could be provided. In particular, information concerning users behind PBXs (called or calling party's availability, capability, and identity) could be made available and be passed through the signaling network.

End-to-end signaling to modify calls in progress and user-to-user data yields further application possibilities. End-to-end uses would open the possibility of user fraud—the high-tech "blue box." Current thinking about that protection focuses on limiting the signaling messages from the PBX to the first exchange, but the topic is still under discussion.

CCITT has a recommendation for the PBX interface. It is entirely concerned with what an E-channel message transfer part should look like (much like a normal message transfer part but with fewer arrangements for detecting failures and restoring service) and how it could be protocol converted from a standard D-channel interface, if necessary. The questions of when and why the E-channel format should be used, and what the user part should be are topics for study and are not included in the latest version of the CCITT recommendations.

The other interface where the user community may come in contact with SS No. 7 protocols is at OSI Level 3, above a LAP-D link layer. The protocols to be used on a basic or primary access D channel are still under consideration by the ISDN specifiers. There is a drive to keep these protocols as simple as possible for low-cost user terminals, but there is also a drive to provide specialized services when required. A simple (but fully compatible) subset of the ISDN user part for simple terminals and a more sophisticated version for full function terminals is a possibility. A fascinating possibility, if the fraud protection approach allows it, is the access to a connectionless data service provided over the signaling network. The datagram service within X.25 has died and there is still an unfulfilled need for a simple transaction-oriented data service, for items such as credit checks. The network service part (that is, message transfer part plus signaling connection control part) of SS No. 7 provides a high performance connectionless service.

Implementation plans

At the current time the lowest three levels of No. 7 (the message transfer part) are fully specified and are probably stable. Implementation work can get underway. The SCCP is also fairly well-defined (although it is notable that there are some differences between the North American version and the CCITT version). Users can plan on having full OSI addressing available.

Above the network service part the situation is somewhat more fluid. It is not clear whether the telephone user part will become widely used outside North America and inside North America there are no plans to use it. The ISDN user part is still in a state of flux, but work is progressing rapidly. Application parts, which use the full OSI layering of the higher protocols, probably will become more prevalent.

The plans of AT&T and Telecom Canada for support of CCIS6 provide one sensible model for moving ahead—using the lower levels and developing a unique upper level, if necessary. The degree of layering that is incorporated in No. 7 allows that flexibility.

One thing is sure, No. 7 cannot be ignored by those who are building or buying telephone equipment that will be used over the next five to ten years. No. 7 will be a central component in the ISDN and, independent of ISDN, No. 7 will be the key to a number of enhanced services for telecommunications users. ■

Walt Roehr heads Telecommunication Network Consulting, a firm he founded in 1984. Roehr has 20 years of experience as a telecommunications engineer, working for a variety of commercial and government clients. He holds a bachelor's degree in electrical engineering from City College of New York and an M. S. E. E. from George Washington University. An expert on Signaling System No. 7, he has presented tutorials on the ISDN signaling technique for Omnicom Inc., an OSI advisory firm in Vienna, Va.

Chapter 2
Local Area Networks (LANs)

The Ethernet and ring networks are commercially successfully in the US and Europe, respectively. An abundance of research publications are available reporting performance of the two networks. This chapter reviews performance issues of existing local area networks in terms of experimental measurements. The FDDI standard for high-speed networks and other proposals are also included. Token passing bus networks are not discussed in this chapter. We point out however, that token bus networks are gaining support due to the availability of higher layer protocol suites of MAP (manufacturing automation protocols).

We omit discussion of the interconnected LANs. There are now two competing proposals for routing algorithms of LAN bridges: source routing and spanning tree (transparent) routing; [Gerla, 1988] provides in-depth coverage of the issue.

2.1 BUS NETWORKS

Ethernet has been a commercially successful bus network. Its medium access protocol is also referred to as 1-persistent CSMA/CD (carrier sense, multiple access with collision detection). The reader is assumed to be familiar with the details of this protocol [Metcalf and Boggs, 1976]. Despite its success, true understanding of the protocol's performance has not been available. Most analytic predictions of Ethernet performance are based on worst-case scenarios, where two nodes at extreme ends of the cable exchange data.

The first article by Boggs *et al.* attempts to diffuse some of confusion arising from earlier worst-cast predictions. Arguments in the article are made from experimental measurements. Improved Ethernet performance is analytically derived in [Kim and Choi, 1989].

Scholl and Coden's article describes a commercial Ethernet based on optical fibers. Combining voice and data on an Ethernet or Ethernet-like bus network has been a popular topic, producing numerous articles. Perhaps the most elegant of these may be Maxemchuk's proposal, which blends the periodic and aperiodic natures of voice and data by a shrewd combination of CSMA and CSMA/CD protocols. The resulting transmission ordering resembles that of the TDM.

2.2 RING NETWORKS

Experimental measurements from a 10 Mb/s ring network (PRONET) are reported in the article by Sventek *et al.* The article examines throughput performance subject to limited receiver buffers and the fairness issue. A new 16 Mb/s token ring product by IBM is briefly described in Strole's article, followed by a description of the ANSI standard for 100 Mb/s FDDI dual-ring protocol. Performance of the FDDI protocol can be found in [Dykeman and Bux, 1988; Karvelas and Leon-Garcia, 1988].

The last article by Zafirovi-Vukotic *et al.* is included to address a commercial success of slotted ring networks in Europe. The article briefly describes a version of the Cambridge ring network at 100 Mb/s speed (called a fast cambridge ring, FCR). Performance issues of slotted and token passing ring networks are examined in the article.

Measured Capacity of an Ethernet: Myths and Reality

David R. Boggs
Jeffrey C. Mogul
Christopher A. Kent

Digital Equipment Corporation
Western Research Lab

Abstract

Ethernet, a 10 Mbit/sec CSMA/CD network, is one of the most successful LAN technologies. Considerable confusion exists as to the actual capacity of an Ethernet, especially since some theoretical studies have examined operating regimes that are not characteristic of actual networks. Based on measurements of an actual implementation, we show that for a wide class of applications, Ethernet is capable of carrying its nominal bandwidth of useful traffic, and allocates the bandwidth fairly. We discuss how implementations can achieve this performance, describe some problems that have arisen in existing implementations, and suggest ways to avoid future problems.

1. Introduction

Local Area Networks (LANs) have become indispensable in the past few years. Many LAN technologies have been designed, and more than a few of these have made it to market; *Ethernet* [8] is one of the most successful. There are many factors that influence a choice between different technologies, including availability, acceptance as a standard, cost per station and per installation, and ease of maintenance and administration. All of these are difficult or impossible to quantify accurately. Performance characteristics, on the other hand, are easier to quantify, and can thus serve as the basis for religious debates even though the potential performance of a LAN technology may have very little to do with how useful it is.

Considerable confusion exists as to the actual capacity of an Ethernet. This capacity can be determined either by measurement or by analysis. Measurements of intrinsic Ethernet performance (its performance in the limiting case) are fairly meaningless because, at least until recently, most feasible experiments could only measure the performance of host implementations and interfaces, not of the Ethernet per se. If no experiment can ''see'' the intrinsic performance of the network, then applications probably cannot see it either, and intrinsic performance therefore does not matter. Analyses, on the other hand, have tended to concentrate on the intrinsic performance; first, because software and interface design is too hard to analyze, and second, because Ethernet performance is ''interesting'' only at very high loads. Therefore, most theoretical studies have examined operating regimes that are not characteristic of actual networks, and their results are also of limited utility in comparing network technologies.

Ethernet works in practice, but allegedly not in theory: some people have sufficiently misunderstood the existing studies of Ethernet performance so as to create a surprisingly resilient mythology. One myth is that an Ethernet is saturated at an offered load of 37%; this is an incorrect reading of the theoretical studies, and is easily disproved in practice. This paper is an attempt to dispel such myths.

We first summarize the theoretical studies relevant to Ethernet, and attempt to extract the important lessons from them. Then, based on measurements of actual implementations, we show that for a wide class of applications, Ethernet is capable of carrying its nominal bandwidth of useful traffic, and allocates the bandwidth fairly. We then discuss how implementations can achieve this performance, describe some problems that have arisen in existing implementations, and suggest ways to avoid future problems.

2. What theoretical studies really say

The literature is full of theoretical analyses of carrier-sense multiple access (CSMA), CSMA with collision detection (CSMA/CD), and related access mechanisms. Most of these analyses rely on simplifying assumptions, either for tractability or simply to avoid drowning in the variety of results that would come from exploring the entire parameter space. In spite of the simplifications, these analyses often require mathematical machinery that render them relatively inaccessible to the average practitioner of computer networks.

This inaccessibility, coupled with potential inapplicability due to simplifying assumptions, has led unwary readers into misunderstanding the theoretical studies. As a result, a substantial mythology has arisen to obscure the true properties of the Ethernet.

In this section we examine the theoretical literature pertaining to the Ethernet, and we attempt to dispel some of the mythology by pointing out the applicability of these studies.

2.1. Parameters affecting performance

The performance of an Ethernet depends upon a variety of parameters, in addition to those fixed by the specification. We assume that the reader is familiar with the principles of operation of an Ethernet, specifically the CSMA/CD mechanism [17].

The fixed parameters of the standard Ethernet are described in the specification [8]. These include the following:

- **Bit rate**: The standard Ethernet operates at a bit rate of 10 Mbits/second. Some studies in the literature describe the "experimental" Ethernet, which operated at 3 Mbits/second [17]. We will assume the standard Ethernet unless otherwise indicated.
- **Maximum propagation delay**: the maximum round trip delay between any two transmitters is defined to be 464 bit times. This is equivalent to about 5 km of coax cable, much more than a transceiver can drive, but a conceivable delay when repeaters and transceiver multiplexors are used.
- **Maximum jam time**: a transmitter which detects a collision continues to transmit for 32-48 more bit times to insure that the other participants reliably detect the collision.
- **Slot time**: an upper bound on the acquisition time of the network. It must be larger than the sum of the maximum round trip propagation time plus the maximum jam time. This is defined to be 512 bit times (51.2 microseconds).
- **Minimum packet length**: The shortest packet must be no shorter than the slot time, so that if there is a collision it is detectable. This leads to a minimum length of 64 bytes, including a 14-byte header and a 4-byte "frame check sequence."
- **Maximum packet length**: This limit, 1518 bytes, bounds the size of buffers that receivers must maintain, and also helps to limit the average access time.
- **Number of hosts**: The Ethernet specification limits the number of hosts per cable segment to 100 (an electrical limit), and the total number of hosts in a multi-segment Ethernet to 1024.
- **Persistence**: The Ethernet is a *1-persistent* CSMA/CD protocol, so-called because a host that becomes ready to transmit when the channel is busy will transmit as soon as the channel is free, with probability 1. Other CSMA/CD protocols have been analyzed; the *non-persistent* protocol waits a random time if the channel is busy. The general case is a *p-persistent* protocol which, after a busy channel becomes free, initiates transmission immediately with probability p, and otherwise waits before trying to transmit. Non-persistent protocols lead to higher delays at low loads but perform better at high loads.

In addition to the fixed parameters, performance depends on several parameters determined by the users of the network. These are not entirely independent variables, since intelligent implementors will choose them so as to make best use of the Ethernet. Further, no single choice of these parameters can be viewed as typical; they vary tremendously based on the applications in use.

- **Packet length distribution**: Within the limits of the Ethernet specification, there is substantial freedom to choose packet lengths. In those networks that have been observed, packet length distributions are strongly bimodal, with most packets either near minimum length or near maximum length [20, 11].
- **Actual number of hosts**: The number of hosts on actual Ethernets varies tremendously, and usually is limited for logistical reasons such as the ease with which a single network can be administered. Typical installations have on the order of tens of hosts to a couple of hundred hosts.
- **Arrival rate of packets**: Although the Ethernet allows a host to transmit a minimal length packet every 51.2 microseconds, most hosts are unable to transmit or receive more than a few hundred packets per second; this effectively limits the arrival rate of packets, and it also places a lower bound on the value of average channel access time that actually affects performance.
- **Actual length of cable**: The actual length of an Ethernet may be far shorter than the specification allows. It is not unusual for installations to use a few hundred meters. This means that collisions are detected far sooner than the worst-case propagation delay would imply; channel acquisition may resume as soon as five microseconds after the beginning of a failed attempt, not the worst-case 51.2 microseconds implied by the specification.

Some of the theoretical analyses make assumptions to provide a simplified model of reality. These may be hard to relate to actual parameters of a real network:

- **Distribution of packet arrivals**: Most theoretical studies assume a simple distribution for the arrival of packets, typically Poisson. Traffic on real networks is seldom so well-behaved; often, there are brief bursts of high load separated by long periods of silence [20].

- **Buffering**: An assumption that hosts have a fixed number of buffers to hold packets waiting for transmission; packets that "arrive" (in the probabilistic sense) for transmission when the buffer is full are discarded. In a real network, flow-control mechanisms limit the arrival rate when the network becomes overloaded; real packets are not discarded, but this is hard to model.
- **Infinite population**: An assumption that even if hosts are blocked waiting for the channel, new packets will arrive for transmission.
- **Slotting**: An assumption that there is a master slot-time clock, with transmissions beginning only at slot boundaries. In a real Ethernet, transmission starts may be separated by less than the slot time; Ethernet is therefore "unslotted."
- **"Balanced star" topology**: An assumption that all hosts are separated from each other by the maximum legal distance; this is essentially impossible in a real installation.
- **Fixed packet size**: It is often much easier to analyze a protocol assuming fixed packet sizes; frequently, the worst-case (minimal packet size) is used.

2.2. How performance is measured

The performance of a network, as with most interesting computer systems, cannot be quantified with a single dimension. Theoretical studies have tended to examine a small set of measures, both for analytic tractability and because it is hard to understand what to measure without knowing what applications are involved.

Performance measures include:

- **Average delay**: the average time it takes to send a packet, measured from the time the host first wishes to acquire the channel.
- **Throughput**: the fraction of the nominal network bandwidth that is actually used for carrying data. Packet headers are considered useful data in calculating this value.
- **Channel capacity**: the maximum achievable throughput for a given set of parameters. Capacity is a function of such parameters as packet length and network length.
- **Fairness**: in a fair network, each host with pending traffic should have an equal probability of acquiring the channel (this is not an equal share of the bandwidth, since hosts use differing packet sizes).
- **Stability**: if the throughput actually drops at high loads then the network is said to be unstable in that region.

2.3. Definitions of offered load

Performance measures are usually described as a function of some parameter of the model; *offered load* is the most common. Several different definitions of offered load are used in the literature.

If one defines the offered load at each host as the fraction of the network bandwidth that the host would use if it had complete access to the network, then G, the offered load on the network as a whole, is simply the sum of the offered loads at each host. Each host's offered load is less than or equal to 1. G can therefore be greater than 1, although the throughput cannot.

Another definition of offered load is the average number Q of hosts waiting to transmit. This concept is what underlies the binary exponential backoff mechanism used to delay retransmission after a collision. If Q hosts are waiting to transmit, one would like each to attempt transmission during a given slot with probability $1/Q$. Essentially, the backoff mechanism is trying to estimate Q.

Although G and Q are both described as the offered load on the network, they are not the same. Q is the offered load as seen by the network, while G is the offered load provided by an idealized customer of the network. Figure 2-1 depicts this schematically; a *generator* is providing packets at some rate, which are being passed via a *buffer* to the *channel*; if the buffer is full when a packet is generated, the packet is discarded. G is measured at the output of the *generator*, whereas Q is measured at the output of the *buffer*, and so does not count the load contributed by the discarded packets.

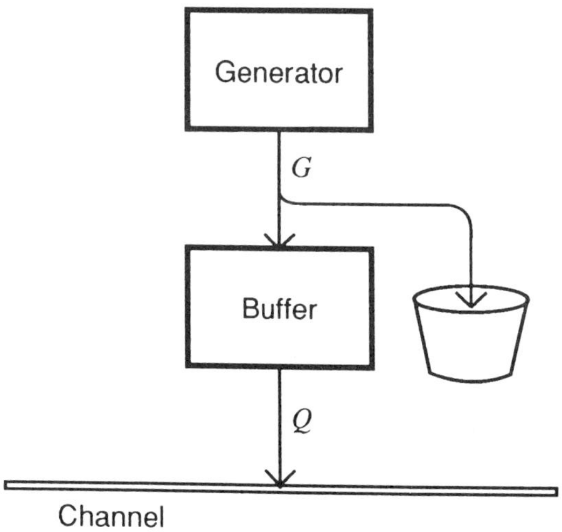

Figure 2-1: Points for measuring load

While G, at first glance, seems to be a more useful measure from the point of view of the user, in most real systems figure 2-1 is not an accurate model, because flow-control mechanisms ensure that packets are not discarded in this way. Flow control between the *buffer* and the *generator* is the norm in most applications, except for those, such as packet voice, that are more sensitive to response time than throughput.

Because it is much harder to model the packet arrival distribution of flow-controlled applications, many of the theoretical studies analyze for response time as a function of G, rather than analyzing for throughput as a function of Q.

2.4. A brief guide to the theoretical studies

What follows is a limited survey of some of the theoretical studies relevant to the Ethernet. This is by no means comprehensive, since formal analyses of multiple-access mechanisms are common enough to rate their own section in *IEEE Transactions on Communications*, but we have attempted to include a representative range, listed in order of publication. We have also attempted to extract from these studies information that might be useful in evaluating the performance of the Ethernet for real applications, or for comparing Ethernet with other technologies.

We do not describe the important early work on CSMA without collision detection, such as the papers by Tobagi and Kleinrock [13, 24], and by Metcalfe [15, 16]. The interested reader may refer to these papers for more information.

2.4.1. Metcalfe and Boggs, 1976

In the original paper on the Ethernet [17], Metcalfe and Boggs provide a simple analysis of the 3Mbit/second "experimental" Ethernet invented at Xerox PARC during 1973 and 1974. They compute the throughput of the network as a function of packet length and Q. The throughput remains near 100% for large packets (512 bytes in this case), even with Q as large as 256, but drops to approximately $1/e$ (37%) for minimal-length packets.

2.4.2. Almes and Lazowska, 1979

Almes and Lazowska [1] continue the analysis of the experimental Ethernet. They present values of response time as a function of G. They show that for small packet sizes, the response times stays under 1 millisecond for G less than 75%; as G increases much past that point, the response time is asymptotically infinite. For larger packet sizes, the knee in the curve comes at a higher offered load, but the low-load response time is worse because a single long packet ties up the network for almost a millisecond. They also point out that as the network gets longer, the slot time gets larger, and the response time is proportionately worse. The performance of the network is thus quite sensitive to the actual slot time.

In addition to a theoretical analysis, Almes and Lazowska simulated the performance of an Ethernet as a way of avoiding some of the simplifying assumptions required to keep the analysis tractable. With fixed packet sizes, the simulation agrees closely with the analytical results. They also simulated a mixture of short and long packets[1], which leads to somewhat worse response times than for a fixed packet size of the same average length. More importantly, the standard deviation of the response time increases much faster than the mean with increasing load, which could be a problem for some real-time applications. As with other performance measures, the variance in response time is quite sensitive to the effective slot time.

2.4.3. Tobagi and Hunt, 1980

Tobagi and Hunt [25] provide a more extensive analysis of the throughput-delay characteristics of CSMA/CD networks. Many of the results in this paper are for non-persistent and p-persistent protocols, and so cannot be blindly applied to Ethernets, but it contains a detailed analysis of the effect of bimodal packet-size distributions on throughput, capacity, and delay.

This analysis starts with three parameters: the lengths of the "short" and "long" packets, and the fraction of the packets that are short. The results show that when only a small fraction of the packets are long, the capacity of the network approaches that which can be obtained using a long fixed packet size. Unfortunately, the fraction of the channel capacity available to short packets drops rapidly as the fraction of long packets goes up; this, in turn, increases the average delay for short packets (which are now waiting for long packets to go by, instead of for collisions to be resolved).

A large fraction of long packets does allow a somewhat higher total throughput before delays become asymptotically infinite. Further, the delays experienced by the long packets decrease, since fewer total packets are being sent and the amount of time spent resolving collisions is lower.

2.4.4. Bux, 1981

Bux [5] attempts a comparative evaluation of a variety of LAN technologies, including bus systems such as Ethernet, and token-ring systems. Such a comparison has value only if the measure used to distinguish the performance of various systems is relevant to actual users. The measure used in this study is the average packet delay as a function of throughput, and it appears to be biased against Ethernet in that it stresses real-time performance over bulk-transfer performance more than measures used in other studies.

Since throughput is always less than the offered load, curves of delay versus throughput are steeper than those relating delay to offered load, and show knees at lower values of the independent variable. Referring to figure 2-1, throughput is measured not at the output of the *generator* or the *buffer*, but on the network itself.

[1]The "long" packets in this study were only 256 bytes long, although the Pup protocol suite [4] used on the experimental Ethernet used a 576 byte maximum packet.

The curves in [5] show knees for CSMA/CD at throughputs of 0.4, instead of the knees at offered loads of about 0.75. To the unwary, this distinction may not be obvious.

Further, the curves displayed in this study are based on unimodal distributions of packet lengths with means of about 128 bytes. Although Bux has attempted to correct for the effects of analyzing a slotted system, the use only of short packet lengths in this study makes it less applicable to bulk-transfer applications. The comparison favors slotted-ring systems over CSMA/CD systems with equal bandwidth; this is valid only for real-time applications using short packets.

Bux does include an interesting figure that demonstrates the sensitivity of CSMA/CD systems to propagation delay. Packet delays are reasonable even for high throughputs so long as packet lengths are at least two orders of magnitude longer than the propagation delay (expressed in bit-times).

2.4.5. Coyle and Liu, 1983 and 1985

Coyle and Liu [6, 7] refine the analysis of Tobagi and Hunt [25] by using a somewhat more realistic model; specifically, rather than assuming an infinite number of hosts generating an aggregate stream of packets with a Poisson distribution, this analysis assumes a finite number of hosts each generating new packets with an exponential distribution. (They analyze non-persistent CSMA/CD, rather than the 1-persistent Ethernet.)

Coyle and Liu define a stability measure, called "drift," the rate at which the number of backlogged packets is increasing. Drift is a function of the number of users (effectively, the load on the system), and the shape of the drift curve depends upon the retransmission rate. For large numbers of hosts and high retransmission rate, CSMA/CD without backoff is shown to be unstable. In other words, the binary exponential backoff mechanism of Ethernet, which adapts the retransmission rate to the load, is crucial for stability. If the packet length is at least 25 times the collision-detection time, the Ethernet *is* stable even at extremely high offered loads.

2.4.6. Takagi and Kleinrock, 1985

Takagi and Kleinrock [22] analyze a variety of CSMA protocols under a finite-population assumption. In particular, they provide an analysis of unslotted 1-persistent CSMA/CD. They solve for throughput as a function of G, and one interesting result of their analysis is that not only does the throughput curve depend on the time to abort transmission after detecting a collision, but that for very small values of this parameter, there is a double peak in the curve. In all cases there is a peak at $G = 1$, corresponding to a load exactly matching the capacity of the network. The second peak, present when collisions are detected quickly, comes at offered loads (G) of 100 or more.

2.4.7. Gonsalves and Tobagi, 1986

Gonsalves and Tobagi [10] investigated the effect of several parameters on performance using a simulation of the standard Ethernet. In particular, they examined how the distribution of hosts along the cable affects performance. Previous studies had assumed the balanced star configuration (all hosts separated by the maximum length of the network). This is not a realistic assumption, and so in this simulation several other configurations were examined.

If hosts are uniformly distributed along the cable, or if they are distributed in more than two equal-sized clusters, those hosts in the middle of the cable obtain slightly better throughput and delay than those at the ends. If the clusters are of unequal sizes, hosts in the larger clusters get significantly better service. This is because the hosts within a cluster resolve collisions quickly, whereas the collisions between clusters, which take longer to resolve due to the higher separation, are more likely to be a problem for members of the smaller cluster (since there are more "distant" hosts for those hosts than for members of a large cluster).

This is an important result because a number of theoretical analyses explicitly assume that the network is fair by design (for example, that of Apostolopoulos and Protonotarios [2]). Users with real-time applications may find position-dependent unfairness of some unusual Ethernet configurations to be an important effect.

2.4.8. Tasaka, 1986

Tasaka [23] analyzes the effect of buffering on the performance of slotted non-persistent CSMA/CD. As the size of the buffer (measured in packets) increases, fewer packets are dropped due to congestion and thus are retained to use up channel capacity that would otherwise be idle. Increased buffer size leads to higher throughput and average delays at lower offered loads, and causes the knees of these curves to appear at lower offered loads.

2.5. Myths and reality

Although many of the theoretical papers accurately describe the limitations of their approaches, it is still difficult to extract the right lessons from these studies. Thus, many people have learned the wrong lessons, and a variety of myths have been propagated.

Pitfalls for the unwary fall into several categories:

- **Not the same protocol**: the standard Ethernet is an unslotted, 1-persistent, carrier-sense multiple-access method with collision detection and binary exponential backoff. Many analyses assume slotted systems, non-persistent or p-persistent systems, or systems without collision detection. These variations can significantly affect real performance: there are reasons why the Ethernet design is as it is.
- **Unrealistic assumptions**: for analytic tractability, many analyses usually assume balanced-star configurations, infinite populations, unimodal or con-

stant packet lengths, small packet lengths, no buffering, etc. Many (if not most) real-world Ethernets include a relatively small number of hosts, distributed randomly over a cable shorter than the maximum length, with flow-controlled protocols that generate a bimodal distribution of packet lengths.

- **Measuring the wrong dependent variable**: if one is building a real-time system, average packet delay (or perhaps the variance of packet delay) is the critical system parameter. If one is building a distributed file system, throughput may be the critical parameter. Since it is not always possible to optimize for both measures simultaneously, the network designer must keep the intended application clearly in mind.
- **Using the wrong independent variable**: when comparing two studies, it is important to understand what the independent variable is. Most studies use some approximation to offered load, but there are real differences in how offered load is defined.
- **Operating in the wrong regime**: Virtually all of the studies cited in this paper examine the performance of networks at high offered load; that is where the theoretically interesting effects take place. Very few real Ethernets operate in this regime; typical loads are well below 50%, and are often closer to 5%. Even those networks that show high peak loads usually have bursty sources; most of the time the load is much lower. Unless real-time deadlines are important, the capacity of the Ethernet is almost never the bottleneck.

The most well-known myth is that "Ethernets saturate at an offered load of 37%." This is a fair summary of what happens for certain worst-case assumptions, but has very little to do with reality. As we show in section 3, an Ethernet is quite capable of supporting its nominal capacity under realistic conditions. In section 4, we look at how implementors can achieve these conditions.

3. Measurements of a real Ethernet

This section presents measurements of the behavior of an Ethernet under varying combinations of packet lengths, network lengths, and number of hosts.

When many hosts are waiting for a long packet to go by on a busy Ethernet, the instantaneous load on the network routinely exceeds its capacity for short periods. Stressing an Ethernet with an unusual steady overload yields insights into how the network handles the usual momentary overloads encountered in normal operation. Therefore, in these tests we attempt to generate a total offered load that continuously exceeds the capacity of the network.

No real Ethernet should be operated this way ("Don't try this at home"). The Ethernet was designed to handle bursty traffic with a high peak-to-average ratio of bit rates, rather than a continuous high load. Most of the time the Ethernet should be idle, but occasionally the load will spike up, and when this happens, the network must handle it fairly. Fairness is important under heavy load conditions; when the load is light and a host can transmit virtually at will, it is not a concern.

3.1. Previous measurements

Several previous studies have measured the performance of an Ethernet.

Schoch and Hupp [19] reported a number of measurements of the 3 Mbit/second experimental Ethernet. They confirmed that for 512-byte packets, the channel utilization at high offered load stayed near 100% even for 64 hosts, whereas for small packets, the utilization approached $1/e$ for large numbers of hosts. They also found that even for moderately short packets, the network is stable at high offered load: the throughput does not drop as the load increases.

Gonsalves [9] made additional measurements of the experimental Ethernet in order to evaluate its potential performance for packet voice applications. He measured both throughput and delay as a function of offered load; he also studied how load affects the rate at which packets must be discarded because they cannot be delivered before a specific deadline.

3.2. Measurement environment

All of the hosts used in this experiment are Titans, prototype RISC personal computers built by Digital Equipment Corporation's Western Research Laboratory. The cycle time of a Titan is 45 nSec; during these tests the machines were executing about one instruction every 1.5 cycles, or about 15 million instructions per second.

The Titan Ethernet controller was designed by one of the authors. It uses the SEEQ 8023/8003 Ethernet chip set with a custom DMA interface to the Titan's 400 MBit/second ECL backplane. The controller's transmitter sends one packet and then must be serviced by interrupt code before it can send again. This means that the interrupt service latency puts an upper bound on the number of packets that can be generated per second.

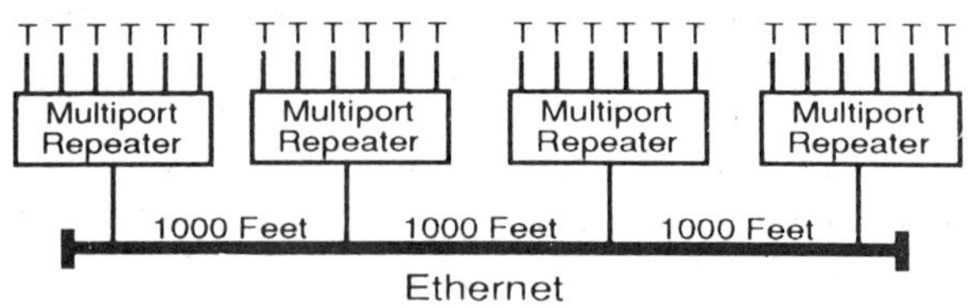

Figure 3-1: Experimental configuration

During the tests, equal numbers of Titans were connected to one of four DELNI multiport repeaters, whose transceivers were attached to a coaxial cable (see figure 3-1).

The test software runs in a light-weight operating system written for diagnostics and bootstrap loading. An Ethernet device interrupt takes about 100 microseconds from assertion of interrupt to resumption of user code; this is about 1500 machine instructions.

3.3. Methodology

A full series of tests with N transmitters requires $N+1$ Titans. One acts as the control machine, coordinating the N generator machines and collecting the data.

Each generator machine listens for a broadcast control packet announcing the beginning of a test and specifying the length of the packets it should generate. After receiving a control packet, it waits 5 seconds, and then transmits packets of the specified length for 20 seconds. During the middle 10 seconds of a test, the Titan measures the performance of its Ethernet transmitter. Next, it waits 5 seconds for the dust to settle and then sends a packet to the control host reporting the statistics gathered during the test. Finally, it goes back and waits for another control packet.

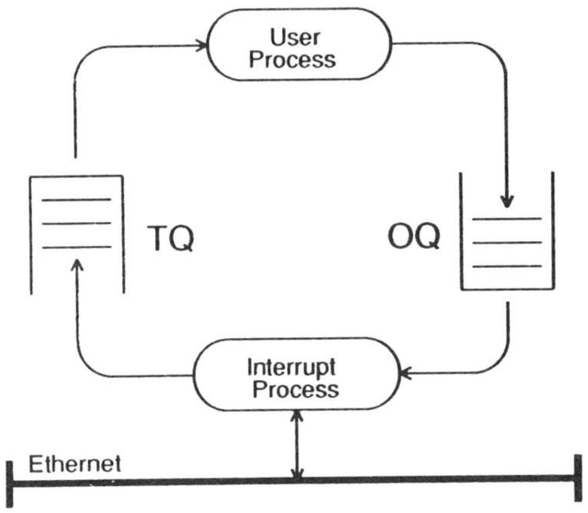

Figure 3-2: Packet generation process

The program that generates packets for the test load is depicted in figure 3-2. Each generator Titan allocates 15 packet buffers, builds Ethernet packets in them, and enqueues them on the *transmitted queue* (TQ). A user-level process blocks while the TQ is empty. When a packet buffer appears on the queue, the user process dequeues it, sets the packet length and enqueues it on the Ethernet *output queue* (OQ). If the transmitter is idle, it is started, otherwise nothing more happens until the next transmitter interrupt. When the hardware interrupts at the end of sending a packet, the interrupt routine dequeues the buffer from the OQ and enqueues it on the TQ. Next, if there is another packet buffer on the OQ, it restarts the transmitter. If the time to send a packet is longer than the time to generate one, then the OQ will usually be full, the TQ will usually be empty, and the user process will be idle some of the time. If the time to send a packet is shorter than the time to generate one, then the OQ will usually be empty, the TQ will usually be full, and the user process will be running all of the time.

During a test, each machine counts the number of bits and the number of packets it successfully transmits, and accumulates the sum of the transmission delays and the sum of the delays squared. Each generator machine sends these four numbers to the control machine at the end of a test.

When calculating bit rate, each packet is charged with 24 extra byte times of overhead to account for the 9.6 microsecond interpacket gap (12 byte times), the 64-bit sync preamble (8 bytes), and the 32-bit cyclic redundancy checksum (4 bytes). This method of accounting yields a bit rate of 10.0 MBits/sec when the network is carrying back-to-back packets with no collisions.

The following statistics are computed from the data collected during a test:

- **Bit rate**: total number of useful bits/second (counting overhead), summed over the entire network.
- **Standard deviation of bit rate**: computed from the per-host bit rates, this is a measure of how unfair the network is.
- **Packet rate**: average rate of successful packet generation, summed over the entire network.
- **Standard deviation of packet rate**: computed from the per-host packet rates, this is another measure of unfairness.
- **Transmission delay**: average delay from beginning of first attempt to transmit a packet to the end of its successful transmission.
- **Standard deviation of delay**: computed from the per-packet delays, this indicates how closely one can predict the response time, for a given load.
- **Excess delay**: the difference between the measured average transmission delay and the ideal delay assuming no collisions, this is a measure of inefficiency.

A series of measurements, involving a test run for each of about 200 combinations of parameter values, takes about three hours. Tests are run after midnight because the Titans are the personal computers that lab members use during the day.

All of the graphs presented in this section show performance measures as a function of the number of hosts, N, involved in the test. For large N or for large packet sizes, the network is the bottleneck, and Q (offered load) is approximately equal to N. When N and the packet size are both small, we could not generate packets fast enough to overload the network, so Q is less than one in these cases.

3.4. Maximum rate attainable between a pair of hosts

We first measured the maximum throughput that could be used for a single "conversation" between a pair of hosts. This was done by setting up one transmitting host and one receiving host, counting the number of packets sent and received, and verifying that all got through. We found that, for all the packet sizes (and thus packet rates) we tried, the receiver could always keep up with the transmitter. In other words, the rate at which a single host can send packets is what limits the throughput between a single pair of hosts (in these tests).

Because receiver performance was not an issue, during subsequent tests each generator host addressed its packets to itself, and disabled its receiver so that it would not waste cycles receiving its own packets.

3.5. Fixed length packets on a long net

The first set of experiments was conducted with 24 hosts sending fixed length packets on 3000 feet (910 meters) of coaxial cable. Six hosts were connected to each of four multiport repeaters spaced at 1000 foot intervals along the cable. This network configuration is similar to the way many buildings are wired for Ethernet: a long backbone segment with repeaters connecting work groups on short segments. The number of hosts per repeater was kept balanced as the total number of hosts in the test was varied. We did tests with nine different packet lengths, including three that exceed the maximum allowed by the Ethernet specification. The curves are labeled with the packet length in bytes.

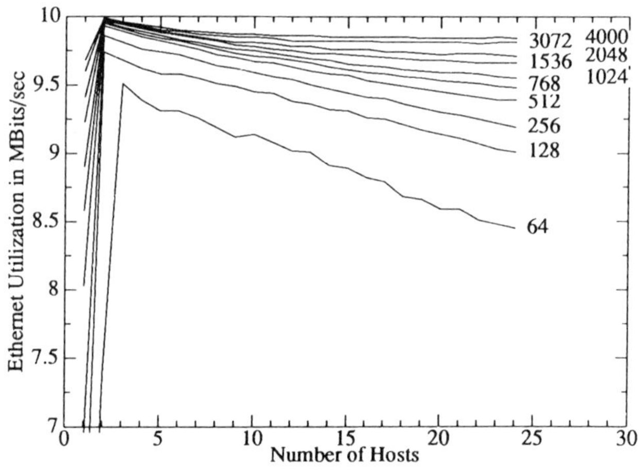

Figure 3-3: Total bit rate

In figure 3-3, notice that the bit rate increases with increasing packet size. This is because for larger packets, there are fewer packets per second and so less time is lost at the beginning of packets colliding and backing off. Notice also that the bit rate decreases with increasing number of hosts. This is because for larger numbers of hosts, there are more collisions per packet. (For small numbers of hosts and small packet size, bit rate first increases until the offered load exceeds the network capacity.)

Figure 3-4 is a measure of the variation in bit rate obtained by individual hosts during a test. (The curves are so hard to distinguish that labelling them would have been pointless.) As the number of hosts increases, the fairness increases. The unfairness for small N is due to the intrinsic unfairness of the Ethernet backoff algorithm. As Almes and Lazowska [1] point out, the longer a host has already been waiting, the longer it is likely to delay before attempting to transmit. When $N = 3$, for example, there is a high probability that one host will continually defer to the other two for several collision resolution cycles, and the measured standard deviation becomes a sizeable fraction of the mean bit rate.

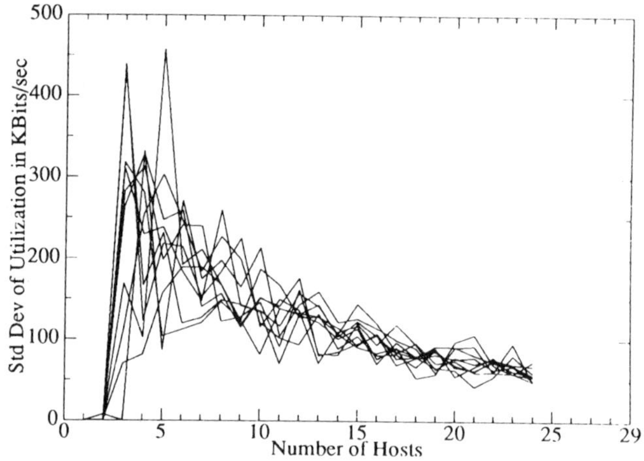

Figure 3-4: Std. dev. of bit rate

As N gets larger, this effect is smoothed out. For 20 hosts, the measured standard deviation is about 20% of mean.

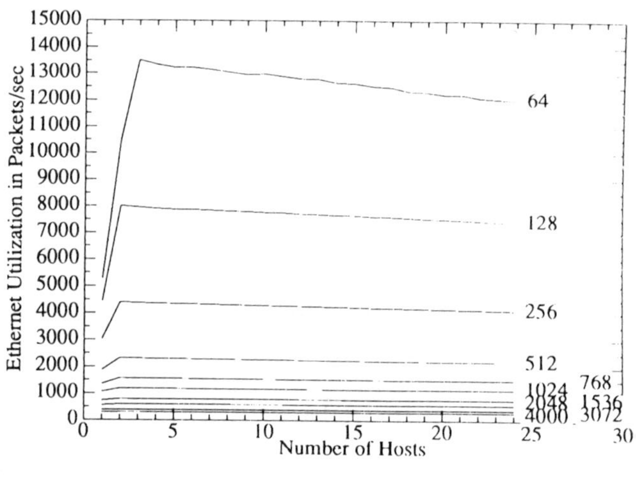

Figure 3-5: Total packet rate

The theoretical maximum rate for 64 byte packets is about 14,200 packets per second. Figure 3-5 shows that we were able to achieve about 13,500 packets per second.

The packet rate peaks at two hosts; no collisions occur until at least three hosts are transmitting. There is obviously no contention for the Ethernet when only one host is sending. Because transmitter interrupt service latency is longer than the interpacket gap, two hosts quickly synchronize: each defers to the other and then transmits without contention. When three or more hosts are sending, there are usually two or more hosts ready to transmit, so a collision will occur at the end of each packet.

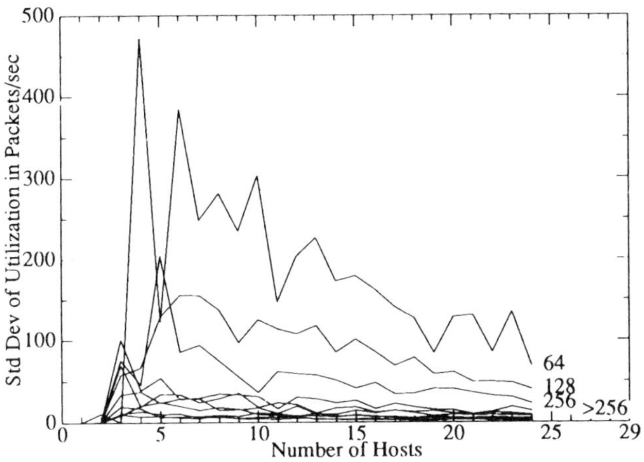

Figure 3-6: Std. dev. of packet rate

Figure 3-6 shows the variation in packet generation rate among the hosts in the tests. Note again that the unfairness decreases as the number of hosts increases. The high variance for 64-byte packets may be an experimental artifact.

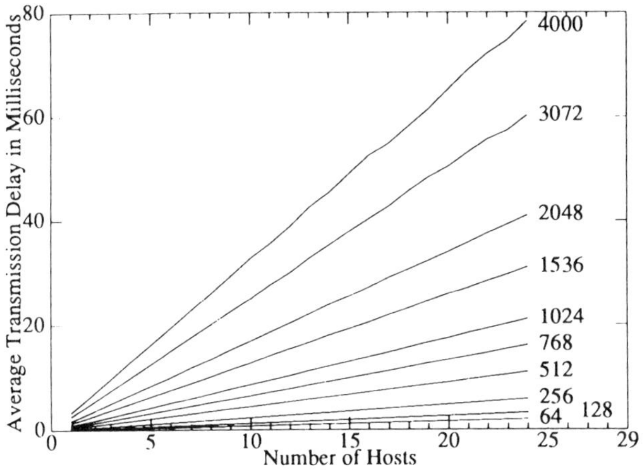

Figure 3-7: Average transmission delay

Figure 3-7 shows that the average transmission delay increases linearly with increasing number of hosts (i.e. offered load). This contradicts the widely held belief that Ethernet transmission delay increases dramatically when the load on the network exceeds $1/e$ (37%).

The standard deviation of the packet delay, plotted in figure 3-8 increases linearly with number of hosts. If there are *N* hosts transmitting, then on average for each packet a host sends, it waits while *N-1* other packets are sent. In the worst case, for large packets and many hosts, the standard deviation is about twice the mean.

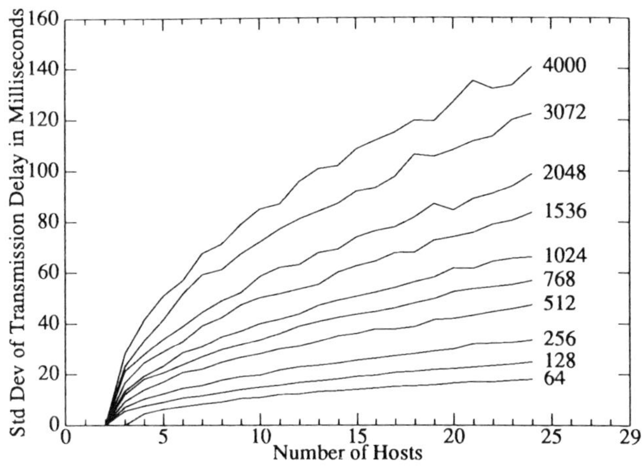

Figure 3-8: Std. dev. of transmission delay

Figure 3-9 shows "excess delay", a direct measure of inefficiency. It is derived from the delays plotted in figure 3-7. The ideal time to send one packet and wait for each other host to send one packet is subtracted from the measured time. The time that remains was lost participating in collisions. Notice that it increases linearly with increasing number of hosts (offered load). When 24 hosts each send 1536-byte packets, it takes about 31 milliseconds for each host to send one packet. Theoretically it should take about 30 mSec; the other 1 mSec (about 3%) is collision overhead. Figure 3-3 agrees, showing a measured efficiency of about 97% for 1536-byte packets and 24 hosts.

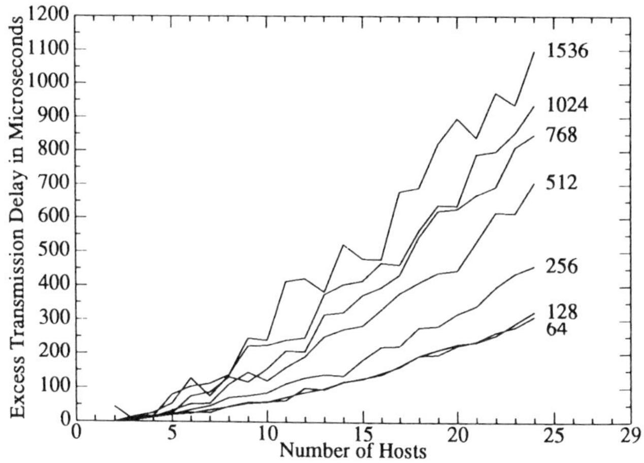

Figure 3-9: Excess transmission delay

3.6. Fixed length packets on a short net

We also ran a set of tests with 23 hosts on a 20 foot Ethernet. Collision resolution time is a function of round trip propagation delay; the quicker that collisions get resolved, the more efficient the net is. The round trip propagation delay for 3000 feet of coaxial cable is about

10 microseconds; it is essentially zero for 20 feet. The data for this experiment look very similar to the previous set; only the bit rate graph is shown.

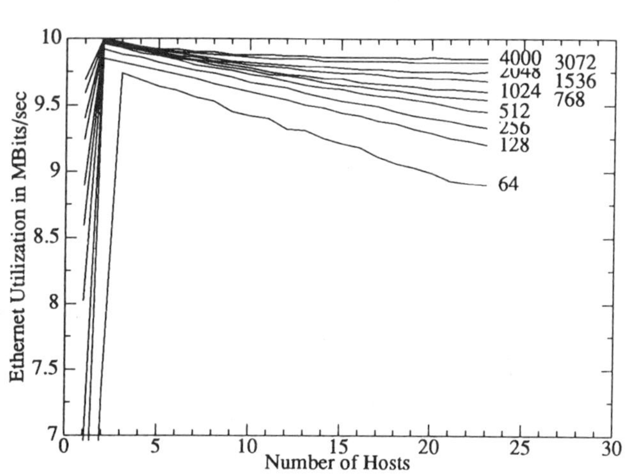

Figure 3-10: Total bit rate (short net)

Comparing figure 3-10 with figure 3-3 shows that network efficiency increases as the collision resolution time decreases. The effect is most pronounced with short packets, where the efficiency drops to only 85% when the packet transmission time is only an order of magnitude larger than the collision resolution time (as in figure 3-3).

3.7. Bimodal distribution of packet lengths

Because a fixed packet length is not characteristic of most actual Ethernet loads, we repeated our measurements using a number of bimodal length distributions. These distributions were composed of minimal-length (64-byte) and maximal-length (1536-byte) packets in varying ratios, with the packet size chosen randomly while preserving the overall ratio. This approximates real Ethernet traffic, which usually consists of many minimum length packets, some maximum length packets, and a few of intermediate size [20, 11].

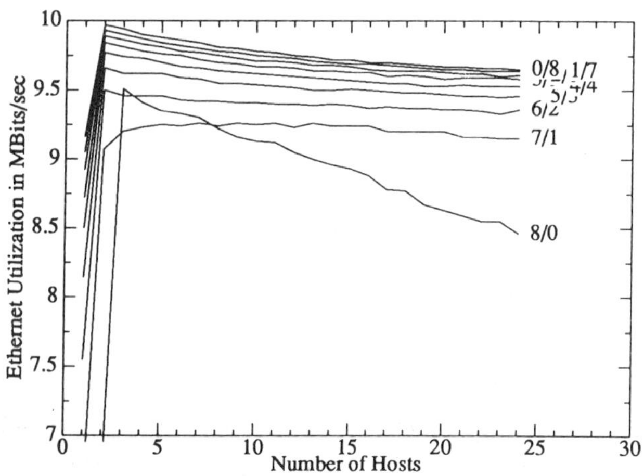

Figure 3-11: Total bit rate (long net; 4 clusters)

Figure 3-11 shows the utilizations obtained with bimodal length distributions. The curves are labeled with the ratio of short to long packets; for example, "6/2" means that there were six short packets for every two long packets.) Notice that when only one out of eight packets is long, the utilization is much higher than when all the packets are short. This is as predicted by Tobagi and Hunt [25].

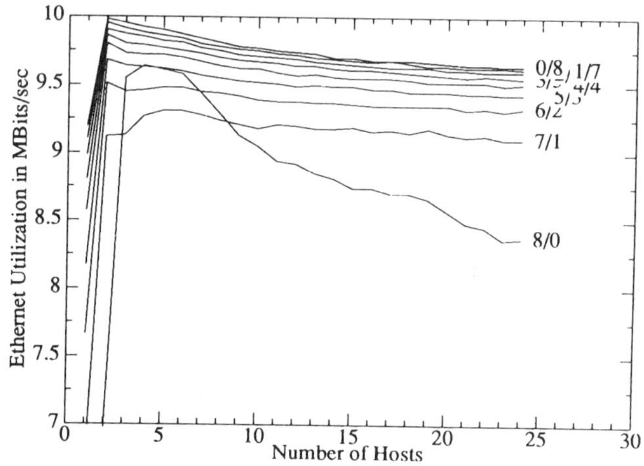

Figure 3-12: Total bit rate (long net; 2 clusters)

In addition to four groups of hosts at 1000 foot intervals as above, we ran a set of tests with half of the hosts at either end of 3000 feet of cable. The average separation between hosts is greater in this configuration than with four clusters of hosts, so the collision resolution time is increased and the efficiency is decreased. For minimal-length packets and 24 hosts, the average delay is about 1.5% higher, and the total utilization, as shown in figure 3-12, is about 1.1% lower, in the two-cluster configuration. For maximal-length packets, there is no appreciable difference between the two configurations.

Gonsalves and Tobagi [10] showed, in their simulation, that unequal-sized clusters increase unfairness; we have not yet attempted to measure this effect.

4. Implications for Ethernet implementations

Between the theoretical analyses, which tell us how to operate the Ethernet as a whole, and our practical experience, which tells us how to obtain good performance from actual implementations and on actual applications, we can draw a clearer picture of how to use Ethernets.

4.1. Lessons learned from theory

For the user who has already decided to install Ethernet technology, the theoretical results provide several guidelines:

- **Don't install long cables**: to cover a large area, break up the cable with bridges or gateways (routers), not repeaters.
- **Don't put too many hosts on one cable**: use gateways to break the network into communities of

interest, trading higher delay for inter-community traffic for better intra-community response time and throughput.

- **Implement the protocol correctly**: proper collision detection and binary-exponential backoff in interface or host software is essential to good performance.
- **Use the largest possible packet size**: this keeps the packet count down, reducing the likelihood of collision and not incidentally reducing overheads internal to hosts.
- **Don't mix serious real-time and serious bulk-data applications**: it is not possible to simultaneously guarantee the lowest delay and the highest throughput (although for moderate requirements both kinds of applications coexist well).

4.2. Prerequisites for high-performance implementations

It is actually fairly difficult to drive an Ethernet into the regime measured in section 3, let alone into the "unstable" regimes predicted by theory. In fact, it can be hard to get good performance even on a lightly-loaded Ethernet.

Some of the pessimism about Ethernet can be traced to poor implementations. For example, one study [18] found that a typical implementation of an Ethernet interface was unable to provide good performance. This study shows packet rates (for minimal-length packets) of 250 packets per second or less, and throughputs (using large packets) of about 150 Kbytes/second or lower. Although more recent implementations can do better than this, it takes careful design and hard work.

For example, Birrell and Nelson [3] were able to obtain 1 millisecond round-trip times for null Remote Procedure Calls (RPCs) even on a 3 Mbit/second experimental Ethernet. Recently, Van Jacobson has reported achieving 8 Mbits/second using TCP over a standard Ethernet [12]. Unlike the superficially more impressive numbers we show in section 3, these results are for real applications.

Networking performance can be limited by any of several weak links. For example, if the backplane or memory system bandwidth is too low, the bandwidth of the network itself becomes less visible. Also, host processor time is precious; per-packet processing time increases latency and, unless it can be completely overlapped with transmission or reception, decreases throughput. The latency costs of packet processing completely dominate the theoretical channel-access times in the light-load regimes typical of most Ethernet installations. Processor speed often has more influence than network bandwidth on useful throughput (see, for example, Lantz, Nowicki, and Theimer [14].)

When the processor, memory, and software are fast enough to support the full network bandwidth, the network interface can be the weak link. In a high-performance Ethernet interface,

- **Transmitter and receiver performance should be matched**: If the transmitter cannot keep up with the receiver, or vice versa, the ultimate performance suffers, as we found when measuring single-host short-packet rates (see section 3.4).
- **The interface should be able to transmit, and to receive and store, several back-to-back packets without host intervention**: otherwise, bandwidth is wasted on lost packets or on channel idle time.
- **The interface should be able to transmit and receive several packets per interrupt**: interrupt latency is often the bottleneck to efficient handling of packets. The ability to receive back-to-back packets implies the ability to handle them in a batch.

4.3. Problems with certain existing implementations

When Ethernets fail, it is almost never because they have been pushed into instability by a high load of useful traffic. This may be because most implementations are incapable of utilizing the bandwidth of the network. Instead, Ethernets usually fail for one of two reasons: hardware failure (including incorrect designs) or accidental high loads.

Transient hardware failures include open or shorted cables, to which all shared-medium LANs are vulnerable, and "jabbering" transmitters (those that send infinitely long packets). Careful hardware maintenance is the only protection against such failures, and it takes considerable preparation and diagnostic skill to locate them; Ethernets are not only "distributed packet switches" but "distributed single points of failure."

Accidental high loads, usually generated by software bugs, are the most common way that Ethernets are pushed into the overload regime analyzed by theoretical studies. One example of such a bug is distressingly common: a broadcast packet is sent but some of the hosts that receive it mistakenly believe that it should be forwarded, as a broadcast, back onto the cable. The recursive nature of this bug leads to exponential growth in the number of such packets, saturating the network. Unless there is a mechanism for aging packets (such as a hop-count field), it may be impossible to recover the network without shutting down every host.

One design bug that can push a highly-loaded network into overload is to use a linear backoff instead of the exponential backoff specified in the standard. At low loads, a linear backoff will give better service to the host that uses it, especially when set against other hosts that play by the rules. At high loads, an incorrect backoff algorithm can drive the network into instability, making it useless for everyone. For this and other reasons, all implementations should be tested at high offered loads, even if they normally would not be used that way.

Host software that assumes low congestion can also push an Ethernet from high load into overload by retransmitting data too aggressively, especially if acknowledgements are being delayed due to collisions. Retransmission delays in higher protocol levels should be backed off to avoid congestive collapse in any network, not just Ethernets. Hosts should not assume that delays on an Ethernet will always be short; even on a network that is normally lightly-loaded, brief intervals of overload are possible.

4.4. Suitability of Ethernet for specific applications

Both the theoretical studies and our measurements support the use of Ethernet for high-bandwidth applications, especially when response time is not closely constrained. Given that the aggregate bandwidth requirements do not frequently exceed 10 Mbits/second, the Ethernet should not be a bottleneck.

When response time is constrained, then Ethernet may still be a reasonable choice, especially if the number of hosts is relatively small and there are few high-bandwidth applications. For example, Gonsalves [9] found that the experimental (3 Mbit/second) Ethernet was "capable of supporting about 40 simultaneous 64-Kbps two-way voice conversations with acceptable quality." Note that this application can accept occasional lost packets; some real-time applications are more constraining.

The Ethernet may not be the appropriate medium for applications requiring low latency and involving a very large number of simultaneous hosts. One such application is distributed programming with fine-grained remote memory access. Spector [21] was able to obtain reasonable performance on an experimental Ethernet with a small number of hosts, but he suggests that only ring networks are able to support a large amount of parallelism.

5. Summary and conclusions

Our measurements, and a careful reading of the theoretical analyses, show that Ethernet is capable of good performance for typical applications, even at high offered load. The performance of CSMA/CD is sensitive to packet size and the time it takes to detect a collision; users must keep these parameters in mind. At higher bit rates or for longer networks, ring topologies may be the only acceptable approach, but experience with Ethernet proves that at 10 Mbits/second, over a kilometer or so of cable, CSMA/CD is quite successful.

Acknowledgements

John DeTreville provided numerous perceptive comments on an early draft of this paper. Michael Fine, Joel Bartlett and Brian Reid also provided important suggestions. The authors are, of course, solely responsible for any errors in this paper.

We would like to thank the members of the Digital Equipment Corporation Western Research Laboratory (WRL) and Western Software Laboratory (WSL) for their patience. During the process of performing our measurements, we took over 90% of the computing capacity, and on several occasions caused network "meltdowns" which made the rest of the computers useless.

References

1. Guy T. Almes and Edward D. Lazowska. The Behaviour of Ethernet-Like Computer Communications Networks. Proceedings of the 7th Symposium on Operating Systems Principles, ACM SIGCOMM, Asilomar, California, December, 1979, pp. 66-81.

2. Theodore K. Apostolopoulos and Emmanuel N. Protonotarios. "Queueing Analysis of Buffered CSMA/CD Protocols." *IEEE Transactions On Communications COM-34*, 9 (September 1986), 898-905.

3. Andrew D. Birrell and Bruce Jay Nelson. "Implementing Remote Procedure Calls." *ACM Transactions on Computer Systems 2*, 1 (February 1984), 39-59.

4. David R. Boggs, John F. Shoch, Edward A. Taft, and Robert M. Metcalfe. "Pup: An internetwork architecture." *IEEE Transactions On Communications COM-28*, 4 (April 1980), 612-624.

5. Werner Bux. "Local-Area Subnetworks: A Performance Comparison." *IEEE Transactions On Communications COM-29*, 10 (October 1981), 1465-1473.

6. Edward J. Coyle and Bede Liu. "Finite Population CSMA/CD Networks." *IEEE Transactions On Communications COM-31*, 11 (November 1983), 1247-1251.

7. Edward J. Coyle and Bede Liu. "A Matrix Representation of CSMA/CD Networks." *IEEE Transactions On Communications COM-33*, 1 (January 1985), 53-64.

8. *The Ethernet, A Local Area Network: Data Link Layer and Physical Layer Specifications (Version 1.0).* Digital Equipment Corporation, Intel, Xerox, 1980.

9. Timothy A. Gonsalves. Packet-Voice Communications on an Ethernet Local Computer Network: an Experimental Study. Proceedings of SIGCOMM '83, ACM SIGCOMM, March, 1983, pp. 178-185.

10. Timothy A. Gonsalves and Fouad A. Tobagi. "On The Performance Effects of Station Locations And Access Protocol Parameters In Ethernet Networks." *IEEE Transactions on Communications 36*, 4 (April 1988), 441-449. Originally published as Stanford University SEL Technical Report 86-292, January, 1986..

11. Riccardo Gusella. The Analysis of Diskless Workstation Traffic on an Ethernet. UCB/CSD 87/379, Computer Science Division, University of California - Berkeley, November, 1987.

12. Van Jacobson. Maximum Ethernet Throughput. Electronic distribution of the TCP-IP Discussion Group, Message-ID <8803100345.AA09833@lbl-csam.arpa>.

13. Leonard Kleinrock and Fouad A. Tobagi. ''Packet Switching in Radio Channels: Part I -- Carrier Sense Multiple Access Modes and their Throughput-delay characteristics.'' *TRANSCOM COM-23*, 12 (December 1975), 1400-1416.

14. Keith A. Lantz, William I. Nowicki, and Marvin M. Theimer. Factors affecting the performance of distributed applications. Proceedings of SIGCOMM '84 Symposium on Communications Architectures and Protocols, ACM, June, 1984, pp. 116-123.

15. Robert M. Metcalfe. Steady-State Analysis of a Slotted and Controlled Aloha System with Blocking. Proceedings of the Sixth Hawaii Conference on System Sciences, January, 1973. Reprinted in the *SIGCOMM Review*, January, 1975.

16. Robert M. Metcalfe. *Packet Communication.* Ph.D. Th., Harvard University, December 1973. Massachusetts Institute of Technology Project MAC TR-114.

17. Robert M. Metcalfe and David R. Boggs. ''Ethernet: Distributed Packet Switching for Local Computer Networks.'' *Communications of the ACM 19*, 7 (July 1976), 395-404.

18. Jose Nabielsky. Interfacing To The 10Mbps Ethernet: Observations and Conclusions. Proceedings of SIGCOMM '84, ACM SIGCOMM, June, 1984, pp. 124-131.

19. John F. Schoch and Jon A. Hupp. Measured Performance of an Ethernet Local Network. Proceedings of the Local Area Communications Network Symposium, Mitre/NBS, Boston, May, 1979. Reprinted in the *Proceedings of the 20th IEEE Computer Society International Conference* (Compcon '80 Spring), San Francisco, February 1980.

20. John F. Schoch and Jon A. Hupp. ''Measured Performance of an Ethernet Local Network.'' *Communications of the ACM 23*, 12 (December 1980), 711-721.

21. Alfred Z. Spector. Multiprocessing Architectures for Local Computer Networks. Tech. Rept. STAN-CS-81-874, Stanford University, Department of Computer Science, August, 1981.

22. Hideaki Takagi and Leonard Kleinrock. ''Throughput Analysis for Persistent CSMA Systems.'' *IEEE Transactions On Communications COM-33*, 7 (July 1985), 627-638.

23. Shuji Tasaka. ''Dynamic Behaviour of a CSMA-CD System with a Finite Population of Buffered Users.'' *IEEE Transactions On Communications COM-34*, 6 (June 1986), 576-586.

24. Fouad A. Tobagi and Leonard Kleinrock. ''Packet Switching in Radio Channels: Part IV -- Stability Considerations and Dynamic Control in Carrier Sense Multiple Access.'' *IEEE Transactions on Communications COM-23*, 12 (December 1977), 1400-1416.

25. Fouad A. Tobagi and V. Bruce Hunt. ''Performance Analysis of Carrier Sense Multiple Access with Collision Detection.'' *Computer Networks 4*, 5 (October/November 1980), 245-259.

Passive Optical Star Systems for Fiber Optic Local Area Networks

FREDERICK W. SCHOLL AND MICHAEL H. CODEN

***Abstract*—The first commercial passive star-based fiber optic Ethernet local area network was described in 1982. Since then several improvements on this system have been made and passive star implementation of three other previously described networks have been developed. The thesis of this paper is that passive optical star systems represent an attractive and versatile architecture for implementation of many fiber optic LAN's. The passive star toplogy, we believe, best achieves the system goals of versatility, reliability, noise immunity, and data security. These networks are built around transmissive optical power splitters that are inherently reliable and contain no electronic components. Each node on the network is served by a duplex fiber cable and an appropriate optical transceiver. Several important technologies have been developed to support the above systems capabilities. The passive optical star topology clearly mandates high optical gain for the transceivers.**

In addition, cost drivers in the market place have led to introduction of integrated circuits, surface assembly technology, high performance LED packages, and improved star coupler manufacturing processes. A variety of system installations in office, government, and factory settings have been completed using the above passive star systems. The largest of these is an Ethernet at Southwestern Bell's 44 story headquarters in St. Louis, MI. This all-optical system has 540 nodes, over 4000 computer terminals, and 92 mi of fiber.

I. Introduction

THE first commercial passive star-based fiber optic Ethernet local area network was described in 1982 [1]. Since 1982 several improvements have been made and passive star implementation of three other previously described networks have been developed: ARCNET, MAP, and PC Net. The thesis of this paper is that passive optical star systems represent an attractive and versatile architecture for implementation of many fiber optic LAN's.

The passive star topology, we believe, best achieves the system goals of versatility, reliability, noise immunity, and data security. These networks are built around transmissive optical power splitters that are inherently reliable and contain no electronic components.

Each node on the network is served by a duplex fiber cable and an appropriate optical transceiver. Noise immunity and data security are ensured by appropriate shielding and packaging of the transceivers. Single fiber breaks result in loss of only the associated terminal, and not in more extensive network failures. Typically, up to 32 nodes can be attached to a single star; larger numbers of nodes can be accommodated using repeaters. Radial length of fiber from star to transceiver can be up to 1 km; again, larger distances can be covered using repeaters. These network sizes are suitable for work station clusters, intrabuilding systems, and campus installations.

Fiber optic networks compatible to several network standards have been built using appropriately designed optical transceivers. The optical transceiver implements the physical layer (layer 1) of the ISO network model. All of the transceivers share a common high output power LED transmitter and a high sensitivity PIN diode burst mode receiver. The fiber optic systems implemented to date are compatible to IEEE 802.3 (Ethernet), MAP and IEEE 802.4, PC Net, and ARCNET.[1]

Our 802.3 compatible fiber optic system was described first in 1983. Since that time over 20 000 optical nodes have been installed. These transceivers are AUI (attachment unit interface) compatible to 802.3; a proposal describing a full fiber optic system is under discussion within IEEE 802 standards groups.

A passive star system developed by NTT is similarly under discussion in the Japanese LAN Standardization Committee. Ethernet transceivers operate at 10 MHz baseband, provide optical transmit, receive, and collision detection.

PC Net was originally a CSMA/CD broad-band system operating at 2 MHz. Our fiber optic version has been developed using passive star technology; the bandwidth available allowed a data rate increase to 10 MHz. MAP, 802.4, and ARCNET all refer to token bus networks, operating at 10 MHz and 2.5 MHz, respectively. We have developed electrically compatible baseband fiber optic systems again using passive star topology.

Several important technologies have been developed to support the above systems capabilities. The passive optical star topology clearly mandates high optical gain for the transceivers. In addition, cost drivers in the market place have led to introduction of surface assembly technology, high performance LED packages, and improved star coupler manufacturing processes. Edge emitting 830 nm LED's with minimum launched power of 0.5 mW in 100 μ-graded index fiber were developed and have been manufactured for the past five years. Reliability measure-

[1]ARCNET is a registered trademark of the Datapoint Corporation.

ments indicate MTTF of 2.0×10^7 h at 25°C. High performance burst mode receivers (packet operation) now have -37 dBm peak sensitivity, 22 dB dynamic range, and operate with 10 MHz Manchester coded data.

A variety of system installations in office, government, and factory settings have been completed using the above passive star systems. The largest of these is an Ethernet at Southwestern Bell's 44 story headquarters in St. Louis, MI. This all optical system has 540 nodes, over 4000 computer terminals, and 92 mi of fiber.

II. Passive Star Topology

Passive optical star-based fiber optic systems have been under development for over 15 years. Early work was on passive star LAN's for avionics systems was carried out by Biard [2] while Rawson and Metcalf [3] later sought to apply passive star technology to the emerging Ethernet LAN standard. More recent work on passive optical star networks [1], [4]–[8] has continued to emphasize application to standards-based LAN's as these emerge as the systems of choice. At present, two separate proposals for passive star systems standards are being reviewed, respectively, by IEEE 802 Committee and by the Japan LAN Standardization Committee.

The physical topology of the passive optical star system is shown in Fig. 1. This network can operate as a logical broadcast bus as in Ethernet CSMA/CD applications or as a token bus in 802.4 or ARCNET applications. The hardware components consist of an optical transceiver at each node, a passive optical star power splitter at the network hub, and sections of duplex fiber optic cable interconnecting each transceiver to the hub. The passive star contains no electronics and serves to optically divide the power from each transceiver to the others in the network.

The size of this network segment is limited by flux budget, dynamic range, and timing considerations for the network protocol being implemented. Flux budget and dynamic range are also affected by the number of nodes attached to a particular segment. Two simple equations describe the optical flux budget and dynamic range. First, define the following parameters:

TX = Peak transmitter power launched into system fiber, dBm.
RX = Peak receiver sensitivity, dBm.
L = Radius of fiber segment, $L_{min} \leq L \leq {}_{max}$, km.
α = Optical fiber loss, dB/km.
M = System optical power margin, dB.
LS = Star coupler loss, including bulkhead optical connectors at star, dB.
FB = Flux budget, or allowable loss between transmitter and receiver, dB.
DR = Receiver optical dynamic range, dB.

Using these parameters the optical flux budget is expressed as follows:

$$FB = TX - RX = LS + 2L_{max}\alpha + M. \quad (1)$$

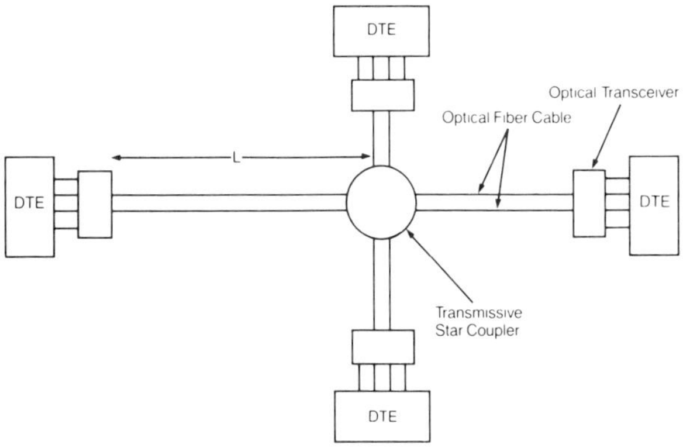

Fig. 1. General layout of a passive star fiber optic network. Individual nodes are optically interconnected by duplex fiber optic cable and a transmissive star coupler. The optical transceivers provide transmit and receive function and other information particular to each network. The case of 802.3 networks is illustrated where the four lines to the DTE include transmit, receive, collision presence, and electrical power.

The receiver dynamic range is described as follows:

$$DR \leq FB - LS - 2\alpha L_{min}. \quad (2)$$

Given a particular receiver, transmitter, and star coupler, these equations can be used to determine L_{min}. For maximum installation flexibility, we want $L_{min} = 0$.

These equations can be applied to a 10 MHz Ethernet CSMA/CD system as an example. The following parameters are assumed:

TX = -7 dBm peak power (100 μ core fiber).
RX = -37 dBm peak power.
α = 5 dB/km.
M = 3 dB.
DR = 22 dB.

Therefore, the relation between star coupler loss and network radius is

$$30 = LS + 2 \times 5 \times L_{max} + 3$$

$$27 = LS + 10\, L_{max}.$$

Using the equation for receiver dynamic range

$$FB - DR \leq LS + 2\alpha L_{min}$$

$$8 \leq LS + 2\alpha L_{min}$$

$$8 \leq LS + 10\, L_{min}.$$

In Fig. 2, these relations are shown graphically using the above parameters and others of interest. For a typical case using a 16-port biconic-fused star coupler, a network radius of 1 km can be achieved. This is equivalent to 2 km distance between optical transceiver or attached DTE's. It is significantly greater than the 500 m maximum segment length that can be attained with conventional coaxial Ethernet. Using biconic-fused couplers the receiver dynamic range is sufficient to permit operation to $L = 0$ for all but the four-port star coupler. An alternate family of star couplers is based on mixing fiber technology (described in Section IV). As shown in Fig. 2, these

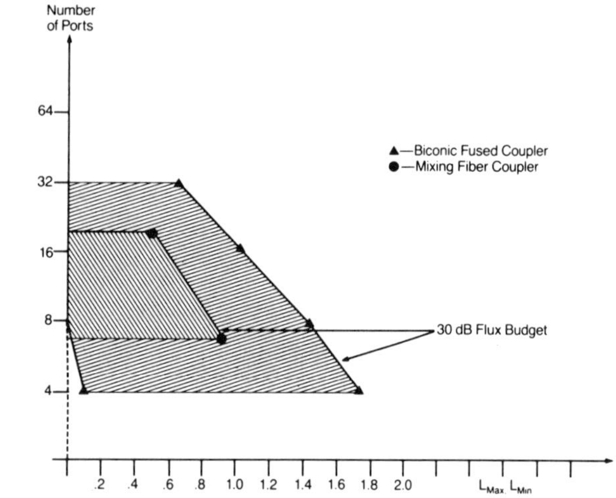

Fig. 2. Operating range for single segment passive star networks. Parameters used: 30 dB flux budget, 5 dB/km fiber, 3 dB margin, and 22 dB dynamic range. Operating range is within the shaded areas, limited on the right by flux budget and on the left by dynamic range.

couplers operate with seven and 19 ports and permit operation to $L = 0$.

Passive star fiber optic systems fill a well-defined application niche in an overall LAN marketplace that is served by coaxial and other wire systems, as well as other fiber optic LAN systems. Some useful points of comparison can be made by limiting consideration to 10 MHz CSMA/CD networks. In comparison to coax 802.3 system, the above discussion shows that larger segments can be built using fiber. Up to 100 nodes can be accommodated per segment using 802.3 coax technology, while a single passive star will presently support up to 32 ports using available LED and PIN receiver technology. A recent survey, however, indicated that 44 percent of all LAN's contain 20 nodes or less [9]; a single passive star segment will adequately support most work group clusters. A second key consideration involves network reliability.

From Fig. 1, it is clear that any cable break or damage will affect only the DTE locally connected to the damaged cable. Connection of additional terminals does not involve intrusion or disruption of the existing network. These features have proved to be very important for networks installed in factory and financial institutions where network availability is critical. Two other fiber optic CSMA/CD networks have been developed and commercialized in addition to the passive star. They are described in [11] and [12]; the topologies adapted are, respectively, an active star and a ring.

Practical networks may involve more than the 32 terminals supported by a single passive star hub; these may be distributed as well over a campus-like installation of many separate buildings.

For example, the backoff algorithm of 802.3 permits up to 1024 nodes per network, while 802.4 token bus has a virtually unlimited number of nodes (limited only by 48 bit address field). These extended networks are created by joining together passive star segments using repeaters. Repeaters also allow interconnection of coaxial systems and fiber optic star segments. Each repeater will regenerate and retime the signals on an input port and transfer these to the output port. An 802.3 repeater is a one-way valve that transmits in one direction at a time unless a collision (simultaneous transmission by two or more nodes) occurs on the network. In this case, the repeater will notify both sides of the network by transmitting a jam signal. A three layer hierarchical topology using passive star hubs is shown in Fig. 3.

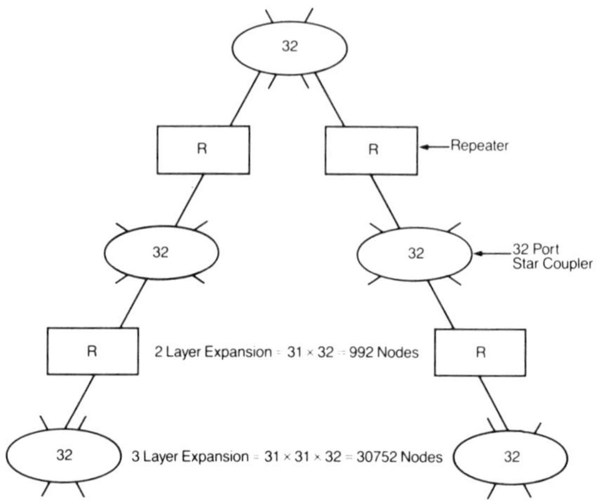

Fig. 3. Three layer expansion of network size beyond the 32 ports of a single segment.

Two layers will accommodate 992 nodes while 3 layers permits up to 30 752 (Ethernet limit is 1024, however). The geographical size of these CSMA/CD networks is limited by the roundtrip propagation delays including cables and optical transceivers. The passive star introduces no propagation delay. Reference [9, Appendix A] describes an 802.3 compatible network of 4.5 km distance from DTE to DTE.

Fig. 4 illustrates the four types of optical repeaters that have been developed to interconnect passive star systems into hybrid coax–fiber networks. Further discussion of repeater function will be given in Section III to follow.

III. Fiber Optic Passive Star Systems

Our emphasis to date has been to develop fiber optic implementations of previously defined network standards. Passive star systems developed to date include the first Ethernet [1], MAP, and PC Net fiber optic implementations as well as a passive star ARCNET compatible system. This approach makes fiber optic LAN's available to users now; other fiber optic networks such as FDDI [13] will likely play an increasingly important role in the future. To date the most widely implemented fiber optic LAN is Ethernet compatible; our discussion will emphasize these systems. Table I summarizes some basic properties of the four network protocols.

IEEE 802.3 compatible Ethernet hardware is shown in Fig. 5. This hardware includes a stand alone optical trans-

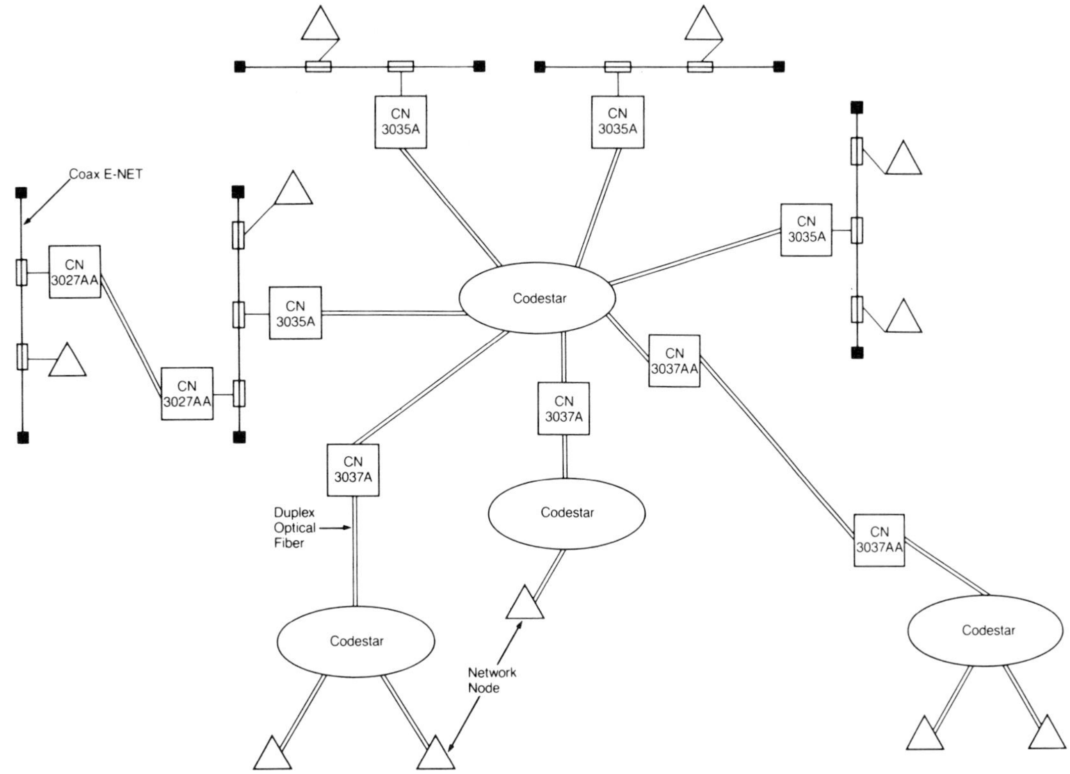

Fig. 4. Optical hybrid network. Repeaters are used to extend the geographical extent of the network as well as to interconnect coax and fiber media.

TABLE I
SUMMARY INFORMATION ON NETWORK SYSTEMS AVAILABLE IN PASSIVE STAR IMPLEMENTATION

Network	Access Protocol	Line Rate	Maximum No. of Stations	Maximum Network Size	Optical Line Coding
802.3 (Ethernet)	CSMA/CD	10 MHz	1024	4.5 km	Manchester
802.4	Token Bus	10 MHz			Manchester
ARCNET	Token Bus	2.5 MHz	255	6.4 km	Miller
PC Net	CSMA/CD	10 MHz	1000	4.8 km	Manchester

ceiver, a board level optical transceiver that interfaces to a common personal computer bus, a passive optical star power splitter, and network repeaters, both electrical to optical and optical to optical. The stand alone optical transceiver implements the physical layer of 802.3 including transmit, receive, and collision detection. The board level optical transceiver implements in addition, the MAC sublayer within the data link layer (layer 2). A gate array implements the interface to the PC bus. System design is facilitated by making use of the optical design parameters listed in Table II.

An important feature of all CSMA/CD networks is collision detection. A collision occurs if two or more nodes attempt to transmit simultaneously. The goal of transceiver design is 100 percent detection of collisions. In practice, a small collision detect inefficiency will not affect network performance. For example, a bit error rate of 1×10^{-9} will cause a certain percentage of frame check errors and missed packets. This percentage will range from 1.2×10^{-5} for the maximum frame length of 12 144 bits to 0.5×10^{-6} for minimum frame length of 512 bits. An undetected collision will also cause a certain number of lost packets depending on the product of the probability of missed collision and the probability of a collision. Since a heavily loaded network may experience large numbers of collisions, a collision detect inefficiency of $< 1 \times 10^{-5}$ would seem to be a suitable practical limit. Several techniques have been proposed for collision detection in passive star systems; see [6], [9], [14]–[19].

We have successfully implemented the pulse width error detection method described in [17]. In this technique, a timing circuit is used to detect the presence of pulses greater than a preset threshold t_{CD}. If only one station is attempting to transmit, then the 10 MHz Manchester code prescribed by 802.3 allows only 50 or 100 ns pulses on the optical fiber. The presence of pulses greater than $t_{CD} > 100$ ns indicates that two or more stations are attempting to transmit. A collision between two packets is shown in Fig. 6. This signal is obtained by connecting an optical waveform analyzer to an unused port on the star coupler and observing the resulting electrical signal on a LeCroy 9400 digital oscilloscope. The 5 MHz signal in the expanded scale trace shows the preamble of the first packet.

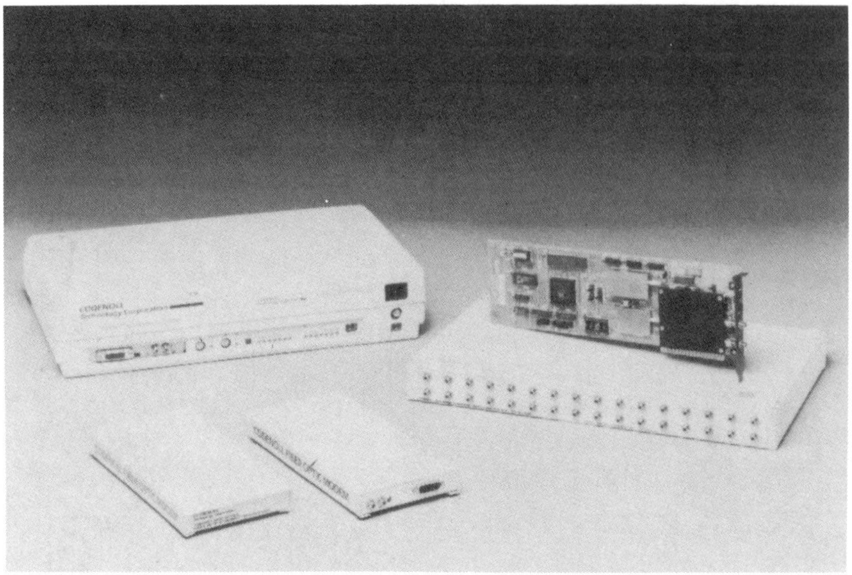

Fig. 5. Typical 802.3 hardware consisting of (clockwise, from foreground): stand-alone optical transceiver, coax-to-optical repeater, transceiver card, and 16-port passive star coupler.

TABLE II
OPTICAL DESIGN PARAMETERS FOR PASSIVE STAR NETWORKS

Transmitter power	7 dBm (peak) (100 micron core, 0.29 na)	
Operating wavelength	830 nm	
Receiver Sensitivity	37 dB (peak)	
Receiver Dynamic Range	22 dB	
Typical fiber attenuation	5 dB/km	
Star coupler losses (maximum)	4 port	10 db
	8 port	13 dB
	16 port	17 dB
	32 port	21 dB
	7 port	18 dB
	19 port	22 dB

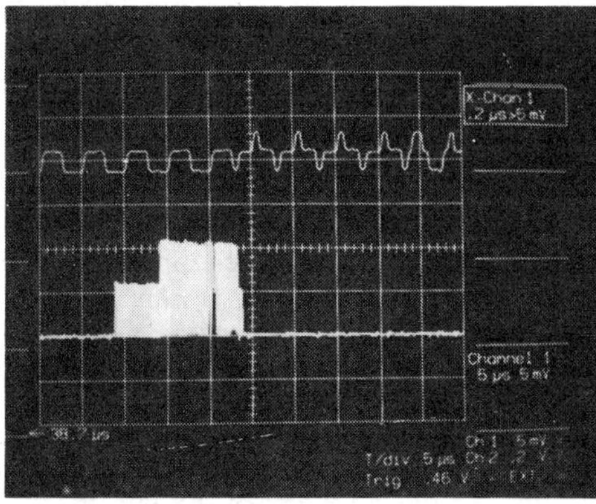

Fig. 6. Optical collision between two packets. Upper trace shows collision event to expanded waveform at 200 ns per division; lower trace shows entire packet at 5 μs per division, with collision occurring 5 μs from start of first packet. While preamble of one packet has only 100 ns between transitions, two colliding packets generate wave with 200 ns between transitions.

The collision with the second packet produces a composite waveform with pulses about 200 ns in width. This collision results from optical pulses with nearly equal amplitude. For large differential optical signal levels the strong signal begins to dominate the weaker, and collisions may not be detected successfully.

These phenomena have been studied in the experimental setup shown in Fig. 7. This three-node test bed for CSMA/CD networks was developed to measure collision detection efficiencies under a variety of real network traffic and topology configurations. Three programmable traffic generators are connected through optical transceivers to a passive star hub. The multiport transceiver units and internetwork bridge units are used for experimental convenience. A logic circuit was developed to monitor, for each optical transceiver, the total packets transmitted, the total sensed collisions (as indicated by collision detect output from the transceiver), the actual collisions occurring (simultaneous transmission by the transceiver in question and one or more of the other transceivers), and false collisions (the transceiver in question signals that a collision has occurred when in fact it is the only unit transmitting). Fig. 8 shows collision detect efficiency as a function of optical power difference. The loopback power was adjusted to be equal at the start; then transceiver $T1$'s power was increased in 1 dB increments. The differential power measured at a particular receiver would be higher by the variations in the star coupler (typically 1–2 dB).

We have conceived and analyzed an improved collision detect method [20] that predicts collision detection inefficiency of less than 1.9×10^{-6} over 6 dB optical differential signal level. The two principal features of this technique are transmission of large amplitude optical pulses and random encoding of 8 of these pulses within the first

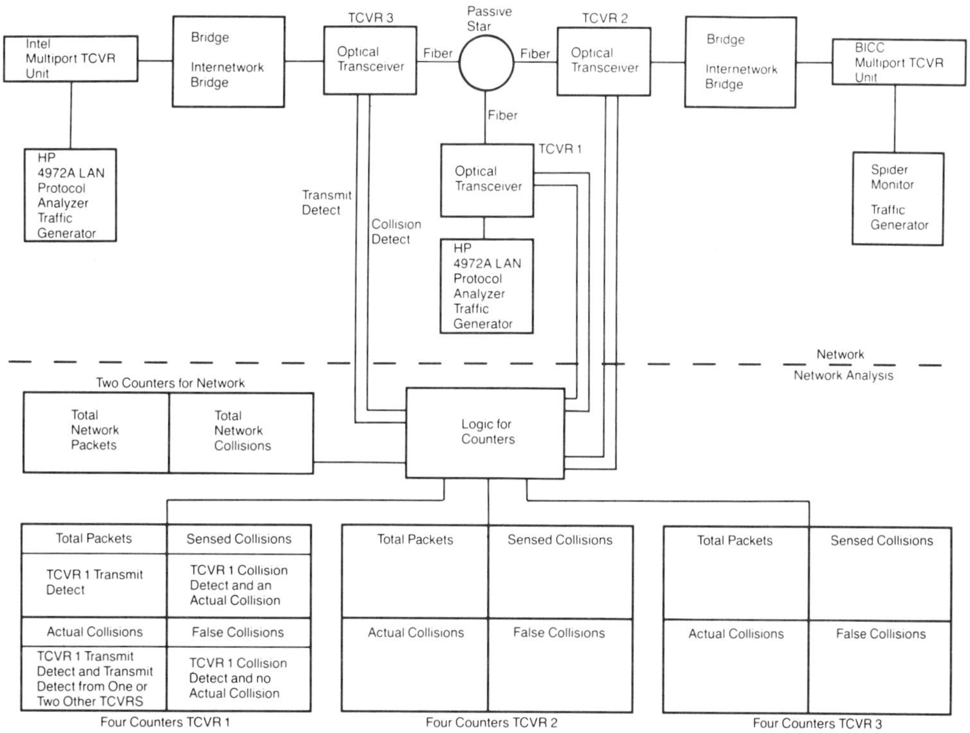

Fig. 7. Testbed for optical networks.

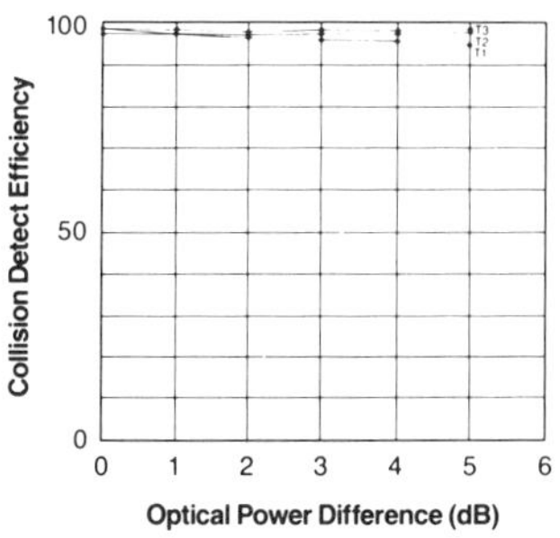

Fig. 8. Collision detection efficiency measured on network testbed. Fiber length was 500 m to star for each transceiver. Transceivers set to generate no false collisions. 64 byte packets were used with 18 percent total network traffic.

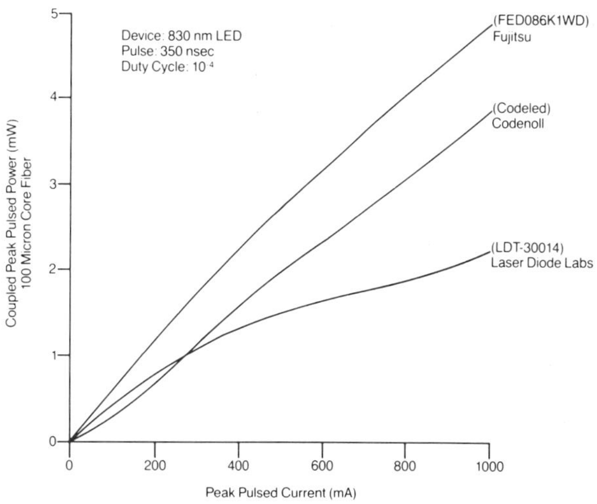

Fig. 9. Pulsed power output of three commercially available LED's. "Codeled" device is edge emitting, the others are surface emitters.

16 bits of the preamble. The $4x$ amplitude large optical pulses are detectable in the presence of a second optical data stream of 6 dB ($4x$) larger amplitude. Since these pulses are 25 ns in width, they are distinguishable from the normal 50 ns and 100 ns intervals. A collision is counted if any transceiver, upon receiving a packet, counts more than eight 25 ns pulses. The probability of a missed collision can be calculated to be $< 1.9 \times 10^{-6}$. The large amplitude optical pulses are generated by edge emitting and surface emitting LED's; Fig. 9 shows pulsed power output characteristics of several commercially available LED's.

An alternate collision detection technique has been reported using coding rule violations detected at the output of a partial response circuit [6], [15], [29]. In [29], a measured collision detection inefficiency of 10^{-7} with optical power difference of 8 dB is reported. Fig. 10 shows experimental data [29] for collision detection time. Future work in this field will involve continued testing of the above described wide dynamic range collision detection techniques, also needed is an international standard using one of these methods.

Optical repeaters were mentioned in Section II and are illustrated in Fig. 4. The repeaters designated 3035A and 3037A are used for locally interconnecting two optical LAN's or an optical and a coaxial LAN. For both re-

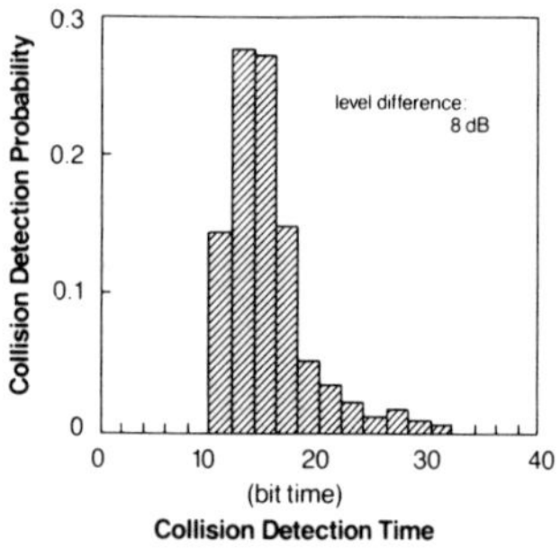

Fig. 10. Collision detection time reported for a CRV technique. Total collision detection inefficiency was reported to be 1×10^{-7}.

peaters, the optical transceiver is packaged internal to the repeater housing. For remote interconnection of coaxial or optical LAN's, the 3027AA or 3037AA repeaters are used. The point-to-point fiber link between the repeater units has additional collision detection via detection of simultaneous transmission and reception on the fiber cables. A recent draft standard describes the fiber optic signaling on remote repeater links [21]. This will ensure interoperability of vendor equipment.

A second CSMA/CD network is IBM PC Net [22]. As originally developed this was a broad-band coaxial cable system operating at 2 Mbits. Our optical transceiver replaces the RF modem with an optical transceiver. The board level transceiver is compatible to the IBM PC bus interface and measures $13\frac{1}{8}'' \times 3\frac{7}{8}''$.

This transceiver contains not only the physical layer interface (layer 1) but also four higher level interface layers: data link (layer 2), network (layer 3), transport (layer 4), and session (layer 5). Layers 3–5 are implemented in firmware executed by an on board 80188 microprocessor.

Board level optical modems essentially compatible to the IEEE draft standard 802.4 have been developed. IEEE 802.4 is a token bus system that originally was based on coaxial cable; the new revision [23] contains two chapters describing fiber optic systems options and fiber optic media. The new revision allows active and passive systems as shown in Table III.

Present designs allow a 20 dB flux budget with a 62.5 μ core fiber. Electrical signals supplied by the optical modem to the MAC layer hardware on a separate card include; receive data, receive clock, and three lines of receive data type information. The MAC layer provides the optical modem with: transmit data, transmit clock, and three lines of transmit data type information. Optical modem cards are compatible to VME bus and multibus backplanes.

A passive star fiber optic ARCNET transceiver card[2] was recently described [24]. The baseband coaxial cable system was developed in 1977 and has more recently been described in detail [25]. This system uses a bipolar line code transmitted over RG 62*A*/*U* cable. An active hub approach is used. Our ARCNET transceiver card mea-

TABLE III
SUMMARY OF FIBER OPTIC SPECIFICATIONS IN 802.4, DRAFT H

Topologies	Passive Star Active Star Active Bus Active Headend	
Signalling	5, 10, 20 Mbs Differential Manchester Coded Packet Transmission	
Optical Media	Networks may be implemented in any multimode fiber provided all parameter specifications are met. A standard test fiber (62.5/125, na – .275) is defined for measurement of fiber optic parameters.	
Optical Connector	As specified by ANSI X3T9.5/84-88	
Emitter Characteristics	Center Wavelength FWHM Launch Power Rise/Fall Time	800-910 nm · 60 nm 9 · 2 dBm defined by wave shape template in draft standard (Fig. 16.3)
Receiver Characteristic	Moderate Sensitivity 11 dBm to 31 dBm High Sensitivity 21 dBm to 41 dBm	
Optical Power Budget	Moderate Sensitivity 20 dB Receiver High Sensitivity 30 dB Receiver	

sures $4.2'' \times 13.3''$ and is compatible to the IBM PC bus. A number of points of comparison exist between the fiber system and the coax system. First, the maximum hub-to-node distance for the coax system is 610 m limited by cable losses. The optical fiber system can support a node to star distance of 1 km limited by roundtrip propagation time of the optical signal. For 1 km this is $2 \times 1\ \mathrm{km}/2 \times 10^8\ \mathrm{m/s} = 10\ \mu\mathrm{s}$. The optical receiver output is blanked for 10 μs following end-of-transmission so that a particular station does not see its own signal returning. This blanking interval is compatible to the minimum 12 μs gap between ARCNET packets.

The coax system uses a bipolar line code with average voltage of zero. For the fiber system, we have adopted a Miller code with constant average duty cycle of 50 percent. The maximum network size of 6.4 km (with repeaters) is the same for fiber and coax system and is limited by internode timing considerations of the ARCNET protocol.

IV. TECHNOLOGIES FOR PASSIVE STAR LAN SYSTEMS

The passive star LAN systems' needs have driven optoelectronic and optical technologies in three main areas; LED transmitter output power, burst mode receiver sensitivity and dynamic range, and optical star coupler excess loss and loss uniformity.

We have developed high brightness 830 nm edge emitting LED's with 0.5 mW output power in a pigtail configuration and 0.2 mW output power in an active device mount package. Sensitivity for burst mode 10 MHz Manchester operation has been improved to -37 dBm peak optical power. Low cost seven and 19 port star couplers have been fabricated using large core mixing fibers and etched cladding design to reduce insertion loss. These technologies will be described in what follows.

[2]Joint development of Codenoll Technology Corporation and Standard Microsystems Corporation.

TABLE IV
BURST MODE RECEIVER SPECIFICATIONS

Peak Optical Power Sensitivity	37 dBm
Dynamic Range	22 dB
Power Supplies	+ 12 (20MA) 12 (Photodiode bias) 5.2 (70 mA)
Design	Si FET with high Z transimpedance
Output	Analog with limiting

TABLE V
SPECIFICATIONS FOR 19 PORT MIXING FIBER STAR COUPLER

Minimum loss	17 dB
Maximum loss	22 dB
Input/output fiber	100 micron core, 0.29 na
Mixing fiber	600 micron core, .22 na, step index
Outside Diameter of etched fibers	120 micron

High power edge emitting LED's were designed, (US patent applied for), to couple maximum power into 100 μ core fiber. This is the most commonly used fiber size for passive star systems. Fig. 11 shows a BeO package configuration suitable for direct fiber pigtailing. Launched powers are routinely 0.500 mW measured through a bulkhead connector in 100 μ, 0.29 nA fiber.

More recently an active device mount (ADM) package has been developed for these LED's. An approach utilizing active alignment and GRIN lens has been chosen. Figs. 12 and 13 show the package design and the actual hardware. Coupled powers of 0.200 mW into 100 μ fiber and 0.130 mW in 62.5 μ fiber are obtained at 100 MA current.

Burst mode receivers are essential parts of all passive star packet communications systems. The specifications for our most recent receiver design are shown in Table IV. This receiver is currently implemented in the low cost surface mount technology as shown in Fig. 14. Board size is 1.2″ × 3.15″; the PIN detector is mounted in an ADM package.

Development of passive star couplers has been motivated by needs to minimize insertion loss, minimize port-to-port variations in loss while maintaining low manufacturing costs. The biconic fused coupler [26] meets only the first two goals well. We have chosen to pursue the development of the mixing rod approach following the earlier developments of others. To date, seven port and 19 port stars have been fabricated with losses for the seven port star of 16.5 dB $\pm$ 1.5 dB and for the 19 port star, 19.5 $\pm$ 2.5 dB. These losses include bulkhead connector loss at the input and output of star. The seven port is not optimized for lowest loss. The 19 port star is optimized by chemically etching the fiber cladding to reduce packing fraction loss. Specifications for the 19 port star are shown in Table V.

Further work is continuing to reduce the loss and improve port-to-port uniformity. The advantage of the mixing fiber approach is that only two straightforward and easily automated manufacturing technologies are needed: connector polishing and etching of silica.

A third approach to multimode coupler technology has recently been described [27]. This technology uses a monolithic glass substrate with multimode waveguides fabricated by ion exchange. For the 1 × 8 coupler reported, excess loss $\leq$2.5 dB, port-to-port uniformity = $\pm$0.75 dB.

V. INSTALLATIONS OF PASSIVE STAR LAN'S

Installations of passive star Ethernet systems have been carried out since 1983—this network system currently dominates over all other types of fiber optic LAN's.

These systems are installed to provide assured network availability, to provide network security, to provide a communication backbone suitable for migration to higher data rates, and to provide a communication medium free from EMI problems. Three installations will illustrate the application of passive star fiber optic LAN's in office and factory settings.

The largest passive star Ethernet LAN is located at the 44 story Southwestern Bell facility in St. Louis, MI [28]. This system contains 540 nodes, 92 miles of fiber, and supports over 4000 computer terminals and workstations.

The architecture uses a three-layer star hierarchy with three stars on the 22nd floor fiber distribution center, eight port stars serving groups of five floors (two spare ports), and each floor served by two 16 port stars. The fiber optic system was installed for the reasons mentioned above and, in addition, because it made the best use of the limited under floor duct space.

Two other installations illustrate the use of passive fiber systems in factory LAN's The layout of the first is shown in Fig. 15. This grinding compounds manufacturer had initially attempted to network equipment in six buildings by threading an Ethernet coaxial cable through them. The exposed portions of this network were very susceptible to electrical storms. In several instances, equipment damage resulted from lightening strikes. The fiber optic implementation shown in Fig. 11 solved this problem.

Coaxial Ethernet segments were retained in three of the buildings connected to a central fiber optic star coupler through repeaters. Individual computers in the remaining buildings connect directly to the star coupler.

A second factory LAN application is illustrated by Fig. 16. This system was installed in an airplane manufacturing plant where noise problems had been encountered with a coaxial cable LAN. This, together with the unique requirement that some computers needed to physically move over distances beyond that afforded by coaxial cable, led to implementation of the hybrid coax/fiber network depicted.

Here a centralized VAX computer is connected to a number of Codenet coax-to-fiber repeaters through a

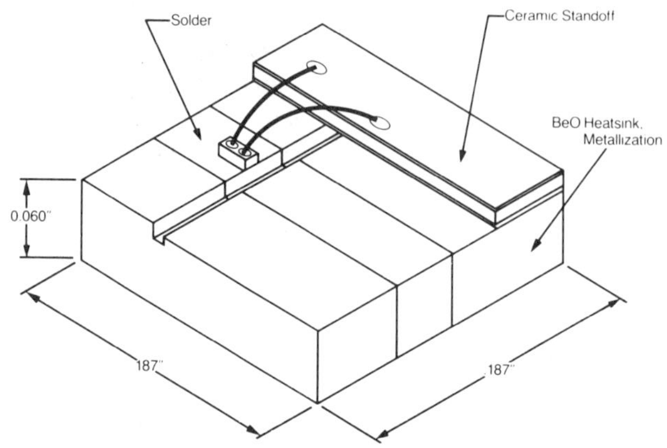

Fig. 11. Physical structure of LED packaged on heat sink. Step in ceramic heat sink is provided for fiber alignment.

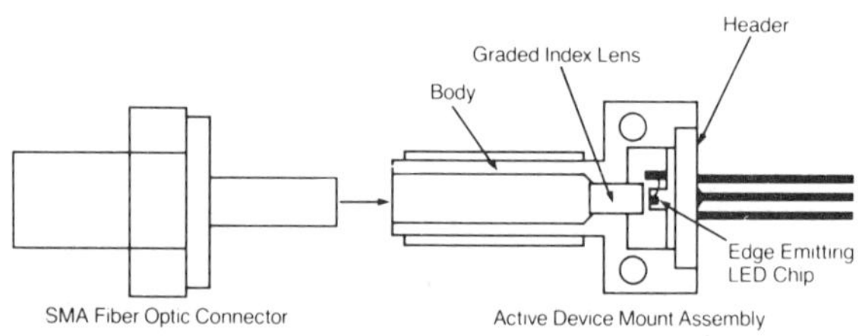

Fig. 12. Active device mount (ADM) design. This package makes use of compact disk laser header and graded index (GRIN) index lens.

Fig. 13. Active device mount located on optical transceiver card.

Fig. 14. Surface mount PC board technology provides low capacitance carrier for *Si* FET burst mode receiver.

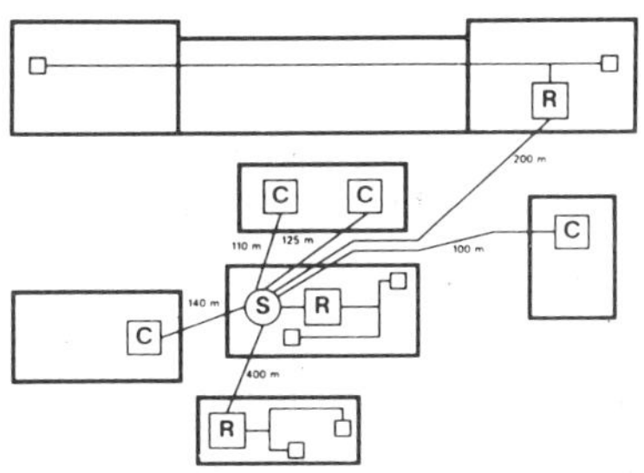

Fig. 15. Fiber optic network implementation for campus-like factory application.

coaxial cable LAN segment. Beyond that point, the network is fiber optic consisting of several passive star segments to which are attached the peripheral computers.

VI. Conclusion

Passive star-based LAN systems have been successfully installed since 1983. They provide an attractive alternative to wire systems for both CSMA/CD and token bus LAN's. In the future, we can expect to see increasing use of integrated circuits and ultimately OEIC's to reduce the manufacturing costs of the optical transceivers. We expect to see fiber optic transceivers integrated directly

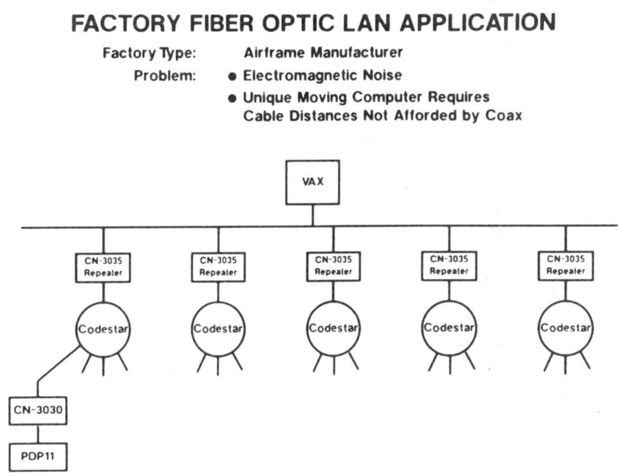

Fig. 16. Fiber optic network implementation for airframe manufacturing application.

into work stations much as RS 232 modems are a standard accessory at present.

Introduction in the future of international standards covering passive star systems will make these systems more accessible to users and installers of LAN's. Application driving these installations will be: backbone

LAN's, individual work station connections where duct space or EMI is an issue, and factory LAN's of all types.

ACKNOWLEDGMENT

Major contributions to the research and development reported herein were made by S. Anderson, W. Hatfield, B. Dutt, J. Helbers, and H. Jayawickrama.

REFERENCES

[1] J. R. Jones, J. S. Kennedy, and F. W. Scholl, "A prototype CSMA/CD local network using fiber optics," presented at Local Area Networks '82, Los Angeles, CA, Sept. 1982.
[2] J. R. Biard, "Optoelectronic aspects of avionic systems," AFAL-TR-73-164, June 1973.
[3] E. G. Rawson and R. M. Metcalf, "Fibernet: Multimode optical fiber for local computer networks," *IEEE Trans. Commun.*, vol. COM-26, pp. 983–990, July, 1978.
[4] S. Moustakas and H. H. Witte, "Passive fiber optic star bus for local area networks: System design and performance," *Siemans Forsch.*, vol. 12, pp. 182–187, 1983.
[5] E. E. Bergmann, "Fiber optics as a physical layer for MAP," Presented at IEEE 802.4H Stand. Meet., Irvine, CA, Feb. 17, 1986.
[6] K. Ozawa, E. Doi, and N. Yokota, "CSMA/CD fiber optic LAN," Presented at IEEE 802.8B Stand. Meet., Vancouver, B.C., July 1987.
[7] W. B. Hatfield, F. W. Scholl, M. H. Coden, R. W. Kilgore, and J. H. Helbers, "Fiber optic Ethernet in the factory automation environment," presented at SPIE O-E/Fiber '87, San Diego, CA, Aug. 1987.
[8] C. H. DeGennaro, "1553 compatible multiple speed fiber optic data bus," Presented at SAE-9C Committee, San Diego, CA, Apr. 13, 1987.
[9] M. H. Coden, W. B. Hatfield, and R. W. Kilgore, "Proposal for passive fiber optic star configured LAN standard for IEEE 802.3 CSMA/CD," Dec. 23, 1986.
[10] *Office Products Analyst*, vol. 11, no. 4, New York, Apr. 1987.
[11] "Joint proposal for active star configured fiber optic LAN standard," Presented to IEEE 802.8A Committee, July 3, 1987.
[12] "Fiber optic CSMA/CD ring network architecture," Presented to IEEE 802.8A Committee, Dec. 22, 1986.
[13] J. Hutchinson and D. Knudson, "Developing standards for a fiber optic LAN–FDDI," *SPIE*, vol. 715, pp. 72–79, 1986.
[14] F. W. Scholl, "Method and apparatus for detecting the collision of data packets," U.S. Pat. 4 560 984, Dec. 24, 1985.
[15] K. Oguchi and Y. Hakamada, "New collision detection techniques and its performance," *Electron. Lett.*, vol. 20, pp. 1062–1063, Dec. 1984.
[16] N. Tokura, Y. Hakamada, and K. Oguchi, "Burst signal receiving apparatus," U.S. Pat. 4 462 582, Dec. 31, 1985.
[17] F. W. Scholl and M. H. Coden, "Data receiver," U.S. Pat 4 561 091, Dec. 24, 1985.
[18] J. W. Reedy and J. R. Jones, "Methods of collision detection in fiber optic CSMA/CD Networks," Advances in Local Area Networks, pp. 264–276, 1987.
[19] J. Helbers, "General requirements for collision detection using large pulse technique," Codenoll Technol. Corp., unpublished memorandum, Mar. 1986.
[20] F. W. Scholl, "Passive star Ethernet systems," Presented at 802.8A Committee, Chicago, IL, June 2, 1987.
[21] "Medium attachment unit and baseband medium specification for a vendor independent fiber optic inter repeater link," IEEE 802.3, draft H., Aug. 1987.
[22] T. Sammons, S. C. T. Schnetlagy, and J. Head, "An inside look at IBM's LAN," *PC Mag.*, vol. 136, Feb. 5, 1985.
[23] "Token passing bus access method and physical layer specification," IEEE 802.4. draft H., Aug. 1987.
[24] G. Karlin and C. Tucker, "ARCNET on fiber: A viable factory automation LAN," presented at SPIE O-E/Fibers '87, San Diego, CA, Aug. 16, 1987.
[25] A. J. Malinger, "The ARCNET local area network design and implementation of a local area network," presented at FOC/LAN 83, Atlantic City, NJ, Oct. 10, 1983.
[26] E. G. Rawson and A. B. Nafarrate, "Star couplers using fused biconically tapered multimode fibers," *Electron. Lett.*, vol. 14, p. 274, 1978.
[27] E. Paillard, "Recent developments in integrated on moldable glass," presented at Fiber Opt. '87 Conf., London, England, Apr. 29, 1987.
[28] D. E. Stein, "High-rise optical communications networks," presented at FOCLAN '85, San Francisco, CA, Sept. 18, 1985.
[29] N. Yokota, "Proposal for IEEE 802.3 AUI Compatible CSMA/CD fiber optic passive star local area network," Presented to IEEE FOSTAR group, Irvine, CA, Jan. 14, 1988.

Frederick W. Scholl received the Ph.D. degree in 1974 from Cornell University, Ithica, NY, where he developed and characterized new ternary nonlinear optical materials.

From 1974 to 1977, with Rockwell International Science Center, he developed new GaAsSb avalanche photodiodes and AlGaAs photodiodes for study of ps phenomenon. From 1977 to 1979 at Optical Information System Division of Exxon Enterprises, he was responsible for semiconductor laser research and production. Following a year at Columbia University, in 1980, he joined Mr. Coden in founding Codenoll. In addition to his responsibilities as Senior Vice President and Director of Codenoll, he has been Adjunct Associate Professor at Polytechnic Institute of New York, Brooklyn. He also has 18 publications and eight patents.

Michael H. Coden received the Bachelor's degree in electrical engineering from the Massachusetts Institute of Technology, Cambridge, the Master's degree in business administration from Columbia University, New York, NY, and the Master's degree in mathematics from the Courant Institute of Mathematical Sciences at New York University, NY.

He has taught Computer Science at Columbia University Graduate School of Business (1974–1975) and Semiconductor Physics at Fairleigh-Dickinson University, Department of Electrical Engineering (1979).

He has been Chairman of the Board, President and Chief Executive Officer of Codenoll since its incorporation in 1980. From 1975 to 1979, he had been employed by Exxon Enterprises, Inc. From 1977 to 1979, he was the Manager of Exxon Enterprises Optical Information Systems, a manufacturer of semiconductor laser and related products, and from 1975 to 1977, he was Program Manager with responsibility for evaluating venture capital investments and providing management assistance. Prior to 1975, he held management positions at Digital Equipment Corporation and Hewlett-Packard Company, where he managed the design and marketing of computer hardware and software and was an Officer of Maher Terminals, Inc., where he designed and implemented on-line computer control of all operational and financial functions of the corporation.

Mr. Coden has received an Award of Merit from the American Chemical Society and The Laser Institute of America. He has presented more than 30 papers (20 of which have been published) in computer science, local area networks, fiber optics, electro-optical technology, and high technology management. He has obtained three patents. He is a member of Beta Gamma Sigma Business Honorary Society and professional societies. He is also listed in American Men and Women of Science, Who's Who in Business and Finance, International Business Mens Who's Who, and Men of Achievement.

A Variation on CSMA/CD That Yields Movable TDM Slots in Integrated Voice/Data Local Networks

By N. F. MAXEMCHUK

(Manuscript received January 6, 1982)

A variation on the carrier-sense, multiple-access/collision-detection (CSMA/CD) protocol for local-area, random-access broadcast networks is presented. The new protocol supports any mix of voice and data traffic, and has an upper bound on the delay of periodic traffic. In conventional CSMA/CD systems, statistical fluctuations in the transmission delay may cause disruptions in reconstructed voice waveforms. This does not occur with the new protocol. With this protocol, data sources use conventional CSMA/CD techniques to access the channel, while voice sources use a subset of these protocols, and appear to acquire a time-division multiplexed (TDM) slot. As in TDM systems, the entire system capacity can be used by the periodic sources. Unlike TDM systems, periodic slots are repositioned when other transmissions interfere with them, timing discrepancies between sources can be tolerated, and a centralized controller is not required to assign slots. Simulations are used to compare the delays in this system with those in conventional CSMA/CD systems. It is shown that at link utilizations of 0.9, voice sources can operate in a time-assignment, speech-interpolation mode. The effect of periodic traffic on conventional CSMA/CD protocols is also determined.

I. INTRODUCTION

There are basic differences between the requirements for data and voice transmission. There are also differences between the capabilities of local networks and the functions that should be performed on these networks, and general global networks. A protocol is described for integrating data and voice in a local-area, random-access broadcast network. It enables a single system, with similar interfaces for voice and data, to be used for both. The protocol satisfies the transmission requirements of both media, and takes advantage of the characteristics of local networks. This protocol can be applied to any system with periodic and aperiodic sources, as long as all of the periodic sources have the same transmission requirements.

In general, transmission capacity is less expensive in a local environment than in the global environment. This has resulted in significant differences in the types of networks implemented for data transmission in the two situations. In global networks, expensive access and switching techniques have been used to minimize the network capacity required to effect communications, whereas in local networks simpler switching techniques and less costly access nodes have been used. Several global networks, which use store-and-forward and packet-switching techniques to reduce the transmission requirements, are described in Chapters 1 and 2 of Ref. 1. A survey of techniques used in local networks, and an extensive bibliography is given in Ref. 2. Many of these local networks are configured as loops or random-access channels. Switching is simplified, but high-bandwidth communications channels are required.

The proximity of nodes in local networks leads to smaller propagation delays than those that occur in global networks. This has resulted in the development of more efficient random-access broadcast techniques for local networks[3,4] than have been developed for global networks (see Ref. 5 for a survey of random-access techniques and additional references).

Most proposals for integrated digital voice and data networks have applied to global networks. A survey of this work and an extensive bibliography can be found in Ref. 6. These proposals use variable-rate speech-coding techniques,[7,8] packet-switching techniques,[9,10] or complex strategies for combining circuit and packet switching.[11] These techniques are expensive to implement and should not be considered for local area communications. As in local data networks, integrated networks should take into account the reduced transmission costs and propagation delays in a local environment. In addition, only simple speech-coding techniques should be considered.

The requirements for digital voice transmission, both in terms of capacity and delay, are significantly different from those of data. The distribution of data traffic on local networks is typically bimodal,[12] comprising short interactive messages and long file transfers. The traffic from these sources arrives sporadically. If the message is divided into packets, a variance in the packet delay can be tolerated, providing the entire message delay is not excessive.

Uncompressed, digitized telephone calls require a large number of bits to be transferred, with much more stringent delay requirements than data sources have. Using a 32-kb/s speech coder, and transmitting only during active speech intervals, over 4-½ Mb of data must be transmitted during a 3-minute telephone call. The data cannot be accumulated over the entire call and then transmitted as a large file transfer between computers, because the participants interact. The maximum delay allowed in current telephone connections is on the order of a few hundred milliseconds. However, in a local network, the maximum delay must be significantly less than this, since the connection may also use an outside facility.

In a packet voice system, overhead bits must be transmitted in addition to the information from the digitized voice source. The more voice samples included in this packet, the higher the ratio of information to total bits in the packet, and the higher the transmission efficiency of the channel. However, the more voice samples included in a packet, the greater the delay between the time a sample is generated and the time it is delivered to the receiving telephone. As a compromise, packets consisting of several tens of milliseconds of speech are used in the systems described in this paper.

The variance of the delay in digital speech systems must also be constrained. The digital-to-analog converter at the receiver uses samples at a fixed rate. If a packet of samples is delayed to the extent that the previously transmitted samples are completely used up before it arrives, a discontinuity occurs in the speech. The probability of this occurring can be reduced by delaying the first packet of voice samples that arrives at the receiver and buffering future packets until they are needed. If the maximum delay is not constrained, this technique can reduce, but not eliminate, the problem. This delay adds to the overall delay between the speaker and the listener, and must be kept small.

In the global networks referenced earlier, packets of voice that do not arrive in time are lost. It is argued that if a small percentage of voice packets are discarded at random, the resultant distortion is tolerable. This same argument is used in proposals to integrate voice and data on local networks.[13,14] It may be valid to assume that voice packets are lost at random when aperiodic data sources generate most of the network requirements, or when the periodicity of speech sources is reduced by variable-rate speech-coding techniques. However, this assumption will probably not hold in a local network.

Based upon the measured data utilization of local networks[12] and the anticipated transmission requirements of digital voice, voice packets will most likely dominate integrated local networks. Since voice packets are generated periodically, if packets from voice sources collide, they are likely to continue colliding on successive transmissions. Therefore, successive delays from the same source will be correlated. Voice sources that do not contend with other sources may have a small average and variance of delay, while those that contend with a large number of voice sources may have a large average and variance of delay. If networks are designed based upon an acceptable average level of lost packets, and these packets are concentrated among a small number of connections during a small period of time, rather than being distributed randomly, the resultant service may not be acceptable.

Instead of reducing the periodicities to achieve fair packet losses, the periodicities can be used to eliminate lost packets entirely. This is accomplished by a variation on the carrier-sense, multiple-access/collision-detection (CSMA/CD) transmission protocol defined in Section II. This protocol transmits data by conventional CSMA/CD techniques, but uses only a subset of these techniques to transmit voice. The periodic sources do not detect collisions. In addition, periodic packets are tailored as described in Section III, the length of data packets is constrained, and periodic packets are given a higher retransmission priority. In Section IV, we show that this protocol limits the delay of voice packets to one data packet transmission time.

Periodic sources using this protocol operate as if a time-division multiplexed (TDM) channel has been assigned to each source. The difference between this channel and standard TDM channels is that it is not locked solidly into a time slot. The time slot may be shifted slightly back in time. When this occurs, the voice samples that arrive during the shift are transmitted in an expanded information area. An interesting characteristic of this system is that a periodic source can gain access to a system that does not appear to have the capacity to handle another periodic source. The system does not fail, but begins to operate as a completely utilized TDM system, with a slightly longer slot period. This phenomenon is explained in Section V. Another result of the time slot mobility is that timing discrepancies can exist between periodic sources without time slots being overwritten. This type of operation is described in Section VI.

In this system, if a periodic source and an aperiodic source are waiting to use a busy channel, the periodic source has a higher access priority and acquires the channel first. Typically, in queueing systems, when the delay of one class of users is reduced by raising its priority, the delay of the remaining users must increase. A simulator, which is described in Section VII, has been written to study the delays in this type of system. The model for the sources in the simulations is explained in Section VIII. The delays of periodic and aperiodic sources in this system are compared with the delays incurred when all sources use the same access protocol. These results are reported in Section IX. In Section X, alternative protocols are described.

II. TRANSMISSION PROTOCOL

In this protocol, all of the packets from an aperiodic source, and the first packet from a periodic source that is beginning to transmit, use a CSMA/CD protocol. Before transmitting, the source listens to the channel and refrains from transmitting if the channel is busy. While transmitting, the source also listens to the channel and stops transmitting if a collision with another source is detected. If the channel is busy, or if a collision occurs, the source tries again after the channel becomes idle. This protocol is not dependent upon the specific retry strategy these sources use to resolve contention. Any of the conventional strategies will work.

After the first transmission, a periodic source transmits all of the data it has accumulated whenever it acquires the channel. The source schedules its next transmission attempt at a fixed time T_p after the last successful transmission. In the first description of the transmission protocol, it will be assumed that T_p is the same for all sources. The effect of timing errors will be investigated in Section VI.

A periodic source listens before transmitting and defers transmission rights to terminals that are currently transmitting. However, this source does not listen while transmitting, and never terminates transmission prematurely. Instead, the packet structure for a periodic source is designed to allow aperiodic sources to detect a collision and terminate transmission before the periodic source begins transmitting useful data. When busy channels are encountered, the periodic source begins transmitting as soon as the channel becomes idle. This protocol, in conjunction with a constraint on the packet size from aperiodic sources, prevents packets from periodic sources from colliding.

As periodic traffic enters a system, the system becomes a TDM system. A periodic source acquires the channel and periodically uses a slot starting at this time until it is delayed by aperiodic traffic, or another periodic source starting to transmit. At this time, the periodic slot scheduled for the periodic source is shifted slightly. Additional data are transmitted in the first delayed slot to compensate for the shift.

III. PACKET STRUCTURE

The packets from aperiodic sources, Fig. 1a, consist of overhead bits and data bits. The overhead bits contain synchronization bits, source and destination addresses, packet-length counts, packet sequence numbers, error-control bits, and any other functions required by the communications protocols. These are variable-length packets. They are constrained to be less than or equal in length to the packets from the periodic sources.

Packets from periodic sources, Fig. 1b, consist of preempt bits, overhead bits, data bits, and overflow bits. During the preempt interval, the periodic source places a signal on the transmission media but

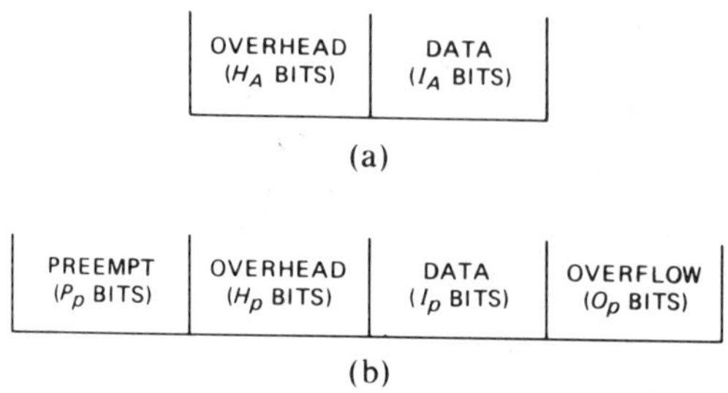

Fig. 1—(a) Packets from aperiodic sources. (b) Packets from periodic sources.

does not send information. This interval is long enough for an aperiodic source to detect a collision, stop transmitting, and have the effects of the transmission removed from the system before the periodic source begins transmitting useful data.

The length of the preempt interval, τ_p, is

$$\tau_p = 2\tau_t + \tau_{on} + \tau_I + \tau_{off},$$

where

τ_t is the maximum one-way propagation delay in the medium,

τ_{on} is the time required for a signal to be detected once it has propagated to a position on the channel,

τ_I is the time before the hardware reacts to a collision,

and

τ_{off} is time for a signal that is turned off to stop affecting a receiver, after the end of signal has propagated to the receiver.

Times τ_{on} and τ_{off} take into account electronics delays in the receivers and distortion in the received waveform caused by finite bandwidth channels which may not be correctly terminated. The preempt interval in a 3-Mb, 1-km simulated system is 38 bits long.

The overhead segment for periodic packets will be smaller than the overhead segment for aperiodic packets. An error-control strategy may not be used for periodic sources, since retransmitted packets will probably arrive too late to be used, and a greater error rate can be tolerated in sampled voice than in data. Packets will not arrive out of sequence; therefore, sequence numbers are not necessary. In addition, the number of bits of data in packets from these sources is deterministic, so that the length field is not required. In the simulations, H_A = 100 bits and H_p = 40 bits.

When a periodic source acquires the channel, it transmits all the data it has accumulated. The source schedules its next transmission T_p seconds after a successful transmission. If the channel is not busy at this time, the samples it has collected fit in the data area. If the channel is busy, it waits before transmitting. The samples that arrive during this time are transmitted in the overflow area. Even when there are no overflow samples to be transmitted, the source transmits carrier during the overflow time. This guarantees that a periodic source takes no more time to transmit when it is delayed than when it acquires the channel immediately.

The size of the overflow area is determined by the maximum delay a periodic source can experience. This transmission protocol guarantees that the maximum delay for a periodic source will not exceed a packet transmission time. In the system simulated, the periodic sources generate 8000 4-bit samples per second, and T_p is 30 ms. A maximum of four samples arrive during a packet transmission interval; therefore, the overflow area is 16 bits.

The first packet from a periodic source may be shorter than successive packets since it does not include a preempt interval or an overhead interval. However, the same packet size will be maintained for this packet as for all other periodic packets. This guarantees that the scheduled interval between the next packet from this source and a packet from another periodic source is at least a periodic packet transmission time, X_p.

IV. CHANNEL CONTENTION

Delay is incurred in CSMA/CD networks when the channel is busy and when transmitting sources collide. Systems with periodic and aperiodic sources are being considered. Therefore, a periodic source can be delayed because

(*i*) the channel is busy transmitting an aperiodic packet,
(*ii*) a collision with an aperiodic source occurs,
(*iii*) the channel is busy transmitting a periodic packet, or
(*iv*) a collision with a periodic source occurs.

In this section, it is shown that only the first and third of these four mechanisms delay periodic sources, and that the maximum delay is less than X_p. It is assumed that every periodic source has the same T_p. The effects of timing inaccuracies are investigated in Section VI.

Whenever a periodic source and an aperiodic source collide, the aperiodic source detects the collision and stops transmitting before the periodic source begins transmitting useful data. Therefore, a periodic source is not delayed by a collision with an aperiodic source.

Whenever a periodic source and an aperiodic source are waiting for an idle channel, the periodic source obtains the channel. The periodic source begins transmitting as soon as the channel becomes idle. If an aperiodic source waits, it detects a busy channel and does not transmit. If an aperiodic source begins transmitting immediately, it detects a collision and stops transmitting. Therefore, a periodic source can only be delayed by an aperiodic source whose transmission is already in progress. This delay is at most the packet transmission time for an aperiodic source, X_a, which is constrained to be less than X_p.

Consider a sequence of k periodic sources scheduled to transmit at times

$$t_{1,1},\ t_{2,2},\ \cdots,\ t_{k,k}.$$

Let

$t_{i,j}$ be the time the transmission from source i is scheduled to appear at the location of source j ($t_{i,j} - t_{i,i}$ is the propagation delay from the ith to the jth source),

$t'_{i,j}$ be the time the transmission from source i actually appears at location j (this takes into account the channel acquisition delay for the ith source),

and,

$D_i = t'_{i,j} - t_{i,j}$ be the delay of the ith source. (Since the propagation time is independent of the delay, the delay between the scheduled and actual arrival is the same for each destination j, and the second subscript is not needed.)

The transmission from each periodic source lasts X_p, and each periodic source schedules its next transmission T_p after its last successful transmission. Therefore,

$$t_{i+1,i+1} - t_{i,i+1} \geq X_p.$$

If $D_i = 0$, the ith periodic source does not delay the $(i + 1)$th periodic source, and, as long as

$$D_i < X_p$$

these two sources will not collide.

Periodic sources cannot be delayed by periodic sources which have not been delayed. Therefore, the first periodic source to be delayed must be delayed by an aperiodic source. The delay incurred by the first periodic source is less than X_p. The effect of this delay may propagate and effect a sequence of periodic sources. In a general sequence of periodic sources, if

$$D_i < X_p,$$

then,

$$t'_{i,i+1} < t_{i+1,i+1},$$

and, the ith and $(i + 1)$th do not collide. The transmission time required by the ith source is X_p, even though it is delayed, and must transmit more samples. If

$$t'_{i,i+1} + X_p \leq t_{i+1,i+1},$$

then the $(i + 1)$th is not delayed by the ith periodic source. This source may be delayed by an aperiodic source, and start a new sequence of delayed periodic sources, but the delay it incurs will be less than X_p. If

$$t'_{i,i+1} + X_p > t_{i+1,i+1},$$

the delay incurred by the $(i + 1)$th source is

$$D_{i+1} = t'_{i,i+1} + X_p - t_{i+1,i+1}.$$

Since the $(i + 1)$th source is waiting for the channel, this delay cannot be increased by an aperiodic source. The delay can be written as

$$D_{i+1} = D_i + t_{i,i+1} - t_{i+1,i+1} + X_p.$$

Since

$$t_{i,i+1} - t_{i+1,i+1} + X_p \leq 0,$$

$$D_{i+1} \leq D_i.$$

Therefore, the delay incurred by a sequence of periodic sources is a nonincreasing function, the maximum delay incurred by a periodic source is less than X_p, and periodic sources do not collide.

V. OVERLOAD TRAFFIC

Consider a system operating in a mode in which the channel capacity is nearly completely used by periodic sources. Assume that a time gap remains that is large enough for another source to begin transmitting, but not large enough to transmit an entire packet. Let another periodic source acquire the channel at this time. Is a failure mode created? Do delays build up until periodic sources start colliding? Is data lost? No. Instead, the system begins to operate without time gaps. The period between channel acquisitions increases, and some or all of the overflow bits in every packet are always used. Whenever a periodic source can acquire the channel, its ability to transmit is guaranteed.

Let a periodic source begin transmitting in a small time gap, such as that described in the hypothetical example. The periodic source it delays is delayed less than X_p. This source transmits the data accumulated during the delay in its overflow area and schedules its next transmission T_p seconds after it successfully acquires the channel. Successive periodic sources are delayed by an amount less than or equal to the delay incurred by the preceding source, as shown in the previous section.

The original interfering source becomes just another source in the sequence of interfering sources. It can be delayed by no more than the delay it originally caused, and can delay the source following it by no more than it did originally. Since the delay is a nonincreasing function, and it cannot go to zero for the over-utilized channel, it must stabilize at some positive time, ϵ, which is the same for all sources. The delay, ϵ, is equal to X_p, minus the sum of the idle channel times for a period T_p before the overflow source entered the channel. Using the terminology from the preceding section:

$$\epsilon = X_p - \sum_{t_{i,i+1} \epsilon T_p} [t_{i+1,i+1} - (t_{i,i+1} + X_p)].$$

When the stable situation occurs, each periodic source transmits a packet every $T_p + \epsilon$ seconds. It transmits the samples that have arrived in this period of time in the data and overflow area. At the end of each transmission, there is a periodic source that has been waiting ϵ seconds. This source acquires the channel before an aperiodic source or the first packet from another periodic source can. Until one of the sources ends its transmission, and channel capacity becomes available, the system operates as a TDM system with a slot period of $T_p + \epsilon$ seconds. No data are lost, and the slot delays do not grow indefinitely.

VI. TIMING CONSIDERATIONS

In a sampled data communication system, it is necessary for the transmitter and receiver to be frequency locked, so that samples are used at the same rate at which they are generated. In broadcast networks, this can be achieved by sending a clock signal outside of the data band or by using a modulation rule with a clock component. The former technique provides accurate timing, but requires that one unit be responsible for inserting the clock on the system. In the latter technique, there is no centralized control, every transmitting unit is identical, but timing discrepancies may exist between the transmitters, particularly when very little data are being transmitted.

Timing discrepancies result in the periodic terminals having different estimates of the interpacket interval T_p. Let the interpacket interval for the ith periodic terminal $T_{i,p}$ be within ϵ of T_p, so that

$$|T_{i,p} - T_p| \leq \epsilon.$$

Let the ith and $(i + 1)$th periodic terminals transmit at

$$t_{i,i+1} = t,$$

and,

$$t_{i+1,i+1} = t + X_p,$$

so that there is no separation of the packets at the $(i + 1)$th source. The next packets from these terminals are scheduled at

$$t_{i,i+1} = t + T_{i,p},$$

and,

$$t_{i+1,i+1} = t + X_p + T_{i+1,p}.$$

These two times may be separated by as little as $X_p - 2\epsilon$. If the first packet in this sequence is delayed by a packet from an aperiodic source, it may be delayed until

$$t'_{i,i+1} < t + T_{i,p} + X_a.$$

With the constraint $X_a \leq X_p$, it is possible that both periodic sources will be waiting for the channel and will collide. This situation cannot be resolved by the protocol. It can be prevented by constraining the length of aperiodic packets to

$$X_a \leq X_p - 2\epsilon.$$

With this constraint, it can be shown, as in Section IV, that sequences of periodic sources do not collide, and that the delay of a periodic source does not exceed X_p.

This length constraint cannot be used to solve all of the problems that arise because of timing inconsistencies. The first packet from a periodic source must have length X_p to reserve an entire slot, but must enter the system like a periodic packet. This can be resolved by allowing periodic sources to recognize and preempt the first packet from a periodic source that begins transmitting within 2ϵ of the periodic source's scheduled transmission. Now, the maximum delay caused by the first packet from a periodic source is not greater than that caused by a packet from an aperiodic source.

VII. SIMULATION

A general-purpose simulator has been written to determine the characteristics of broadcast networks. In this simulator, the protocols used by sources, the traffic generated by a source, the physical position of a source on a channel, the parameters that define a transmission channel, and the interconnection of channels can be varied.

In the simulator, time jumps between the occurrence of significant events, rather than continuously moving forward by small increments. Functions such as wait for idle channel and listen while transmitting are implemented by giving the channel the characteristics of an active device. The channel notifies a source when the state of the channel, at the location of the source, has changed. Actually, the channel in this type of system is passive and sources that perform these functions examine the channel continuously. However, the interval of time in which the state of the channel can change and a significant event can occur is small compared to the average time between significant events. A simulator that models the continuous operation of the sources would have to examine the channel at small intervals, and would be much less efficient, in terms of processing, than the technique selected.

In this simulator, the channel maintains lists of sources that are transmitting, listening while transmitting, and listening for an idle channel. When the state of the channel changes, channel maintenance routines notify the terminals in the lists at the appropriate times. These routines take into account the propagation delays between terminals, and the detection time of terminals.

Queues of messages and packets waiting to be transmitted from the sources are maintained without creating physical queues. Instead, the clocks associated with packet and message generation are allowed to run independent of the simulation clock. Each source keeps track of the time the packet and message it is transmitting were generated. When the transmission is complete, this time is updated to determine the next packet or message arrival at this source. If this time is less than the simulation time, this object has been waiting in a queue, and can be transmitted immediately. If it is greater than the simulation time, the queue is empty, and the message or packet is scheduled to arrive in the future.

The periodic terminals approximate voice terminals that only transmit during active speech intervals. For the parameters selected in the simulations, the speech interval durations are more than three orders of magnitude greater than the packet transmission time, and more than six orders of magnitude greater than certain of the channel-related events. The simulator tracks the fine structure of channel-related events. Because of the different time scales, the simulations could not be conducted for a long enough time to guarantee that all combinations of active and inactive intervals occur with the proper probabilities. To compensate for this, a different random-number generator is used to generate active and silent intervals than is used to generate the other random events in the simulation. This allows different protocols for periodic sources to be compared when the same active and silent intervals occur.

The systems simulated have the following characteristics:

(*i*) Sources are uniformly distributed along a single cable.

(*ii*) Channel length is 1 km.

(*iii*) Transmission rate is 3 Mb/second.

(*iv*) Time to detect the presence and removal of carrier, τ_{on} and τ_{off}, is 1 μs.

(*v*) Time to detect interference, τ_I, is 4 μs.

(*vi*) Time the channel must remain idle between transmissions is one-third of a microsecond.

These parameters do not correspond to a particular hardware realization of a system, but are indicative of what might be expected.

VIII. SOURCE MODELS

Aperiodic and periodic terminals are simulated. They correspond to a number of data sources sharing a store and forward node, and a two-way voice conversation. Some of the parameters associated with these terminals are selected to study specific characteristics of the system or to facilitate programming, rather than to accurately model the source. The results of the simulations provide an indication of how the system operates and not precisely how many terminals of a specific type can be supported.

The model for the sources in this system is shown in Fig. 2. In the remainder of this paper, the packets from the various sources are referred to as follows:

Type A—all of the packets from aperiodic sources.

Type F—the first packets in active intervals from periodic sources.

Type P—all other packets from periodic sources.

Type X—a composite of all of the packets from periodic sources.

Type T—a composite of all of the packets from all sources.

The protocols for periodic sources are referred to as follows:

Π_A—Periodic sources use the protocol proposed in this paper.

Π_L—Periodic sources use the same protocol as data traffic and transmit enough overhead bits to have the same packet size as Π_A.

Π_S—Periodic sources use the same protocol as data traffic, but transmit fewer overhead bits than Π_L.

Π_X—Either Π_L or Π_S. This symbol is used when describing the intersection set of these two protocols.

The doublet (x, y) refers to packets of type x, when the periodic sources use protocol y.

In the model, all of the sources test the state of the transmission medium before transmitting (LBT). If the channel is currently being used (B), they wait until it is idle (WIC), then schedule a transmission retry according to a source-specific strategy (SRB). Transmission priorities can be established by using different strategies for different sources. For instance, if one type of source can retry sooner than another, these sources are more likely to find the channel idle and will have a smaller average delay. For this set of simulations, all of the sources begin transmitting as soon as the channel becomes idle. If the channel is not heavily used, there is seldom more than one source waiting, and this will lead to the smallest possible delay. If more than one source is waiting, they collide and the collision retry strategy determines which source obtains the channel.

Transmitting sources may or may not listen while transmitting. In this set of simulations, sources listen while transmitting type A, (F, Π_A), and (X, Π_X) packets. When a collision is detected, a source-

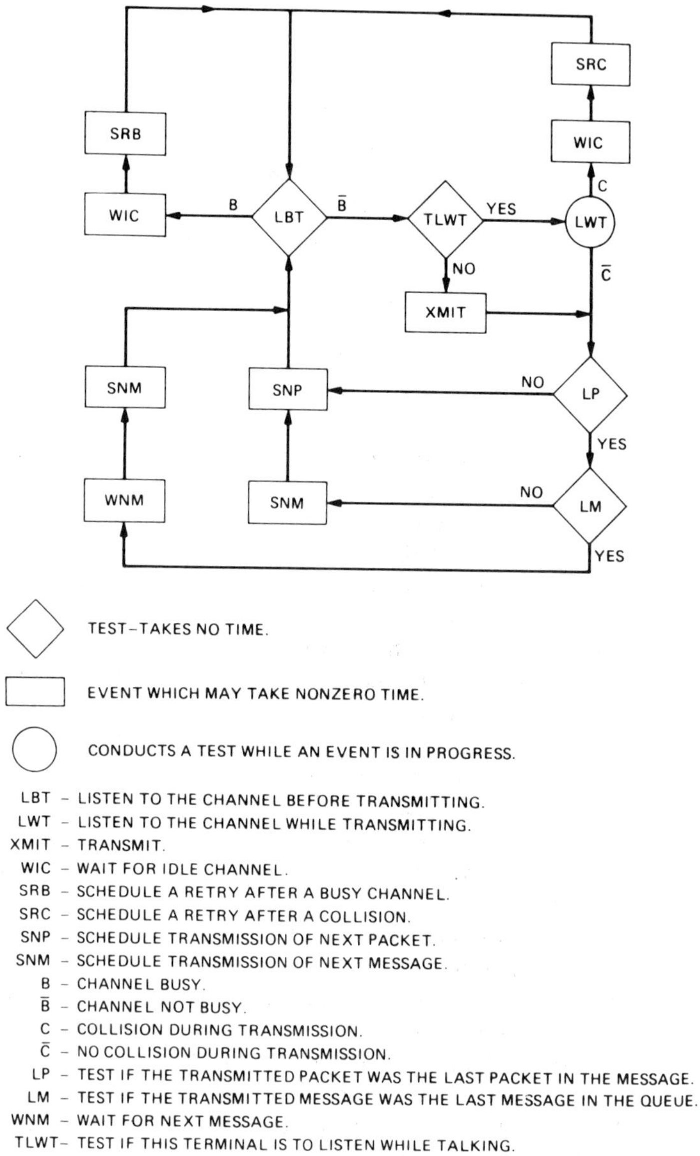

Fig. 2—Model of the transmission strategies used by the sources in the simulations.

dependent retry strategy (SRC) is initiated. The sources do not listen while transmitting type (P, Π_A) packets.

The algorithm, SRC, provides another opportunity to assign transmission priorities. Sources generating type A and (X, Π_X) packets distribute their retrys uniformly over an interval $2^i\Delta\tau$, where i is the number of collisions this packet has experienced, and $\Delta\tau$ is two round-trip transmissions delays. The value of i is limited to ten, so the retry interval varies from two to 1024 round-trip delays. Sources generating type (F, Π_A) packets distribute their retrys uniformly over an interval Δ_T. Under heavy-load conditions, these souces retry sooner than those generating type A packets and have a smaller average delay. The periodic sources model voice terminals that only transmit during active speech intervals. Type (F, Π_A) packets are given a higher priority because they must acquire the channel in less than 50 ms more than 98 percent of the time.[15,16] It is shown in Section IX that this requirement is met at channel utilizations of 0.9.

In this model, after a packet is successfully transmitted the source determines if this is the last packet in a message. If it is not, the next packet transmission is scheduled. If it is, the next message is scheduled. For the periodic sources, a message corresponds to an active interval during which packets are generated. After the last packet in an active interval is transmitted, there is never a message waiting. Instead, a silent interval is generated, and lasts until the next message arrives. Type (P, Π_X) packets are generated every T_p seconds after the first packet successfully acquires the channel. These packets are queued until they can be transmitted. Type (P, Π_A) packets are generated T_p seconds after the last successful channel acquisition.

The aperiodic sources generate variable-length messages. The messages arrive independently and are queued until they can be transmitted. The message length determines the number of packets in the message and the length of the last packet. The source transmits all of the packets in a message before determining if another message is waiting. The packet scheduling algorithm (SNP) waits before transmitting successive packets from this source. This enables other sources to acquire the channel.

The aperiodic sources in these simulations have the following characteristics:

(*i*) The message interarrival process is a negative exponential distribution with a mean of 33.47 ms.

(*ii*) The packet size is 1050 bits, 90 overhead bits, and 960 data bits.

(*iii*) The message length is deterministic—it is 960 bits. (Each message generates a single maximum-size packet.)

(*iv*) The minimum time between successive packets from the same source is four round-trip transmission delays.

These characteristics do not correspond to any known data source. They are selected to allow comparisons between systems with different mixes of periodic and aperiodic sources. The average number of bits transmitted by one of these sources is equal to the average number of bits transmitted by a periodic source. This allows the network utilization to remain constant, while the mixture of periodic and aperiodic sources is varied. From this, the effect of periodic rather than aperiodic requirements on a broadcast network is determined. Using these characteristics, type A packets are almost the same size as type (X, Π_A) packets. This causes the largest possible delay to be incurred by the periodic sources.

The periodic sources have the following characteristics:

(*i*) For all of the sources, T_p is set to 30 ms. This results in 960 data bits per packet.

(*ii*) The duration of silent and active intervals are exponentially distributed with means of 0.185 and 1.31 seconds, respectively.

(*iii*) For type (X, Π_A) packets, $H_p = 40$ bits, $O_p = 16$ bits, and $P_p = 38$ bits. This results in 1054-bit packets.

(*iv*) For type (X, Π_L) packets, $H_p = 94$ bits. This retains the same link utilization as aperiodic sources and periodic sources using the new protocol.

(*v*) For protocol Π_s, $H_p = 40$ bits. This shows the effect of increasing the link utilization in the new protocol.

The mean values of the distributions are selected so that the total active time, the average active interval, and the number of active intervals correspond to those in a two-way telephone conversation.[17] This model assumes that the two speakers in a conversation do not talk simultaneously. It ignores the 0.07 probability of double talk. This model is easier to implement than one which allows double talk and should provide a reasonable indication of the system delays.

IX. SIMULATION RESULTS

Simulations have been conducted for protocols Π_A, Π_s, and Π_L. The fraction of the requirements generated by periodic sources was set to 0, 10, 25, 50, 75, 90, and 100 percent of the total requirements and the nominal offered utilization, S, to 0.7, 0.8, and 0.9. For protocols Π_A and Π_L, the actual values of S are very close to the nominal values. For protocol Π_s, the utilization generated by aperiodic sources and the number of periodic packets is the same as for the other protocols. However, the overhead associated with periodic packets is reduced. Therefore, the actual value of S is less than the nominal value.

Statistics were taken for 10 seconds of elapsed time after a 2-second initialization period. As the utilization varied from 0.7 to 0.9, the total number of packets in the simulations varied from about 20,000 to 25,500. The time between successive starts of active intervals for the periodic traffic is on the order of a second or two. It is unlikely that all combinations of active and inactive periodic terminals are witnessed in a 10-second interval. However, the same active intervals occurred for each of the protocols, and the data show definite trends. The average and standard deviation of the channel access delay are plotted as a function of the percentage of the load generated by periodic sources in Figs. 3 and 4.

For Π_L, the periodic and aperiodic sources use the same transmission protocol and have the same packet lengths. For this protocol, the network utilization remains constant as the load shifts from aperiodic to periodic sources. All of the curves (T, Π_L) display the same characteristic. For a small fraction of periodic load, the delay measures increase. As the fraction of the load generated by periodic sources becomes large, the delay measures decrease.

This behavior can be accounted for by characteristics of periodic sources. As the periodic sources change between the active and inactive mode, the short-term utilization changes. The delay versus utilization curve for this type of system is concave upward so that the delay averaged over a number of utilizations is greater than the delay associated with the average utilization. This accounts for the initial increase in the curves. The periodic sources transmit packets at a multiple of T_p after the first successful transmission. This separates sources so that a large number of sources do not collide. Since it takes longer to resolve contention between larger numbers of sources, the average delay decreases as the number of sources that may be separated increases. This accounts for the decrease in the delay measures when the fraction of the load generated by periodic sources exceeds ten percent.

Curves (T, Π_A) show the delay characteristics for all of the packets in a system that uses the new protocol. The total utilization is the same as for Π_L. The delays for (T, Π_A) are less than those for (T, Π_L). Protocol Π_A guarantees that periodic sources do not collide with each other and that they successfully acquire the channel when they collide with aperiodic sources. Since there are fewer collisions, and fewer of the collisions that occur result in unsuccessful transmissions, less of the channel capacity is wasted. The decrease in delay is a result of the reduced-channel contention.

Comparing Π_A and Π_L demonstrates the favorable characteristics of the new protocol but not the unfavorable characteristics. Protocol Π_A requires additional bits to be transmitted in the preempt and overflow areas of the message. To show the effect of these bits, Π_s is examined. This protocol is the same as Π_L, except the overhead in periodic packets is reduced. As the fraction of the load generated by periodic sources increases, the average channel utilization decreases. Curves (T, Π_s) are below (T, Π_L), but are still above (T, Π_A). The latter result is due to the ratio of data bits to packet length in the system simulated. The preempt header is dependent upon the propagation delay in the system. As the channel becomes longer, or the transmission rate

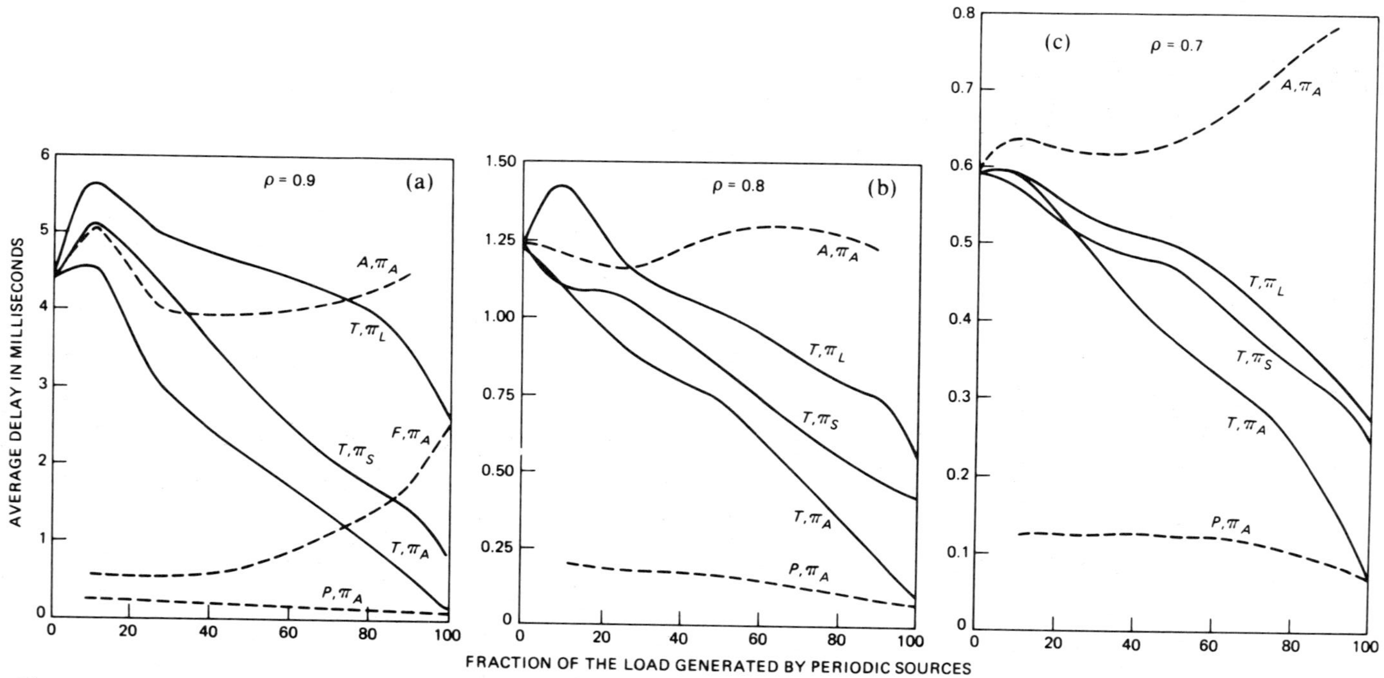

Fig. 3—Average delay versus the fraction of the load generated by periodic sources for several types of sources with protocols at the following utilizations: (a) $\rho = 0.9$; (b) $\rho = 0.8$; (c) $\rho = 0.7$.

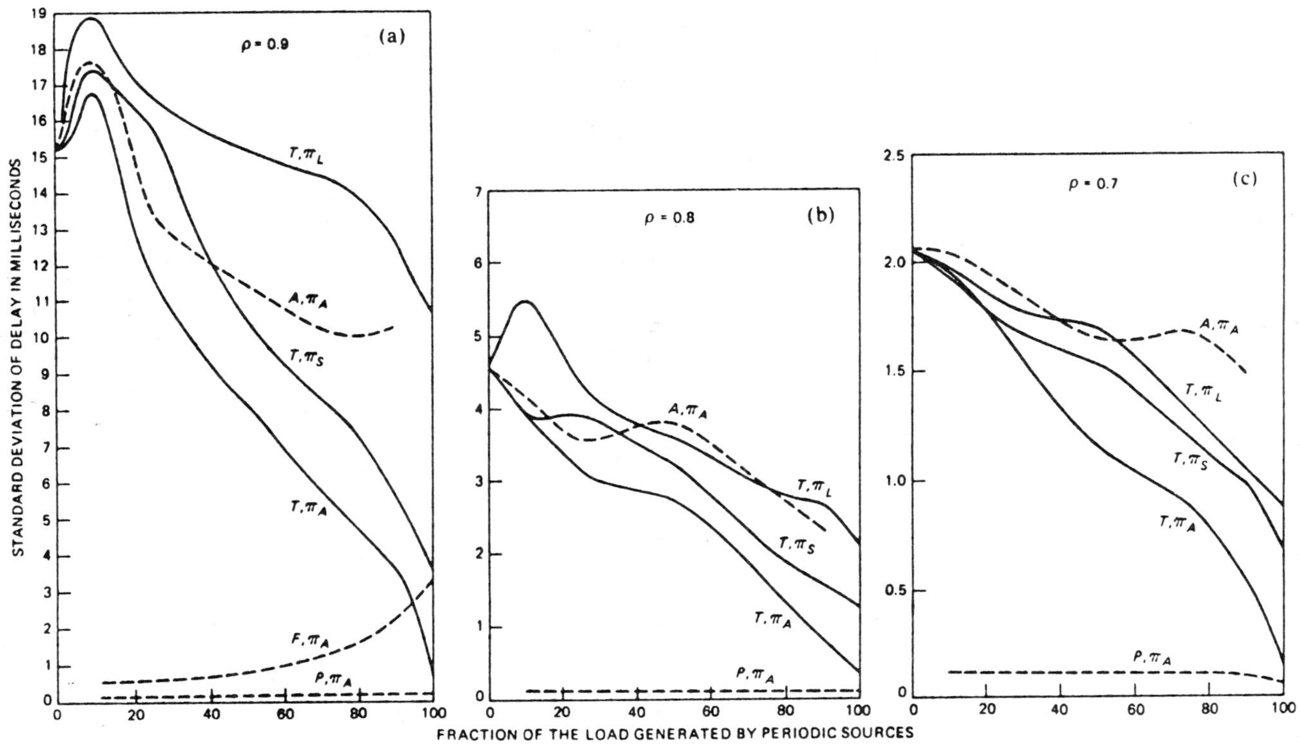

Fig. 4—Standard deviation of delay versus the fraction of the load generated by periodic sources for several types of sources with protocols at the following utilizations: (a) $\rho = 0.9$; (b) $\rho = 0.8$; (c) $\rho = 0.7$.

increases, this will comprise a larger fraction of the packet in Π_A, and the operation of Π_s will improve with respect to Π_A. A similar improvement will occur if T_p is decreased.

Protocol Π_A establishes an upper bound on the delay experienced by periodic traffic by giving this traffic a higher priority than aperiodic traffic. The dashed curves (A, Π_A) and (P, Π_A) represent the components of (T, Π_A) generated by aperiodic and periodic traffic, respectively. As expected, curves (A, Π_A) are above (T, Π_A), while (P, Π_A) are below. However, there are some interesting effects noted on these curves. As the fraction of the load generated by periodic sources increases, the number of sources that have a higher priority than the aperiodic sources increases. The delay experienced by the aperiodic sources is expected to increase. However, at utilizations of 0.8 or 0.9, there is not a significant increase in the average delay, and the standard deviation of the delay decreases.

From curves (T, Π_L), it can be concluded that it is definitely beneficial for aperiodic sources to share a channel with periodic sources. From curves (A, Π_A), it may even be beneficial for aperiodic sources to share a channel with higher priority periodic sources.

It has been stated that speech-activated systems should have a channel-acquisition delay less than 50 ms at least 98-percent of the time. From curves (A, Π_A), it should be possible to meet this constraint up to utilizations of 0.9, even if the first packet of speech is given the same transmission priority as aperiodic traffic, assuming that the 98-percent point corresponds to two standard deviations beyond the mean. In Π_A, the first packet from a periodic source is given a higher priority than packets from aperiodic sources. Using this strategy, it should be possible to lose no samples, not even those used to detect the presence of energy. For instance, if the transmitter stores the last 30 ms of speech samples, and it takes 10 ms to detect the presence of energy, no active samples are lost when the channel is acquired within 20 ms. At utilizations of 0.9, this appears to be well within the capability of the system.

X. ALTERNATIVE PROTOCOLS

Several other protocols for periodic traffic were considered. These protocols did not operate as well as the protocol described here and were not completely characterized. However, it is instructive to describe two of these protocols and their perceived shortcoming.

One possible approach to reduce the average and variance of the delay for periodic traffic is to give the periodic traffic a higher retransmission priority than aperiodic traffic. This approach is used in Π_A to reduce the delay for the first packet in a periodic sequence. Note in Figs. 3 and 4 that this strategy works best when the high-priority traffic is a very small fraction of the total traffic. Priority systems do not reduce contention between high-priority users. In a local network designed to handle data and voice requirements, the voice transmission requirements are expected to far exceed the data requirements. A priority system will not provide significant improvements in this type of an environment.

In addition, the priority system may reduce delay problems, but it does not eliminate them. The delay is still an unbounded statistical variable. The probability of exceeding a specified delay is nonzero, and packets will be lost. This probability may be acceptable for moderate utilizations, but as the utilization approaches one, it will increase and become significant. Because of the high transmission requirements of voice, it is expected that a voice and data system will have to operate at much higher utilizations than data only systems.

Another protocol, which was tried, gives periodic traffic absolute preemptive right to the network. The periodic packets have a preempt header. They begin transmitting without listening to the network and interrupt any terminals that are transmitting. The preempt header is long enough for the nonpreemptive terminals to detect a collision and stop transmitting before useful data are transmitted by the preemptive terminals. This protocol results in zero transmission delay for periodic packets once they have acquired a slot. They acquire the slot for the first packet by behaving like an aperiodic terminal, as in the recommended protocol.

This protocol requires exact timing synchronization between periodic terminals and does not operate well at high-channel utilizations. Since the periodic terminals do not listen to the channel before transmitting and transmit periodically at what they perceive to be T_p seconds, if there is a discrepancy between various terminals' estimates of T_p, they will eventually interfere with one another. In addition, when the utilization for periodic sources is slightly greater than 50 percent, it is possible for these sources to be spaced in such a way that no new sources can obtain a time slot.

When this protocol was simulated, it was found that when the channel utilization from periodic sources exceeded 0.7, the channel efficiency decreased rapidly and the access time for the first packets from periodic sources and packets from aperiodic sources became intolerably long. The channel efficiency is the fraction of time the channel is being used to transmit data that are successfully received, over the total time that data are being transmitted. The unsuccessfully transmitted data occurs when packets are interrupted by collisions. The propagation delay in local networks is small compared with the packet transmission time, and terminals listen before transmitting. Therefore, for the previously defined protocols, collisions only occur near the beginning of transmission. In this protocol, a packet can be interrupted at any time. As the fraction of time that preempt terminals use the system increases, the probability of interrupting a terminal that has completed a significant portion of its transmission increases and the channel efficiency decreases. As the channel efficiency decreases, the fraction of the channel capacity remaining for successful transmissions decreases and the delays increase. When the channel utilizations generated by periodic sources approached 0.7, the delay experienced by first periodic transmissions and aperiodic transmissions ranged from several tenths of a second to a second. This is an intolerable delay.

XI. CONCLUSIONS

When periodic and aperiodic data are transmitted on the same network, using conventional CSMA/CD protocols, the average and variance of the delay can be reduced by the periodicities in the data. By requiring periodic sources to begin generating periodic packets after they have successfully acquired the channel, the periodic requirements are separated in time, and collisions are reduced. This results in a decrease in the network delay. Channel contention is further reduced by requiring periodic sources to transmit all of the data they have acquired whenever they transmit and to schedule their next transmission a fixed time after their last successful channel acquisition.

If, in addition, periodic sources are only delayed by aperiodic sources with transmissions in progress, the size of packets from aperiodic sources is less than that from periodic sources, and the size of packets from periodic sources is not increased by the delay, then packets from periodic sources do not collide. The maximum channel acquisition delay experienced by periodic sources in this mode of operation is a packet transmission time.

By delaying the use of the first packet from a periodic source by a packet transmission time, the next packet is always received before all of the samples in previous packets are used. There are no discontinuities caused by the late arrival of samples in the reconstructed waveform, and packets of samples are never discarded because of excessive transmission delays.

In this mode of operation, periodic sources appear to have a dedicated TDM channel. There is a difference between the bandwidth allocation in this type of system and that of a true TDM system. In a TDM system, the time slots for a particular channel always carry the same number of bits and are fixed with respect to the beginning of a periodically recurring frame. In this system, the time slot for a particular channel may be moved back with respect to a hypothetical start of frame when the channel is busy. When this occurs, the number of bits of data being carried by the slot increases. In the extreme, when the channel is overutilized, the slot is shifted back by the same amount every time it occurs. The slot period and the amount of data in the slot increases.

The mobility of time slots, in conjunction with the access protocol, allows all of the excess capacity that is not being used to transmit periodic packets to be used by aperiodic sources, even if the excess capacity is dispersed in small intervals throughout the TDM frame. The system can shift between any mix of periodic and aperiodic traffic without a central controller reallocating the bandwidth.

In conventional TDM systems, all of the transmitters must be able to accurately determine the location of their assigned slots to avoid transmitting simultaneously with other sources. In this system, a periodic source is willing to wait for a communication in progress to be completed. Therefore, timing inaccuracies between periodic sources can be tolerated, as long as the inaccuracy is not long enough to cause two periodic sources to simultaneously initiate communications.

Finally, the simulation results indicate that periodic sources can acquire a channel quickly enough to operate in a time-assignment, speech-interpolation mode. Storage and delay in the system may be used to reduce, and possibly eliminate, clipping which normally occurs at the beginning of active intervals in this type of system.

REFERENCES

1. M. Schwartz, *Computer-Communication Network Design and Analysis*, New York: Prentice-Hall, 1977.
2. J. F. Hayes, "Local Distribution in Computer Communications," IEEE Communications Magazine, *19*, No. 2 (March 1981), pp. 5–14.
3. R. M. Metcalfe and D. R. Boggs, "Ethernet: Distributed Packet Switching for Local Computer Networks," Commun. of the ACM, *19*, No. 7 (July 1976), pp. 395–404.
4. R. C. Crane and E. A. Taft, "Practical Considerations in Ethernet Local Network Design," Hawaii Int. Conf. Syst. Sciences, (January 1980), pp. 166–75.
5. F. A. Tobagi, "Multiaccess Protocols in Packet Communication Systems," IEEE Trans. Commun., *COM-28*, No. 4 (April 1980), pp. 468–88.
6. T. Bailly, A. J. McLaughlin, and C. J. Weinstein, "Voice Communication in Integrated Digital Voice and Data Networks," IEEE Trans. Commun., *COM-28*, No. 9 (September 1980), pp. 1478–89.
7. T. Bailly, B. Gold, and S. Seneff, "A Technique for Adaptive Voice Flow Control in Integrated Packet Networks," IEEE Trans. Commun., *COM-28*, No. 3 (March 1980), pp. 325–33.
8. D. Cohen, "A Protocol for Packet-Switching Voice Communication," North-Holland Publishing Co., Computer Networks, *2*, No. 4–5, (September–October 1978), pp. 320–31.
9. G. J. Coviello, "Comparative Discussion of Circuit vs Packet Switched Voice," IEEE Trans. on Commun., *COM-27*, No. 8 (August 1979), pp. 1153–9.
10. O. A. Mowafi and W. J. Kelly, "Integrated Voice/Data Packet Switching Techniques for Future Military Networks," IEEE Trans. Commun., *COM-28*, No. 9 (September 1980), pp. 1655–62.
11. M. J. Fisher and T. C. Harris, "A Model for Evaluating the Performance of an Integrated Circuit- and Packet-Switched Multiplex Structure," IEEE Trans. Commun., *COM-24*, No. 2 (February 1976), pp. 195–202.
12. J. F. Shoch and J. A. Hupp, "Measured Performance of an Ethernet Local Network," Commun. of the ACM, *23*, No. 12 (December 1980), pp. 711–21.

13. J. F. Shoch, "Carrying Voice Traffic Through an Ethernet Local Network—A General Overview," IFIP WG 6.4 International Workshop on Local-Area Computer Networks, Zurich (August 1980), pp. 429–46.
14. D. H. Johnson and G. C. O'Leary, "A Local Access Network for Packetized Digital Voice Communication," IEEE Trans. Commun., *COM-29,* No. 5 (May 1981), pp. 679–88.
15. S. I. Campanella, "Digital Speech Interpolation," Comsat Tech. Rev., *6,* No. 1 (Spring 1976), pp. 127–58.
16. K. Bullington and I. M. Fraser, "Engineering Aspects of TASI," B.S.T.J., *38,* No. 2 (March 1959), pp. 353–64.
17. P. T. Brady, "A Technique for Investigating On-Off Patterns of Speech," B.S.T.J., *44,* No. 1 (January 1965), pp. 1–22.

Token Ring Local Area Networks: A Comparison of Experimental and Theoretical Performance *

J. Sventek, W. Greiman,
M. O'Dell ** and A. Jansen ***
Computer Science and Mathematics Department, Lawrence Berkeley Laboratory, Berkeley, CA 94720, USA

Many existing performance models for local area networks are based on the assumptions of random traffic patterns and infinite buffer capacity. All real network implementations violate the infinite capacity assumption, while an increasingly popular class of local area network applications violates the randomness assumption. This paper describes the experimental measurements of the characteristics of a commercially available token ring network. It is shown that the violation of both assumptions invalidates model predictions, and that the utility of all access control mechanisms for such applications must be re-evaluated. Finally, it is shown that congestion is a very real problem even in small "simple" local area networks. Schemes for controlling interface congestion may require rather novel approaches to the problem.

Keywords: Token Ring Local Area Networks, Performance Models, Infinity Capacity, Randomness, Access Control, Congestion, Interface Congestion

Joseph S. Sventek has been with the Computer Science Research Department at Lawrence Berkeley Laboratory since receiving his Ph.D. in 1979; since 1982 he has lead the Distributed System research group. His primary interests are in the areas of computer network performance and distributed application design.

* This work was supported by the Applied Mathematical Sciences Research Program of the Office of Energy Research of the U.S. Department of Energy under Contract DE-AC03-76SF00098.

** Current address, Group L Corporation, 481 Carlisle Drive, Herndon, VA 22070, USA.

*** Current address, EDP-Center, University of Oslo, Pb 1059, Blinden, Oslo 3, Norway.

1. Introduction

The design and analysis of local network protocols has largely concentrated on "link access" protocols like Ethernet [6], with analytic analysis and simulation providing powerful tools for exploring protocol behavior. The general procedure is to create an analytic model of the protocol under study, derive an approximate analytic solution, and then verify it by comparison with simulation studies [2], [4], [11]. These modeling efforts have shown that critical user-visible characteristics of a local network (e.g. stability and fairness)

William H. Greiman has been at Lawrence Berkeley Laboratory since 1969, the year he received his Ph.D. in theoretical physics from Iowa State University. For three years he did research in high energy physics. Then he moved to the Mathematics and Computing Department. From 1972 to 1976 he developed operating systems and communication front ends. Since 1976 he has been working as a Computer Scientist in the Computer Science Research Department. His research and interests have focused on computer systems performance evaluation and computer networks.

After receiving an M.S. in Computer Science from the University of Oklahoma in May, 1980. **Michael D. O'Dell** joined the Computer Science Research Department at Lawrence Berkeley Laboratory. He is currently a member of the engineering staff of Group L Corporation. His primary interests are in the areas of operating systems and computer networks.

After receiving an M.S. degree in Computer Science, **Arild Jansen** has held a variety of positions at the EDP Center, University of Oslo, Oslo, Norway. He is currently a manager of the EDP Center. His primary interests are in high level protocol standardization and the sociological impact of electronic inter-user communication.

depend strongly upon the access control mechanism used by the interconnection technology.

Most network protocol models, whether for analytic or simulation purposes, make several simplifying assumptions. Two of these assumptions are of particular interest to this discussion:

(1) Each receiver on the network is assumed to have infinite buffer capacity and no processing dead time, implying that a receiver can actually receive all messages successfully placed onto the medium. This permits the models to focus on the fairness and stability of the access control mechanism with respect to each transmitter's ability to place messages on the communication medium.
(2) The distribution of packet lengths is assumed to be exponential; the distribution of packet arrivals is assumed to be Poisson, which results if there is no correlation between the packets. This absence of correlation is essential to permit closed form analytic solutions for the characteristics of a network.

In any modeling effort, it is necessary to validate the models by comparison with the performance of actual networks. In particular, it is necessary to show that either the model assumptions hold true, or that the network performance is independent of the assumptions for the application regime which dominates the use of the communication medium. The measurements described in this paper attempt of compare the model predictions with a real token ring network, where the dominant application of the network is to provide communication between several disk-less work stations and a file server.

The choice of this application regime was motivated by several factors:

(1) An increasing number of "disk-less workstations" are becoming commercially available.
(2) The packet size distribution of such applications is expected to be highly non-random.
(3) The use of the network for paging/swapping by the client systems, coupled with the expected persistence and optimized network access of the client operating systems in such cases, should present a heavy, non-random (in time) load to the network.

The remainder of the paper describes the performance measurements conducted for a commercially available 10-Mbit token ring for use in such an environment. Section 2 contains a description of an idealized token ring. The specifics of the measured hardware are outlined in section 3, followed by the experimental design in section 4. Section 5 consists of the results of the measurements. In section 6, we summarize the results of the measurements and describe their impact upon local area networks in general. Section 7 offers some proposals for addressing the problems discovered in the course of this work.

2. Properties of an Ideal Token Ring

In a token ring, access to the transmission channel is controlled by sequentially passing a permission token around the ring. When a station wishes to transmit, it waits until it receives the token, at which time it transmits its message before sending the token on to the next station. Each station is responsible for removing its own packets from the ring.

The performance model which describes this operation is that of a multiqueue system with a non-exhaustive cyclic server [5]. The basic features of such a system are:

(1) The throughput versus offered-load curve (see fig. 1) increases linearly until 100 percent channel utilization is reached. Beyond 100 percent the curve remains constant. This latter feature illustrates that the token ring is stable when confronted by more than 100 percent offered-load.

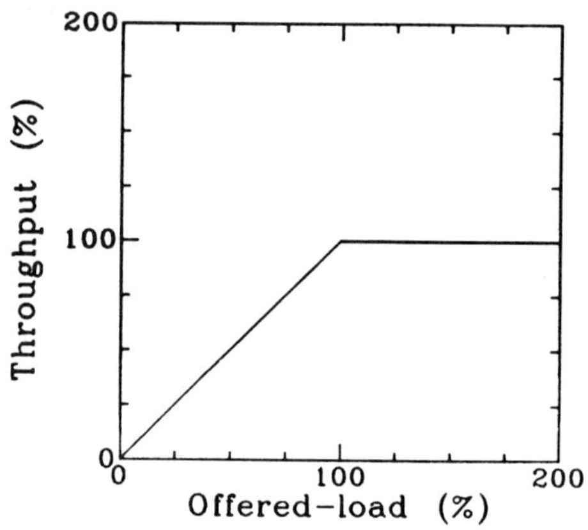

Fig. 1. Throughput versus offered-load for an ideal token ring.

(2) The cyclic polling system enforced by the access method guarantees that each node has equal access to the medium. As a result, the access method is fair.
(3) The polling nature of the access method tends to eliminate randomness on the ring when the offered load is high.

3. The PRONET Local Network Interface

The ring interface hardware used in these experiments is the PRONET local network interface [8], manufactured by Proteon Associates, Inc. When two or more hosts are interconnected via these interfaces and a wire center, a star-shaped ring [9] results. Such a network provides one-way baseband communication between nodes at 10 Mbits/second.

Some of the salient features of the interface are:

(1) Full-duplex Direct Memory Access is provided by the device, thus permitting concurrent transmit and receive DMA operations.
(2) The receiver portion of the device is singly buffered. After a packet has been received, the interface is busy until the packet has been transferred to host memory.
(3) The transmitter portion of the device is singly buffered, i.e. DMA cannot overlap transmission on the ring.
(4) The transmit buffer must be reloaded to retransmit a packet.
(5) If a packet is not accepted by its destination because the receiver buffer is not yet empty, a bit indicating "message refused" is set at the end of the packet. This bit is returned to device state registers permitting software to inspect the outcome of the transmission attempt.

4. Experimental Configuration

The test configuration for measuring the performance of the PRONET interfaces consisted of a single DEC PDP-11/34 and two DEC VAX-11/780s. The PDP-11/34 was configured with core memory and a memory cache. The PRONET Unibus interface accesses Unibus memory directly on the 11/34, while it accesses VAX memory through the intervention of the "buffered data paths" in the VAX Unibus Adapter (UBA).

Measurements of raw hardware event timings were obtained by placing probes on the Unibus backplane and measuring the time intervals on a calibrated oscilloscope.[1] For the measurements requiring software, stand-alone programs (written in assembly language) were used. The network interface was the only device active in each machine during the tests. The programs busy-waited for each transfer to complete, thus avoiding any interrupt latency. When transmitting, each system simply transmitted uniform size packets as quickly as possible, thus producing a deterministic arrival distribution similar to the continuously queued sources of [10]. When receiving, each system continuously waited for a packet to arrive. Upon arrival, the message was copied into memory, and the device was re-enabled for the next message. The DMA copy to memory was initiated by the device and the program looped on a status bit until the receive-copy composite was finished.

Additional capabilities were provided in the software on the 11/34 to drive some of the measurements:

(1) The software could optionally delay a specified time after successfully completing each transfer. This permitted the code to simulate a fixed processing interval required for each packet.
(2) When receiving packets, a history of the source host of each packet was kept to permit the fairness of the system to be determined.

5. Performance Measurements

5.1. Throughput

Since the interface is singly buffered in each direction, with no overlap of DMA and transmission on the ring, the time to transmit (or receive) data is

software time + DMA time + ring transfer time.

The ring transfer time is about 0.80 μsec/byte for large messages (10 Mbits/sec with small header

[1] Proteon Associates graciously provided the logic and timing diagrams necessary for locating test points and understanding detailed interface operation.

Table 1

Machine	Direction	Time (μs/byte)
11/780	Transmit	2.23
11/780	Receive	2.23
11/34	Transmit	1.17
11/34*	Receive	0.99

and trailer). By placing probes on the Unibus backplane, the DMA times were measured for transfers to memory (receive) and from memory (transmit). The results are given in Table 1. Note that the VAX has a very slow DMA rate even though the buffered data path was used. The VAX receive and transmit times are equal since the buffered data path masks memory access delays. On the PDP 11/34, receive and transmit times are unequal since there is no buffering between memory and device and different bus requests are made for read/write access to memory. As one can easily see, the DMA times are greater than the ring transfer time in all cases.

The maximum transfer rate can be calculated from the above measurements. Assuming no software overhead and large packets, the maximum transmit rates are listed in Table 2. These rates are upper bounds since no software overhead was assumed. The rates of Table 3 were measured sending 512 byte messages using the test programs described above.The 11/34 receive data rate is estimated from the zero software overhead values because we were unable to drive the 11/34 to a receive data rate limit. Two points are worth noting about this data:

(1) The software times in these test programs are small by design. In real networks, the software times dominate.
(2) For both types of CPUs, the maximum bandwidth achievable by a single machine does not remotely approach total utilization of the medium. This inability to receive each message on a loaded network represents a clear violation of the infinite capacity assumption.

Table 2

Machine	DMA + Ring Time (μsec/byte)	Transmit Rate (Mbits/sec)
11/780	3.03	2.64
11/34	1.97	4.06

Table 3

Machine	Transmit Data rate (Mbits/sec)	Receive Data Rate (Mbits/sec)
11/780	2.56	2.56
11/34	3.84	4.23 (est)

5.2. Stability

Measurements of throughput versus offered-load were made for the 11/34 sending to one of the 11/780s (see fig. 2). Note that the throughput curve experiences a sharp dip at the point where the 11/34 exceeds the maximum receive rate of the 11/780. From this point on, every other message is missed by the receiver, since the receiver's ring interface is busy DMA'ing the last message into host memory.

The dip in the throughput curve indicates that this ring implementation is not stable with respect to a single transmitter and receiver. While several transmitters would be able to originate a constant amount of traffic once the medium was completely saturated, consistent with the usual model predictions, the successful traffic to any particular receiver would experience the type of instabilities seen in fig. 2. Such instabilities can only be avoided if each receiver is able to receive every message on the medium.

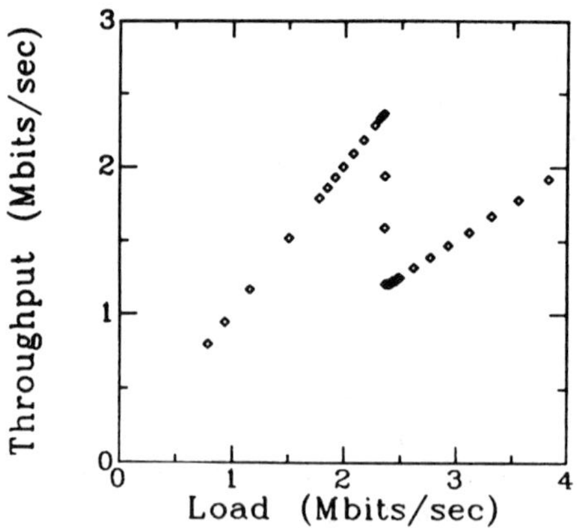

Fig. 2. Throughput versus offered-load for a single transmitter and single receiver pair. The severe transition at 2.3 Mbits/sec is due to the transmitter overrunning the receiver.

5.3. Fairness

In an attempt to measure the fairness of the ring, both 11/780s transmitted 512-byte packets continuously to the 11/34. A measurement run consisted of starting the 11/780s and then receiving 10,000 packets as counted by the 11/34. The software in the 11/34 kept a history of the sender of each packet received (indicated by the second byte of the message, as inserted by the transmitter's hardware). Several measurement runs were conducted, with fixed delays of varying sizes introduced after each received packet in the 11/34. The delay models a fixed processing overhead for each packet. If the access method is truly fair, then one would expect that for a given run, 50 percent of the packets should originate from each host, independent of the processing delay. A quick perusal of figs. 3 and 4 will indicate that this is not so for a large number of delay values, where one of the transmitters is always unsuccessful in having its messages received.

The observed starvation behavior can be explained in the following manner. The polling nature of the ring's access method causes packets from the 11/780s to alternate. The upper portion of fig. 3 shows this synchronization of packets ("convoys") to scale. The singly buffered nature of the network interface causes the receiver to be

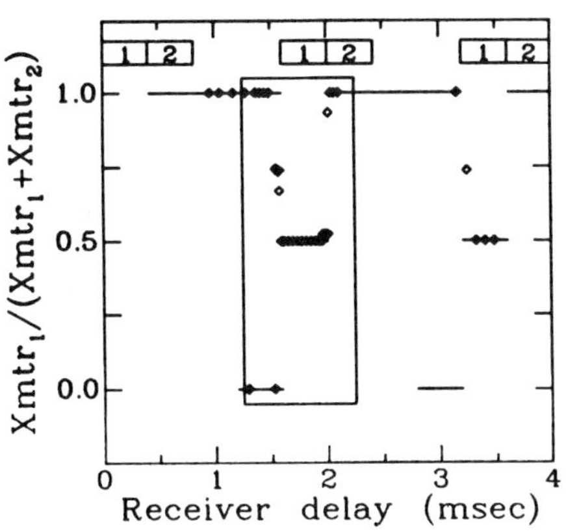

Fig. 3. Fraction of packets received from transmitter 1 versus the receiver's processing delay for each received packet. The solid curves are the expected values from analysis of the convoy pattern displayed at the top of the figure. The multi-valued nature of the function prior to each transition region is due to the initial conditions of the receiver for a particular run, as described in the text.

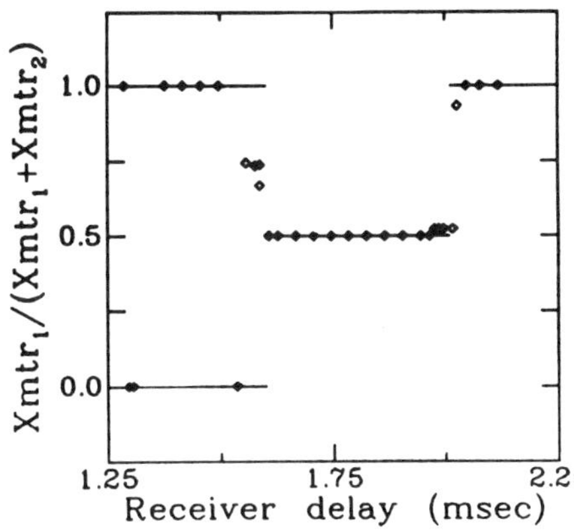

Fig. 4. An expansion of the first transition region of fig. 3.

unable to read the second message in the convoy, since it is busy copying the first message into host memory. Defining the transmission time of a packet on the wire as t and the delay between packets from a particular host as d, then the only receiver delay regions (D) in which a fair distribution will be detected are those when

$$(n * d) < D < ((n * d) + t) \quad n = 1, 2, \ldots$$

i.e. when the delay is just long enough to cause the receiver to miss the start of the message from the previously successful transmitter, but just short enough to pick up the start of the trailing message of the other transmitter. This description easily extends to cover the situation when the number of transmitters is larger than 2, or the number of receive buffers is larger than 1.

One should also note the multi-valued nature of the function just preceding each transition region in fig. 3. The lower branch of the function in these cases results when the initial read for a particular run locks onto the second packet of a convoy. The delay values in these regions are such that the receiver will continue to receive the second packet in the convoy, thus resulting in complete starvation of the first transmitter. Since the position of the convoy with respect to the receiver is random for the first read in a particular run, one would expect the relative population of the two branches to be

$$(d - t)/t.$$

As a result, performance predictions for a system operating in these regions have very little value.

6. Discussion

The data presented above indicate that very counter-intuitive behavior occurs when the assumptions of infinite buffer capacity and random traffic are violated. The following discusses these results in the context of the specific ring interface measured, as well as for networks in general. A solution to these problems for token rings is proposed in section 7.

6.1. Finite Capacity

When discussing the finite capacity problem, two parameters determine the behavior of the network when these is sustained traffic: the receiver delay (D), which is the time necessary for the interface to be ready for another message from the network, and the transmit time for a packet on the wire (t). Two particular operating situations can be characterized:

$D < t$: In this situation, it takes the transmitter longer to originate the message than it does for the receiver to copy it to memory and be ready for another message. By providing two receive buffers and permitting DMA to overlap message reception, a receiver can see all messages on the ring. Previous low-speed local networks (e.g. 3 Mbit Ethernet) fall into this category.

$D > t$: Since the reception DMA is longer than the transmission time, no amount of buffering will permit the receiver to handle sustained traffic. The ring interfaces measured fall into this category. Note that if the traffic can be characterized as bursty as opposed to sustained, then an optimal number of buffers can be provided to smooth out the traffic distribution (on the average).

While the current interfaces could be pushed into the $D < t$ category via engineering changes, the finite buffer problem scales with the current trends in local networks. The attractiveness of higher speed local nets guarantees that the network speed will be faster than the host channel speeds – i.e. $t \ll D$.

The messages lost due to finite capacity will generally trigger timeout and retransmission mechanisms in higher level protocols. Moreover, the coupling between protocol layers can be quite non-linear. Most importantly, no amount of upper-level protocol effort can produce throughput in the face of low-level starvation. See [3] and [7] for studies of these problems.

It should also be apparent that this particular problem is independent of the access control mechanism. All that is required is that the traffic be bursty enough to more than exhaust a receiver's message buffers. Independent measurements of such anomalous behavior for 10 Mbit Ethernet interfaces are reported in [7].

6.2. Unfair Traffic Distributions

The grossly unfair distributions displayed in fig. 3 can be attributed to two causes:

(1) lack of multi-buffering in the receiver interface;
(2) the non-random ordering of the messages on the ring.

The previous discussion has shown that no amount of buffering will guarantee stable performance under all traffic conditions. It is therefore necessary to concentrate on the non-random ordering of packets on the ring.

The regularity of the traffic offered by the continuously queued servers, coupled with the synchrony of the ring access control mechanism, forces the messages in a convoy to be identically ordered. Since limited buffering on the receiver's interface will only permit n consecutive messages to be received from the convoy, the remaining messages in the convoy will always be missed by the receiver. This situation leads to the extreme starvation seen for many values of receiver delay seen in fig. 3.

Broadcast mechanisms, with the randomization inherent in their collision resolution protocols, should not be quite so susceptible to this pathology. It is feasible, however, that several identical, continuously queued servers could synchronize in time such that the collision resolution protocol would not be invoked, but the resulting traffic could still overflow the limited buffering of a single receiver.

For example, assume that the total offered load to the wire is ~ 50 percent of capacity, and that the resulting load is > 100 percent of the capacity of a single receiver. The number of collisions (randomizing events) would be expected to be low, resulting in constant ordering of messages as they pass the receiver. If the interface is subject to finite buffering, then the messages from the same hosts will be missed each time, in analogy with the ring situation. Again, it is important to note this interface congestion anomaly is not induced by high total load, but by high loads with respect to a single receiver.

7. Proposals for Improving Performance

Please note the problems seen in these experiments are not an indictment of the specific hardware we used. While some problems do result from specific hardware design decisions, these decisions did not, by themselves, produce the anomalies seen. On the contrary, we believe these results show that congestion control is a real, serious problem in the design of even "simple" local area networks.

7.1. Receiver-Access Protocols

Historically, most local network design, analysis and modeling has been done for *link-access* protocols. While the value of such efforts cannot be overemphasized, this study shows that stable link access characteristics, while necessary, do not guarantee adequate network performance in the face of interface congestion. Therefore, we propose that after a new protocol design has been shown *link-access stable*, efforts be applied to designing an accompanying *receiver-access* protocol to be *receiver-access stable*. Receiver-access stability implies transmitters will not see suddenly-decreasing throughput or delivery unfairness in the presence of highly asymmetric (congested) conditions like those in this experiment. A receiver-access protocol embodies the congestion control strategy used by each transmitter to optimize the chances that its packets successfully placed onto the medium will actually be successfully received. This is a somewhat different view in that congestion control is usually seen as a global issue instead of relating to a single host. In most local networks, congestion of the network medium is handled by the fairness and stability of the basic link-access protocol. Interface congestion is controlled with the receiver-access protocol.

Analyzing receiver-access protocols implies modeling both the receiver implementation and the strategy used by the transmitter for providing "best effort." Note that this is *not* a call for guaranteed-reliable link-level protocols, but it is an acknowledgement of a specific kind of unreliability and the desire to reduce its impact. Receiver-access protocols will quite likely complicate models to the point of analytic intractability, thereby requiring simulation and measurements of real implementations. Only in this way can reasonable guarantees be made about the behavior of a real network. As a first attempt, we now present an analysis of the experimental system from the receiver-access point of view.

7.2. An Example of a Poor Receiver-Access Protocol

The initial receiver-access protocol implemented in the driver software for these tests and the production local network which uses the ring interfaces is very simple-minded, and consequently very poor. Recall the "refused" bit indicates that the recipient of a packet is an active member of the ring, but did not take the packet off the wire because its input buffer was not empty. The first-order strategy implemented is to simply, and immediately, retransmit a packet when it returns marked "refused." We thereupon discovered one design flaw in the interface: the transmit buffer must be reloaded via DMA to do the retransmission. A holding register for the initial byte count would allow retransmission of a refused packet without the overhead of the DMA. The critical error, however, is the retransmission strategy, not the need for the additional DMA.

By repeatedly attempting to transmit the packet until the number of refusals reached some threshold, we guaranteed the creation of the convoy traffic pattern which induces the anomalies described above.

7.3. A Better Receiver-Access Protocol

Provided with the knowledge that some formal mechanism is needed to stabilize receiver access and control interface congestion, we examined the problem to see how it can be accomplished in the framework of the interfaces at hand, and how the interface design could be improved to make this easier. Some of the following discussion could apply to an Ethernet-style network, but the scheme described below relies critically upon the back-channel signal provided by the "refused" bit.

Assuming the absence of message priorities, which will be discussed later, the main problem is one of fairness, or insuring transmitters all have equal likelihood their messages are missed. This requires a mechanism for "stirring" the convoy so transmitters trade positions in fair ways.

One challenging characteristic of a token ring is that each transmitter's observed frequency of message origination varies with the load on the ring. This is not surprising, but the extrema of these values differ by a factor of over 200 for large rings. To reorder the convoy, each sender must wait a random time interval based upon one convoy rotation time, or the original packet order will be preserved. This means a simple scheme based upon a fixed convoy rotation time would wait far too long for small rings and reduce the available throughput. Providing the interface with a mechanism which can provide software with a measurement of the "speed" of the last convoy rotation would allow more accurate estimation of the waiting times necessary to reorder the convoy. Schemes for providing this mechanism are currently under study.

Other improvements would arise from relocating the "refused" bit. In the current ring implementation, the refused bit is at the end of a message; relocating it to the front of the packet would allow large packets to be terminated immediately upon receipt of the refused bit. This relocation helps in two ways: less bandwidth is used for useless packets, and the abrupt termination helps break the synchrony of the convoy.

The receiver-access protocol which evolves from this analysis is essentially the Ethernet CSMA/CD algorithm with some definitions changed [6]. The "refused" bit is taken to mean "collision" at the receiver. The convoy rotation timer in the interface is used to estimate the loading of the ring and is used in formulating the initial random backoff interval, with successive "refusals" exponentially scaling the back-off interval. Providing a countdown timer in the interface to process these delays would make the software simpler.

At present, this receiver-access protocol has not yet been implemented; a software implementation is being designed to support further experiments.

7.4. Congestion Control and Multiple Buffers

We now return to the issue of multiple buffering in receivers. Many current interface designs subscribe to the idea that having "enough" buffers in the receiver will alleviate the problems described in this study. While having more buffers allows longer convoys to be received without incapacitating congestion, disaster is merely postponed. Given a network with no receiver-access protocol, when the service times of the packets queued in receiver interface buffers exceed their transmission times, the receiver *will* fall behind with the eventual result being lost packets from interface congestion. Packets lost because the receiver is too congested to accept it are no different from those lost because the checksum was corrupted by failing hardware. Providing massive buffering may well reduce the likelihood of debilitating congestion to levels acceptable for *some* applications, but buffering alone *cannot* be the basis for meaningful guarantees about network behavior.

One other possibility to consider is that excessive buffering could exacerbate interactions with higher-level protocols by preserving stale data packets in an already congested interface. For instance, many datagram-based or real-time protocols (packet voice, for instance) want the most recent version of a retransmitted packet, and not one which has been delayed in an over-buffered receiver interface due to congestion. Designers of systems with low latency requirements should evaluate the receiver buffering issue with extreme care.

7.5. Implications for Rings with Message Priorities

Several ring designs include provisions for message priorities [1] which are intended to provide a better quality of service for high priority messages. The results of this study cast some doubt on the implementability of such differentiated service. Serious problems arise when the traffic rate of low priority messages into one interface becomes sufficient to saturate the finite internal buffering of the interface. This could easily produce starvation of a high priority source with little possibility of resolution. The only fix we see at present would be to design the interfaces to discard low priority messages from the internal buffers when a higher priority message arrives. Even then, there are no absolute guarantees, but the likelihood of prompt delivery is improved.

8. Conclusions

This study revealed the unsetting fact that current performance models do not appear to predict the performance of real-world network implementations, particularly in regimes prone to interface congestion. The reasons for the discrepancies were analyzed, and schemes for improving modeling, congestion control protocols, and interface hardware architecture were presented. We believe these proposals indicate directions for further fruitful research.

References

[1] W. Bux, F. Closs, P. Janson, K. Kummerle and H.R. Muller, A Reliable Token-Ring System for Local-Area Communication, Report RS 1095 (#39517) (IBM Zurich Research Laboratory, 1981).

[2] W. Bux, Local-Area Subnetworks: A Performance Comparison, IEEE Trans. Commun. 29 (October 1981) 1465–1473.

[3] J.E. Donnelley and J.W. Yeh, Interaction Between Protocol Levels in a Prioritized CSMA Broadcast network, Computer Networks 3 (1979) 9–23.

[4] H. Jafari, T.G. Lewis and J.D. Spragins, Simulation of a Class of Ring-Structured Networks, IEEE Trans. on Computers 29 (May 1980) 385–392.

[5] P.J. Kuehn, Multiqueue Systems with Nonexhaustive Cyclic Service, Bell Syst. Tech. J. 58 (March 1979) 671–698.

[6] R.M. Metcalfe and D.R. Boggs, Ethernet: Distributed Packet Switching for Local Computer Networks, Communications of the ACM 19 (July 1976) 395–404.

[7] J. Nabielsky, Interfacing to the 10 Mps Ethernet: Observations and Conclusions, Mitre Corporation Report (March 1983).

[8] Proteon Associates, Inc., Operation and Maintenance Manual of PRONET UNIBUS HSB (August 1982).

[9] J.H. Saltzer and K. Pogran, A Star-Shaped Ring Network with High Maintainability, Computer Networks 4 (1980) 239–244.

[10] J.F. Shoch and J.A. Hupp, Measured Performance of an Ethernet Local Network, Communications of the ACM 23 (December 1980) 711–721.

[11] F.A. Tobagi, Carrier Sense Multiple Access with Message-based Priority Functions, IEEE Trans. Commun. 30 (January 1982) 185–200.

Norm Strole, IBM Corp., Research Triangle Park, N. C.

Inside Token Ring Version II, according to Big Blue

An insider discusses IBM's new 16-Mbit/s Token Ring. Phone wiring can't handle it, but multimode fiber fits right in.

Most informed analysts agree that IBM's enhanced Token Ring is likely to have a profound effect on the future of LANs. Because of this, and because certain questions regarding the LAN's design, applicability, and integration could best be answered by those who developed it, Data Communications *invited IBM to address these issues in an article. IBM accepted, and this article is the result. Furthermore, because the 16-Mbit/s Token Ring will most impact Ethernet as a leading LAN technology, a counter-perspective from a key Ethernet proponent, Digital Equipment Corp., was also solicited—and the offer accepted. The DEC article immediately follows this one.*

The second generation of IBM Token Ring network products was announced this past November at the Comdex trade show in Las Vegas. The new LAN provides a fourfold increase in the basic Token Ring transmission rate, from 4 Mbit/s to 16 Mbit/s. Additional levels of fault tolerance and more efficient LAN-to-LAN and host connectivity are supported. IBM has also worked to provide users with a smooth migration path between its existing and new Token Ring products.

To assist in migration, new Token Ring adapter boards operate at either the current 4-Mbit/s data rate or the faster 16-Mbit/s data rate. This feature allows the new adapter boards to be used initially in an existing 4-Mbit/s Token Ring, and later in 16-Mbit/s installations as required.

Rate selection is supported using software parameters on new adapter cards for PS/2 computers equipped with the Micro Channel bus. The Micro Channel's Programmable Option Select feature, in conjunction with memory on the PS/2 motherboard, makes software-controlled rate selection possible. Dual-rate adapter boards for PCs, either with the PC or the AT bus, employ a hardware switch.

IBM's existing wiring concentrators, called Multistation Access Units (MAUs), will also support either 4- or 16-Mbit/s data rates. This is possible since there are no active components within MAUs, such as logic or analog elements. An MAU contains a series of relay elements that are activated via power derived from the attaching station(s).

Shielded cable only

LAN users with unshielded twisted pair installing 16-Mbit/s products will have to upgrade office-to-wiring closet connections to data-grade media. At 16 Mbit/s, signal filtering (which, while required to meet Federal Communications Commission standards for radiation emission, degrades signal quality) severely limits the number of workstations and the lobe length that can be supported on a LAN employing unshielded, telephone-type twisted pair. For 4-Mbit/s operation, users still have the option to use unshielded twisted pair.

Optical-fiber media are now also required in the majority of cases for links between wiring closets. This is because maximum wiring closet-to-wiring closet separation supported by existing 4-Mbit/s products cannot be obtained at 16 Mbit/s using data-grade media (although data-grade media are acceptable for 16-Mbit/s operation between wiring closets that are proximate, such as closets that are vertically positioned in a multistory building).

Use of optical fiber for links between wiring closets is supported through the new 8220 dual-rate (16- or 4-Mbit/s) optical-fiber converter. The converters are deployed in pairs, with one converter located at each of the wiring closets that are being interconnected. Physical separation of these closets may be up to approximately 2 kilometers, provided high-quality fiber is employed.

Multimode fiber sizes that are supported by the 8220 include 50/125, 62.5/125, 85/125, and 100/140 micron. The 62.5/125-micron fiber is more common in the United States, while 50/125-micron fiber is more common in Japan and some European countries.

Optical-fiber media are also required to take full advantage of the features provided by the optical-fiber converter, including its support for fault-tolerant "dual-ring" network installations.

In a dual-ring configuration, a fiber pair forms the connection between two optical-fiber converters. One of the fibers is designated as the primary, the other is designated the secondary (corresponding to the primary and secondary rings of the dual Token Ring LAN).

1. LAN interconnection. *The major use for IBM's 16-Mbit/s Token Ring LAN will be as a backbone for a number of slower-speed ring or bus networks. Existing 4-Mbit/s LANs will be best for individual or departmental LANs.*

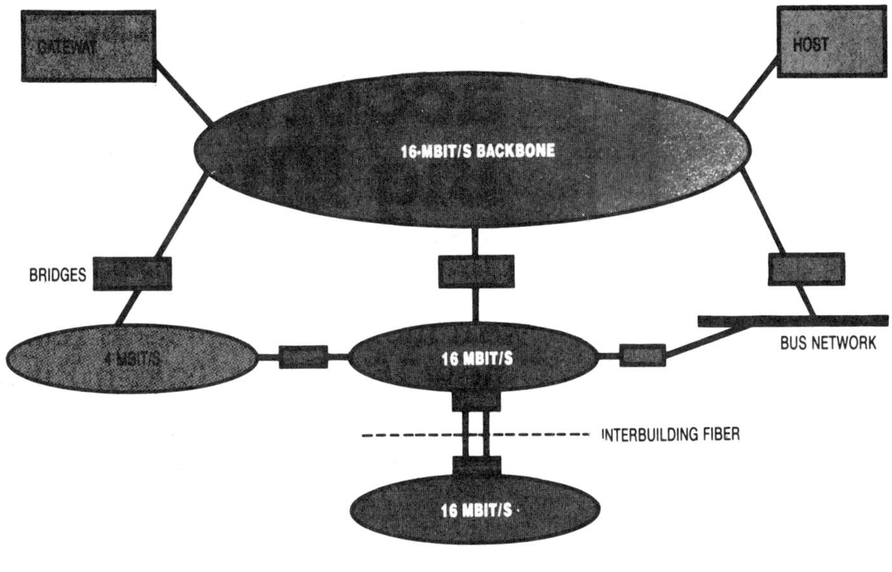

Within the pair of converters, the network adapters constantly monitor the optical signal on both the primary and secondary fiber links. In the event that the signal path is disrupted, either by a break in the fiber or by manual disconnection, both converters automatically perform an internal wrap to the secondary link, which isolates the failed segment from the network. Loss of power in either one of the converter pairs will also result in an automatic wrap at both converters.

The dual-ring capabilities of the converters make use of error-detection and recovery functions of built-in adapter boards. A converter can participate in polling and beaconing like other LAN stations on the ring and exchanging messages with IBM's LAN Manager network management software.

Users can also designate a converter as a "critical resource" in a large network (this is accomplished using LAN Manager software). The converter will then maintain a continuous logical session with the LAN Manager software.

Enhanced bridge connectivity

The introduction of IBM's 16-Mbit/s products enables the practical implementation of hierarchical LAN topologies—where a number of slower-speed ring or bus networks are interconnected via a faster backbone (Fig. 1).

There are a number of advantages offered by 4-Mbit/s data rates, such as operation over unshielded twisted pair and lower attachment costs. Segmenting or partitioning an overall network by constructing smaller networks connected via a backbone also facilitates network expansion and maintenance. Also, the physical separation provided at the bridges prevents the overall system from being affected by a localized failure.

Interconnection of LANs in a campus environment is best accomplished through the use of an optical-fiber backbone connected through wiring closets using the 8220 converter. Interbuilding fiber links do not require the surge suppressors that copper links must have.

Routing of frames within a complex network of Token Ring LANs is accomplished by IBM's existing Source Routing protocol. With Source Routing, a Routing Information (RI) field within the frame header carries specific ring and bridge identifiers that indicate the path for the frame to follow to reach the destination station. The route is established when one station initiates a link to the other.

Source Routing protocols allow considerable flexibility in designing network topologies. For example, workstations can be moved from one location to another without requiring hardware or software changes. Also, Source Routing allows redundant network components to be installed to maximize system availability. For instance, the network backbone can be duplexed to ensure that service is never disrupted. The maximum number of stations within an enterprise-wide LAN is essentially unbounded since frame forwarding at intermediate bridges is based solely on the content of the RI field and does not require that bridges maintain station address tables. In fact, IBM's adapter board implements the RI compare logic in VLSI (very large-scale integration) circuitry.

In a multiring system, single route broadcast paths through parallel bridge configurations are automatically determined via a new function included in IBM's Network Bridge Program (Version 2.1). The new version accommodates interconnection of 4- and 16-Mbit/s LANs using

PC/ATs or PS/2s (based on the Intel 80286 or 80386 microprocessors).

With the single-route transmission features, packets that are designated for single-route broadcast will follow a designated route, even when active parallel bridges are present. Version 2.1 of the bridge program accomplishes this by only forwarding frames received with a "single route broadcast" bit that is set to the "on" position. This prevents a target network from receiving duplicate frames. A user can employ the LAN Manager software to instruct a bridge to forward or not forward frames based on the position of the single-route broadcast bit. Stations may also send "general broadcast frames" that will flow through all bridges. This is often used within the route-selection protocol during link establishment.

Version 2.1 also allows two remote 4- or 16-Mbit/s LANs to be interconnected at speeds from 9.6 kbit/s to 1.344 Mbit/s. This interconnection scheme is implemented through a pair of remote bridges connected to each other over leased lines (Fig. 2).

Routing of frames through the remote bridges is based on the Source Routing protocol. Users can add software (documentation gives sample programs) to add filtering in order to restrict certain stations (based on station address) from accessing the remote LAN. Maximum frame sizes vary, depending on the line speed of the remote link. The normal end-to-end Logical Link Control (LLC) and Netbios protocols operate across both LANs and the intermediate link for recovery of lost frames.

Parallel remote bridge links may also be active simultaneously, either to provide additional bandwidth or as additional backup. Although cascaded remote bridges are permitted, such installations are not recommended with the slower-speed links because of end-to-end propagation delays that may be encountered due to queuing delays.

Commonly used methods from IBM and third parties for interconnecting remote 4-Mbit/s LANs entail the use of dedicated gateways. These often operate in a store-and-forward mode at the message level. The throughput limitations of this mode of operation could contribute to longer response times than would be experienced on the LAN. In contrast, the remote bridge forwards packets as they arrive, without the intervention of higher-layer protocols, thus offering significant throughput improvements over conventional gateways.

A number of network management features are incorporated within Version 2.1 of the bridge program, such as detection and reporting of link outages and collection of performance statistics and status information.

Additional bridge interconnection of 4- and/or 16-Mbit/s Token Ring LANs can be accomplished via a channel within a broadband system by installing the IBM Broadband PC Network hardware and software along with Token Ring hardware within a PS/2 or IBM PC/AT. The IBM PC Network Bridge Program is required. The broadband channel data rate supported by the PC Network family of adapters is 2 Mbit/s. As with the other bridging schemes, the Source Routing protocol is implemented to minimize bridge latency and to achieve greater ring-to-ring throughput.

A key aspect of any LAN is the connectivity options that allow intelligent workstations to gain direct access to one or more host computers.

Options for mainframe connectivity that will be made available include new 16- and 4-Mbit/s adapter features for the IBM 3745 Communications Controller, as well as the IBM 3174 Subsystem Controller, both enabling direct access to an IBM mainframe.

The IBM 8232 LAN Channel Station supports the Transmission Control Protocol/Internet Protocol (TCP/IP) connectivity to an IBM mainframe.

Midrange attachment for the Enterprise System/9370 will allow this system to be used as a server, an application host, or as a gateway to another computer system. Also, the IBM Series/1 Token Ring attachment card will support the 16-/4-Mbit/s attachment for S/1 systems with a 4956 processor. Direct attachment of the IBM Application System/400 is also planned.

Network management within a Token Ring environment remains unchanged with the introduction of the 16-Mbit/s products. As before, each adapter is capable of detecting failures that occur within itself or those caused by other stations on the ring.

Available user application interfaces for Token Ring environments also remain unchanged with the introduction of 16-

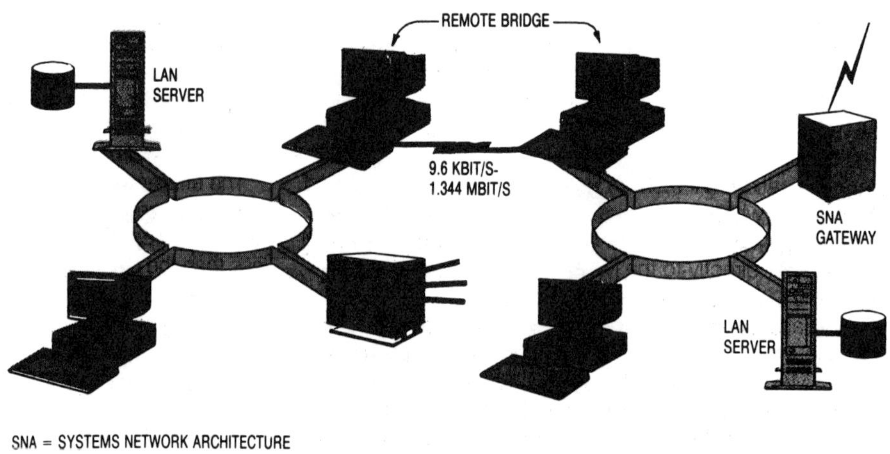

2. Bridge to 1.344 Mbit/s. *IBM's Source Routing protocol, as implemented in a new version of the LAN Bridge Program for PS/2 computers, allows the interconnection of multiple LANs over leased telephone lines.*

3. Interface stacks. *The new 16-Mbit/s Token Ring products support the same APIs as existing 4-Mbit/s products. For users writing custom software, LLC and MAC level interfaces are provided.*

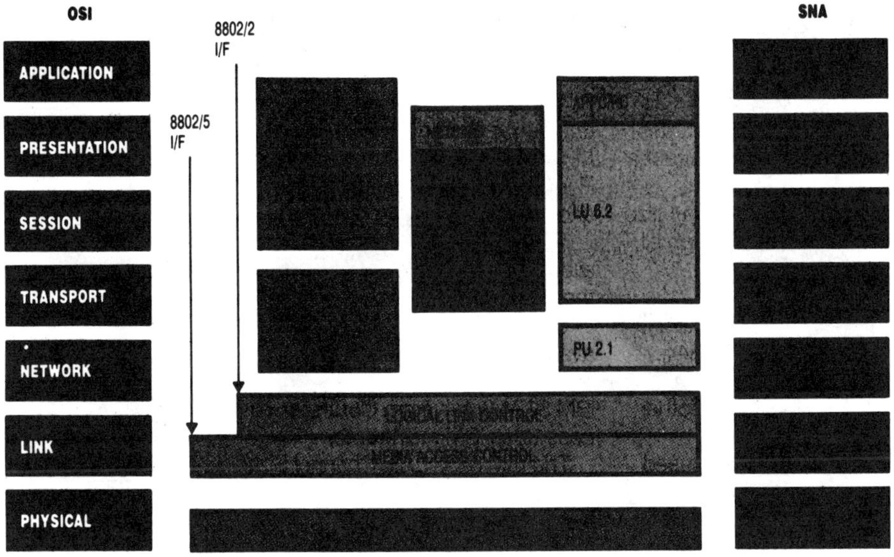

APPC = ADVANCED PROGRAM-TO-PROGRAM COMMUNICATIONS
OSI = OPEN SYSTEMS INTERCONNECTION
TCP/IP = TRANSMISSION CONTROL PROTOCOL/INTERNET PROTOCOL
SNA = SYSTEMS NETWORK ARCHITECTURE

Mbit/s adapters (Fig. 3). The Netbios networking protocols have gained widespread acceptance in the industry and continue to provide multivendor interoperability at the application level. The Netbios protocols can provide both reliable end-to-end message delivery and datagram services.

Application interfaces

TCP/IP and IBM Systems Network Architecture protocols are also supported. The TCP/IP networking protocols provide intra- as well as inter-LAN communications services. (TCP/IP has a broad base of support in both university and industrial networks. A number of network products and applications now rely on TCP/IP to provide the communications base for connectivity via a wide range of physical interconnection schemes.)

IBM Systems Network Architecture (SNA) applications in the workstations can link directly to SNA host applications via the PC implementation of Advanced Program-to-Program Communications (APPC/PC), with LU 6.2 and PU 2.1 support.

For users who want to customize their LAN software, two additional interfaces are provided: one to the LLC level and another directly to the Media Access Control (MAC) level. The LLC protocols provide reliable end-to-end packet delivery and also implement a dynamic window flow control scheme, which relieves congestion at intermediate bridge nodes. MAC-level protocols provide only basic packet-delivery services.

A number of performance enhancements have been incorporated into the new line of products. The amount of random access memory (RAM) on the 16-/4-Mbit/s Token Ring adapter cards has been increased to 64 Kbytes from the previous maximum of 16 Kbytes. Thus, the maximum frame size that a 16-Mbit/s station can transmit and receive is variable up to approximately 18,000 bytes. When operating at 4-Mbit/s, the maximum frame size for the new adapters increases from 2,048 bytes to approximately 4,500 bytes.

Support for larger frames is especially beneficial to those applications that require movement of large files, such as imaging or system-to-system information transfer. The additional adapter memory also allows centralized applications, such as LAN servers, to support up to 254 concurrent sessions.

IEEE 802.5 Standard Options for the 16-Mbit/s Token Ring include Early Token Release (ETR). This capability can be selected by users to enhance the performance of new Token Ring products when operating in 16-Mbit/s mode. The new boards, when operating at 4 Mbit/s, are compatible with existing Token Ring boards.

ETR is under consideration by the IEEE 802.5 committee as an extension to the Token Ring architecture standard. The committee may also provide for the use of ETR at 4-Mbit/s operation. It can boost performance at 4-Mbit/s when rings are physically long (30 kilometers or more).

The performance benefit derived from ETR is its ability to compensate for the inherent delay, or latency, within a Token Ring. This delay is due to the signal propagation time over the media and the delay as intermediate nodes repeat the signal.

Delay becomes more of a problem as the physical ring length, number of active stations, and/or the ring transmission rate are increased. This is because these three factors increase the number of bytes required to "fill" the ring (Fig. 4A).

The efficiency of the Token Ring protocol may be affected if a high percentage of the frames are shorter than the latency of the ring, resulting in an overall average frame size that is less than the ring latency. This is due to the idle characters that must be inserted by a transmitting station until it has recognized its source address in the header of the returning frame—and subsequently released a new token.

ETR allows the transmitting station to release a new token as soon as it has completed frame transmission, whether or not the frame header has returned to that station. This optimizes the use of the available ring bandwidth when the average frame size is less than the ring length and decreases the token wait time at subsequent stations. As illustrated in Figure 4B, the frames from two (or more) different stations can be on the ring simultaneously with this enhancement to the token-access protocol. The maximum ring efficiency with ETR could exceed 95 percent regardless of ring length for most applications, thus making more efficient use of total bandwidth for information transfer.

The elimination of ring latency as a limiting factor in maximum ring efficiency will be a major benefit of ETR. The latency for a typical 4-Mbit/s LAN with 100 active stations and 5 kilometers of cable is approximately 50 bytes. A higher-speed, 16-Mbit/s LAN would more likely be used in a backbone to link several LANs over a wider area. Assuming over 100 local LANs were connected via a 16-Mbit/s backbone with 25 kilometers of cable, the ring latency is approximately 300 bytes. From these examples it is apparent that ETR is most applicable to 16-Mbit/s systems that span a large area.

One side effect of ETR operation is that it may alter the behavior of the priority protocol, as the station may not have received the header of the transmitted frame prior to releasing the token. When this occurs, the priority of the next token is the same as the token that was captured by the station. Any priority reservations in the header of the returning frame are ignored.

4. Early Token Release. *In (A) ETR reduces network delay by eliminating the idle characters that transmitting stations previously had to insert until they could release a new token, and (B) it allows frames from two or more stations to be on a LAN simultaneously.*

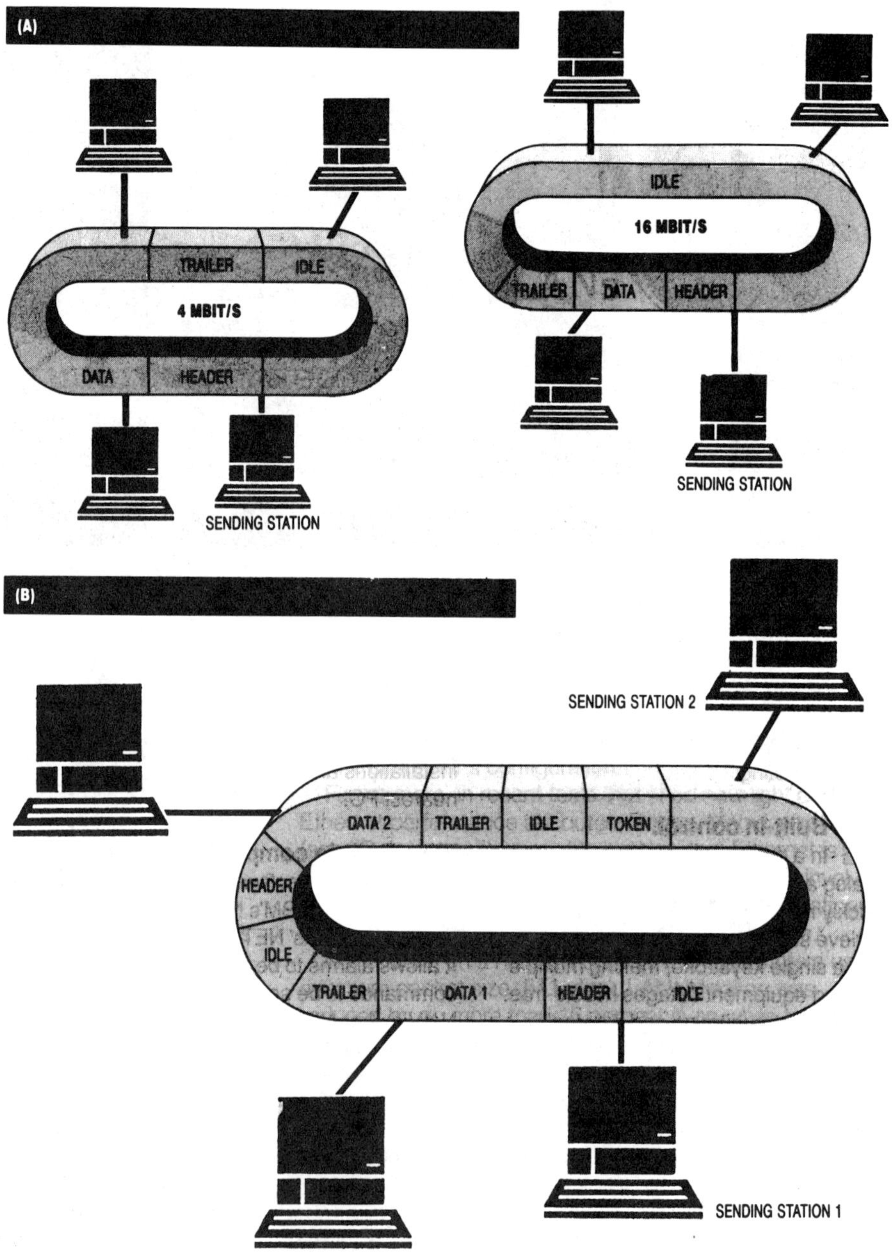

Performance modeling has shown that a mix of "long" and "short" frames within an application is sufficient to ensure that the priority stations gain access during periods of heavy traffic with little noticeable impact.

This second generation of IBM Token Ring products is the first implementation of the latest IEEE 802.5 Token Ring standard. In the future, the emerging Fiber Distributed Data Interface standard will extend Token Ring transmission rates to 100-Mbit/s. Thus, Token Ring networks will continue to play a major role in LAN systems for the 1990s. ■

Norm Strole is an advisory engineer at IBM's Research Triangle Park laboratory. Strole received a B.S. degree in electrical engineering from North Carolina State University in 1973. He received M.S. and Ph.D. degrees in electrical engineering from Duke University in 1975 and 1980, respectively. He is a Senior Member of the IEEE.

Sunil Joshi and Venkatraman Iyer, Advanced Micro Devices, Sunnyvale, Calif.

New standards for local networks push upper limits for lightwave data

Recent activities in ANSI committees foretell a different yardstick for high-speed local fiber networks.

While other standards bodies now specify top-end data rates of 10 Mbit/s for local networks, the American National Standards Institute (ANSI) is developing network standards for data rates an order of magnitude higher—around 100 Mbit/s. These are not futuristic speeds. Most companies involved in developing these standards plan to introduce products supporting such data rates during 1985 and are pushing to finalize the standard by mid-1984. Called the Fiber Distributed Data Interface (FDDI), this proposed new ANSI standard (currently in committee X3T9.5) specifies a token-passing ring architecture for local networks using optical fiber cable.

There are two reasons for the need for high data rate networks: a dramatic increase in computer processing power over the last few years and an enormous increase in the volume of stored or processed data. As a result, it is the lower-speed local networks that quickly become the weak link between devices needing to transfer huge amounts of data as quickly as possible.

Clearly, data rate requirements depend on the services supplied and the network's application. For example, back-end networks, which connect computers to other storage devices and peripherals, require high-speed data transfer. These networks have a fairly small number of nodes (frequently fewer than 50), and span relatively small distances—usually within a computer room. In general, a backbone network has to be at least as fast as the devices on it in order to minimize buffering constraints. As the speeds of hard disks and optical disks increase beyond 40 or 50 Mbit/s, back-end networks need to be even faster. In addition, protocols for such networks must provide for "streaming" operations, in which several data packets are sent end-to-end in a single network access. This is essential for disk transfers because, with streaming, disks can send entire tracks of data in one access, eliminating the wait for a complete disk revolution between network access opportunities.

Backbone

An organization wishing to connect its token bus, token ring and Ethernet networks would use a special bridge or gateway network to do so. Known as a backbone network, these interconnecting links can span distances of up to several kilometers, often connecting many floors of a building or even several buildings together. Since internetwork traffic is funneled through the backbone, it has to be able to accept fairly high data rates. Also, devices requiring high throughput can be connected directly to the backbone. Large mainframes in different buildings, for example, may have direct connections to the backbone. Or, a distributed PBX architecture can be implemented by connecting the local PBXs in each building via the backbone network, which likewise should be fast enough to carry real-time voice information.

Unlike back-ends, front-end networks typically connect computers to devices that are more user-interactive, such as terminals, word processors, workstations, and printers. Front-end networks can have hundreds of nodes spanning a few kilometers. The IEEE 802.3, 802.4, and 802.5 standards (see "Local network standards") are essentially designed for front-end applications.

Until now users have been satisfied with data rates of 10 Mbit/s or so for front-end networks. But if it were available—and affordable—most network users would welcome higher data rates and their services.

For example, a 10-Mbit/s data rate is not sufficient to support many real-time voice conversations, much

Local network standards

Numerous standards committees in the United States are standardizing various layers of the International Organization for Standardization (ISO) seven-layer Open System Interconnect (OSI) model. The most notable are the IEEE and ANSI committees, the primary focus of which has been ISO's physical and data link layers (table).

The IEEE is comprised of professional engineers, whereas ANSI membership is comprised of companies. ANSI is also the official body representing the United States in ISO. Hence, ANSI's organization closely parallels ISO's in terms of technical committees and subcommittees. Indeed, ANSI standards could eventually become ISO standards.

The IEEE defines standards for data rates of approximately 40 Mbit/s and below. ANSI focuses on data rates greater than 40 Mbit/s. Both committees have liaison members who represent them at the other committees' meetings—mainly to share expertise and to avoid duplication of effort.

The IEEE 802 committee is specifying a series of local network standards that define four access technologies for a variety of physical media. The four standards are described briefly as follows:

■ IEEE Standard 802.3. This is essentially the same as the Ethernet standard, which uses carrier sense multiple access with collision detect (CSMA/CD) as the access method. The standard is defined at a 10-Mbit/s data rate on a coaxial cable in a bus.

■ IEEE Standard 802.4. This bus standard uses a token-passing mechanism to determine network access. The primary application of this scheme is in factory automation where the advantages of bus topology are combined with the priority potential and deterministic properties of token access.

■ IEEE Standard 802.5. This defines a token-passing ring network. The medium is coaxial cable and the interconnection is a physical ring topology with a provision for passive bypass that uses relays.

■ IEEE Standard 802.6. This is a project, now under study, to put together a metropolitan area network.

ANSI activities

The ANSI X3T9 Committee has the charter to define I/O interfaces. Several subcommittees are working on standards.

■ ANSI X3T9.2 Small Computer Systems Interface (SCSI). Popularly called "scuzzy," this interface defines a scheme to interconnect low-end disk drives and other peripherals to computers.

■ ANSI X3T9.3 Intelligent Peripheral Interface (IPI). IPI defines a scheme to interconnect somewhat higher-end peripherals to host adapters over a parallel bus. At the link layer, IPI sends data in the form of packets, much like a local network.

■ ANSI X3T9.5 Local Area Networks: Two subcommittees within X3T9.5 are defining different local network standards: Local Distributed Data Interface (LDDI) is in the process of modifying and adopting a 70-Mbit/s coaxial cable network proposed by Digital Equipment Corp. (DEC) based on their CI-network. It uses a star topology. And Fiber Distributed Data Interface (FDDI) defines a 100-Mbit/s fiber-optic ring, which uses a token-passing access scheme.

Standards activity in networking

APPLICATION	SC16	WG5	X3T5.5	—	—	—	ISO COMPATIBLE	—
PRESENTATION	SC16	WG5	X3T5.5	—	—	—	ISO COMPATIBLE	—
SESSION	SC16	WG6	X3T5.6	—	—	ECMA-75	ISO COMPATIBLE	—
TRANSPORT	SC16	WG6	X3T5.6	—	802	ECMA-72	ISO COMPATIBLE	—
NETWORK	SC6	WG2	X3S3.3	X.25	802	—	—	—
DATA LINK	SC6	WG1	X3S3.2 X3T9	X.21	802	ECMA-82	—	TR30.2 TR40.2
PHYSICAL	SC6	WG3	X3S3.1 X3T9	X.21	802	ECMA-81	—	TR30.1 TR40.1

TC = TECHNICAL COMMITTEE
SC = SUBCOMMITTEE
WG = WORKING GROUP
ISO = INTERNATIONAL ORGANIZATION FOR STANDARDIZATION
ANSI = AMERICAN NATIONAL STANDARDS INSTITUTE
ECMA = EUROPEAN COMPUTER MANUFACTURING ASSOCIATION
CCITT = CONSULTATIVE COMMITTEE FOR TELEPHONE & TELEGRAPH
IEEE = INSTITUTE OF ELECTRICAL & ELECTRONICS ENGINEERS
NBS = NATIONAL BUREAU OF STANDARDS
EIA = ELECTRONIC INDUSTRIES ASSOCIATION
TR = EIA SUBCOMMITTEE

1. Inner ring. *Primary and secondary fiber rings transmit lightwave data in opposite directions. Class A stations—either mainframes or wiring concentrators—connect to both the primary and secondary rings. Class B stations connect only to the primary ring. Fiber cable is terminated with bulkhead connectors.*

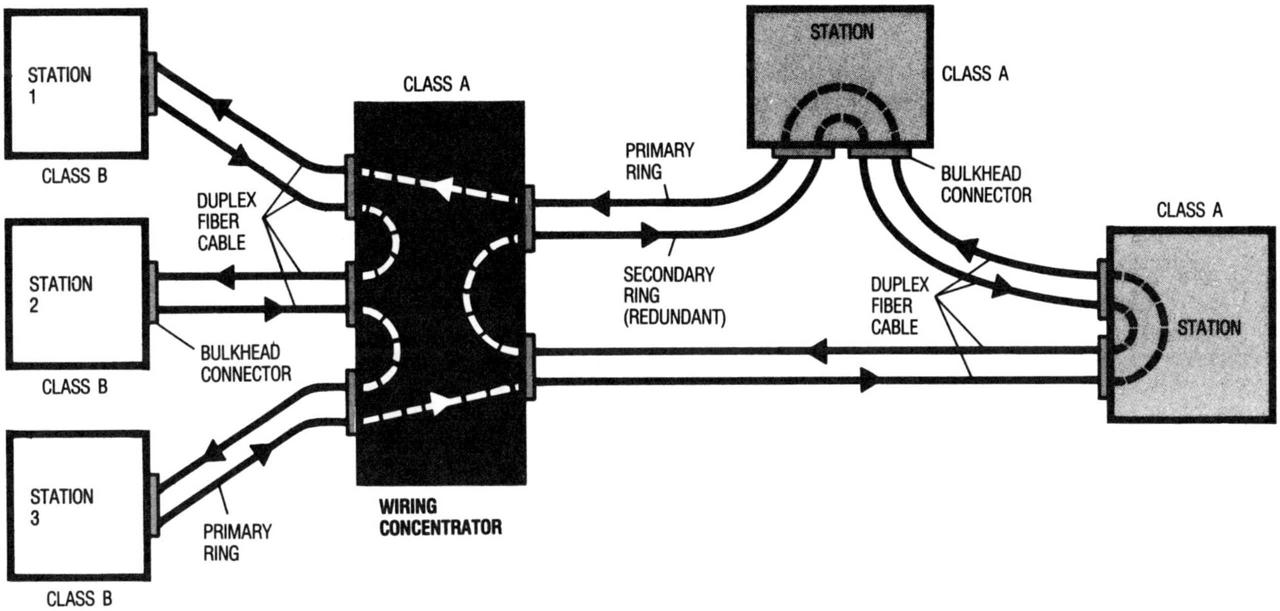

less video traffic. But as the need for these features increases with developments such as teleconferencing, higher data rates seem much more attractive. On a more practical note, low-speed networks are usually adequate for terminals and printers, but for engineering workstations, which require many huge file transfers daily, a high-speed front-end network is almost essential.

Designers have imagined many and varied fiber optic applications for years, but expense has kept many of those plans the stuff of dreams. Now, however, fiber costs are often comparable to those of coaxial cable and are expected to keep dropping. The breakthroughs have come from improvements in technology and the potential for mass production. Until recently, for example, it would have required a laser diode costing several hundred dollars to drive a fiber at over 100 MHz. This speed can be achieved now by using LEDs, priced at less than $10. In fact, one of the objectives of the ANSI FDDI specification group is to define the standard so that it can be implemented with fair-priced components within the reach of today's technology.

When cost is no longer a concern, fiber optics is a clear winner for a number of reasons. Four are:

- *Bandwidth.* Fiber-optic cables allow high data rates—in the range of several hundred megabits per second.
- *Attenuation.* Lightwaves experience relatively low signal loss, resulting in efficient communication over tens of kilometers, without repeaters.
- *Noise susceptibility.* Fiber cables transmit information as light and neither generate nor are affected by electromagnetic interference (EMI). This makes it ideal for use in high EMI environments such as factory floors.
- *Security.* Because it is difficult to tap a fiber cable without interrupting communication, fiber is more secure from interception.

The light-ring standard

As mentioned, the FDDI is a 100-Mbit/s token-passing physical ring using fiber-optic cable. The ANSI X3T9.5 committee specifies this as a local network standard. And various corporate architects of FDDI plan to use it for their own products because the standard includes features that support the many applications required for high marketability.

For example, some would use FDDI for back-end computer room applications where the network need only span several meters. Others intend to use FDDI-based products for connections with circumferences of over 100 kilometers. The FDDI specification places no lower bounds on the number of nodes and the distance between them; at the same time, the upper limits are reasonably large, permitting a wide range of implementations. Some of the main limits are:

- Up to 1,000 nodes on the ring
- Up to 2 kilometers between two nodes
- Up to 200 kilometers ring circumference.

With the maximum 1,000-node configuration, the average node separation will be 200 meters in order to limit the ring circumference to 200 kilometers. Yet several nodes may be separated by up to 2 kilometers as long as the average separation is 200 meters.

These limits are imposed to minimize latency, or the time it takes a signal to travel around the ring. The maximum ring latency is an important parameter for some real-time network applications. And with the proposed FDDI standard, it is held to only a few milliseconds.

***2. Fault tolerance.** As shown in A, an FDDI network can sustain a single fault and still maintain the integrity of the entire network. All stations simply reconfigure on the primary ring. If a network similar to that shown in B sustains two faults, it can split into two smaller rings which remain operative internally.*

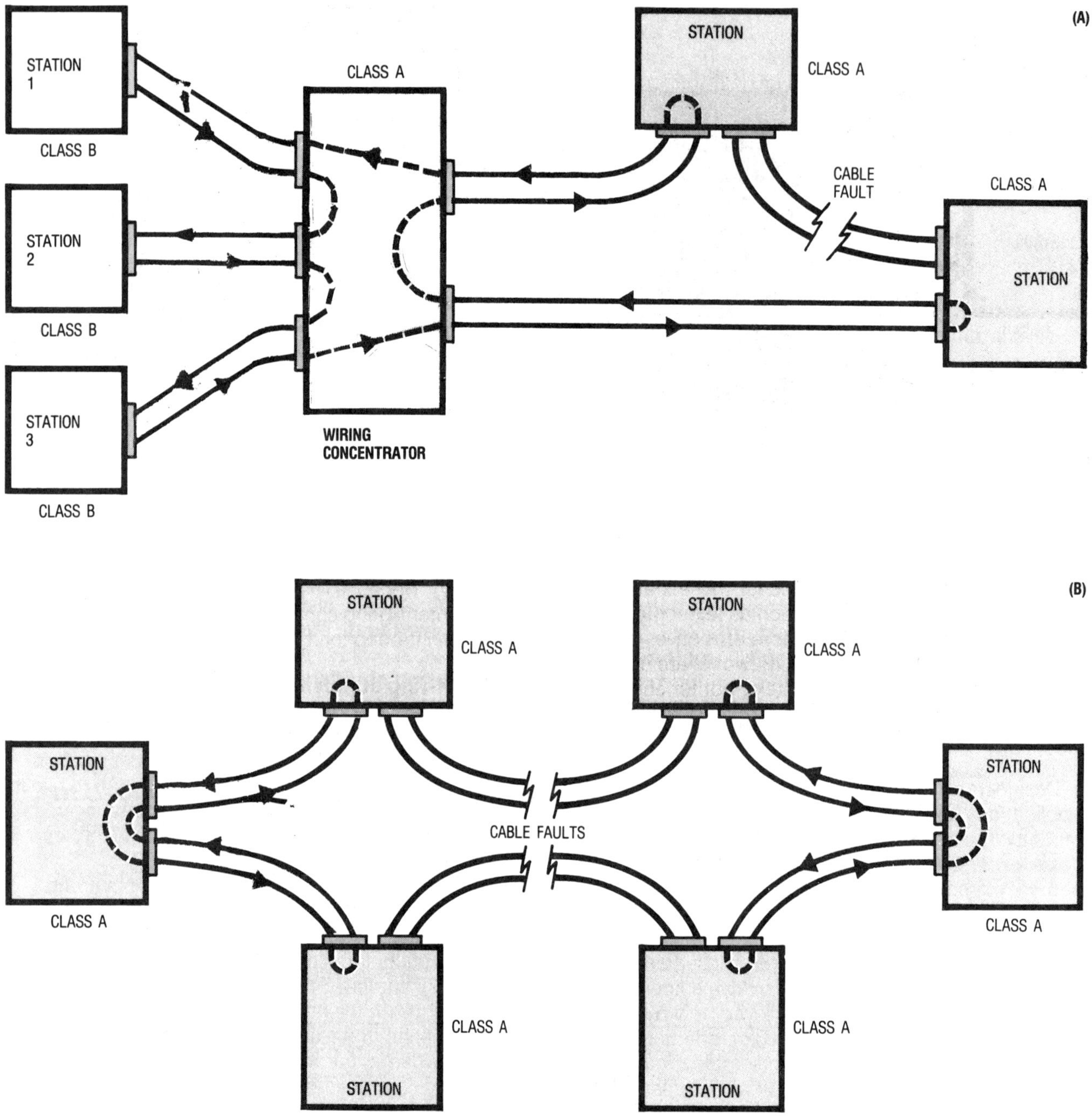

The FDDI ring is a combination of two independent counter-rotating rings, each running at 100 Mbit/s. If both rings operate simultaneously, the effective throughput is 200 Mbit/s. An FDDI scheme can actually use a special case of this where one ring connects all the nodes, and the second counter-rotating ring only a selected few.

Figure 1 illustrates a possible FDDI network configuration with fiber-optic cables forming the inner and outer rings. The paths through which the data travels around the ring are also shown. The ring that reaches all of the nodes is called the secondary ring and carries data in the opposite direction of the primary ring. The primary ring connects only the class A stations (an explanation of station classes can be found below). Such a concentric scheme is useful during ring reconfiguration. If the outer ring fails, for instance, the network can continue operating on the inner ring and still keep the intact portions of the outer ring (Fig. 2a).

In an FDDI network, nodes or stations can be of two

categories, either A or B (Fig. 1). Class A nodes connect to both the inner and outer rings; Class B nodes connect to only the outer one.

The two node classes allow users to tailor network complexity to meet cost objectives. For example, Class B stations need to connect to only one ring so they can be implemented for a lower cost. They are, however, isolated if that ring fails. On the other hand, while Class A stations may require additional hardware to connect to two rings, they offer a level of fail-safeness by being able to swap and continue operations on the good ring should the other fail. Typically stations that need more fault tolerance will be configured as Class A. Less critical stations can be Class B.

Another Class A station is the Wiring Concentrator (WC) illustrated in Figure 1. As its name implies, this device acts as a hub node through which several other stations can be connected. A WC allows a physical ring to be easily maintained, since, like a star network, failing nodes may easily disconnect. The WC can be a service point, with short connections taken out to individual stations, with several WCs so configured in a network. The WC is always a Class A station, with other Class A and B stations connecting to it.

An appealing package

The packaging of the fiber cable also makes the concept of WCs very attractive. As implied in Figure 1, the fiber cable in FDDI would contain two physical fibers packaged in one jacket with a bulkhead connector on either end. For example, for the dual ring connecting the Class A stations, each jacketed cable would contain one fiber for the primary ring and one for the secondary ring. For the Class B stations on the primary ring, the same cable would carry the incoming and outgoing data. Hence, when a Class B station is connected to a WC, only one physical cable need be routed between the two devices. Since the cable carries two fibers, a ring path is physically established.

Although implementing combined media, such as fiber and coaxial cable, is not specified in the FDDI proposal, it is easily accomplished. For example, a dual ring scheme connecting Class A stations can be built using fiber, but the shorter connections between a WC and the Class B stations can be coaxial cable. As long as the data rate is maintained and the token-ring protocol is adhered to, the existence of multiple media in a ring is transparent to the network.

The most typical faults in a network are caused either by failed components or broken cables. Even if a connection is not fully broken, there may be a substantial degradation that shows up as increased bit error rate. Figure 2A shows how an FDDI ring would reconfigure its data paths if a link—or a pair of links—in a dual ring cable became inoperative. The stations would sense the breakage and use the appropriate paths on the secondary ring to keep the network running. In FDDI this reconfiguration would happen automatically within a few milliseconds; the proposed standard defines a station management interface to facilitate this. Also, when a broken link is restored, the station management handshake will allow the ring to automatically return to its original state.

4B/5B coding

Q	00000	QUIET	
I	11111	IDLE	
H	00100	HALT (FORCED BREAK)	
J	11000	1ST OF SEQUENTIAL SD PAIR	
K	10001	2ND OF SEQUENTIAL SD PAIR	

		HEX	BINARY
0	11110	0	0000
1	01001	1	0001
2	10100	2	0010
3	10101	3	0011
4	01010	4	0100
5	01011	5	0101
6	01110	6	0110
7	01111	7	0111
8	10010	8	1000
9	10011	9	1001
A	10110	A	1010
B	10111	B	1011
C	11010	C	1100
D	11011	D	1101
E	11100	E	1110
F	11101	F	1111
T	01101	ENDING DELIMITER (USED TO TERMINATE DATA STEAM)	
R	00111	DENOTING LOGICAL ZERO (RESET)	
S	11001	DENOTING LOGICAL ONE (SET)	

V	00001	NOTE: THESE CODE PATTERNS ARE INVALID BECAUSE THEY ALLOW MORE THAN THREE CONSECUTIVE ZEROS IN A ROW.
V	00010	
V	00011	
V	00101	
V	00110	
V	01000	
V	01100	
V	10000	

Figure 2B shows the effect of a second failure. In this case, the networks split into two smaller independent networks, both of which remain operative internally.

Nuts and bolts

If a fault occurs in the cable going to a Class B station, that station is cut off from the network and the WC bypasses the node. Situations in which stations connected to the ring lose electrical power or are turned off can be handled by optical bypasses. Electrically operated relays provide pathways that allow the fiber transmissions to bypass a node—so that when power is lost, a mirror directs the lightwaves through an alternative path. A bypass can be provided in either Class A or Class B stations. Most Class A stations that include WCs will probably implement bypasses to get the additional level of fault tolerance. A Class A node without an optical bypass will appear as if all its links are broken when powered off. Optical bypasses (or even electrical bypasses) may be used in WCs at their Class B interfaces to mediate situations where the Class B stations fail.

The specification for FDDI physical layer includes the optical and mechanical characteristics of the fiber and the connectors, the power budgets, the dynamic range required in the receivers, and the encoding/-decoding scheme used.

FDDI specifies three different options for fiber: 100/400 micron diameter fiber (for example, 100 is the diameter of the fiber and 140 is the outer diameter of the cladding or jacket); 62.5/125 micron diameter fiber; and 85/125 micron diameter fiber. For all three, an 850-nanometer wavelength is specified. Bulkhead connectors specify the mechanical interface (Fig. 1).

Several of the low-speed standards, including the IEEE's, use Manchester encoding for baseband transmission. With this method, data and clock information are encoded together so each bit cell or bit period is divided in half—with line transitions occurring mid-way in each bit period. Each time a positive-going transition occurs, a "1" is sent; each time a negative-going transition occurs, a "0" is sent.

Unfortunately, Manchester encoding is only 50 percent efficient. Had it been used with FDDI at 100 Mbit/s, the encoding scheme would have required 200-Mbit/s bandwidth. Accordingly, the LEDs and PIN receivers would have had to operate at 200 MHz.

As a result, FDDI proposes a more efficient encoding scheme, called 4B/5B, to keep the baud rate down. In 4B/5B, the encoding is done on four bits at a time to create a five-cell "symbol" on the medium. The efficiency is 80 percent: to transmit 100 Mbit/s requires a bandwidth of only 125 MHz. The advantage is that inexpensive LEDs and PIN diode receivers, which operate at only 125 MHz, can be used.

The table shows the symbol encoding used in 4B/5B. It is an encoding scheme between 4-bit data "nibbles" and 5-baud symbols. The symbols are transmitted in a Non-Return to Zero Inverted (NRZI) line transmission format. The assignments between the four-bit nibbles and the five-cell symbols are chosen so that a transition is present at least once in three cells on the medium. Given an NRZI format, no more than three zeros could occur in a row, since, with NRZI, the absence of a line transition indicates a "0."

There are eight symbol combinations that violate this requirement, which should never occur in a properly operating ring. There are 32 possible code groups (or symbols) and 8 are invalid, so 24 are left for other interpretations. Sixteen of these correspond to the 16 data-values for a nibble. The remaining symbols are assigned special meaning as control characters at the physical layer. Some of the control symbols are used to derive line states such as "quiet," "idle," or "halt." Others are used as special delimiters for packets or fields within a packet. These include symbols denoted as J, K, T, R, and S. In FDDI these have been assigned specific meanings. For instance, J and K always occur in pairs and act as start delimiters for a packet (Fig. 4). The symbol T is used in conjunction with R and S as an ending delimiter.

Aside from delineating packets, these special symbols play a major role in station management handshake during ring reconfigurations. FDDI uses a token-passing ring consisting of stations serially connected by a transmission medium to form a closed loop. The packets are transmitted sequentially from one station to the next, where they are retimed and regenerated before being passed on to the "downstream" station. (Idle stations can either be bypassed or configured to function as active repeaters.) The addressed station copies the packet as it passes. Finally, the station that transmitted the packet strips it off the ring.

A station gains the right to transmit when it has the "token," which is a special packet that circulates on the ring behind the last transmitted packet. A station wanting to transmit captures the token, places its packet(s) on the ring, and then issues a new token, which the next station can capture for its transmission.

Taking turns

The access scheme in a network determines when a node has permission to transmit. The FDDI access scheme is based on a token packet which represents a permission to transmit. However, to support real-time needs of certain applications it is desirable to ensure that a node obtains a transmission opportunity (that is, assurance of getting the token) within a certain time.

The FDDI access scheme is called timed-token because it has a provision for specifying the time it takes the token to make a cycle of the ring. In essence each node on the ring measures the time it takes for the token to come back to it and compares that time with a prenegotiated target time. If the token comes back sooner than the target time, there is probably a light transmission load on the network. The node can then initiate packet transmissions on the ring as long as it does not exceed the target time. When the token comes back later than the target time, a heavy load on the ring is implied. In such cases the node is only allowed to transmit a high-priority class of packets called "synchronous packets." The node has to wait until the token rotation time (TRT) gets shorter than the target time before it transmits any of its more mundane packets that are not synchronous.

In addition, the FDDI uses the timed-token scheme to allocate the ring's bandwidth. This scheme assumes that the TRT, or the time between two successive token sightings at a station, has a linear relationship to the load on the ring (Fig. 3). The operative TRT T_{opr} is the same at all nodes on the ring. It is guaranteed in a normally functioning ring that TRT will never exceed 2 T_{opr}. The TRT would normally be less than or equal to T_{opr} transient conditions it may exceed it. Hence, nodes that have synchronous traffic (voice, for example) are assured at least one transmission opportunity in every 2.T_{opr} period. This is an upper bound on TRT; the TRT itself may vary from one revolution to the next. Thus, synchronous stations need some minimal buffering to support their traffic. The node that needs the lowest TRT determines the T_{opr} for the entire ring through a bandwidth allocation algorithm to be specified. The access scheme is independent of data rate and pro-

3. Upper limits. *TRT is a direct function of loading. T_{opr} is a prenegotiated length of time in which each node normally expects the token to return.*

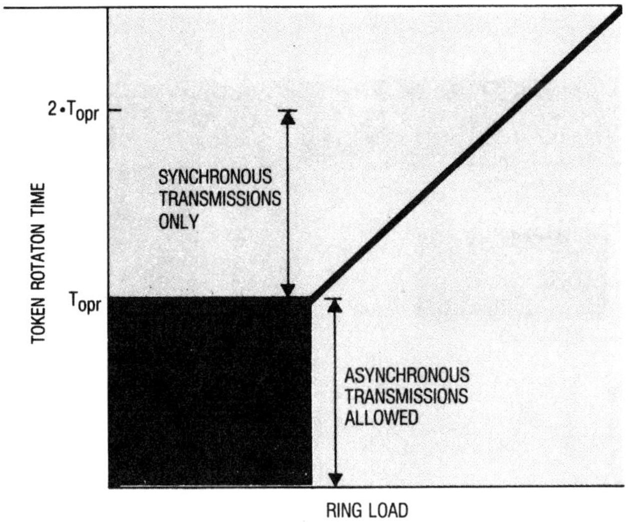

T_{opr} = OPERATIVE TOKEN ROTATION TIME, A PRENEGOTIATED EXPECTED TIME BETWEEN TOKEN SIGHTINGS ON THE RING.

vides an easy transition to higher performance rings. The total synchronous load is the sum of synchronous traffic at the synchronous nodes. Thus, the maximum synchronous load that can be supported, as shown in Figure 3, sets up the configuration limits on the ring.

A TRT less than T_{opr} implies that the load on the ring is light. Asynchronous traffic (file transfers, keyboard commands) can be supported and prioritized at such times. Low-priority packets can be transmitted only when the ring load is light—that is, when TRT is less than a certain threshold for that priority level. In this fashion, the timed-token scheme provides a simple but powerful way to support synchronous and asynchronous traffic on the same ring, with an additional capability of dynamically prioritizing asynchronous traffic as a function of ring load.

Error recovery

Ring malfunctions such as noise, babbling nodes, or configuration violations are reflected in the TRT. And, if the TRT exceeds $2T_{opr}$, then ring malfunction is implied. (FDDI also specifies an early warning timer which times out before $2T_{opr}$ and initiates recovery.) All the nodes then act in concert to recover the ring. Each starts transmitting "Claim Token" packets that contain the TRT value needed by that particular node to support its synchronous traffic. The node with the lowest TRT wins. However, in the case of a broken ring, this process does not go to completion as indicated by a TRT exceeding a threshold value. The nodes start sending out beacon packets, indicating serious ring failure, and any node that receives a beacon repeats it. This enables the fault to be localized and triggers the reconfiguration procedure, thus enabling the ring to function again—albeit at reduced performance. The FDDI standard also specifies the handshake procedures for such things as a station that is either getting onto a running ring or going to a preconfigured setup.

Figure 4 shows the format for an FDDI packet; the token is a special packet and is also shown.

The packet preamble (PA) consists of at least idle symbols. The start delimiter (SD) consists of the symbols J,K (table). The frame control (FC) is 1 byte, and indicates frame type. The destination/source address (DA/SA) can be either 16 or 48 bits long. The address length is specified in the FC field. The information field (INFO) can have from 0 to approximately 9,000 data symbols. An autodin cyclic redundancy check (CRC) comprises the frame check sequence (FCS). End-of-frame (EDF) is followed by frame status (FS), which is provided to the transmitting node by the nodes through which the packets pass.

Comparing

The IEEE 802.5 standard specifies a 4-Mbit/s token-passing ring. When the FDDI subcommittee began work it tried to maintain the services and facilities of 802.5 but modified other sections to obtain data rates up to 100 Mbit/s. A comparison between the two standards will illustrate the factors to be considered when going to higher data rate protocols. Specifically, the physical layer specification is crucial at high speeds. For a data rate of 100 Mbit/s, the group encoding limits the bandwidth to 125 MHz on the medium; as mentioned, a Manchester encoding scheme would have required an increased bandwidth of up to 200 MHz for the same 100-Mbit/s data rate.

FDDI's decentralized clocking scheme requires designing for clock variations between two stations instead of designing for the sum of clock variations on the whole ring. At high speeds, where the bit time is very small, the latter scheme would put tough (and therefore expensive) requirements on the clocking circuitry. Both schemes require latency buffers. However, the buffer in FDDI is divided among the nodes. In IEEE 802.5 each node has a buffer large enough to compensate for all clock variations on the ring. At high speeds the number of bits of buffer goes up and so the 802.5 scheme becomes impractical. The 802.5 centralized clock scheme supports packets of any length. FDDI packets are limited to around 4,500 bytes.

In the 802.5 standard a transmitting node flips a bit in the token—making the bit a start delimiter—and then appends its own packet behind the token. This means that a packet has to be ready and set up for transmission when the token appears. This also means increased hardware and costs. In FDDI the token is a special packet that is stripped off the ring by the node that wants to transmit. When a node takes the token, it puts idles on the ring while the packet is getting set up for transmission. Symbol level manipulation is sufficient in FDDI, unlike in 802.5 where bit-level manipulation is necessary.

From the above it appears that the IEEE scheme is more efficient. However, at high speeds it is the ease of processing, not the efficiency of data transmission, that causes the bottleneck. To ease the speed, size, and complexity requirements on the electronics, the price of a few bytes is an easy tradeoff. In FDDI the token is

4. Data bars. *Information is passed around the ring in frames. The PA field synchronizes the frame with the node's clock. Other fields indicate a frame's beginning, length, destination, source, contents, data integrity, and end. The token is similar to an information packet, but with no information field.*

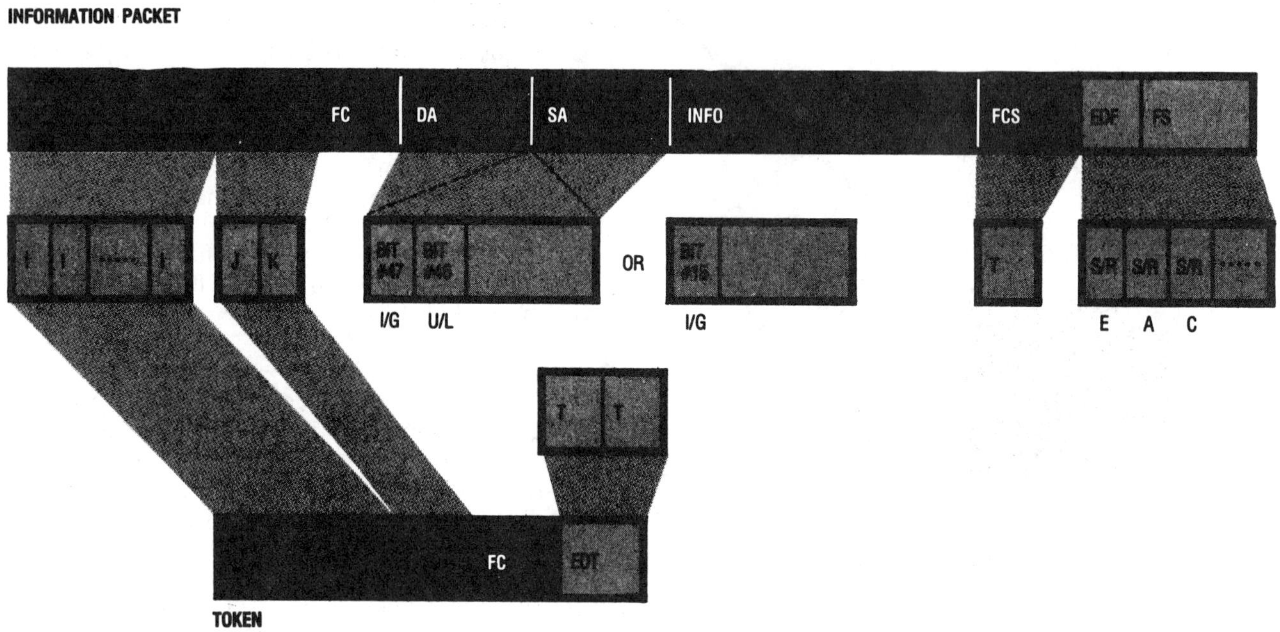

PA = PREAMBLE
SD = START DELIMITER
FC = FRAME CONTROL
DA = DESTINATION ADDRESS
SA = SOURCE ADDRESS
INFO = INFORMATION
FCS = FRAME CHECK SEQUENCE
EDF = ENDING DELIMITER (INFORMATION)
EDT = ENDING DELIMITER (TOKEN)
FS = FRAME STATUS
I/G = INDIVIDUAL/GROUP ADDRESS
U/L = UNIVERSAL/LOCAL ADMINISTRATION OF ADDRESS
E = ERROR DETECTED
A = ADDRESS RECOGNIZED
C = COPIED
= DATA SYMBOLS
= SPECIAL SYMBOLS (I, J, K, T, R, S; SEE TABLE 1)

transmitted immediately after the packet. The 802.5 configured ring waits for the transmitted packet to come back before giving out the token, putting out fill (null data) in the meanwhile. The FDDI scheme is thus more efficient, especially in large rings with a lot of latency.

The FDDI timed-token scheme sets up a traffic framework through a bandwidth allocation scheme. This framework is maintained until some node wants it changed. Synchronous traffic is easy to support on the FDDI ring, and the TRT is the single parameter that controls the ring. In 802.5 each frame has three priority bits (to indicate one of eight levels of priority) and three reservation bits to indicate the priority desired. By suitable manipulation of these fields, synchronous traffic can be supported. Note, however, that traffic is regulated on a per-packet basis.

The dual-ring option specified in FDDI provides the additional reliability and fail-safeness essential for high-performance applications. Because the 802.5 standard is aimed at the low-end market, the dual ring is not specified, though it could be supported.

The success of FDDI will depend on the amount of support it receives from component manufacturers, and the cost reductions achieved as users and manufacturers progress along the curve involved with learning how to implement the standard.

Toward that end, Advanced Micro Devices (AMD) plans to introduce the first chip in its Supernet family of high-speed networking products in mid-1985. The first chip in the family, the Link Protocol Processor (LPP), does most of the buffer management functions, and the FDDI ring-access scheme can be implemented as a module to work with it. ■

Sunil Joshi, section manager for AMD's packet networks division, is involved in product planning for high speed local networks. Venkatraman Iyer is a product planning engineer for that division.

Performance Analysis of Slotted Ring Protocols in HSLAN's

MIRJANA ZAFIROVIC-VUKOTIC, IGNAS G. NIEMEGEERS, DURK S. VALK

Abstract—**The performance of a number of slotted ring protocols supporting integration of synchronous and asynchronous traffic in HSLAN's is evaluated in this paper as follows: the Cambridge Fast Ring, a variant of the Cambridge Fast Ring, and Orwell. The performance of their basic access mechanisms is compared and contrasted to the multiple token ring. The effect of a uniframe scheme for supporting synchronous traffic is examined. The influence of system parameters and performance of the integrated services slotted ring protocols is evaluated.**

I. Introduction

THE slotted ring protocols originate from the Cambridge Ring [8]. They are suitable for high speed LAN's (HSLAN's) because of their good performance at high transmission rates [20]. Furthermore, the synchronous transmission scheme makes the implementation at high transmission rates in fiber optics easier. In this paper, the performance of the following MAC layer protocols or access mechanisms (AM's) for slotted rings is evaluated: the Cambridge Fast Ring (CFR) [17], Orwell [4], and a variant of the CFR (CFRV) [25]. Similar protocols can be found in other slotted rings, e.g., FXNET [3] and Upperbus [6]. Our analysis is applicable to these systems as well. Transmission rates in excess of 100 Mbits / s are assumed.

The performance of the basic slotted ring AM's is analyzed. This includes an analysis of the sensitivity to the configuration and workload parameters, and a comparison to the multiple token ring protocol and the uniframe scheme. Furthermore, integration of synchronous (e.g., voice, video) and asynchronous traffic (e.g., file transfer) is studied. The influence of the system parameters under both symmetric and asymmetric loads is evaluated. The performance analysis is conducted using simulation and analytical models. In the analytical modeling, the MAC service provider is regarded as a multiple cyclic server system with a limited or exhaustive service discipline.

The performance of slotted ring protocols in high speed networks has been studied in [4], [5], [12], and [13]. However, except for [13], these analyses concern telephone traffic only. Furthermore, no comparison of the various slotted ring protocols except for a limited study in [13] is available yet. In this paper, we present such a comparative analysis for both synchronous and asynchronous traffic. The sensitivity to configuration and load parameters, the influence of the system parameters, and the effect of asymmetric loads is also studied. The analytical models and the stability conditions for the slotted rings on which the analysis is mainly based have recently been developed by ourselves in [22], [24], and [25].

The paper is organized as follows. Protocols are described in Section II. Issues related to the integration of synchronous and asynchronous traffic are presented and an introduction to the performance analysis is given in Section III. Notation is introduced in Section IV. The workload model is presented in Section V. Configuration and system parameters are chosen in Section VI. A sensitivity analysis and a comparison to the token passing AM and a study of the uniframe scheme is done in Section VII. The influence of the system parameters is examined in Section VIII. A comparative performance analysis of the slotted ring protocols supporting integrated services is presented in Section IX. Section X contains the conclusions.

II. Slotted Ring Protocols

The ring is partitioned into equal length slots (Fig. 1). We assume that this is achieved by introducing a latency register located in a monitor station to virtually lengthen the ring to a multiple of the slot length. Let us call a LLC_PDU a packet and a MAC_PDU a minipacket.

Slots circulate around the ring and can be empty or full. A full slot is occupied by a minipacket. Stations are actively coupled to the ring: they repeat or modify the slots. A sending station may fill an empty slot by a minipacket. A full slot circulating around the ring reaches the destination station which reads it and passes it on to a higher layer.

There are two classes of basic slotted ring AM's depending on which station empties a full slot: the source station or the destination station. Furthermore, after emptying the slot there are two possibilities: the slot can be used by the station that empties it or it must be passed to the next downstream station.

If the source station releases a slot which was full (Fig. 2), there are two variants of this AM depending on the maximum number of slots that can be used by one station at a time. Either at most one slot at a time can be occupied by a station or more than one slot at a time can be carrying

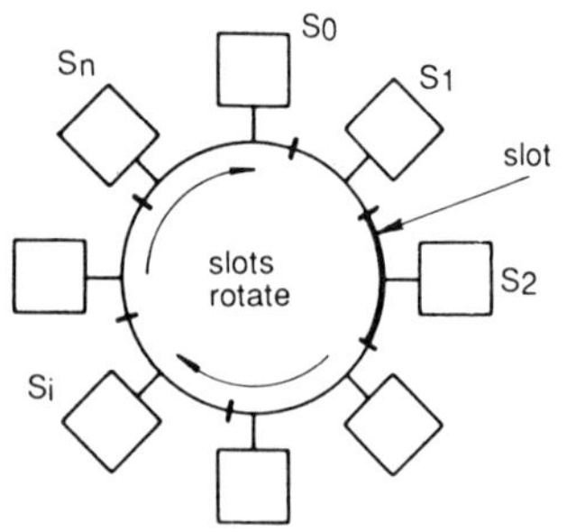

Fig. 1. The slotted ring structure.

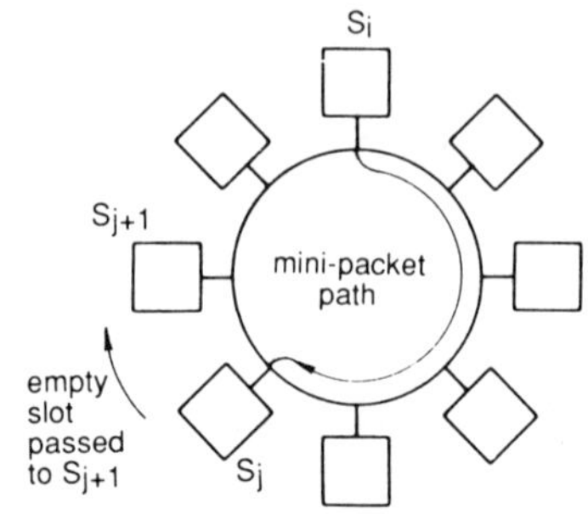

Fig. 3. Path of a minipacket sent from S_i to S_j in the Orwell basic AM.

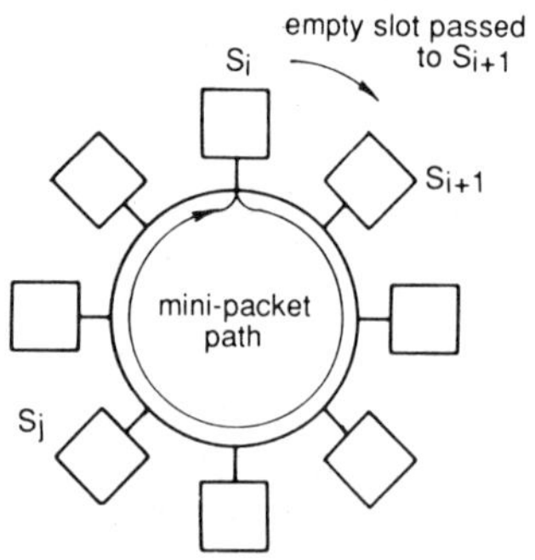

Fig. 2. Path of a minipacket sent from S_i to S_j in the CFR and the CFRV basic AM.

minipackets from the same source. An example of the former protocol can be found in the CFR. Examples of the latter protocol are the CFRV, FXNET, and Upperbus. In all these protocols, the source station after emptying the slot has to release it and pass it on to the first downstream station. An exception are the channel slots in the CFR and in the CFRV which can be reused by the source.

If the destination station releases a slot which was full (Fig. 3) typically more than one slot at a time can be carrying minipackets from the same source. An example of such a protocol can be found in Orwell. After emptying the slot a destination station has to pass it on to the next station. Fairness is not guaranteed and precautions have to be taken to prevent a station from hogging the ring.

In the case of source release, an acknowledgment from the destination to the source station is provided using the same slot. If only one slot at a time can be used by a station an immediate retransmission of an erroneous minipacket is possible. In the case of destination release, no acknowledgment is provided at the MAC layer. In this paper, we do not study acknowledgment or retransmission mechanisms. We assume that each station is capable of using every empty slot that arrives and of reading every slot destined to itself.

Let us now describe the AM's of the CFR, the CFRV, and Orwell.

The CFR distinguishes two types of slots: normal and channel slots. Both kinds of slots are subject to source release. A normal slot must be passed by the source to the first downstream station. This way a periodical access of each station to this slot is ensured. A channel slot, on the other hand, can be reused by the source. Channel slots are intended for large bandwidth or bursty sources. In the CFR, only one slot at a time can be used by a station. When we relax this condition and permit that more than one slot at a time can be used by a station, we obtain the CFRV. The CFR is a relatively inexpensive and simple LAN which has been designed for short ring lengths and moderate transmission rates, e.g., up to 100 Mbits / s. We compare it to the other slotted rings working at high transmission rates.

In Orwell, slots are released by the destination. It offers two service classes: class 1 for delay sensitive services, requiring a guaranteed bandwidth, e.g., synchronous traffic, and class 2 for delay tolerant services, e.g., asynchronous traffic. Access is organized in cycles to ensure fairness. Each station S_i in the ring has a counter D_i [4], indicating the number of packets allowed to be sent during a so called reset interval or a cycle. During a cycle each station can send up to $D_i - D_{\min}$ packets of class 1 (we denote this value by $D_{i,1}$) where $D_{\min}$ provides a guaranteed (safeguard) bandwidth for class 2 packets (we denote this value by $D_{i,2}$). Since each station has independent counters, the throughput can be regulated according to the station's needs. After all the stations have used their counters or have nothing to send, a reset is issued which resets all the counters to their initial value and starts a new reset interval.

We distinguish three basic AM's which correspond to cases where in the CFR and in the CFRV only normal slots are used, and in Orwell it is assumed that reset intervals do not exist, i.e., $D_{i,1}$ and $D_{i,2}$ are infinite.

III. Scope of the Performance Analysis

The main characteristics of slotted rings in HSLAN's compared to their use in medium speed LAN's (transmission rate around 10 Mbits / s, e.g., IEEE P802.3, P802.4, and P802.5), which are relevant for modeling and performance analysis, are the following ones: 1) synchronous traffic sources and large bandwidth sources create most of the load [11]; such sources create a different type of load than the smaller bandwidth sources in medium speed LAN's, 2) the information field of a slot contains more bits in a HSLAN than in a medium speed LAN, so the number of slots per packet is typically smaller in HSLAN's than in medium speed LAN's for the same application, and 3) typically, there are more slots in a HSLAN than in a medium speed LAN given the same configuration.

Synchronous traffic also demands a different quality of service (QoS) than asynchronous traffic. Small packet delays and a regular access to the medium have to be provided. There are two basic ways in which synchronous and asynchronous traffic can be integrated in a LAN: 1) by packetizing the synchronous traffic (e.g., a 64 kbit/s source produces 512 bit packets every 8 ms), and 2) by sending the synchronous traffic using a PCM-like transmission, i.e., one byte every 125 μs. The former solution demands an upper bound on the delays for the synchronous traffic at the MAC layer and the latter one called a "uniframe scheme," requires a frame structure to support both PCM-like circuit switched channels and asynchronous traffic packet switching.

The CFR, CFRV, and Orwell implement the former way of integration. In Orwell, access is organized in cycles to provide fairness and ensure fulfillment of the QoS requirements for synchronous traffic. In the CFR and the CFRV, such a mechanism does not exist at the MAC layer. However, in these protocols there is a minimum bandwidth available to each station so the QoS given to the synchronous traffic can be guaranteed, provided synchronous traffic uses less than this minimum bandwidth and provided that it is given priority over asynchronous traffic at transmit time.

Examples of a uniframe scheme are FDDI II [14], Σ Network [7], and FXNET [3]. Let us now briefly describe it. In a frame of duration 125 μs, there are a number of bytes used for synchronous transmission. Some are also used for (protocol) control information. Other bytes are available for asynchronous traffic using a LAN protocol: in our case, a slotted ring protocol or a token passing protocol. If a slotted ring AM is used, we assume as it has been done in FXNET that an integer number of slots fit into the remaining part of the frame.

In this paper, we first study the performance of the basic slotted ring AM's. Then we study the first type of integration of the two traffic classes in more detail and compare it to the uniframe method. A reason to study in detail the first type of integration is that the slotted ring protocols perform very well with short packets which are typical and desirable for synchronous traffic. Note that the merits of the two integration schemes depend not only on the performance of the MAC layer, but also on the technical and economical feasability.

Only a few papers concerning performance analysis of slotted ring protocols in HSLAN's are available in the literature. In [4], it has been shown that the delays in Orwell are much smaller than in the token ring for voice only traffic. In [5], the performance of the Cambridge ring, used also only for telephony, has been evaluated. In [12], the blocking probability in Orwell for telephone traffic has been approximated. A scheme for integration of voice and data based on the Orwell basic AM has been proposed in [9]. The scheme has not been described in enough detail to be used as a basis for further study. A limited performance comparison in [13] of the CFR, the CFRV basic AM's, and a simple (not a multiple) token passing ring operating at 100 Mbits/s shows that the slotted rings perform well with short packet lengths.

We define packet delay as the duration of the MAC layer service per packet, i.e., the packet delay from arrival at S_i till its complete delivery at the destination. We obtain the estimates of the expected packet delay at S_i. Simulations provide also the maximum packet delay which is the maximum packet delay observed during a simulation.

Let us now introduce the simulation and analytical models.

The simulation models are detailed models of the MAC protocols. They are written in SIMULA and are documented in [10] and [18]. Ninety-five percent level confidence intervals have been obtained on estimates of the expected delay except when the correlation between the batches was too large (see [18] and [2]).

The analytical models have been developed and tested by us in [22] for Orwell, in [24] for the CFR, and in [25] for the CFRV. An overview of analytical models is given in [21]. They provide exact stability conditions (though without a proof) and approximations of the expected packet delays. The accuracy of the models has been tested by comparing to simulation results for a number of cases thought to be representative of HSLAN's. The analytical models provide estimates that have a deviation up to 20 percent of the point estimates obtained by the simulations over a range of utilizations which go from 0.4 to 1.3 for Orwell, from 0.0 to 0.8 for the CFRV, and from 0.0 to 0.7 for the CFR (deviation even less than 10 percent for the CFR). At high utilizations the errors are larger, however, the approximations are sufficiently good and indicative for a comparative analysis. The exact solution of the token ring model as a cyclic server system with an exhaustive service discipline has been given in [16]. The same models are used for uniframe schemes. The load due to synchronous traffic and uniframe control information is accounted for by decreasing the available capacity for asynchronous traffic. This has been done by decreasing the transmission rate in the model. The error in delay estimates caused by this is less than a uniframe duration, i.e., 125 μs.

IV. Notation

Let us introduce the following notation: n—number of stations minus one; S_i—ith station in the ring, $i = 0, 1, \cdots, n$; w—transmission rate (Mbits/s); σ—duration of a slot (μs); s—number of slots in the ring; λ_i—packet arrival rate at station $S_i (1/\mu s)$, $i = 0, \cdots, n$; γ_i—the expected number of minipackets a packet is split into at S_i, $i = 0, \cdots, n$; ρ—relative load, or the expected number of minipackets arriving in the system during σ time units, such that

$$\rho = \sum_{i=0}^{n} \lambda_i \gamma_i \sigma; \qquad (1)$$

p_{ij}—an element of the packet communication source_to_destination matrix, $\| p_{ij} \|_{n+1 \times n+1}$ which represents the relative traffic intensity from source S_i to destination S_j, and

$$0 \le p_{ij} \le 1 \quad \text{and} \quad \sum_{j=0}^{n} p_{ij} = 1,$$
$$i, j = 0, \cdots, n; \tag{2}$$

τ_{ij}—propagation time from S_i to S_j including the latency at station $S_j(\mu s)$, $i, j = 0, \cdots, n$; τ—slot rotation time or τ_{ii} for all $i(\mu s)$, such that

$$\tau = s\sigma. \tag{3}$$

V. Workload Model

Let us now specify a workload model representative for a HSLAN. It includes the arrival process of packets and of minipackets, the distribution of their lengths and the load specification.

Packets are segmented and MAC protocol control information (PCI) is added to form a number of minipackets. The information field of a minipacket in a slotted ring protocol has a constant length. The length of a minipacket is a constant and equals $\sigma \times w$ (bits) i.e., a slot length.

A study of the scenarios and workload models for a HSLAN conducted in [11] shows that it can be expected that in an office environment a great deal of the total load consists of synchronous traffic. We assume that synchronous traffic represents 80 percent and asynchronous traffic 20 percent of the total offered load at each station. Let us now specify the arrival process and the packet length distribution for synchronous and asynchronous traffic.

1) Synchronous Traffic: We denote this traffic by $T0$. If synchronous traffic is sent using a uniframe scheme, only a specification of the intensity is needed. If synchronous traffic is packetized, we use the following model. We assume that each packet has a constant length and fits into one slot. In our case, packet lengths are 512 bits and are equal to the slot information field length. Use of a Poisson arrival process for synchronous traffic cannot be justified (see [23]). We assume a bulk deterministic arrival process where the interarrival times are constant and determined according to the total load. In reality, the arrivals of synchronous traffic packets have a random but a periodic behavior, i.e., given a number of connections, the number of packets arriving in a packetization interval is a constant (possibly with small deviations). The deterministic arrival process has been chosen to model the regularity of the arrival process of synchronous traffic. The bulk arrival process represents in a conservative way the fact that not all the arrivals are equidistant in time. The bulk size is chosen to be 4 which is a realistic value but could be subject to discussion.

2) Asynchronous Traffic: We assume a Poisson arrival process and a bimodal packet length distribution which has been argued to be a good one for LAN's [19]. So we distinguish two packet lengths. We denote packets belonging to the group with shorter average packet length by $T1$ and those belonging to the group with longer average packet length by $T2$. $T2$ represents, e.g., file transfers and $T1$, e.g., management messages, electronic mail, etc. We assume the following packet lengths: 1024 bits for $T1$, and 8192 bits for $T2$. The relative load (in terms of offered bits per time unit) is taken as follows: 13 percent of $T1$, and 87 percent of $T2$.

Two load types and traffic patterns have been used: a symmetric and an asymmetric one. In the symmetric case, the load intensity at each station S_i, $i = 0, \cdots, n$, is the same and

$$p_{ij} = \begin{cases} \frac{1}{n}, & j \neq i \\ 0, & j = i \end{cases} \qquad i, j = 0, \cdots, n. \tag{4}$$

In the asymmetric case, one station S_0 is taken to be different from the others. In this case, the load at S_0 is $1/3$ of the total load. At all other stations S_i, $j = 1, \cdots, n$ it is equal. p_{ij} take the following values:

$$p_{ij} = \begin{cases} \frac{1}{2}, & j = 0 \\ \frac{1}{2} \cdot \frac{1}{n-1}, & j \neq i, j \neq 0, \\ 0, & j = i \\ i = 1, \cdots, n, j = 0, \cdots, n, \end{cases} \tag{5}$$

and

$$p_{0j} = \begin{cases} \frac{1}{n}, & j \neq 0 \\ 0, & j = 0 \end{cases} \qquad j = 0, \cdots, n. \tag{6}$$

This asymmetric load has been taken as a representative of such type of load, not as a model of a particular application. However, S_0 could represent, e.g., a high throughput gateway.

VI. Configuration and System Parameters

Configuration and system parameters typical for HSLAN's have been taken as follows: *configuration:* cable length = 5 km, number of stations = 40 or 10, *system parameters:* transmission rate = 140 Mbits/s, slot information field length = 512 bits, overhead in slot = 48 bits, and latency register at each station = 24 bits. An equal distance between the neighboring stations has been assumed, so $\tau_{i,i+1} = \tau_{j,j+1}$, $i, j = 0, \cdots, n$. We assume that the preamble and the MAC_PCI length in a token ring add up to 200 bits. We assume that the length of the overhead in a uniframe scheme equals 84 bits.

A common slot information field length of 512 bits has been chosen for all the slotted rings. In the CFR [17], the slot information field is 256 bits and in Orwell [4] 128 bits. We do a comparative performance analysis of these protocols and take the same information field length for all the protocols. The overhead (MAC_PCI) length is taken to be 48 as in the original protocols.

An analysis concerning the system parameters has been done for the case with 10 stations. The reason is that the protocols are more sensitive to the values of system parameters if the number of stations is smaller (see [22], [24], and [25]). However, the case with 40 stations is a more realistic one for the office environment. A delay analysis and a sensitivity analysis for the configuration and workload parameters of these protocols have been done for the case with 40 stations.

VII. A Comparative Analysis of the Basic Slotted Ring AM's and the Token Ring

The basic slotted ring AM's can be modeled as multiple cyclic server (Fig. 4), as has been done in [22], [24], and [25]. A slot corresponds to a server. The service discipline is limited, so only one minipacket can be served per visit to the queue. The switchover time is constant and equals $\tau_{i,i+1}$. The service time in the CFR and CFRV is also constant and equals τ. The Orwell basic AM has been studied in [22] under a symmetric load only, i.e., for

$$\lambda_i = \lambda_j, \quad \gamma_i = \gamma_j, \quad i, j = 0, \cdots, n. \tag{7}$$

In Orwell, the service time is a random variable X_i such that

$$P\{X_i = \tau_{i+1,j+1}\} = \frac{1}{n}, \quad j \neq i, \quad j, i = 0, \cdots, n.$$

In Orwell, the order of visiting the queues by the servers is random and in the CFRV it is a regular one. In the CFR, the service is performed only if there is no other server serving the same queue.

Let us present now the exact stability conditions which have been proposed in [22], [24], and [25]. We will use them to compare the capacity of the protocols. They are independent of the interarrival time and packet length distributions except for their first moments. For the CFR, it has been shown [24] that the system may have a few stable patterns when all the queues are instable. The necessary and the sufficient condition for all queues to be stable are different. The necessary condition is the following one:

$$\lambda_i \gamma_i \sigma < 1 - \rho, \quad i = 0, \cdots, n, \tag{8}$$

and the sufficient one is

$$\lambda_i \gamma_i \tau < 1 - \rho, \quad i = 0, \cdots, n. \tag{9}$$

For a symmetric load relations, (8) and (9) become, respectively,

$$\rho < 1 - \frac{1}{n+2} \tag{10}$$

and

$$\rho < 1 - \frac{s}{s+n+1}. \tag{11}$$

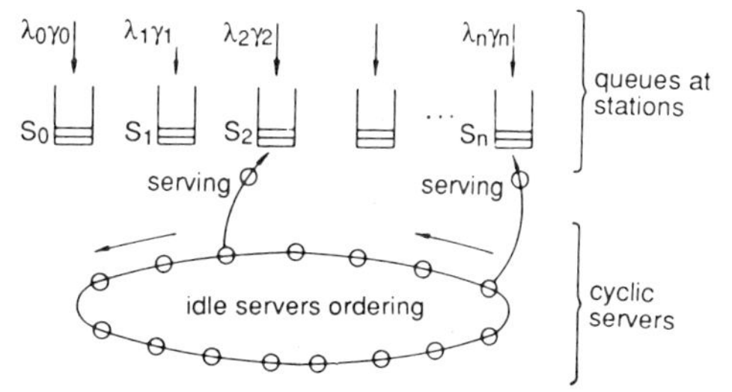

Fig. 4. A multiqueue multiple cyclic server model of a slotted ring.

For the CFRV relations, (8) or (10) represent both a necessary and a sufficient stability condition. For the Orwell basic AM under a symmetric load, the following necessary and sufficient stability condition holds:

$$\rho < 2 - \frac{4}{n+3}. \tag{12}$$

Relation (12) shows that under a symmetric load the capacity of Orwell is close to twice the transmission rate. It is also about two times larger than the one of the CFRV [see (10)]. As expected, the CFRV is more efficient than the CFR. This is apparent from the fact that the sufficient stability conditions are different. Note that the stability conditions depend on ρ and n but not on s except for the sufficient condition of the CFR.

Let us now do a comparative sensitivity analysis of the delays of the basic slotted ring AM's by means of the analytical models developed in [22], [24], and [25]. The results are compared to the multiple token ring with an exhaustive service discipline, i.e., the target token rotation time is infinite. Exact results from [16] have been used for the token ring modeling. A comparative analysis of a uniframe integration scheme will also be presented.

The workload introduced in Section V yields an average packet length of 621 bits. If we choose the packet length of the synchronous traffic equal to the slot information field length, no bits are wasted. Synchronous traffic represents 80 percent of the total load. So, we determine the average packet length as the average packet length observed on the medium, i.e., it is calculated starting from the number of minipackets appearing in the ring. An exponential packet length distribution is assumed and a configuration with 40 stations is assumed.

The *packet delay versus load* is depicted in Fig. 5. The Orwell basic AM has the best performance. The CFRV basic AM performance is the next best and has very low delays at loads <80 Mbits/s. The CFR basic AM has larger delays than the CFRV due to the fact that only one slot at a time can be used by a station which increases the waiting times for access to the medium. The token ring has the worst performance. We attribute this to its efficiency for short packet lengths, e.g., less than 1000 bits (see also Fig. 9).

The *packet delay versus the number of slots*, or the ring length, is shown in Fig. 6. The load is 80 Mbits/s. The delay increase for the CFRV and the Orwell basic AM is linear and equals the increase in the propagation delay.

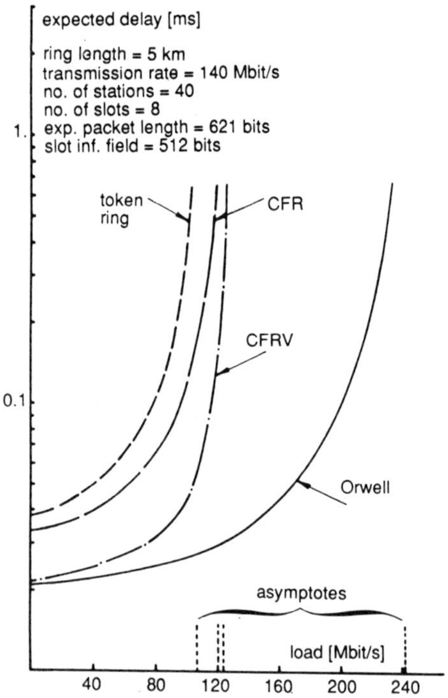

Fig. 5. Packet delay versus load in the basic AM's.

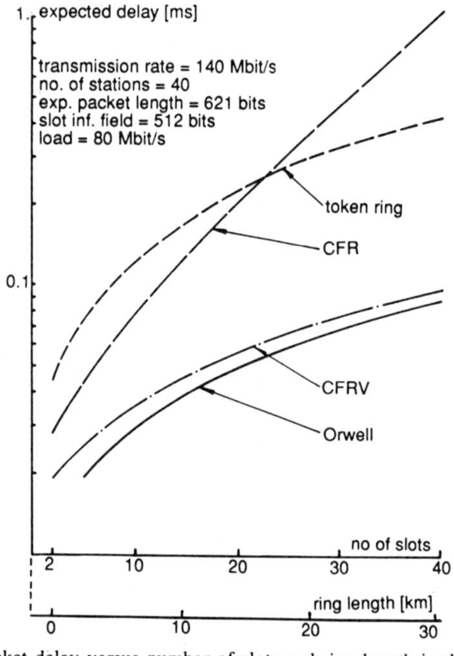

Fig. 6. Packet delay versus number of slots and ring length in the basic AM's.

So, these two AM's perform well with different numbers of slots. The token ring has a nonlinear and much larger increase in the delays than the CFRV and Orwell because, in addition, longer ring delays cause longer switchover times. In the slotted rings, these are compensated by introducing new slots which increase the capacity (i.e., the number of servers). The CFR has a degradation of performance if the number of slots increases (i.e., the ratio between the number of stations and the number of slots decreases) due to the fact that only one slot at a time can be used by a station.

The *packet delay versus the number of stations* is presented in Fig. 7. The load is 80 Mbits / s. The ring delay is kept constant. So, the addition of a new station does not introduce extra latency. The CFR is the most sensitive to the number of stations. This is because of the change of the ratio between the number of stations and the number of slots. The others are sensitive only for a small numbers of stations, e.g., less than 15. All delays converge to horizontal asymptotes. This is a property of cyclic servers systems (see [16], [22], [24], and [25]). In the slotted rings, the packet delays decrease with an increasing number of stations and in the token ring they increase. We attribute this to the different service disciplines, namely, the limited discipline in the slotted rings and the exhaustive discipline in the token ring. If the number of stations is smaller, the probability that the server is serving the queue where a packet arrives is larger and the expected waiting time is smaller in case of an exhaustive discipline. With a limited service discipline a more regular service is provided resulting in smaller delays for a larger number of stations.

The *packet delay versus transmission rate* is shown in Fig. 8. The relative load is held constant at 0.625, e.g., 80 Mbits / s load at 140 Mbits / s transmission rate. Above 200 Mbits / s, the CFR has increasing delays, mainly due to the increase of the number of slots. In the CFRV, Orwell and the token ring delays decline. The Orwell and the CFRV basic AM's have much smaller delays than the token ring in the case studied. The delay curves for the CFRV and Orwell eventually have the same asymptote at the expected propagation time from source to the destination, i.e., $\tau/2$ in the limit. The token ring converges to an asymptote which equals the sum of the expected propagation time from source to destination and a nonzero waiting time for the token. In case of the slotted ring, increasing the transmission rate implies more slots in the ring, i.e., more servers. In a token ring, the service time (i.e., transmission time) of a packet tends to zero, however, due to a constant relative load the token is held for a nonvanishing time at each station.

The *packet delay versus average packet length* is shown in Fig. 9. The load is 80 Mbits / s. The increase in the delays of the CFRV and Orwell is caused by longer transmission times. The transmission time is dependent on the number of minipackets per packet (i.e., on the packet length and the slot information field length) and on the slot duration. Note that the queueing delays (packet delays excluding the transmission delay) are the same for short and long packets given the load intensity and the average packet length. The CFR has much larger delays than the CFRV if the packet lengths are larger because in addition to the effect of longer transmission times only one slot at a time can be used by a station. The delays in the token ring decrease with increasing packet length till about 1300 bits and then they increase. If the average

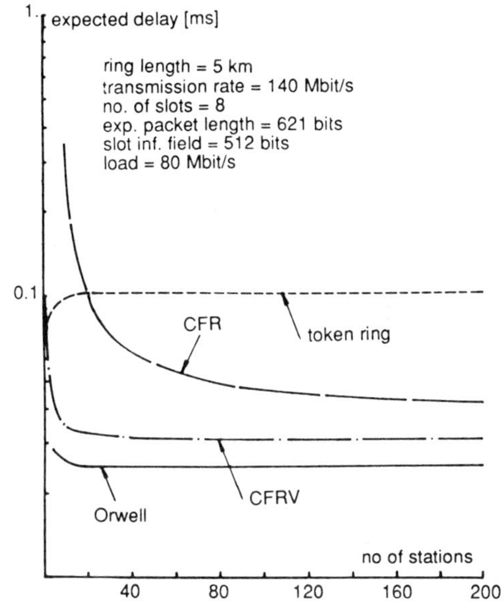

Fig. 7. Packet delay versus number of stations in the basic AM's.

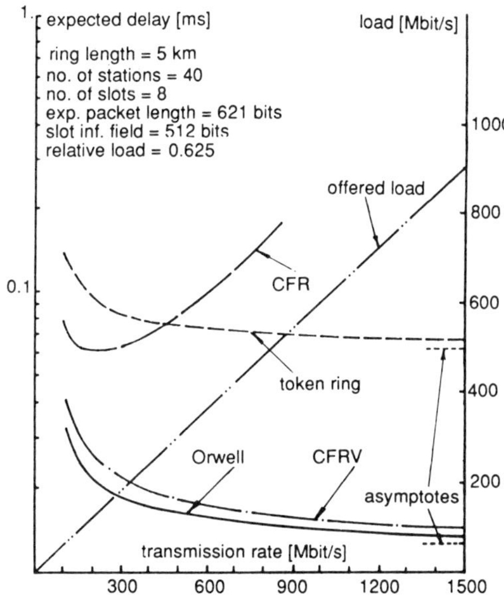

Fig. 8. Packet delay and offered load versus transmission rate in the basic AM's.

packet length is short then the token passing AM has a small throughput and a bad performance. This can be illustrated as follows. Assume that all the packets have a constant packet length and a layout which is the same as for the minipackets. The token passing AM with an exhaustive service discipline is worse than the CFRV provided that on the average the transmission at a station is shorter than the ring latency. In addition to this, if the packet length is short the ratio between the preamble and PCI length (assuming an asynchronous transmission scheme) and the data information field length in a MAC_PDU is large and the token passing AM inefficient. On the other hand, if the average packet length grows, the transmission time also grows and so do the delays. The

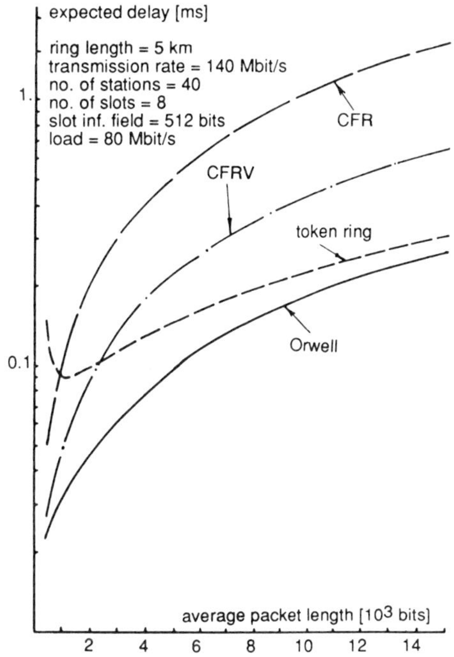

Fig. 9. Packet delay versus average packet length in the basic AM's.

token ring has a worse performance than the CFRV if the average packet length is small relative to the slot information field length, e.g., in our case less than about 2300 bits. If the average packet length is large relative to the slot information field length, e.g., 15 000 bits in Fig. 9, the delays in the token ring are comparable to those in the Orwell basic AM. This demonstrates that in typical HSLAN's (without a uniframe) which are expected to carry a lot of synchronous traffic in short packets, slotted rings have an advantage in terms of throughput and expected delay over a token ring. This advantage is very pronounced for the CFRV and Orwell.

If a *uniframe scheme* is used, very small and constant delays are ensured for synchronous traffic. The *delays of the asynchronous traffic versus asynchronous traffic load* are depicted in Fig. 10. An average packet length of 4319 bits has been taken. This is according to the workload of Section V. An exponential packet length distribution is assumed. Again, we determine the average packet length as the average packet length observed on the medium, i.e., it is calculated starting from the number of minipackets appearing in the ring. The synchronous traffic load is 80 Mbits/s if full duplex connections are assumed. The delays in the CFR basic AM are much larger than in the other protocols. This is because the CFR performs badly for large packet lengths. The token ring performs better than the CFRV basic AM because of the relatively long average packet length. An additional reason is that some capacity is wasted in the slotted ring protocols since an integer number of slots must fit in a frame leaving some unused capacity. In token passing rings this does not occur. At low loads delays of the token ring are larger than in the CFRV due to waiting for access to the medium,

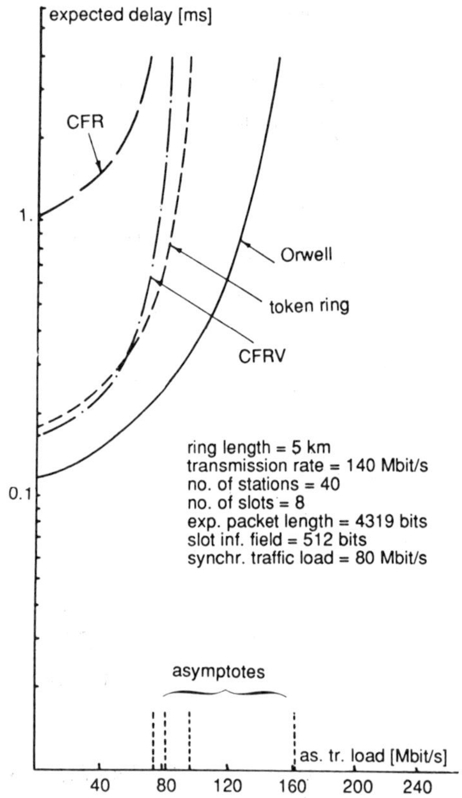

Fig. 10. Packet delay asynchronous traffic versus asynchronous traffic load in the basic AM's with a uniframe scheme.

i.e., waiting for a token. The maximum capacity of Orwell stays approximately the same, i.e., 242 Mbits/s. This is mainly because of the destination release of the slots in the Orwell basic AM which already almost provides a full duplex connection if the number of stations is large relative to the number of slots. Note that if the number of stations is small, e.g., 4 the usage of uniframe over Orwell may improve the performance significantly [26]. For all the other basic AM's when a uniframe is used the maximum carried load increases, e.g., 162 versus 124 Mbits/s for the CFRV and 175 vesus 107 Mbits/s for the token ring (see also Fig. 5). Hence, usage of a uniframe does improve performance both for synchronous and asynchronous traffic for the token ring and the slotted ring protocols. It has the most effect on the token ring. The price for using a uniframe scheme is protocol complexity since it has to integrate both PCM-like circuit switched channels and a MAC protocol for data.

The slotted ring protocols are clearly advantageous if the average packet length is small and that happens when synchronous traffic is packetized. We present further on a performance analysis of these protocols for that case.

VIII. Influence of the System Parameters

In this section, we study by means of simulation the three slotted ring protocols under both symmetric and asymmetric loads in the cases when 1) the number of normal and channel slots changes in the CFR and the CFRV and 2) the D_i allocation changes in Orwell. We use a configuration with 10 stations which implies seven slots in the ring. A queue prioritizing the synchronous traffic has been assumed.

A. Number of Normal and Channel Slots

Normal slots can be viewed as cyclic servers with a limited service discipline meaning that after carrying a minipacket from a particular station another station must be offered the service. Channel slots can be viewed as cyclic servers with an exhaustive service discipline. So, they serve a queue until it becomes empty. Only then the service is offered to another station. The fairness of the limited service discipline is provided at the expense of ceasing the service at a queue which is not empty. On the other hand, an exhaustive service discipline permits an unlimited possession of a server by a station and thus hogging of the ring. This property is undesirable. Since a queue is served exhaustively, this discipline, for a Poisson arrival process, provides smaller expected packet delays than the limited service discipline (see, e.g., [16] for a single cyclic server system). At low relative loads, smaller maximum packet delays are expected, but at larger relative loads, the unfairness of the service may result in too large maximum delays.

A symmetric load of 100 Mbits/s has been chosen. Results of the simulations for the CFR are presented in Table I and for the CFRV in Table II.

Let us analyze the CFR. With an increasing number of channel slots in the ring, the expected packet delays of the three traffic types ($T0$, $T1$, and $T2$) decrease. The maximum packet delays also decrease for all traffic types except for $T1$ if the number of channel slots equals 7 where it gets slightly larger again. Let us analyze the CFRV. With an increasing number of channel slots in the ring both the expected and the maximum packet delays for $T1$ and $T2$ decrease. The number of channel slots does not influence the performance of $T0$ very much. The reason for the observed behavior is the same as was given in the introductory paragraph of this section.

All the expected and maximum packet delays are smaller in the CFRV than in the CFR. The only exceptions concern the maximum packet delays which are larger in the CFRV for $T1$ and $T2$ in the case of 0 channel slots, and for $T0$ in the case of 7 channel slots. The ratio between the expected packet delay and its maximum can be used as an indication of the variance of the delays. For example, a larger expected value but a smaller maximum value of the packet delay in the CFR than in the CFRV, indicates that the variance of the delay in the CFR is much smaller. Both in the CFR and the CFRV under a symmetric load and if the number of channel slots equals 4, the performance is very good for all traffic types. Introducing more channel slots does not improve it. More channel slots increase the risks associated with unfairness.

The delays are much larger under an asymmetric load for the CFR than for the CFRV, so we take different load

TABLE I
PACKET DELAYS VERSUS NUMBER OF CHANNEL SLOTS IN THE CFR UNDER A SYMMETRIC LOAD OF 100 Mbits / s

traffic type	no of chan. slots	exp. packet delay [ms]	95% conf. int. [ms]	max packet delay [ms]
T0	0	0.145	0.003	1.052
	3	0.100	-	0.921
	4	0.097	0.001	0.842
	7	0.092	0.001	0.549
T1	0	0.289	-	4.131
	3	0.142	0.010	1.744
	4	0.132	0.009	1.488
	7	0.122	0.009	1.507
T2	0	0.825	-	3.481
	3	0.544	0.010	1.866
	4	0.526	0.008	1.843
	7	0.508	0.008	1.798

TABLE II
PACKET DELAYS VERSUS NUMBER OF CHANNEL SLOTS IN THE CFRV UNDER A SYMMETRIC LOAD OF 100 Mbits / s

traffic type	no of chan. slots	exp. packet delay [ms]	95% conf. int. [ms]	max packet delay [ms]
T0	0	0.087	-	0.938
	3	0.062	-	0.634
	4	0.067	-	0.715
	7	0.072	-	0.783
T1	0	0.273	-	4.840
	3	0.137	0.013	1.364
	4	0.119	0.008	1.257
	7	0.098	-	0.716
T2	0	0.633	-	5.291
	3	0.279	0.015	1.940
	4	0.241	0.012	1.595
	7	0.182	0.007	0.897

TABLE III
PACKET DELAYS VERSUS NUMBER OF CHANNEL SLOTS IN THE CFR UNDER AN ASYMMETRIC LOAD OF 40 Mbits / s

station	traffic type	no of chan. slots	exp. packet delay [ms]	95% conf. int. [ms]	max packet delay [ms]
S0	T0	0	0.129	-	0.768
		3	0.114	0.002	0.552
		4	0.112	-	0.548
		7	0.109	0.002	0.548
	T1	0	0.071	-	0.670
		3	0.065	-	0.515
		4	0.062	-	0.500
		7	0.058	-	0.493
	T2	0	0.569	-	1.107
		3	0.462	-	0.838
		4	0.457	-	0.850
		7	0.452	-	0.839
S5	T0	0	0.195	-	0.771
		3	0.144	-	0.545
		4	0.141	-	0.550
		7	0.139	-	0.536
	T1	0	0.764	-	6.171
		3	0.212	-	1.509
		4	0.211	-	1.513
		7	0.205	-	1.509
	T2	0	1.004	-	4.098
		3	0.611	-	1.919
		4	0.602	-	1.896
		7	0.595	-	1.770

TABLE IV
PACKET DELAYS VERSUS NUMBER OF CHANNEL SLOTS IN THE CFRV UNDER AN ASYMMETRIC LOAD OF 100 Mbits / s

station	traffic type	no of chan. slots	exp. packet delay [ms]	95% conf. int. [ms]	max packet delay [ms]
S0	T0	0	0.055	-	0.553
		3	0.058	0.003	1.087
		4	0.070	0.005	0.585
		7	0.055	-	0.621
	T1	0	0.067	-	0.805
		3	0.090	-	0.562
		4	0.092	-	0.923
		7	0.087	-	0.454
	T2	0	0.360	-	1.209
		3	0.073	-	1.136
		4	0.232	-	0.743
		7	0.178	-	0.750
S5	T0	0	0.160	0.015	0.862
		3	0.071	0.003	0.794
		4	0.073	-	0.804
		7	0.061	-	0.421
	T1	0	103.330	-	180.282
		3	0.125	-	0.930
		4	0.097	-	0.754
		7	0.088	-	0.794
	T2	0	59.751	-	476.384
		3	0.272	-	1.224
		4	0.223	-	1.088
		7	0.169	-	0.517

values. We take a 40 Mbit / s load for the CFR and 100 Mbits / s for the CFRV. Results of the simulations for the CFR are given in Table III and for the CFRV in Table IV. Note that the delays at the lightly loaded stations S_i, $i = 1, \cdots, n$, do depend on the position of the station with respect to the heavily loaded station S_0. The results are given for station S_0 which has a heavy load and for station S_5 which has a light load.

Let us examine the CFR. The delays are large already at a load of 40 Mbits / s. The different traffic types experience different delays, e.g., both the expected and the maximum delays are even 10 times larger for $T1$ at S_5 than at S_0 and the expected delays for $T2$ are about 1.1–1.5 times larger. Having 3–7 channel slots improves the performance.

Let us now analyze the CFRV. The delays are smaller at S_0 than at S_5 except for $T2$ in 4 and 7 channel slot configuration. For the same traffic types the performance is even better under the same total load (see also Table II) in the asymmetric case than in the symmetric case, except for the case with zero channel slots. The performance improves if the number of channel slots increases. The performance is much better already with 3 channel slots compared to zero.

Under an asymmetric load, the performance in both the CFR and in the CFRV is the best if the number of channel slots equals 7, but having, e.g., 4 channel slots already improves the performance significantly.

Both in the CFR and in the CFRV the delays are larger at a lightly loaded station than at a heavily loaded station, except for the case with 7 channel slots in the CFRV. In a single cyclic server case with an exhaustive service discipline, the delays are smaller at the heavily loaded station [16]. This property is attributed to the large probability that customer arriving to the heavily loaded queue finds the server serving that very queue. It has been shown in [1] that this property also holds for a single cyclic server system with a limited service discipline and zero switchover times. For nonzero switchover times, no conclusions concerning this property could be deduced. The results of the simulations in Tables III and IV show that this property also holds with a large difference in the delays for the case with only normal slots. This property does not hold in the CFRV with only channel slots but the differences in the delays are not large.

B. D_i Allocation in Orwell

Let us now study the influence of D_i allocation. The analysis of the Orwell basic AM corresponds to a D_i value of infinity. The intent is not to give a complete analysis of the sensitivity to the D_i value, but to merely demonstrate that the D_i values have indeed a large impact on the performance of all traffic types. A number of experiments have been done. First a symmetric load of 140 Mbits/s has been used.

In the first set, the ratio $D_{i,1}/D_{i,2}$ is kept constant at 4, corresponding to the share of synchronous and asynchronous traffic, and $D_{i,1}$ is varied. The packet delays are shown in Table V and the reset intervals in Table VI. The performance of $T0$ improves if $D_{i,1} = 8$. Larger values do not improve the performance of $T0$ but $T1$ and $T2$ get better performance. The reset intervals increase with increasing D_i.

Next $D_{i,1}$ is kept constant at 8 and $D_{i,2}$ is changed. Results are shown in Tables VII and VIII. Table VII shows that the expected and the maximum delays of $T1$ and $T2$ decrease with increasing $D_{i,2}$. The delays of the synchronous traffic increase in that case. $D_{i,1} = D_{i,2} = 8$ provides a good compromise for all traffic types. The expected duration of the reset intervals does not change with $D_{i,2}$ contrary to Table VI, but the maximum does.

Results for $D_{i,1}/D_{i,2} = 1$ and different $D_{i,1}$ values are shown in Tables IX and X. $D_{i,1} > 5$ yields small delays. For $D_{i,1} = 4$ the performance for $T0$ is bad. The reset intervals also increase a lot if $D_{i,1}$ is changed from 4 to 5.

The performance for an asymmetric load of 120 Mbits/s is shown in Table XI and Table XII. $D_{i,1}/D_{i,2} = 1, i = 0, \cdots, n$ and $D_{0,j}/D_{i,j} = 5, j = 1, 2, i = 1, \cdots, n$. The expected and the maximum packet delays are smaller if $D_{i,1}$ and $D_{i,2}$ are larger, e.g., 20 at S_0 and 4 at other stations. If D_i is infinite, the delays at the heavily loaded station S_0 are smaller than at the lightly loaded station S_5, which is the property also noticed for the CFR and the CFRV.

We can conclude that if the reset interval has to be small with respect to the total load, the Orwell protocol is very sensitive to the D_i values. If inadequately chosen they can degrade the performance strongly. In these cases, sudden changes of the loads can strongly influence the performance. On the other hand, if the reset intervals do not have to be small relative to the load, the performance is

TABLE V
PACKET DELAYS VERSUS $D_{i,1}$ IN ORWELL UNDER A SYMMETRIC LOAD OF 140 Mbits/s WHERE $D_{i,2} = 4 * D_{i,1}$

traffic type	Di,1	exp. packet delay [ms]	95% conf. int. [ms]	max packet delay [ms]
T0	4	0.128	0.005	0.628
	8	0.086	0.001	0.404
	64	0.085	0.001	0.810
	∞	0.081	-	0.817
T1	4	26.623	-	86.005
	8	27.188	-	86.743
	64	11.381	3.272	54.381
	∞	0.078	0.004	0.989
T2	4	24.045	18.403	86.149
	8	24.044	19.191	87.121
	64	13.707	3.874	54.405
	∞	0.256	0.007	1.352

TABLE VI
RESET INTERVAL DURATION VERSUS $D_{i,1}$ IN ORWELL UNDER A SYMMETRIC LOAD OF 140 Mbits/s WHERE $D_{i,2} = 4 * D_{i,1}$

Di,1	exp. reset int. [ms]	95% conf int. [ms]	max reset int. [ms]
4	0.173	0.007	1.640
8	0.349	-	2.216
64	3.096	-	8.835
∞	16.399	-	67.185

TABLE VII
PACKET DELAYS VERSUS $D_{i,2}$ IN ORWELL UNDER A SYMMETRIC LOAD OF 140 Mbits/s WHERE $D_{i,1} = 8$

traffic type	Di,2	exp. packet delay [ms]	95% conf. int. [ms]	max packet delay [ms]
T0	2	0.086	0.001	0.404
	4	0.092	0.002	0.458
	8	0.116	0.010	0.860
	10	0.123	0.010	0.892
	12	0.135	0.012	0.921
	100	0.176	0.021	1.344
T1	2	27.188	-	86.743
	4	0.702	0.138	7.526
	8	0.167	0.146	2.607
	10	0.112	0.009	1.798
	12	0.101	0.006	1.579
	100	0.076	0.003	0.383
T2	2	24.405	19.191	87.120
	4	1.806	-	9.097
	8	0.563	0.018	2.730
	10	0.494	0.013	2.104
	12	0.438	0.010	1.693
	100	0.177	-	0.544

TABLE VIII
RESET INTERVAL DURATION VERSUS $D_{i,2}$ IN ORWELL UNDER A SYMMETRIC LOAD OF 140 Mbits/s WHERE $D_{i,1} = 8$

Di,2	exp. reset int. [ms]	95% conf. int. [ms]	max reset int. [ms]
2	0.349	-	0.443
4	0.398	-	2.153
8	0.387	-	1.850
10	0.394	-	1.910
12	0.393	-	1.528
100	0.393	-	2.153

TABLE IX
PACKET DELAYS VERSUS $D_{i,1}$ IN ORWELL UNDER A SYMMETRIC LOAD OF 140 Mbits/s WHERE $D_{i,2} = D_{i,1}$

traffic type	Di,1	exp. packet delay [ms]	95% conf. int. [ms]	max packet delay [ms]
T0	4	6.125	-	14.171
	5	0.201	0.025	1.184
	8	0.116	0.010	0.860
	16	0.100	0.004	0.834
T1	4	0.188	0.017	3.174
	5	0.167	0.014	2.773
	8	0.167	0.146	2.607
	16	0.141	0.124	2.458
T2	4	0.779	0.020	3.139
	5	0.812	0.017	2.884
	8	0.563	0.018	2.730
	16	0.334	0.018	2.582

TABLE X
RESET INTERVAL DURATION VERSUS $D_{i,1}$ IN ORWELL UNDER A SYMMETRIC LOAD OF 140 Mbits/s WHERE $D_{i,2} = D_{i,1}$

Di,1	exp. reset int. [ms]	95% conf. int. [ms]	max reset int. [ms]
4	0.173	0.007	1.640
5	0.257	-	1.401
8	0.349	-	2.216
16	0.739	-	2.970

TABLE XI
PACKET DELAYS VERSUS $D_{i,1}$ IN ORWELL UNDER AN ASYMMETRIC LOAD OF 120 Mbits/s WHERE $D_{i,2} = D_{i,1}$

D0,1	exp. reset int. [ms]	95% conf. int. [ms]	max reset int. [ms]
10	0.156	-	0.244
15	0.193	-	2.478
20	0.235	0.007	0.442
30	0.245	0.009	0.556
∞	0.300	0.021	6.529

very good both under symmetric and asymmetric loads, and the Orwell basic AM also has a very good performance and a low sensitivity to the system and configuration parameters. This indicates that another way of integration of synchronous and asynchronous traffic may be needed to provide a better performance in these cases. Some attempts in that direction are proposed in [9]. The problems reported here concerning the D_i allocation are comparable to the problems of selecting values of certain system parameters which regulate the fairness of the access protocol in other LAN's. For example, it has been reported in [15] that the choice of TTRT in FDDI also has a large influence on the performance of FDDI.

IX. DELAY ANALYSIS OF THE INTEGRATED SERVICES SLOTTED RINGS

Let us now examine the integration of synchronous and asynchronous traffic. We assume a configuration with 40 stations implying 8 slots and we limit ourselves to symmetric cases.

TABLE XII
RESET INTERVAL DURATION VERSUS $D_{0,1}$ IN ORWELL UNDER AN ASYMMETRIC LOAD OF 120 Mbits/s WHERE $D_{i,2} = D_{i,1}$, AND $D_{0,j} = 5 * D_{k,j}$, $k \neq 0$

station	traffic type	Di,1	exp. packet delay [ms]	95% conf. int. [ms]	max packet delay [ms]
S0	T0	10	45.707	-	96.884
		15	0.609	-	3.335
		20	0.170	-	1.961
		30	0.137	-	1.623
		∞	0.058	0.001	0.508
	T1	10	0.274	-	4.173
		15	0.204	-	3.839
		20	0.187	-	3.697
		30	0.196	-	3.377
		∞	0.053	-	0.396
	T2	10	1.413	0.044	4.632
		15	1.242	-	4.037
		20	1.020	-	3.794
		30	1.059	0.036	3.512
		∞	0.155	0.005	0.635
S5	T0	2	26.736	-	96.842
		3	0.376	-	3.278
		4	0.130	-	1.651
		5	0.103	-	1.242
		∞	0.067	0.002	0.437
	T1	2	0.177	-	2.412
		3	0.267	-	2.178
		4	0.140	-	1.948
		5	0.134	-	1.549
		∞	0.115	-	1.000
	T2	2	0.970	0.136	2.692
		3	0.866	0.108	2.479
		4	0.708	0.087	2.287
		5	0.728	-	2.230
		∞	0.240	-	1.050

The share of each traffic type is varied. In accordance with the results of Section VIII, 6 channel and 2 normal slots are used for the CFR and the CFRV. In Orwell, the total D_i allocation per station is 10. This ensures that the total D_i allocation is not too small, so reset intervals do not appear too often. $D_{i,1}/D_{i,2}$ is chosen equal to the ratio of synchronous and asynchronous traffic. Although this ratio is not the optimal one, it has been used since it provides a fair share of capacity to all traffic types and, on the other hand, obtaining the optimal ratio is difficult.

The delays of different traffic types versus asynchronous traffic load are shown in Fig. 11 for the CFR, in Fig. 12 for the CFRV, and in Fig. 13 for Orwell. The synchronous traffic load is used as a parameter. The results have been obtained by simulations.

In the CFR and the CFRV the delays increase rather slowly with the load until the utilization is about 0.8. From there on they rise sharply. Random fluctuations in load intensity may cause a sudden degradation of performance by significantly increasing the delays. If the load intensity of both synchronous and asynchronous traffic in the CFRV is controlled, bounded delays and a good QoS for both traffic classes can be provided. A guaranteed synchronous service can also be provided by merely controlling the amount of synchronous traffic itself (given queueing priority). However, this would not be sufficient when a large share, e.g., 80 percent of the load consists of synchronous traffic. The performance of the CFR is

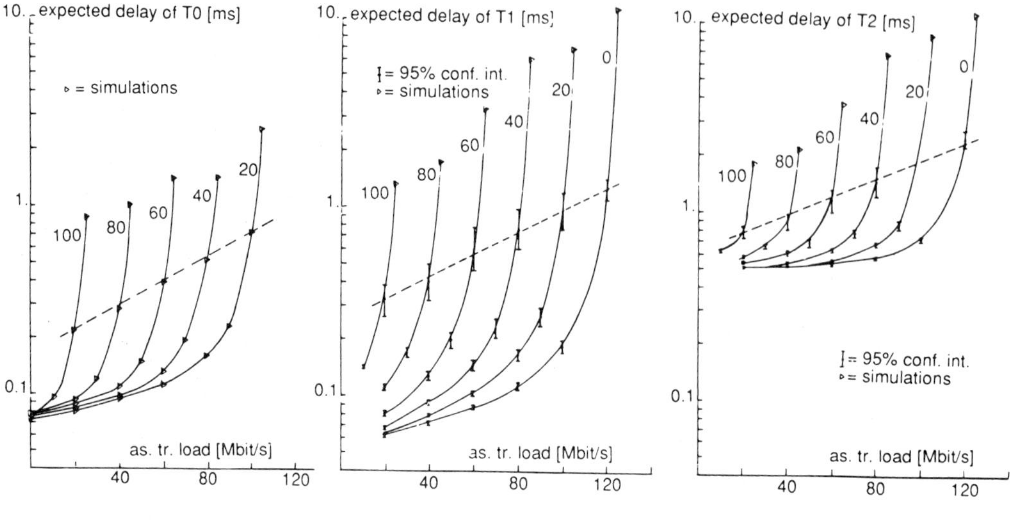

a) Delays of T0. b) Delays of T1. c) Delays of T2.

Fig. 11. Packet delays versus asynchronous traffic load given a synchronous traffic load in the CFR.

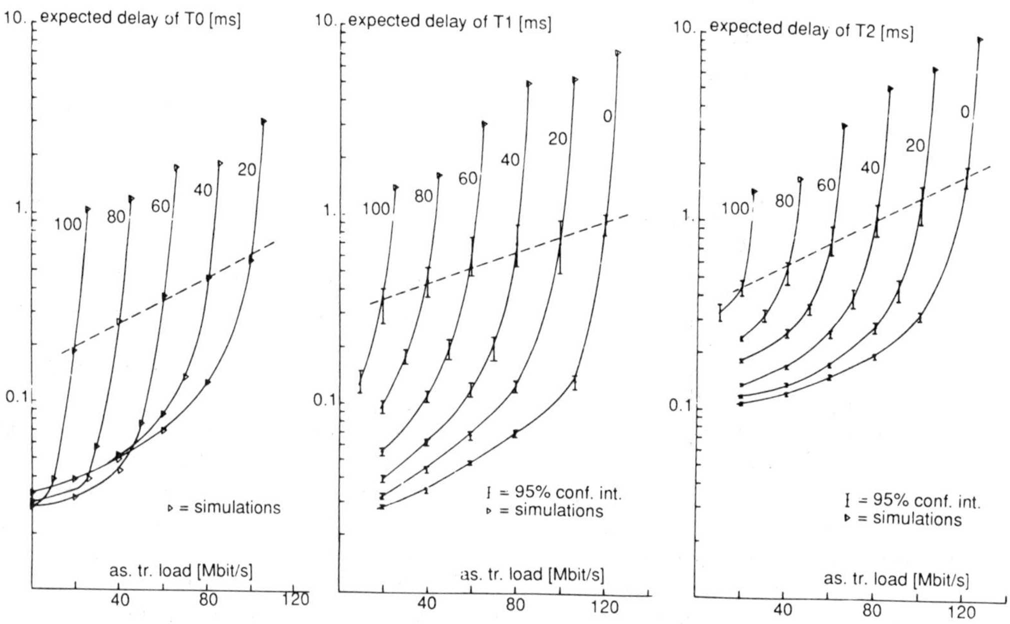

a) Delays of T0. b) Delays of T1. c) Delays of T2.

Fig. 12. Packet delays versus asynchronous traffic load given a synchronous traffic load in the CFRV.

worse than the CFRV. In the CFR and the CFRV, the delays increase if the share of asychronous traffic increases given a constant total load. This is because of the following two reasons: the larger variance of the packet length distribution of asynchronous packets, and the larger variance of the asynchronous traffic interarrival time due to a Poisson arrival process. The dotted lines in Figs. 11 and 12 represent a constant total load curve of 120 Mbits / s. The delays of $T1$ and $T2$ vary by a factor 4 in the CFR and by a factor 3 in the CFRV when the share of asynchronous traffic changes from 20 to 120 Mbits / s.

In Fig. 12(a), a counterintuitive behavior of the CFRV appears: if there is no asynchronous traffic, the expected delay decreases when the load increases. This is due to the regularity of the deterministic arrival process of synchronous traffic. Note, however, that the difference in the delays due to this is small, less than 5 μs.

Orwell exhibits a sharp increase of the asynchronous traffic delays already starting at a utilization of about 0.5 while the increase of the synchronous traffic delays is much slower. The difference is due to the D_i allocation. The delays in Orwell do not increase smoothly because of

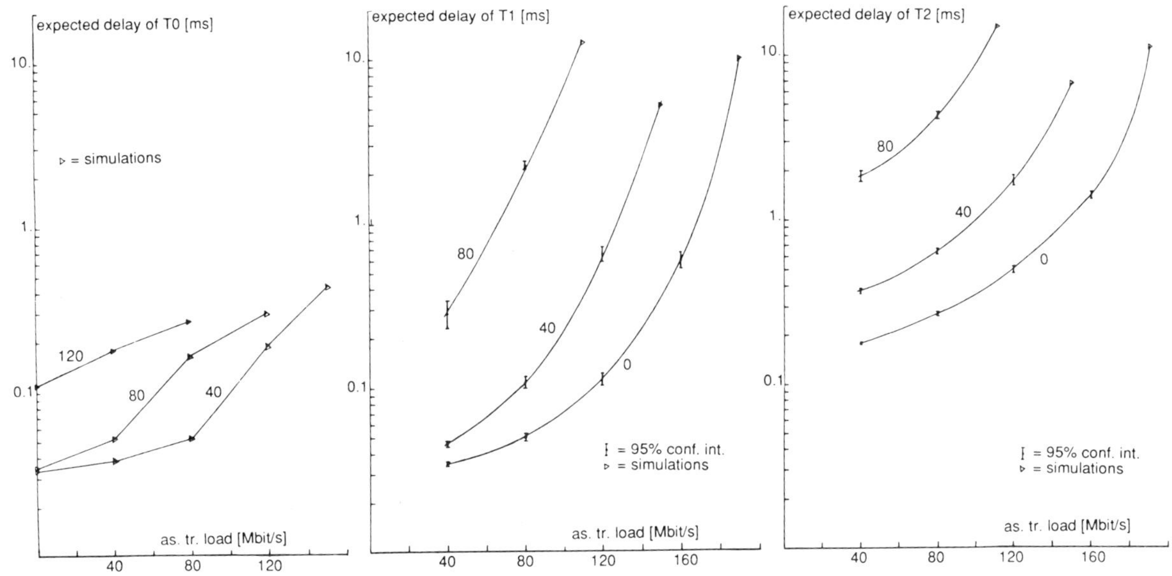

a) Delays of T0. b) Delays of T1. c) Delays of T2.

Fig. 13. Packet delays versus asynchronous traffic load given a synchronous traffic load in Orwell.

the high sensitivity to the choice of D_i: the $D_{i,1}/D_{i,2}$ ratio does not follow exactly the ratio of synchronous and asynchronous traffic. Note that the delays for synchronous traffic are small. This happens at the expense of the asynchronous traffic delays. This is in agreement with the results in Table V.

X. Conclusion

Slotted ring protocols are suitable for optical fiber HSLAN's because of their efficiency at high transmission rates and the synchronous transmission scheme. However, packet segmentation and reassembly may impose large processing demands if the packet lengths are large relative to the slot information field size. Let us now summarize the performance analysis results we obtained for these protocols. The multiple token passing ring has been used in this paper for comparison to the slotted ring basic AM's.

The Orwell and the CFRV basic AM's perform very well with respect to the number of slots and the number of stations. Their performance also improves if the transmission rate increases. The Orwell basic AM has the largest capacity due to the destination release of slots. Its capacity is almost twice the transmission rate under a symmetric load. The CFRV has about half of that capacity. The CFR basic AM has a much larger sensitivity to the system parameters and a much worse performance than the other two slotted ring protocols. Token passing is inferior to the CFRV and the Orwell basic AM's unless the average packet length is much larger than the slot length (e.g., >2300 bits). The performance of the token passing is improved most of all by the uniframe scheme. This scheme also improves the performance of the CFRV. It does not improve the performance of the Orwell basic AM in a configuration with a large number of stations relative to the number of slots. The asynchronous traffic delays in the CFR with a uniframe are large, e.g., >1 ms.

The CFR has larger delays than the other slotted rings. So, the CFR has worse performance than the CFRV and Orwell with either long cable lengths, a small number of stations, or asymmetric loads. By increasing the transmission rate to more than, e.g., 200 Mbits/s the performance does not improve. This is the price for MAC layer acknowledgment and the possibility of immediate retransmission of erroneous minipackets.

The CFRV can be used successfully over a wide range of parameters. Combining the normal and the channel slots improves its performance especially under asymmetric loads. This corresponds to combining two service disciplines: limited and exhaustive. It performs very well under asymmetric loads. If the load intensity of both synchronous and asynchronous traffic on the network is controlled, bounded delays and a good QoS for both traffic classes can be provided. Furthermore, the CFRV allows for an acknowledgment at the MAC layer.

The good properties of the Orwell basic AM indicate that it can be used successfully over a wide range of parameters. Some improvement of the performance of Orwell is achieved if a larger $D_{i,2}$ allocation is taken for asynchronous traffic than what corresponds to the share of asynchronous traffic. The performance is good under asymmetric loads.

Because of the very good performance of Orwell and the CFRV protocol with long cable lengths and a large number of stations, they are suitable for metropolitan area networks, as well.

ACKNOWLEDGMENT

The authors would like to thank R. van den Brink for doing the simulation runs in Section VIII, which have been documented in [2].

REFERENCES

[1] O. J. Boxma and B. Meister, "Waiting time approximations for cyclic-service systems with switch-over times," *Performance '86/ACM SIGMETRICS 1986*, 1986.

[2] R. v. d. Brink, "Influence of system parameters on performance of the slotted rings: Simulations," University of Twente, Rep. 080-87-46, Oct. 1987.

[3] L. M., Casey, R. C. Dittburner, and N. D. Gamage, "FXNET: A backbone ring for voice and data," *IEEE Commun. Mag.*, vol. 24, pp. 23-28, Dec. 1986.

[4] R. M. Falconer and L. Adams, "Orwell: A protocol for an integrated services local network," *Brit. Telecom. Tech. J.*, vol. 3, no. 4, pp. 27-35, Oct. 1985.

[5] R. M. Falconer, J. L. Adams, and G. M. Walley, "A simulation study of the Cambridge ring with voice traffic," *Brit. Telecom. Tech. J.*, vol. 3, no. 2, pp. 85-91, Apr. 1985.

[6] W. K. Giloi, P. Behr, and G. Zuber, "Upperbus: A high speed backbone for metropolitan area networks," in *Proc. EFOC/LAN'86*, Amsterdam, The Netherlands, June 1986, pp. 286-289.

[7] K. Hiyama, H. Narisawa, and H. Satou, "An integrated services optical fiber local area network 'Σ Network,'" *Hitachi Rev.*, vol. 32, no. 4, pp. 165-170, 1983.

[8] A. Hopper, "The Cambridge ring—A local network," in *Advanced Techniques for Microprocessor Systems*, Hanna F. K., Ed. Stevenage, UK: Peter Pergrinus, Ltd., 1980, pp. 67-71.

[9] J. O. Limb and S. Kandiyoor, "Improved scheduling of traffic for a high-speed slotted ring," *Electron. Lett.*, vol. 23, no. 15, pp. 806-765, July 1987.

[10] K. v. d. Lande, "Design and implementation of a Cambridge fast ring variant simulation model, University of Twente, The Netherlands, Rep. 080-87-06, Feb. 1987.

[11] L. Lamberelli, A. Luvison, D. Roffinella, and M. Sposini, "Service integration in wideband local area networks: Problems and system solutions," presented at Int. Tirrena Workshop Digital Commun., Tirrena (Pisa), North Holland, 1985.

[12] I. Mitrani, J. L. Adams, and R. M. Falconer, "A modelling study of the Orwell ring protocol," in *Teletrafic Analysis and Computer Performance Evaluation*, O. Boxma, J. Cohen, and H. Tijms, Eds. New York: Elsevier, 1986, pp. 429-438.

[13] N. N. Pedersen and R. Sharp, "Analysis of channel access schemes for high-speed LAN's," in *Proc. ACM SIGCOMM'84: Commun. Architect. Protocols*, Montreal, P.Q., Canada, June 1984, pp. 164-171.

[14] F. E. Ross, "FDDI—A tutorial," *IEEE Commun. Mag.*, vol. 24, pp. 10-23, May 1986.

[15] A. Schill and M. Zieher, "Performance analysis of the 100 Mbit/s optical token ring," in *IFIP WG 6.4 Workshop High Speed Local Area Networks*, A. Danthine and O. Spaniol, Eds. Aachen, W. Germany: North Holland, 1987, pp. 53-74.

[16] H. Takagi, *Analysis of Polling Systems*. Cambridge, MA: The MIT Press, 1986.

[17] S. Temple, "The design of the Cambridge fast ring," in *Ring Technology Local Area Networks*, I. N. Dallas and E. B. Spratt, Eds. New York: Elsevier, 1984, pp. 79-88.

[18] D. Valk, "Performance analysis of integrated services slotted ring protocols in gathering HSLAN's," University of Twente, The Netherlands, Rep. 080-87-08, Mar. 1987.

[19] E. Vyncke and A. Danthine, "A realistic simulation of a wideband backbone network," in *Proc. ICCC'86*, North Holland, pp. 176-181, 1986.

[20] M. Zafirovic-Vukotic and I. G. Niemegeers, "An evaluation of high speed local area network access mechanisms," in *Communication in Distributed Systems*, N. Gerner and O. Spaniol, Eds. Aachen, W. Germany: Informatik-Fachberichte no. 130, Springer-Verlag, Feb. 1987, pp. 426-440.

[21] —, "Analytical models of the slotted ring protocols in HSLAN's," in *IFIP WG 6.4 Workshop High Speed Local Area Networks*, A. Danthine and O. Spaniol, Eds. Aachen, W. Germany: North Holland, 1987, pp. 115-134.

[22] —, "Performance modelling of the Orwell basic access mechanism," in *Proc. ACM SIGCOMM'87: Frontiers Comput. Commun. Technol.*, Stowe, VT; also in *Comp. Commun. Rev.*, vol. 17, no. 5, Special Issue, Aug. 1987, pp. 35-48.

[23] —, "Applications, workload models, benchmarks and configurations for HSLAN's," University of Twente, Rep. 080-86-33, Oct. 1987.

[24] —, "Performance modelling of the Cambridge fast ring protocol," in *Proc. 1988 Int. Zurich Sem. Digital Commun.*, Zurich, Switzerland, Mar. 1988, pp. 163-170.

[25] —, "Performance modelling of a HSLAN slotted ring protocol," presented at Proc. ACM SIGMETRICS'88, Santa Fe, NM, May 1988.

[26] —, "HSLANs for communication within a gateway: A performance evaluation," in *Proc. IFIP WG 6.4 Workshop HSLAN'88*, A. Danthine, O. Spaniol, Eds. Liege, Belgium: North Holland, Apr. 1988, to be published.

Mirjana Zafirovic-Vukotic was born in Belgrade, Yugoslavia, in 1958. She received the B.S. degree in mathematics in 1981 and the M.S. degree in operations research in 1984 from the University of Belgrade. She is also a Ph.D. candidate. She worked as a Research Assistant in operations research at the University of Belgrade from 1982 to 1984. Since 1984 she has been working as a Research Assistant at the University of Twente, The Netherlands, on performance analysis of integrated services high speed LAN's. Her interests are in computer communication systems and performance analysis.

Ignas G. Niemegeers (S'70–M'78) was born in Gent, Belgium, in 1947. He received the M.S. degree in electrical engineering from the Rijksuniversiteit Gent in 1970. In 1972 he received the M.S. degree in computer engineering and in 1978, the Ph.D. degree from Purdue University, West Lafayette, IN.

From 1978 to 1981 he was a System Design Engineer at Bell Telephone Mfg. Cy., Antwerp, Belgium, where he was involved in the design of the DPS 1500 system, a high capacity X.25 packet switching network. From 1981 to 1986 he was with the Department of Electrical Engineering, University of Twente, Enschede, The Netherlands. Since 1986 he has been a Professor of computer science at the same university. His present research interests are architectures for local integrated communication infrastructures, parallel processing in communication switches, performance analysis, and performability of fault tolerant systems.

Durk S. Valk was born in Leeuwarden, The Netherlands, in 1964. He received the Ir. degree in electrical engineering from the University of Twente, The Netherlands in 1987.

He joined Logica B.V., The Netherlands, in 1987. He is currently also working on performance analysis tools. His interests are in computer and communication performance analysis and communication systems.

Chapter 3
Metropolitan Area Networks (MANs)

Both MAN and broadband ISDN (the topic of the next chapter) attempt to provide wideband fiber optical access to residential and business premises. The scope of a MAN is limited to one metropolitan area, whereas broadband ISDN spans a wide area network. Recognizing common interests, the IEEE 802.6 subcommittee, responsible for MAN standards, is expected to cooperate with evolutionary developments of broadband ISDN standardization at the CCITT.

In 1988, the IEEE 802.6 subcommittee approved the standards for MAN. The medium access control (MAC) standard of the MAN has been referred to as QPSX (queued packet, synchronous exchange), and more recently as DQDB (distributed queue, dual bus). The DQDB protocol is based on a dual-bus architecture and has two excellent properties: the self-healing capability and superb performance characteristics over a wide range of network loads and sizes.

The first article by Newman *et al.* describes the details of the protocol. The next five articles present alternative approaches to the MAN architecture. Limb's article describes Fasnet, which is the basis of Bellcore's METROCORE prototype. Fasnet has a dual-bus structure as does the DQDB protocol. Fasnet's performance with combined voice-data traffic is studied in Mark and Limb's article.

The Manhattan Street Network (MSN), discussed by Maxemchuk, possesses a regular mesh topology, which has physical bearings on streets in a metropolis. MSN is also attractive for its simple node structure, requiring a limited buffering. For reasons of reliability, nationwide carriers install fiber optic trunk lines in a similar mesh structure. MSN thus has a potential for applications to nationwide fiber networks as well.

Rocher's article raises an interesting interface issue analogous to arguments on ac *versus* dc electrical outlets. Although interface issues are thoroughly discussed, issues of network structures are left open.

Finally, a chordal (braided or daisy-chained) ring topology is included. The chordal ring network is particularly attractive due to its capability of dynamic reconfiguration without disrupting the existing network. Although details of the protocol amenable to MAN applications are to be worked out, the chordal ring structure can be a valuable alternative in a MAN where additions or removals of nodes are frequent.

The QPSX Man

R. M. Newman
Z. L. Budrikis
J. L. Hullett

Introduction

Public communications are catching up with demands set by information technology developments and will soon lead them. A number of standards forums, the CCITT, T1, the IEEE, the ISO and others are working on interrelated projects to define interfaces, networks and protocols that promise to make data communications over a wide spectrum as ubiquitous and as universally switchable as the telephone. More than that, from the user's prospective all communications will be fully integrated flowing over the one interface and access loop, with telephone just one of the many services.

The CCITT has already defined the integrated services digital network (ISDN) and is defining the broadband ISDN, or B_ISDN. T1 is defining a synchronous optical network (SONET), capable of carrying the large traffic within the network. ISO is standardising protocols and procedures so that different equipments can communicate with each other. The IEEE in P802 is bridging the gap between private and public networks. It is defining local area networks (LANs) suitable for customer premise or campuses, and metropolitan area network (MAN) that will interconnect LANs and provide other integrated services as part of the public network. In fact, the IEEE defined MAN can be a significant stepping stone on the path to a full B_ISDN.

The IEEE MAN is being defined in 802.6, a Working Group of P802. The project is actively supported by Telcos which gives assurance that it will measure up to public network requirements. A technology proposal that has particularly wide Telco support is one put forward from Australia. It is the Dual Bus QPSX MAN [1],[2].

QPSX (the queued packet and synchronous switch) is a distributed switch/network that will fulfill the requirements of a public MAN. The switch architecture of QPSX is based on two contra-directional buses as shown in Fig. 1. This allows full duplex communications between each pair of nodes of the switch. The communications on QPSX are integrated, providing isochronous (circuit switched) and non-isochronous (packet switched) services. The two switch modes function independently and the total capacity of the switch is shared flexibly and completely between them.

The bus architecture of QPSX is inherently highly reliable. The operability of the network is to all intents independent of the operation of the individual nodes. Furthermore, fault handling mechanisms have been developed that permit the isolation of potentially catastrophic faults with no service degradation in the case of a single failure, and minimal disruption even with multiple failures.

The access of packets to the buses of QPSX is controlled by a unique MAC (media access control) protocol referred to as distributed queueing [3]. The distributed queueing protocol provides near perfect access characteristics independent of network size and speed. In particular, the protocol supports 100 percent throughput and can provide average packet access delay equal to that of a perfect scheduler, the ultimate queueing system.

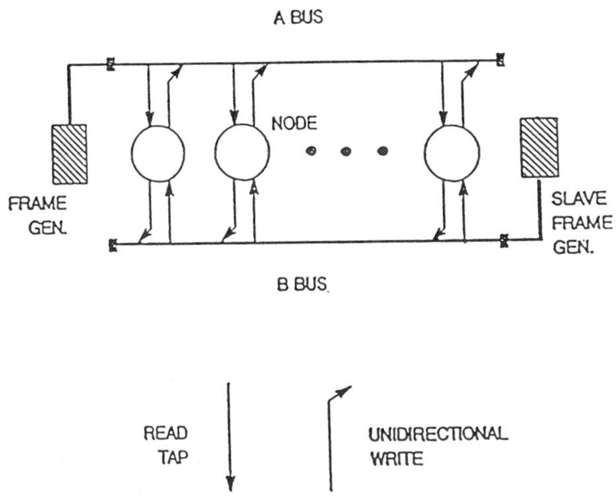

Fig. 1. QPSX dual bus architecture.

The operation of the protocol is based on two control bits, the busy bit which indicates whether a slot on the network is used and the REQ (request) bit which is sent whenever a node has a packet waiting for access. Each node by counting the number of REQs it receives and non-busied packets that pass it, can determine the number of packets queued (i.e., in line) ahead of it. This counting establishes a single ordered queue across the network for access to each bus. With such queued access, levels of priority can be established by operating a number of queues, one for each level. Within each level the perfect performance characteristics are maintained: the packets will gain access as soon as capacity becomes available, but priority is always given to packets in higher level queues.

QPSX as a Public Network MAN

The P802 forum started with development of LAN standards and much of the MAN work has evolved from this base. However, the MAN has very much greater functionality and a different role. The increased functionality includes the ability to cover much larger geographical areas, to operate at much higher speeds, and to provide switching for a mix of traffic services. These capabilities take the MAN well beyond the realm of in-house or building networks and into the domain of the public network.

QPSX MAN is particularly well suited to application in the public network. It has the necessary features of outstanding reliability and independence of network size and speed, to meet the requirements of all public network services. As well, it has the features, such as security, network administration, bandwidth control and charging facilities, which are also essential in the public network. A further advantage in relation to its application in the public network is that it has a frame structure that will be compatible with the evolving broadband ISDN. Hence, QPSX can be seen as an evolution path to this future public network technology.

The Dual Bus Architecture of QPSX

The principal components of the QPSX architecture are a head station, two unidirectional buses, and a multiplicity of access units (AUs). These are configured in the unidirectional bus structure shown in Fig. 1. The head station generates the frame synchronization on the forward bus and the end station generates the frame pattern at the same rate on the reverse bus.

Access units attach to both buses via read and write connections. The writing on to the bus is by logical OR of the data from upstream with the data from the unit. The read connection is placed ahead of the write connection and allows all data to be copied from the bus unaffected by the unit's own writing. The digital bus implementation of the AU connection is shown in Fig. 2.

The bus architecture of QPSX has been chosen for its reliability. As can be seen from the Fig. 2, the data bus passes directly by each node in the network. Data does not pass through the node as is the case in ring networks. Hence, in QPSX, nodes can fail or even be removed from the network with no consequence on the operation of the rest of the network. The network does not even need to reconfigure in this case.

The only fault conditions in QPSX that possibly may affect the operation of the network are a "jabbering" node or failure of the bus. The first fault is handled by a fault detection and isolation scheme which may be instituted, again without causing reconfiguration of the network. Failure within the bus is the only condition handled by network reconfiguration. The QPSX reconfiguration scheme is simple and very fast acting.

The Looped Bus Architecture

An extension of the dual bus architecture is the looped bus shown in Fig. 3. The only change in this from the original architecture is that the end points of the buses are co-located. Data do not flow through the head point of the loop. Thus, while the looped bus architecture may appear similar to a ring, it has no log-

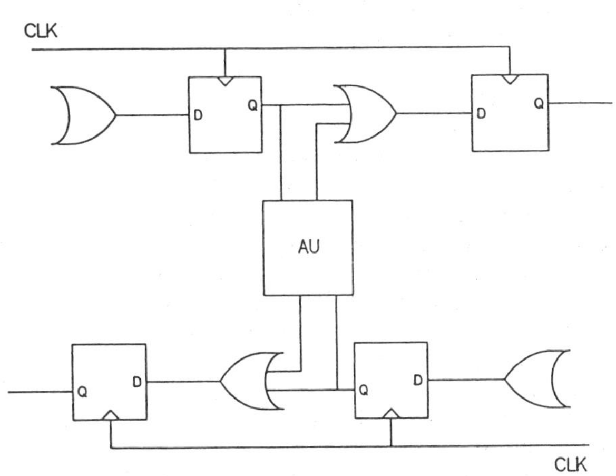

Fig. 2. Digital bus realization.

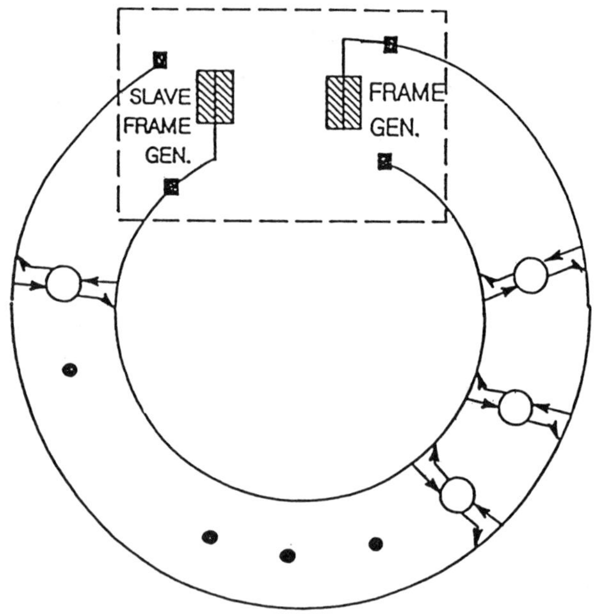

Fig. 3. QPSX looped bus architecture.

ical similarity to it. Hence, it provides all the reliability advantages associated with the QPSX bus architecture.

There are two purposes in looping the QPSX bus architecture. One is that the framing generator for the two buses is common and this point provides the entry point for the PSTN clock into the network. The second and more important reason for looping the buses is that it provides an automatic route diversity within the switch. In the case of a bus fault, the network can isolate the fault and close the data buses through the head point of the loop. Hence, the network is reconfigured and fully operational in the reconfigured state. This is achieved without the need for any redundant transmission hardware, which is in direct contrast with ring networks that require a second redundant ring to institute network reconfiguration.

The Architecture of the Public MAN

The role of the QPSX MAN in the public network is to provide interconnectability for a range of communication services not yet provided. The geographical scope of the MAN is such that it is planned to cover complete cities. In practice it is most sensible if the MAN covers an area that corresponds to the telephone number plan area, NPA. While the dual bus architecture of QPSX is such that a single network would operate over very large geographical areas, it will not be long before the capacity requirements of the attached customers will exceed 2X150Mbits/s capacity of the network. In fact the number of networks required in large cities could well approach the hundreds and even thousands. As a result, there is a need for bridged interconnection of networks to form a single large MAN.

In order to permit a manageable and efficient interconnection of the networks within the MAN, a simple interconnection architecture and bridging algorithm is required. The architecture of the public network based on QPSX is very similar to that of the public telephone network and is illustrated in Fig. 4. It is based on a hierarchy with possible cross links between switches at a common level of the hierarchy. The cross links provide route diversity in the case of failure and also provide a short path between switches which have a high degree of cross traffic.

At the lowest level of the MAN hierarchy is the access network, shown as level one in the Fig. 4. This level provides for the customer's connection into the public network. The protocol operated over the access network is the same QPSX protocol. The access network will go from the customer to the nearest site owned by the public network provider. In terms of utilizing the cabling efficiencies associated with distributed switching, if the network provider has space leased at sites away from the central office, then connection to the public network switch will be at those points. Otherwise, all connections to the public switch will be in the central office. In this latter case, the length of the access network loops generally will be longer.

The point where the access network reaches the office level network is the edge of the public network. This point, called the network connection point (NCP), is responsible for maintaining and administering the customer's connection. The functions performed at NCP are illustrated in the Protocol architecture diagram, Fig. 5. There are two principal sets of functions performed at the NCP. These are routing and edge-to-edge services. The routing layer performs the function of converting the destination E.164 address indicated in packets sent by the customer to determine where the packet is to be routed. The routing layer will translate that address into an internal address used within the QPSX MAN. The edge-to-edge

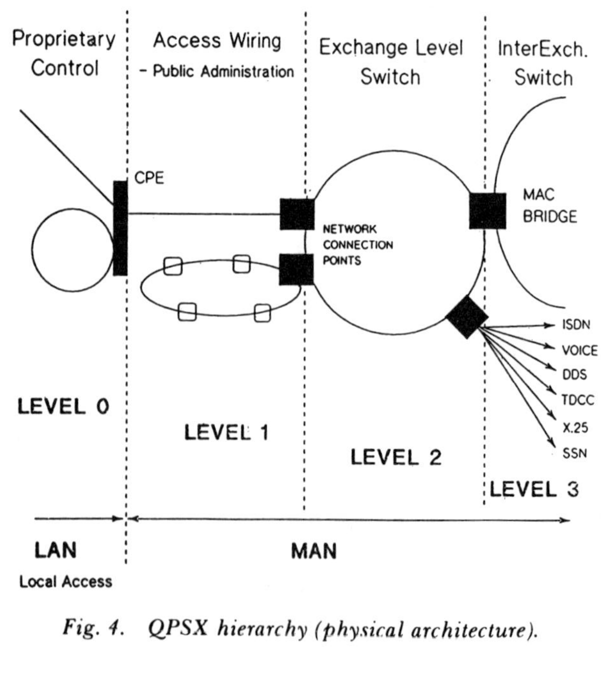

Fig. 4. QPSX hierarchy (physical architecture).

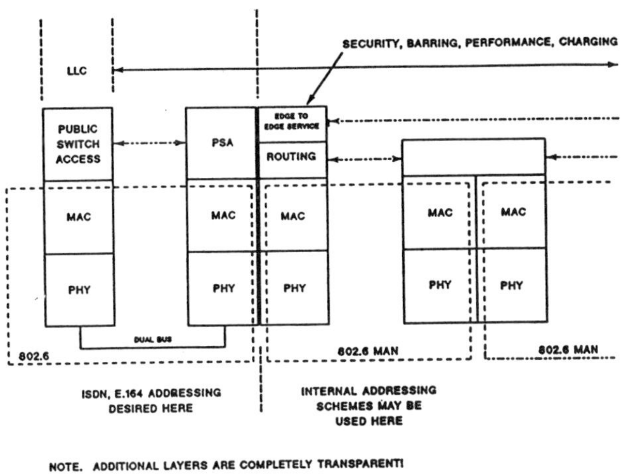

Fig. 5. Protocol architecture.

services layer will perform functions related to the network provider's management of the customer's connection, such as security provision, charging, performance monitoring barring, etc.

The NCP represents the point where the customer is physically and logically connected to the switch hierarchy of the QPSX MAN. The switch hierarchy consists of office level switches at the lowest level, interoffice switches at the next higher level and so on upwards to whatever depth is required. All switches are interconnected by bridges which can provide up to 150Mbits/s throughput. The same system hardware can be used at all levels of the hierarchy, though ultimately higher rate switches will be required at the higher levels. Nodes at all switch levels are connected to the QPSX bus at cluster points, and clusters may be either widely distributed, or adjacent thus forming a centralized switch realization.

Communications across MANs will be routed through the source MAN to a gateway to the wide area network (WAN). This gateway will again perform the necessary routing layer functions using the E.164 address contained in the packet to determine the path through the WAN to the destination MAN. It is not intended, at least in the first realizations of QPSX, for synchronous communications to be carried in the WAN or even in the interoffice switches of the QPSX hierarchy. Rather this traffic will be given over to the existing switch services provided for it.

The routing within the MAN is based on a simple protocol which permits very high throughput in the bridges of the MAN. In fact, a bridge can switch between two networks up to the full capacity of the QPSX dual bus. Route diversity within the bridging algorithm is also provided to circumvent bridge faults.

Frame Format and Synchronization

Communications on the QPSX buses are within the frame format shown in Fig. 6. The frame interval is 125 microseconds, matching that of digital telephony. The frame is subdivided into a fixed number of equal size units called slots. Slots provide the subdivision between the isochronous and non-isochronous traffic. A slot may be allocated to contain either type of communication. Slots are allocated to isochronous use commensurate with the demand for isochronous capacity. The slots not allocated to isochronous use are available for packet communications.

For isochronous communications, each octet within a slot provides a 64kbit/s channel. By using multiple octets per frame higher rate circuits can be constructed. By the use of a multi-frame count in the frame header, circuits at sub-64kbps can also be constructed.

For non-isochronous communications, a slot once claimed for access by a node, is used to transfer a single segment of a packet. Access to slots is handled by the Distributed Queueing protocol. The segmentation of packets into slots and the reassembly of the packet at the destination is by a source-identifier-linked segmentation and reassembly protocol. Both protocols are discussed in later sections.

The frame structure for QPSX is very similar to that which is currently under consideration in CCITT as the structure for B-ISDN. It has already been decided that the B-ISDN frame will consist of fixed sized cells, these cells being a similar concept to QPSX slots. In the standardization effort of QPSX, it is expected that slot size and slot header size will be aligned with that adopted by CCITT for B-ISDN cell size and cell header. Further levels of compatibility such as the format of the header will be a future issue. However, it is expected that IEEE 802.6 will develop a MAN standard well in advance of the B-ISDN standard by the CCITT.

The Distributed Queueing Protocol

Distributed Queueing is a media access control (MAC) packet switching scheme that controls the access of fixed length data segments to the slots on the QPSX bus. The protocol overcomes many of the limitations of other media access protocols. The maximum throughput of the network is not constrained by network speed or size and the protocol operates on the highly reliable dual bus architecture.

The operation of Distributed Queueing is fundamentally different from all existing MAC protocols. In all other access schemes, no continuous record is kept in the nodes that explicitly indicates the state of the network. Hence, when in these protocols a node has a packet to transmit, it must first derive the information from the network as to when to access. This leads directly to a sensitivity in almost all protocols to the size of the network. The larger the network, the longer the delay in obtaining the access information. This limits the size over which the protocols may operate efficiently.

By contrast, with Distributed Queueing a current state record is kept in every node which holds the number of segments awaiting access to the bus. When a node has a segment for transmission, it uses this

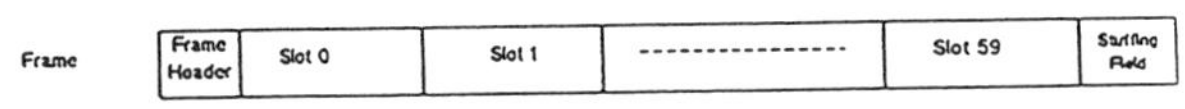

Fig. 6. Frame format for a QPSX man.

count to determine its position in the distributed queue. If no segments are waiting, access is immediate, otherwise deference is given only to those segments that queued first.

With the queueing of segments for the bus, slots are never wasted. This guarantees minimum access delay at all levels of loading right up to the maximum network utilization of 100 percent. This performance is achieved with negligible control overhead, in fact only two bits per slot, and is effectively independent of network size and speed.

The Basic Distributed Queueing Algorithm

The operation of the basic Distributed Queueing algorithm is particularly simple, as illustrated in Fig. 7. The protocol uses just two control bits, busy and REQ, to control access to the slots on the forward bus. An identical, but independent arrangement applies for access to the opposite bus.

The Busy control bit is a marker at the head of each slot which indicates whether the slot is full and hence not available for access. The REQ control bit is the crux of the queueing mechanism and is used to signal when a segment has queued.

When a node has a segment for transmission on the forward bus it will issue a single REQ bit on the reverse bus. This bit will pass to all upstream nodes, where upstream is defined in relation to flow on the forward bus. This REQ bit serves as an indicator to the upstream nodes that an additional segment is now queued for access.

Each node keeps track of the number of segments queued downstream from itself by counting the REQ bits as they pass on the reverse bus, as shown in Fig. 8a). For each REQ passing on the reverse bus, the request (RQ) counter is incremented. One REQ in the RQ counter is canceled each time an empty slot passes on the forward bus. This is done since the empty segment that passes the node will be used by one of the downstream queued segments. Hence, we see that with these two actions the RQ counter keeps a precise record of the number of segments queued downstream.

When a node has a segment for access to the bus, it transfers the current value of the RQ count to its second counter, the countdown (CD) counter. This action loads the CD count with the number of downstream segments queued ahead of it. This, along with

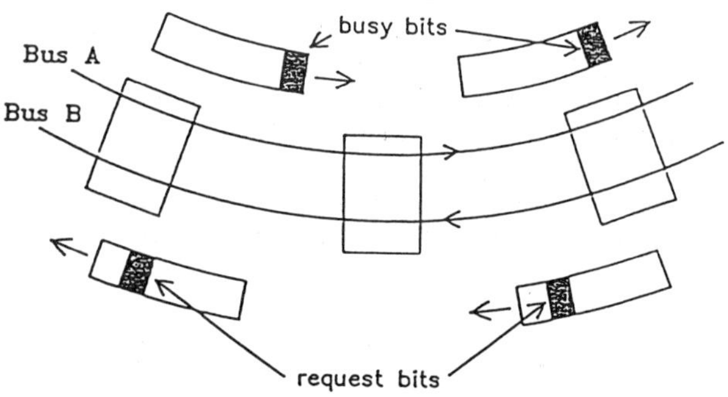

Fig. 7. Distributed queueing. Queue formation on Bus A.

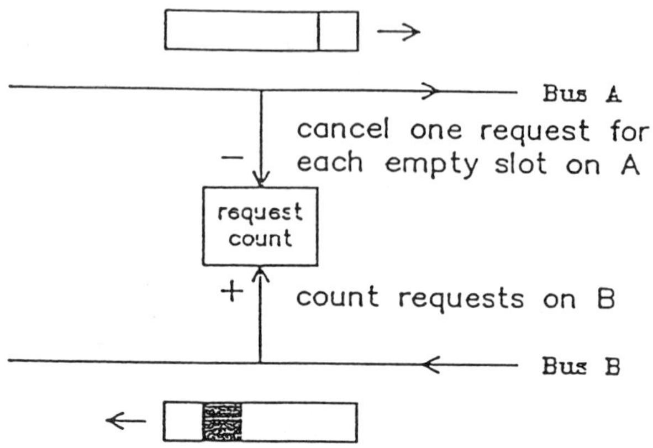

Fig. 8a). Distributed queueing (Bus A). Node not queued to send.

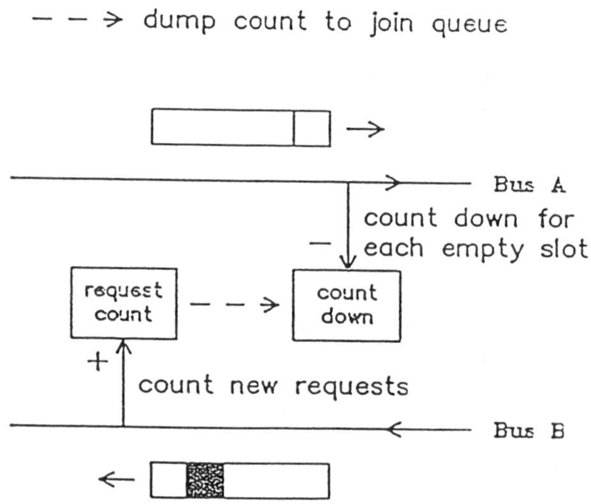

Fig. 8b). Distributed queueing (Bus A). Node queued to send.

the sending of the REQ for the node's segment, effectively places the segment in the distributed queue.

To ensure that the segments registered in the CD counter gain access before the newly queued segment in the given node, the CD counter is decremented for every empty slot that passes on the forward bus. This operation is shown in Fig. 8b). The given node can then transmit its segment in the first empty slot after the CD count reaches zero. The claiming of the first free slot ensures that no downstream segment that queued after the given segment, can push in and access out of order.

During the time that the node is waiting for access for its segment, any new REQs received from the reverse bus are added to the RQ counter, as shown in Fig. 8b). Hence, the RQ count still tracks the number of segments queued downstream and the count will be correct for the next segment access.

Thus, with the use of two counters in each node, one counting outstanding access requests and the other counting down before access, a first-in-first-out queue is established for access to the forward bus. The queue formation is also such that a slot is never wasted on the network if there is a segment queued for it. This is guaranteed since the CD count in the queued nodes represents the number of segments queued ahead. Since at any point in time one segment must have queued first, then at least one node is guaranteed to have a CD count of zero. It is that node that will access.

Priority Distributed Queueing

The Distributed Queueing protocol can be extended to permit the assignment of priority to segment access. The priority access control is absolute in that segments with a higher priority will always gain access ahead of segments at all lower levels. This is achieved by operating separate distributed queues for each level of priority.

The operation of separate distributed queues for each level of priority consists of the use of a separate REQ bit on the reverse bus for each level of priority and separate RQ counters for each priority level. The counters operate much as before, except that account must be taken of REQs at higher levels. That is, a RQ counter operating at a particular level will count REQs at the same and higher priority levels. Thus the RQ count records all queued segments at equal and higher priorities.

The operation of the CD counter is also slightly altered. A CD counter operating at a particular priority level will, in addition to counting down the passing empty segments, be incremented for REQs received at higher priority levels. This allows the higher priority segments to claim access ahead of already queued segments.

With such a priority scheme, the access performance for high priority traffic is ideal. Immediate access can be gained without any delay due to already queued traffic. Such a feature is very important for network signaling. It would also be important for packet voice, should that be implemented.

Performance

The operation of the Distributed Queueing protocol provides two key advantages over the operation of other existing protocols. One is that, through its use of REQ counters, an ordered queue is established for access where capacity on the network is never wasted. The second advantage is through the use of segmented access, which allows finer granularity and control in the allocation of the bandwidth resource.

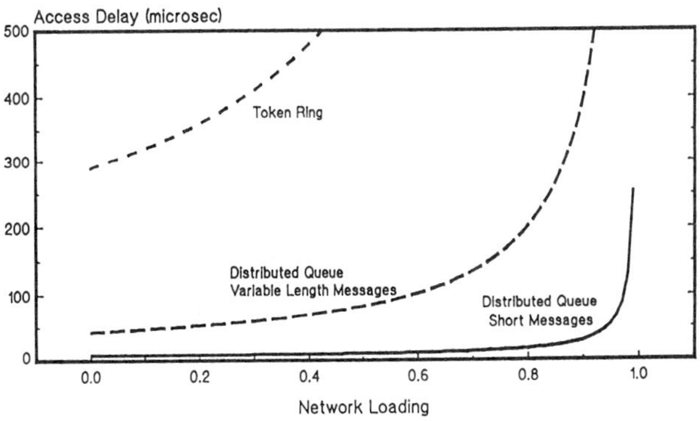

Fig. 9. Delay for distributed queueing and token ring (100 Mbits/s, 100km).

The access performance advantage for the Distributed Queueing protocol is best demonstrated by the graph in Fig. 9. This figure shows the access delay for packets on a typical MAN. To represent the delay best it has been normalized to the average length of the packets. Also, shown on the figure is the token ring protocol. Distributed Queueing has a very clear advantage in terms of access delay.

The advantage demonstrated in Fig. 9 is even further exaggerated when either the speed or the size of the network increases. The normalized delay performance for the Distributed Queue is unaffected by these network parameters, whereas the token ring normalized delay will continue to increase with the network size and speed. The increase in delay is clearly undesirable.

The limitation of the token ring is the delay in the transfer of the token. Even at very low utilization, a node wishing to access must wait for the token to pass to it. On average this delay is at least half of the ring latency at low loading and increases beyond this as load increases. Such a delay eventually limits the usable size of a token ring network.

The second performance advantage associated with distributed queue access is that of segmentation. Packets sent over the QPSX network are not sent as a whole but rather in segments to suit the slot size on the bus. Segmented transfer versus whole packet transfer is much like the comparison of time shared processing versus batch processing.

With whole packet transfer and likewise batch processing, the single processor is allocated to serve each packet unit at a time. Consequently, very large packets cause long delays for all other packet transfers. Most affected by this are short packets and high priority packets. Both must wait while the large packet completes access.

By contrast to this operation, with segmented access such as in the distributed queue the access is shared on a segment by segment basis. This means that even during the transfer of a very large packet, higher priority packets need only wait at most one segment before gaining access. Likewise, other packets at the same priority will be able to obtain immediately their share of the capacity. Demonstration of the access advantage for the short packets is illustrated also in Fig. 9.

Thus, segmentation provides a much finer granularity for bandwidth sharing and aligns communications switching with the performance expectations now commonly accepted in the computer industry.

The Segmentation and Reassembly Scheme

We have noted the performance advantages associated with segmented access of packets. Now we describe the scheme used to control the segmentation of packets at the source and the reassembly at the ultimate destination.

The segmentation process is straightforward and involves the fragmentation of the original packet to match the slot size on the bus. This process is shown in Fig. 10, where the segments of the packet are formed. This includes also the forming of a packet header segment which is necessary in routing the packet and instituting the reassembly process.

All segments consist of a header field along with the actual data as shown in Fig. 11. The header field consists of two sub-fields. The first is a single bit, the C/D subfield, which is used to indicate whether the segment is a control or a data segment. The second field is the source identifier (SI). This is used to provide the logical linking between slots of the same packet.

The SI is a fifteen bit field. Each node in a QPSX dual bus will have one or more unique SI number(s) related to its physical location along the bus. The SI is in addition to, and independent of, any network layer addressing. The allocation of the SI numbers is locally administered and only has significance within a single network. The purpose of the SI is to identify segments of a single packet. This information is used to reassemble the segmented message at the destination. To describe the operation of the packet transfer scheme, the segmentation of the packet at the source is described first and the action of the receiver is considered later.

The train of segments sent by the source is shown in Fig. 10. The first segment of a multi-segment message is a control segment. Within this segment a control field immediately after the segment header will indicate the beginning of message (BOM) code. This code signifies the start of a new packet transfer. The SI field of this segment is set to the source's value and the information field of the segment contains additional control information amounting to a packet header. Part of this information is routing information required to transfer the packet to the destination. All routing within the MAN is based on physical access unit addresses, not on the network layer addresses. Also, contained within the information field of the first segment is the length of the packet.

All segments of the packet until the last are each placed in the information fields of slots following the first control slot. In the header of each of these slots the C/D bit is reset to indicate data and the SI field is the same value as was sent in the first or control segment. The transfer of the multi-segment packet is completed by sending a further control slot which contains the end of message (EOM) code in the control field after the segment header.

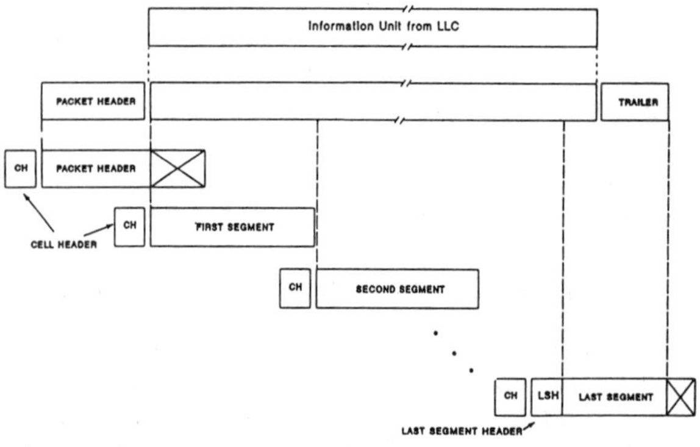

Fig. 10. Packet segmentation scheme.

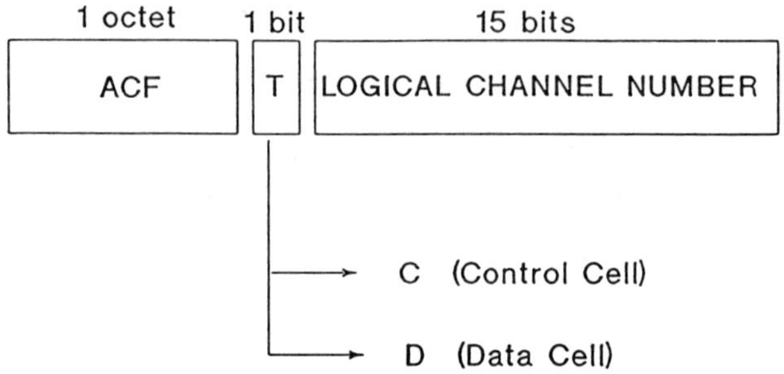

Fig. 11. QPSX segment header.

For the transfer of a message that only requires a single segment, the SSM (single segment message) code is used in the control field of the first segment. The SI is not required in this case, however it is still used for consistency in operation.

To receive the packets segmented as described above, each AU will implement the state machine in Fig. 12. This state machine will handle the receipt of a packet from a single source at a time. To handle concurrent receipt of more than one message, multiple state machines and receiver buffers are used in the AU. In the diagram, the condition for a state transition is shown at the start of the transition line and the action taken is at the arrowhead of the transition line.

The receive machine consists of two states: IDLE and REASSEMBLE. In the IDLE state the machine is not currently receiving any message. Thus, in this state the machine will check for control slots with the control field equal to BOM or SSM. In the case a BOM code is received the machine will check the destination identifier in the first part of the information field. If the destination identifier is its own, then this indicates the packet is intended for that station and the state machine enters the REASSEMBLE state. In the case that SSM is detected in the IDLE state and the destination identifier field matches, the length and information fields are copied and the AU will indicate to the higher layers that a packet is received. The receive

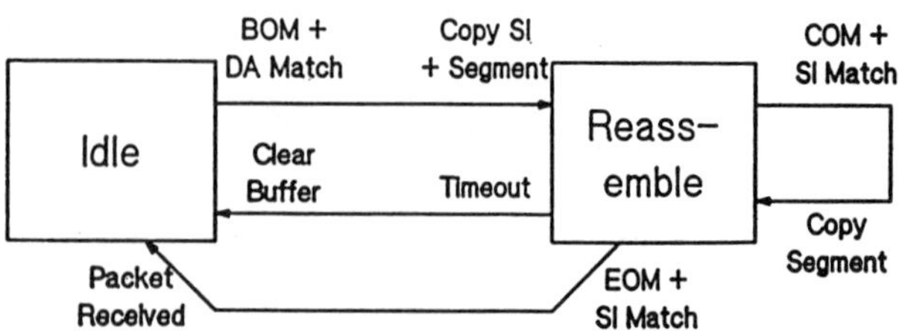

Fig. 12. Reassembly machine.

machine remains in the IDLE state after copying a SSM segment.

The REASSEMBLE state is used to receive the segments following the BOM control segment of a multisegment message. In this state, the segments with the SI equal to that copied from the BOM segment will be received. The information fields of these following segments are appended together to form the complete packet. New packets addressed to the given station in this state are ignored by the state machine. Other idle receive machines are used to receive such messages. When the EOM segment of the packet is detected by the receive state machine, again with the same SI, the machine will pass the reassembled packet to the station and return to the IDLE state. This completes the receipt of the packet.

To guard against the loss of the EOM segment, which would cause the receive state machine to be locked in the REASSEMBLE state, a timer is used. This timer is started after each segment is received. If the timer expires before the next segment is received, a failure in the transfer is assumed. The machine will then clear all copied slots and return to the IDLE state. Loss of segments of the packet can be detected by comparing the length field given in the BOM slot with the number of slots received.

The SI can be reused by the source for its next packet transmission immediately after the EOM segment is sent. Even if the destination of the first packet did not receive its EOM slot, the immediate reuse of the SI does not present a problem. The BOM issued for the new packet can serve to terminate the receipt of slots by the first destination. As a further safeguard, knowledge of packet length can prevent the receipt of further segments not intended for the first destination.

Stations that need to send more than one message concurrently, such as bridges, may be allocated more than one SI. The number of packets that may be sent concurrently equals the number of SIs allocated.

Reliability Features of the QPSX Network

The reliability of QPSX stems both from its architecture and its Distributed Queueing medium access control protocol or MAC. There are two reasons for the inherent reliability of the bus architecture. First, the network nodes are logically adjacent to the bus not serially and logically connected as in a ring. As a consequence data do not pass through the nodes and passive MAC failures have no effect. Secondly, because the node writing function is by logical "OR," i.e., there is no

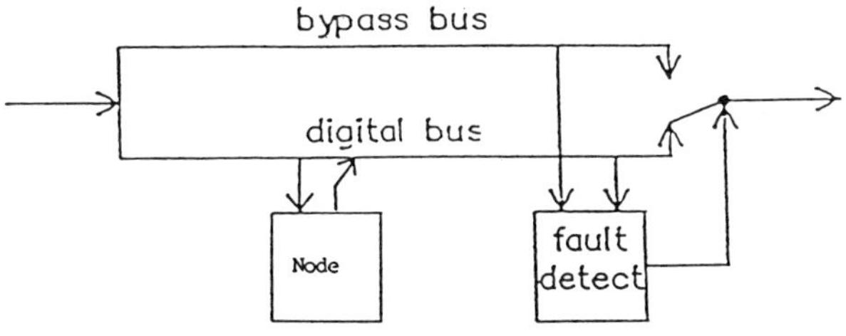

Fig. 13. Node fault detection.

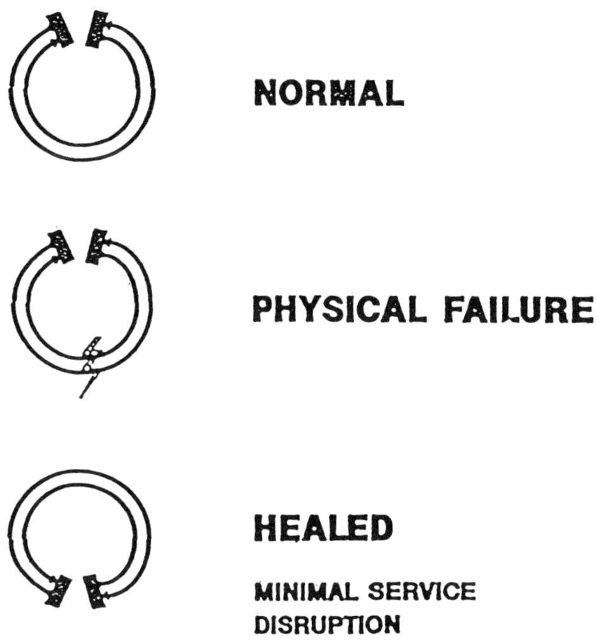

Fig. 14. Healing mechanism. Physical breaks are healed by repositioning the natural break in the loop.

overwriting, active MAC failures and bus failures are easily detected by input to output coincidence checking. An arrangement to implement this is shown in Fig. 13. Such failure points when they occur are bypassed at the node. The operation can be totally hardware based, with immediate response and causes no loss of synchronism, and minimal network disruption. With QPSX no node monitoring functions are required to check the MAC.

In this context a comparison with ring based networks is pertinent. With rings [4] both active and passive node MAC failures are catastrophic in that both invoke healing mechanisms associated with redundant rings. Such action involves considerable network disruption because of the need to reconfigure with attendant loss of synchronization. Any additional node failure can then produce isolated network islands.

With the QPSX architecture those faults involving node failure do not need to invoke any network healing mechanism which reconfigures the network. Faulty nodes are easily and non-disruptively bypassed.

Failures in the physical transport however do involve reconfiguration. But on healing the network captures the same synchronization and therefore there is minimal disruption to all synchronous communications. The healing mechanism is depicted in Fig. 14 and shows how physical faults or line breaks are healed by repositioning the natural break in the loop to the position of the break due to failure. Significantly there is no loss of network capacity involved in network reconfiguration. Here again the contrast with ring type networks is sharp. If a ring network carries traffic on a back-up ring then network reconfiguration, due to either node or transport failure, introduces a 50 percent reduction in packet data capacity. Alternatively if the back-up ring carries no data traffic, there is always a 50 percent inefficiency factor relative to the QPSX bus network. This is a critical factor in relation to cost efficiency, more particularly so in the public environment.

With QPSX it takes a real disaster, such as a backhoe through a physical transport, to invoke network healing. As explained, node failures do not invoke reconfiguration since faulty nodes are excised using very simple low complexity and low cost hardware. With rings however node failure invokes global healing.

Summary

QPSX is being developed as a public metropolitan area network and is in the IEEE Standards arena for 802.6 MAN Standardization. It has the inherent robustness and reliability necessary in a public network and it is being developed with management and healing features that materially enhance that reliability.

Next after reliability, an important issue is performance and in this respect QPSX is again quite outstanding. It uses a unique distributed queueing access control that permits fair access with the smallest possible delay and up to 100 percent loading. Moreover, the performance is independent of network rate or size.

The basic QPSX subnetwork is a pair of oppositely directed buses with station, bridge and gateway access units attached to both. Bus rates are up to 150 Mbits/s, with the streams divided into fixed length slots recurrently at the frame rate of eight kilohertz. Slots can be reserved for synchronous communications and those not reserved are available for packet-switched transfers.

A metropolitan area network will consist of many, possibly hundreds or even thousands, of QPSX subnetworks. These will be interconnected in an approximately hierarchical manner using bridges. MANs will be interconnectable to each other via gateways over a Wide Area Network. Customer premises networks or LANs will connect to a MAN also via gateways. Routing over the MAN and WAN will be by physical addresses of gateways, bridges and stations, related to E.164 addresses. Data packets up to eight kbytes in length will be transferred in segments over virtual circuits, a circuit normally existing only for the duration of a packet. Thus, with respect to packets the network will give a datagram or connectionless service. A connection-orientated packet service will also be possible.

Data transfer in QPSX resembles closely the transfer mode that has been decided by CCITT SGXVIII for B_ISDN making alignment of the two standards feasible. Complete alignment requires that the slot or cell as well as the segment or cell header sizes be equal in the two and that the headers encompass the product set of functions, expressed by identical codes and formats. Thus aligned, QPSX MAN would greatly speed the advent of B_ISDN.

References

[1] Draft proposed, IEEE Standard 802.6 metropolitan area network (MAN), QPSX dual bus media access control and physical layer protocol documents. Submitted July 1987.

[2] Z. L. Budrikis, J. L. Hullett, R. M. Newman, D. Economou, F. M. Fozdar, R. D. Jeffery, "QPSX: A Queued Packet and Synchronous Circuit Exchange," *Proc. 8th Int. Conf. Comp. Comm.*, Munich, 1986, pp. 288-293.

[3] R. M. Newman, J. L. Hullett, "Distributed Queueing: A Fast and Efficient Packet Access Protocol for QPSX," *Proc. 8th Int. Conf. Comp. Comm.*, Munich, 1986, pp. 294-299.

[4] F. E. Ross, "FDDI—a tutorial", *IEEE Comm. Mag.*, vol. 24, no. 5, pp. 10-17, May 1986.

Robert M. Newman was born on September 15, 1963. He received his Bachelor of Engineering with first class honors from the University of Western Australia in 1985, and has completed his Ph.D. research studies at the University of Western Australia. In both or these degrees Mr. Newman was involved in research work that led to the initial concept of QPSX switching and has also performed detailed modelling of the switch's operational characteristics.

Since April 1987, Mr. Newman has been employed as Projects Director, Network Systems for QPSX Communications Pty Ltd. In this role he has been responsible for the development of the initial QPSX switch products and has been actively involved in the standardization of dual bus technology in the IEEE 802.6 Committee. Mr. Newman is the co-inventor of the core QPSX technology.

Zigmantas L. Budrikis is a Director of QPSX Communications Pty Ltd and is on the Electrical Engineering Faculty in the University of Western Australia. He was born in Lithuania on September 29, 1932. He graduated B.Sc. majoring in physics in 1955 and B.E. with first class honors in electrical engineering in 1957, both from the University of Sydney, and Ph.D. in 1970 from the University of Western Australia.

Dr. Budrikis worked in the research laboratories of the PMG's Department (now Telecom Australia) from 1958 to 1960 and in Australia's Aeronautical Research Laboratories in 1961. Since 1962 he has been on the academic staff of the University of Western Australia. He was a visiting lecturer at U.C. Berkley in 1968, a visitor at Vilnius University at Technical University Munich in 1976-1977, an MTS at Bell Laboratories Holmdel in 1972 and 1974, and a consultant on four separate occasions at AT&T Bell Laboratories at Holmdel and Murray Hill over the period 1981 to 1985. He has worked extensively on problems of TV signal coding, human vision modeling, electromagnetic theory foundations, data communications and computer graphics. He is author of some forty papers and holder of several patents.

Dr. Budrikis is a Fellow of the Institution of Engineers, Australia and a member of IEEE, the New York Academy of Science, and the Optical Society of America.

John L. Hullett was born in Perth, Western Australia on December 20, 1942. He received his Bachelor of Engineering with first class honors in electrical engineering in 1965 and his Ph.D. in 1970 from the University of Western Australia.

Dr. Hullett worked in the research laboratories of Telecom Australia from 1969 to 1972 where from 1971 he was head of the Data Systems Division. Since 1972 he has been a member of the academic staff of the University of Western Australia and a member of the faculty of engineering. He has worked extensively in digital, data and optical communications, and networking and has published some forty five papers in these fields. He is a member of the IEEE.

Dr. Hullett is the co-inventor of the core QPSX technology and acts now as the Director of Strategic Planning and the Technical Division of QPSX Communications Pty Ltd. ■

Description of Fasnet—A Unidirectional Local-Area Communications Network

By J. O. LIMB and C. FLORES*

(Manuscript received January 15, 1982)

Fasnet is an implicit token-passing, local-area network aimed at supporting high data rates and carrying a wide mix of traffic (data, voice, video, and facsimile). Transmission is unidirectional with stations attaching to the medium passively via directional couplers. A link consists of two lines, one to carry traffic in each direction. Unidirectional transmission provides the potential for efficient operation at high data rates, while the passive medium provides the potential for high reliability. We describe the physical configuration and the protocol and give channel utilization for the condition of continuously queued sources. Mechanisms to control the access of various traffic types are described. Finally, the interconnection of multiple Fasnets is studied for one particular configuration, a ring.

I. INTRODUCTION

Local computer networks operating at 1 to 10 Mb/s are being commercially offered and appear to adequately meet current demands for computer communications within the office environment. However, future needs stimulated by both a broader range of services than is now available and changes in system architecture (e.g., the trend towards distributed processing) could increase significantly the demand for digital capacity. For example, one would like to be able to handle video information, voice traffic, and facsimile, as well as computer traffic, in a single digital system. The availability of a cheap, high-capacity communication conduit between computers will itself stimulate increased traffic. For example, processing time can be traded for communication capacity; rather than transmitting a text file and

* Department of Electrical Engineering and Computer Science, University of California at Berkeley.

formatting it at a remote location, one may choose to transmit a formatted version or even a bit map. Thus, while today 10 Mb/s may be regarded as an extremely generous bit rate for a local computer network, 200 Mb/s may become limiting for an integrated communications network.

Carrier-sense multiple access with collision detection (CSMA/CD) is reliable and reasonably efficient even under heavy load for most conditions.[1] Shoch and Hupp[2] show that measurements of channel utilization of an Ethernet* yield results that are close to calculations made by Metcalf and Boggs, using a simplified model and assuming that active stations have data continuously queued for transmission. Utilization is reported in Ref. 1: $\eta = \gamma/[\gamma + F(M)]$, where γ is the ratio of packet duration to slot time and $F(M)$ is a slowly varying function of M, the number of active stations. Utilization is plotted as a function of γ in Fig. 1, with M as a parameter. If we assume that slot time is 50 μs,[3] and the transmission rate is 100 Mb/s, then for an average packet length of 1000 bits, $\gamma = 0.2$ and utilization is in the range 7 to 8 percent.† Shorter packets, higher transmission rates, or longer slot times would further decrease efficiency. Also note that the above equation does not incorporate source acquisition and synchronization time which, like slot time, is relatively more significant at higher transmission speeds. Thus, it appears that CSMA/CD is not viable for operation at high data rates.[5]

Fasnet is an implicit token-passing protocol developed to efficiently utilize the channel capacity when the ratio of packet duration to the maximum station-to-station propagation time is small (<1). Information flows in only one direction on the medium, unlike the usual CSMA/CD configurations (although see Refs. 5 and 6), but like CSMA/CD the essential passivity of the medium is retained. The access method is closely related to a ring protocol (e.g., see Ref. 7) and may be regarded as a variant of implicit token passing.

Reliability was an important consideration in the design of Fasnet. Consider both the transmission medium and the control electronics. Reliability of the transmission medium may be enhanced by keeping active electronics in the medium to a minimum. Bus architectures such as Ethernet have occasional repeaters, depending on the length of the signal path. Cable-TV (CATV) type architectures have periodic line amplifiers whose spacing is determined by the number of stations (taps), as well as by cable attenuation. Ring architectures usually have

* Ethernet is a trademark of Xerox Corporation.

† Since Ethernet requires $\gamma \geq 1.0$ for collisions to be reliably detected, utilization for values of $\gamma < 1.0$ were obtained by multiplying the utilization for $\gamma = 1$ by the value of γ. This can result in efficiencies much lower than that obtainable by other CSMA protocols, e.g., p-persistent CSMA.[1]

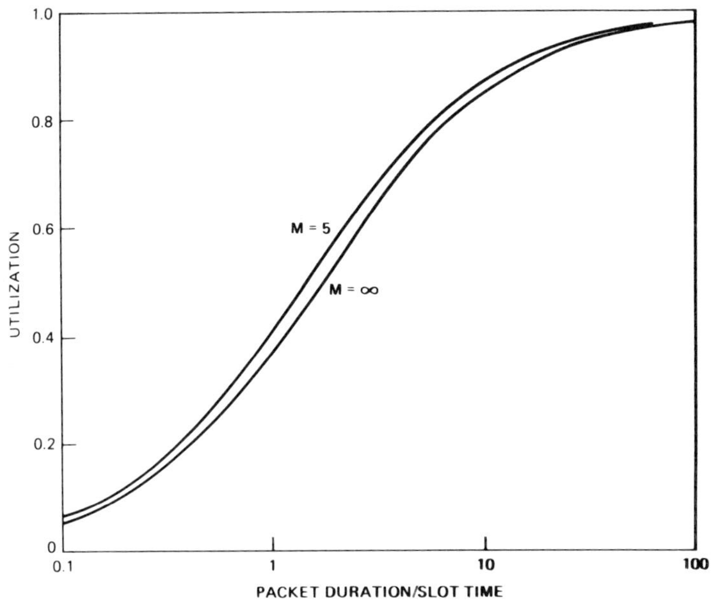

Fig. 1—Plot of efficiency as a function of the ratio of packet duration to slot time. Parameter is the number of simultaneous users. Efficiency changes little as the number of users goes from 5 to ∞.

most electronics in the signal path; digital regeneration is usually provided at each station.[7] Turning to control, reliability considerations tend to favor distributed control. An alternative to a fully distributed system is to permit some stations to perform unique functions but have these functions assumable by any station on the network; however, this can result in a large cost penalty. A further alternative is to have the function performed on a server basis (two or more stations provide the service to all other stations). The Fasnet medium resembles that employed in CATV, and control is primarily distributed with some functions assumable by all stations.

There is considerable attraction in having a single system to handle all forms of traffic in the environment. Indeed, such a system may be regarded as an extension into the local environment of the Integrated Systems Digital Network (ISDN) that is being so vigorously pursued in the long distance and local loop environments by the common carriers.[8] An integrated transmission system simplifies the implementation of services that utilize different types of traffic. Examples are voice-annotated electronic mail[9] and interactive use of voice and facsimile.[10] An integrated transmission system also provides the opportunity to reduce overall transmission needs by taking advantage of the complementary nature of some types of traffic; for example, most electronic mail can be deferred until after the voice busy hour. Further, one would anticipate cost reductions in having one integrated system over a number of separate systems.

While design of an integrated system offers opportunities, it also presents the need for compromises and trade-offs. Consider terminal costs and transmission efficiency. If a system is to provide economical interconnection for telephones and terminals, it must permit construction of an interface that is cost-effective relative to alternative solutions. This may mean that some interfaces have to be tailored to the specific application to make them competitive.

Virtually any type of traffic should be able to exploit the channel efficiently. For example, environments that generate a large number of short messages (e.g., computer terminal traffic), as well as environments that generate a preponderance of long messages (e.g., file transfers), should be able to operate efficiently. This requires that there be a minimum of structure at the lowest common level of the service. For example, a packet structure which mandated a source-address field, while useful for computer traffic, may be unnecessary overhead for a voice channel where call set-up would establish the identity of the source.

The description of Fasnet starts in Section II with the physical loop; the access protocol is described in Section III. The performance of the basic system is given in Section IV, followed by a discussion of some of the system design issues (in particular, the synchronization and signaling procedures) in Section V. Section VI describes variations of the basic system, including methods for improving efficiency, particularly when the number of users is small. Section VII describes mechanisms to support the efficient management and control of mixtures of different traffic types. Section VIII describes the interconnection of Fasnets, with consideration of the impact of the topology on throughput and ability to handle localized sources of traffic. Section IX summarizes the paper.

II. PHYSICAL CONFIGURATION

The basic link, as shown in Fig. 2, consists of two lines. One line passes all stations carrying traffic in one direction and the other line passes all stations carrying traffic in the other direction. For line A, station S_1 is referred to as the head station and S_n the end station. For line B the assignment is reversed. Together the two lines provide a connection between any pair of stations attached to the link. While the lines may be either twisted pair, coaxial cable, or light fibers, we will be primarily concerned with a coaxial cable implementation. Each station makes two connections to each line. A read tap precedes a

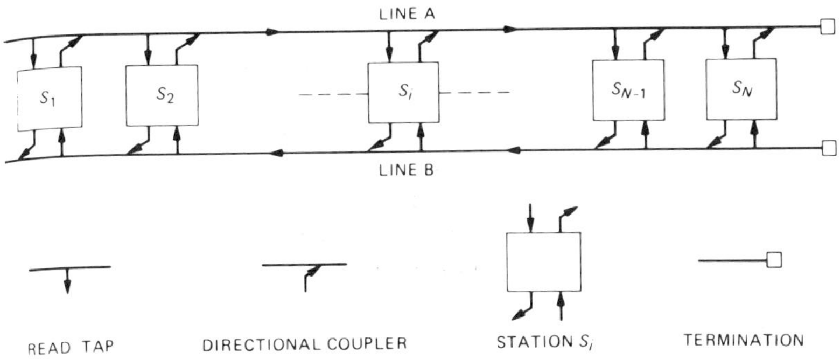

Fig. 2—Physical configuration of a Fasnet link.

passive directional coupler used for writing. The nature of the directional coupler is such that very little energy travels in the reverse direction on the line so that the signal read virtually simultaneously from the read tap will be unaffected by the signal being written on the line via the directional coupler. A station writes on the line by adding energy to the signal already existing on the line. Except for specific fields of the header, the protocol ensures that only one station at a time writes on the line. Thus, once a signal is written on a line, it is not removed or changed by any station. This has certain implications for the line code that is selected (Section 5.2).

Depending on the length of the line, amplifiers are needed to boost and compensate the signal. The technology and design procedures used for CATV systems[11] are directly applicable here, although the noise margin required for a high-quality video signal is somewhat greater than that required for two- or three-level digital transmission.

Links may be joined together to form a network of links. Usually, links will be run in pairs of lines, but this is not always necessary. The advantage of using multiple links is that the traffic-carrying capacity of the network can be increased and reliability may be improved by the use of redundant paths.

An earlier version of Fasnet[12] differs primarily from the system described here in that a link consists of a single unidirectional line that passes each station twice—on the outbound or write side and on the inbound or read side. Each station makes three connections to the line, a read tap for control purposes, and a directional write tap on the write side, and a read tap for recovering data on the read side. The primary advantage of the scheme described here is that the link can carry approximately twice the traffic of the earlier version. A disadvantage is that a station must select the correct line on which to transmit, and this will depend on the relative physical location of the destination.

III. PROTOCOL DESCRIPTION

The data link layer may be divided into two sublayers.[13] One sublayer, the logical link control with which we are less concerned here, provides functions like addressing, windowing, and acknowledgments. The other sublayer, the media access control with which we are more concerned, determines when and how to send information via the physical medium. This is influenced by the media type, the physical configuration, and the technology used.

3.1 Frame format

The frame structure suggested in Ref. 13 and its relation to the data link sublayers is shown in Table I. The information unit is delivered by the network layer. The logical link control appends the source address, the destination address, the link control field for windowing, acknowledgments, and similar functions. We call this unit a packet, and in the work described here we will assume it is of fixed length. The media access control sublayer appends (*i*) the frame check sequence computed on the previous fields for error detection and (*ii*) the access control (AC) field which determines how and when each station may access the physical medium. The main objective in the design of this field is to control access among all active stations in an efficient, reliable, and fair manner. The frame start and frame end delimiters are unnecessary, since the stations are kept in tight bit and frame synchronization (see Section 5.1). The duration of the frame is referred to as a slot.

3.2 Access control

Basic access control for Fasnet is as follows. The head station, S_1, initiates a cycle on line A. After a cycle has been initiated, each active station on the line with packets destined in the right direction is allowed to access the line for one slot. To do this, each station monitors the line. When it senses the line idle, it seizes the line for one slot. It has to wait for a new cycle to be initiated before it attempts to access the line again. The exact manner in which this is done efficiently and

Table I—Protocol and frame structures

	Protocol Structure	Frame Structure
Data link layer	Logical link control sublayer	DA/SA/LC/IU
	Media access control sublayer	FS/AC/DA/SA/LC/IU/FCS/FE
Physical layer	Physical layer signaling	FS/AC/DA/SA/LC/IU/FCS/FE

FS: Frame starting delimiter
AC: Access control field
DA: Destination address
SA: Source address
LC: Link control field
IU: Information unit from network layer
FCS: Frame check sequence
FE: Frame ending delimiter

fairly is described in the next paragraphs. If a station has priority, it is given permission to access the line for an integral number of slots. In this manner, the active stations can access the line for a specified duration in the order in which they are physically located on the line. The operation on line B is identical to that described above with S_N replacing S_1 as head station.

To describe the operation in more detail, let $\{S_1, S_2, \cdots, S_N\}$ be the set of stations in the order of their physical locations as shown in Fig. 2. Let AQ_i and BQ_i be the number of packets queued at station S_i for access to lines A and B, respectively.

When the next packet arrives at S_i from the Network layer interface,
if destination address $j > i$, then AQ_i is incremented by 1;
if destination address $j < i$, then BQ_i is incremented by 1.

The structure of the AC field is shown in Fig. 3. Let t_{fn} be the start of the nth frame. The AC field is from t_{fn} to t_{bn}. Station S_i gains access to line A in the following manner. Let S_i be permitted access for p_{max} packets each cycle. At t_{fn}, the start of the nth frame, the read tap reads the START bit of the AC field. The start of cycle is indicated by START = 1. Because of gate delays in the decision circuitry and propagation delays in the tap cables, the outcome of the read operation is only known at t_{sn}. This additional time of duration τ_{dec}, shown in Fig. 3, is of the order of a few bit times for a 100 Mb/s line and nanosecond logic. Next, the station may simultaneously read the BUSY bit via the read tap and write BUSY = 1 via the directional coupler. Again, the outcome of the read operation is only known at t_{bn} after a delay of τ_{dec}. Nonetheless, the write operation does not alter the BUSY bit if it is already set to 1. The nature of the signaling to achieve this is explained in Section 5.2. At t_{bn}, if BUSY = 1, the station defers until the BUSY bit of the next frame. If BUSY = 0, the station accesses the line for the remaining frame duration.

Station S_i is said to be in one of four states:

IDLE —if it has no packets to transmit, i.e., $AQ_i = 0$.

WAIT—if it is waiting for the start of cycle.

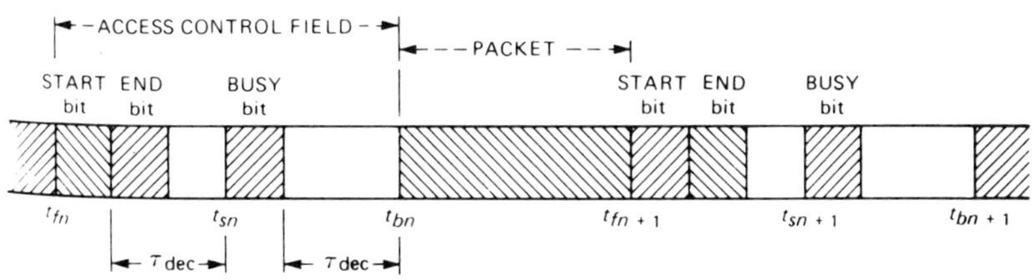

Fig. 3—The frame structure of Fasnet. Each frame consists of (*i*) an access control field containing START, END, and BUSY bits; and (*ii*) the packet as provided by the logical link sublayer.

DEFER —if it is deferring to busy users who are upstream on the line.

ACCESS—if it is accessing the line.

The station makes transitions (denoted as →) between states as follows (Fig. 4): While $AQ_i = 0$, S_i is in IDLE. Upon arrival of a packet for line A, $AQ_i > 0$ and $S_i \rightarrow$ WAIT. The station reads the START bit of every frame. When START = 1 $S_i \rightarrow$ DEFER, and the station simultaneously reads and writes the BUSY bit as described above for every frame. When BUSY = 0 $S_i \rightarrow$ ACCESS. Now it accesses the line for p_{max} frames and also writes BUSY = 1 for each. Then $S_i \rightarrow$ WAIT. The station may cease to access the line earlier if $AQ_i = 0$, whereby $S_i \rightarrow$ IDLE.

Station S_1 initiates cycles by START = 1 in the first frame of each cycle. There is an additional bit, END, in each frame to indicate the end of cycles. This bit can be conveniently located in the blank portion of the frame after the START or BUSY bits. When station S_N reads BUSY = 0 on line A (indicating that all active stations have accessed the line), it sets END = 1 in the next frame on line B. On receipt of this frame on line B, S_1 then initiates a new cycle on line A. Thus, in the worst case, line A will be silent once every cycle for a time equal

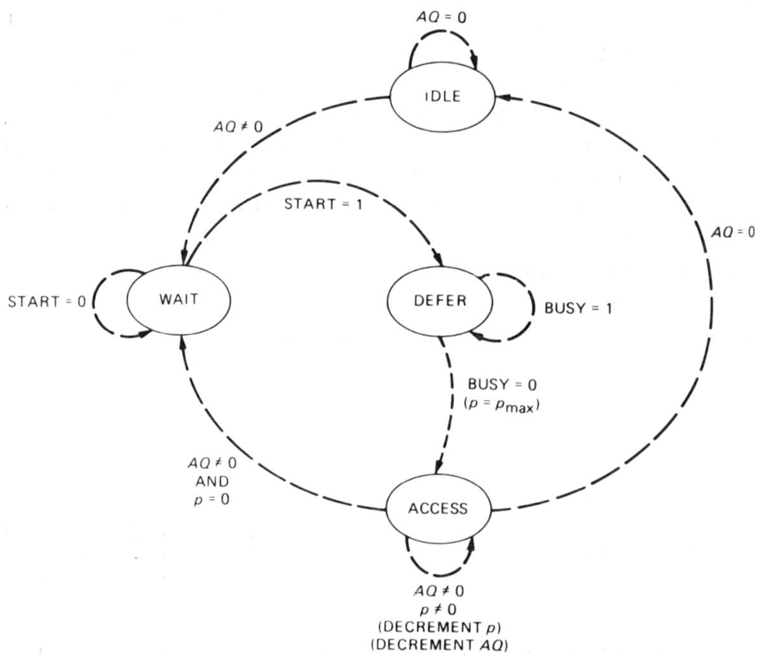

Fig. 4—State transition diagram describing the operation of a Fasnet station.

to twice the end-end propagation delay, plus twice the frame duration, as each end station has to wait until the next frame to set the START or END bits.

The operation on line B is identical, with the roles of S_1 and S_N reversed. Thus, the two lines cycle independently of each other with access being passed between the stations in the same order as their physical locations on each line.

In the protocol described above, the outcome of the read operation on the START bit needs to be known before the BUSY bit is written so that the first frame of a new cycle does not remain idle. Should the START and BUSY bits be adjacent to each other, a station will only learn that START = 1 after the BUSY bit has passed and the frame will not be used. However, for large cycle lengths and short packet lengths, the reduction of one decision interval, τ_{dec}, per frame would be greater than the addition of the extra idle frame.

A further alternative is to have the first frame of each cycle contain only an access field. However, unequal frame sizes complicate synchronization for a very small increase in efficiency.

3.3 Error recovery

The protocol is controlled by the START, BUSY, and END fields. An error in a BUSY field will have no lasting effect; it will result in a packet being overwritten if the busy bit is changed from a 1 to a 0. Alternatively, an empty slot will go unused if the busy bit is changed from 0 to 1. Of more significance is an error in the START and END fields. If a START field is set to 1 in error, two STARTs or a START and an END would be simultaneously present on the loop.

It will be shown that generation of additional STARTs and ENDs will not propagate and have little effect on the operation of the link. We will assume that end stations do not generate STARTs or ENDs that are closer together than the round-trip delay time, τ_r; under normal operation this cannot occur. A false START = 1 will occur either in the active portion of a cycle (including the first empty slot) or in the empty slots occurring at the end of the cycle. If the former, a new cycle will start before, or as, the old one is finishing. Since the additional START = 1 will not generate an END = 1 on the return line (because no transition from busy slot to empty slot is detected), the condition will not propagate. If the false START = 1 occurs in the empty slots, other than the first, the new cycle will start prematurely (actually increasing utilization temporarily). One of two conditions results:

(*i*) The busy part of the additional cycle terminates at least one slot before the next normally occurring START = 1, in which case the end station will detect an end condition. However, because the period since the last END = 1 is less than τ_r a new END = 1 will not be generated.

(*ii*) There is no empty slot before the next normally occurring START = 1. As a result, an additional end condition is not detected by the end station.

Thus, a START = 1 resulting from a fault condition will not produce additional END = 1 bits on the return line. On the other hand, END = 1 faults, unless they are closer together than τ_r, will produce additional START = 1 slots which as just described, have a transient effect on the operation of the link.

Consider the condition where a START or END bit is changed from a 1 to a 0. A new cycle would fail to initiate. After a time-out greater than the longest permitted cycle time, the head station will issue a START = 1, and the link will continue to operate normally. Should a head or end station fail, the station next to the head or end station would assume the functions on detecting loss of timing or after timing out on the arrival of START = 1 or END = 1.

3.4 Fault diagnosis

The independent lines of the Fasnet link provide the opportunity to localize and mitigate some types of faults. Consider first that a line is severed because of some catastrophic event or something trivial like a cable connector failing. The result will usually be either a short or open circuit leading to a gross impedance mismatch. The fault will most likely terminate all effective communication on the upstream side of the fault because of reflections from the mismatch traveling back into station interface units via the read tap. The downstream segment will be affected very little because the directional couplers will propagate little energy in the direction of the mismatch. Thus, a diagnostic program in the end station can determine between which stations the mismatch lies. This is done by having the end station send a query to each station via the intact line and determining which stations respond to the query.

A difficult type of station fault is to have a station continuously write garbage on a line. Diagnostic programs in the end stations, again by querying each station, can determine the faulty station and remove it from service. The head station on the line with the fault, after being informed of the fault by the end station, via the other line, queries each station in turn. If the station fault is confined only to the write circuit, the faulty station will respond. The next station on the downstream side will not respond, since it will not be able to read the query sent by the head station because of the interference from the faulty station. If both read and write circuits in the faulty station are affected, the last correctly responding station will be the station on the upstream side of the faulty station. Thus, the fault is isolated to one of two stations. Both stations may then be disconnected by means of a control signal sent via the functioning line. The faulty station may then be uniquely determined by returning one of the stations to service. If the fault condition resumes, the returned station is faulty and is disconnected; otherwise the other station is faulty.

IV. PERFORMANCE

4.1 Sample operation

Typical operation of Fasnet for lines of 2.5-km individual length, 100 Mb/s bandwidth, and 200-bit frame length is shown in Fig. 5. It shows the time-space relation of the frames on each line. The horizontal axis represents time divided into slots $A_1, A_2, A_3, \cdots$ for line A and $B_1, B_2, B_3, \cdots$ for line B. The vertical axis represents the physical locations of the active stations S_1, S_2, S_3, S_4, and S_5 with S_1 and S_5 serving, additionally, as end stations. The electrical line length is five frames. Station S_1 initiates the cycle in frame A_1, and access passes from S_1 to S_2 to S_3 to S_4. When the end station, S_5, senses BUSY = 0 in frame A_5, it sets END = 1 in frame B_9. Receipt of this frame by S_1 causes it to

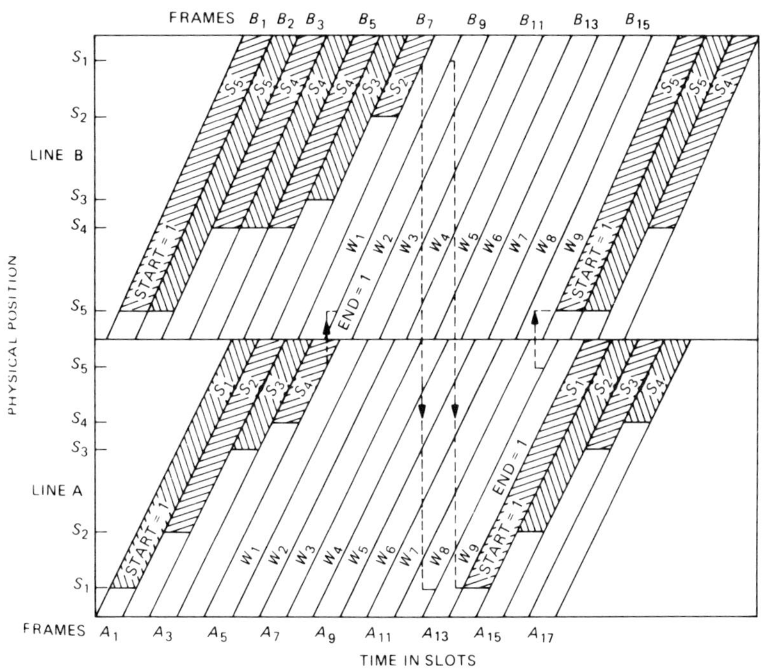

Fig. 5—A graph of activity on a Fasnet link (lines A and B) as a function of time. The dotted lines indicate the flow of information from one line to another.

initiate a new cycle in A_{14}. Similarly, a cycle on line B starts at B Assume that S_5 and S_4 are permitted access for up to two and three packets, respectively. Station S_1 senses BUSY = 0 in frame B_8 and sets END = 1 in A_{13}. Receipt of this frame by S_5 causes it to start a new cycle in B_{17}.

4.2 Utilization

As there are no collisions, no capacity is lost through collision resolution. However, the utilization is not 100 percent as each line is idle at the end of each cycle. The idle period is nine frames, W_1, W_2, ... W_9, on each line in Fig. 4. In the worst case, this is equal to twice the end-end propagation delay, plus twice the slot time (one slot time on average) as each end station has to wait until the start of the next slot to set the START or END bits. If

ν = speed of propagation on the line (m/s)

W = line capacity (b/s)

L = line length (m)

F = frame size (bits)

M = number of busy stations with downstream traffic,

then if each station is allowed access for only a single packet per cycle.

$$\text{cycle length} = \tau_c = M^*(F/W) + 2^*(L/\nu) + (F/W)\,,$$

$$\text{duration of busy frames} = \tau_b = M^*(F/W)\,, \tag{1}$$

and

$$\text{utilization} = \eta = \frac{M^*(F/W)}{M^*(F/W) + 2^*(L/\nu) + (F/W)}\,. \tag{2}$$

The effective utilization is lower since a fraction of each frame is devoted to access control. However, for large F, it results in only a small reduction in utilization. If $\nu = 2.5 \times 10^8$ m/s, $W = 100 \times 10^6$ b/s. $L = 2.5 \times 10^3$ m,

$$\eta = \frac{M^*F}{(M + 1)^*F + 2000}\,.$$

For $M = 100$,

$F = 50$ $\quad\eta = 71$ percent

$F = 100$ $\quad\eta = 83$ percent

$F = 200$ $\quad\eta = 90$ percent

$F = 500$ $\quad\eta = 95$ percent

$F = 1000$ $\quad\eta = 97$ percent.

For the same values of ν, W, and L, with $F = 500$[1,2] and assuming an Ethernet slot time $T = 50 \times 10^{-6}$ s,[3] we can compare the performance of Fasnet and Ethernet as the number of stations is varied as shown in Table II.

Unlike Ethernet, Fasnet has the desirable feature that as the load increases, the utilization also increases. The above figures do not reflect the fact that in practice the length of packets is variable and, consequently, the fixed frames of Fasnet frequently will be only partially filled. The effect on η depends on the distribution of packet size, and to some extent is determined by the system design. For example, in a system designed for large amounts of voice traffic, F could be set equal to the size of a voice packet.

V. IMPLEMENTATION CONSIDERATIONS

The design criteria previously stressed in the introduction affect the implementation in important ways. In particular, the requirement to operate at high speeds and the unidirectional operation of the bus affect the design of the synchronization system; in turn, the type of synchronization and the use of directional couplers impact the choice of the line code that is used.

5.1 Synchronization

Bus systems in which signals travel in both directions on the line require the receiving stations to adapt to the signals transmitted by the sending station because the amplitude, dispersion, and phasing of the received signal vary depending upon the position of the transmitting station on the line. Synchronization can be achieved very quickly when the signaling rate is low relative to the bandwidth of the transmission medium. At higher signaling rates, synchronization needs to

Table II—Fasnet versus Ethernet as a function of number of busy stations

M	Fasnet (in percent)	Ethernet* (in percent)
5	50	4.1
10	67	3.9
50	91	3.7
100	95	3.7

* Note that since the minimum permissible packet length is 5000 bits, η is calculated as 0.1 of η with 5000-bit packets. Other CSMA protocols that do not require collision detection perform better.

be more accurate to achieve good error performance. Ethernet specifies a synchronization preamble of 64 bits and for higher transmission rates an even longer sequence may be required. Thus, for short messages efficiency would be significantly reduced. Using a unidirectional bus, each station can be synchronized to a common clock issued from the head station. Thus, if all stations add signals to the cable in phase with the transmitting clock, stations will receive the signals in correct phase. Similarly, fixed gain and frequency compensation can be employed. The problem of reliability can be overcome by giving each station the ability to supply clock. The clock drive would be inhibited by detection of, and locking to, an incoming clock.

Initial tests have shown that a simple, cost-effective method of synchronization is to synchronize to a continuously injected pilot tone placed at the high end of the signaling band. The synchronizing function then assumes a negligible fraction of the transmission capacity.

In addition to bit synchronization, frame synchronization is also required. This is achieved by sending periodically a synchronizing bit pattern. Design is simplified if this is sent after an integral number of frames, say 64 or 128. With tight bit and frame synchronization, successive frames may be butted together without a gap.

5.2 Signaling

Because synchronization is achieved independently of the data signal, line codes with fewer transitions may be considered. It is particularly convenient if a code is chosen that couples no energy to the line when one of the logic states is continuously transmitted (assume logic 0). Each station at the end of transmission then simply returns to logic 0, and there is no need to "disconnect" the transmitter from the line. The two line codes we are investigating are a bipolar three-level code, Fig. 6a, and a nonreturn to zero (NRZ) two-level code, Fig. 6b. The two-level signal has a greater noise margin; however, one has to contend with the dc signal component.[14]

The Fasnet protocol does not permit subsequent stations to modify a signal already transmitted by an upstream source. In principle, this could be done. For example, a signal from one station could be deleted from the line by a second station writing the complement on the line. In practice, signal levels would have to be matched very accurately for such a scheme to work.

There is one condition in which more than one station may add energy to the same bit in a frame—the BUSY bit of the AC field. As a result, the amplitude of this bit may far exceed the amplitude of the remaining signal. This may lead to errors in adjacent bits of the AC field. To prevent this, guard bands on either side of the BUSY bit

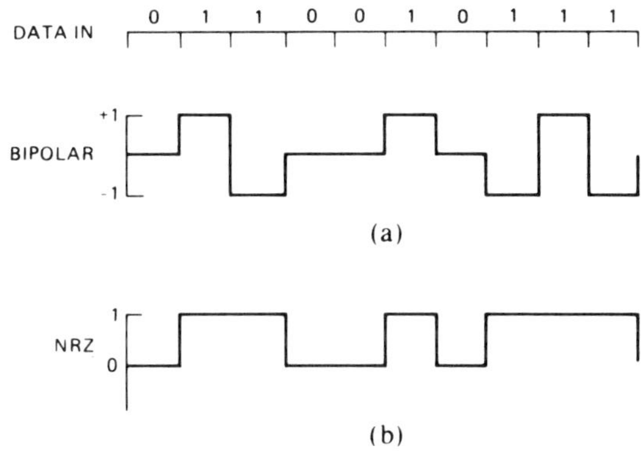

Fig. 6—(a) Bipolar three-level line code. (b) NRZ two-level line code.

should be used. Notice from Fig. 3 that the access field is configured so that the guard bands fall in the intervals τ_{dec} and, in practice, will have a comparable duration.

VI. IMPROVING UTILIZATION

As can be seen from (1), efficiency increases (*i*) as cycle length increases and (*ii*) as the idle period at the end of each cycle (intercycle gap) decreases. At the expense of some increase in complexity, techniques may be devised to improve utilization by increasing cycle length or reducing intercycle gap.

6.1 *Control of cycle length*

Since START = 1 may be read by all stations, the length of the last cycle, τ_c, may be determined by any station. As previously described, each station may transmit up to p_{max} packets per access. Thus, by controlling p_{max}, stations may influence the value of τ_c. Station control of τ_c by manipulation of p_{max} is obviously limited. For example, let us assume that p is fixed at 1 and that we have stations each generating packets at a rate $< 1/\tau_c$. Increasing p will not change the cycle length since packets will be transmitted before a queue can form. On the other hand, increasing p for heavily loaded stations will lead to an increase of τ_c, provided τ_c is less than the accepted maximum.

6.2 *Reducing inter-cycle gap*

Three methods are described for reducing the intercycle gap and hence increasing the line utilization. In the first, stations detecting the END = 1 bit seize empty slots on the other line; in the second, stations use the END field as a request field; in the third, the end station

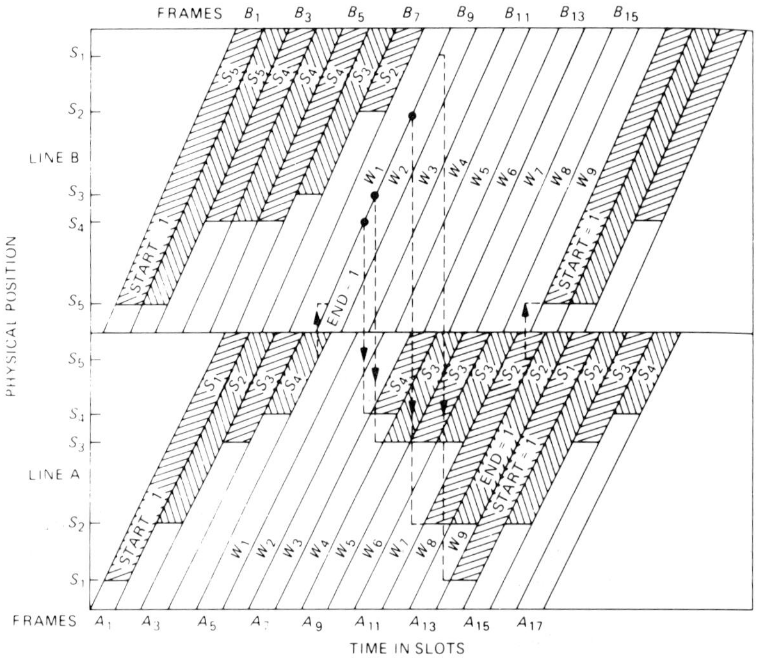

Fig. 7—A graph of activity on a Fasnet link as a function of time. Empty slots encountered on line A by stations that have read END = 1 on line B are utilized. The IDLE period is reduced to three slots. Corresponding use of IDLE slots on line B would also occur but is not shown.

attempts to estimate the end of a cycle, setting END = 1 before BUSY = 0 is received.

Considering the first method, any station S_i in the WAIT state that observes END = 1 may attempt to seize any empty slots on the opposite line.[12] The number of empty slots seized depends on the time the END = 1 frame takes to propagate to the next active station, which then seizes empty slots, thus preempting active stations downstream.* The intercycle gap now depends on the propagation time from the last active station to the end station and back. (The relative timing of the frame starts in the two lines will also affect the gap size). As shown in the example of Fig. 7, the intercycle gap has been reduced from nine slots to three, shown for line A only.

In the second method, stations in DEFER and ACCESS states write REQUEST (REQ) = 1 on the return line (the END field is now

---

* A station may only transmit a single frame at a time because of the possibility of preemption. This will interfere with construction of "superpackets." Superpackets are used to increase efficiency by reducing the effect of packet overhead. The overhead is attached only to the first packet of the superpacket.[12]

replaced by an REQ field). After all stations have been served, the head station will read REQ = 0 and initiate a cycle. To ensure that at least one REQ = 1 occurs in a cycle for which only one station is active, a station in changing from WAIT to DEFER or ACCESS writes at least one REQ = 1. Approximately speaking, the average intercycle gap is now equal to twice the propagation time from the head station to the last active station, plus half a slot time. Notice that this procedure is more distributed in that the end-station function of recognizing the end-of-cycle END condition and writing END = 1 on the return line is now bypassed. Each station now performs an equivalent operation.

A further refinement is to observe that the head-station function can also be distributed.* Each station with traffic to transmit need not wait for the head station to issue START = 1. Rather, after reading REQ = 0 on the return line, it can switch from WAIT to DEFER or ACCESS mode setting the p register to the allowed number of packets that may be transmitted per cycle. The intercycle gap is virtually eliminated under heavy traffic conditions and for a large number of users as in the first method. For two continuously queued stations, the intercycle gap is twice the delay time between the stations, plus one slot time on average. Notice that even though no START is being issued, or is necessary, the loop cycles. However, as in the first method, packets from a source during one cycle will usually not be consecutive. The station protocol is summarized in Fig. 8 by means of the state diagram. As seen in comparison with Fig. 4, the protocol is more complex; however, the algorithm no longer requires the centralized head or end-station functions. The issue of distributed control versus centralized control is particularly important in efficiently accommodating different types of traffic and is discussed further in Section VII.

In the third method, each active station in the DEFER state indicates its desire to transmit by setting an additional request field to 1 (REQ = 1) in the access field of the line on which it wishes to write, as in Ref. 12. The end station estimates the cycle length and transmits END = 1 timed to arrive at the head station as the last slot to be used in the cycle is leaving the head station. If the estimate was too low, the end station will read REQ = 1 in the last frame in the cycle indicating that the estimate of the length of the last cycle was too short. Therefore, the estimate of the length of the next cycle would be increased. If the estimate is too high, there will be empty slots prior to the arrival of the next START = 1, and the estimate of the length of the next cycle would be decreased. If the estimate is correct, then the

* Suggested by Z. L. Budrikis in connection with the earlier single-line version of Fasnet.

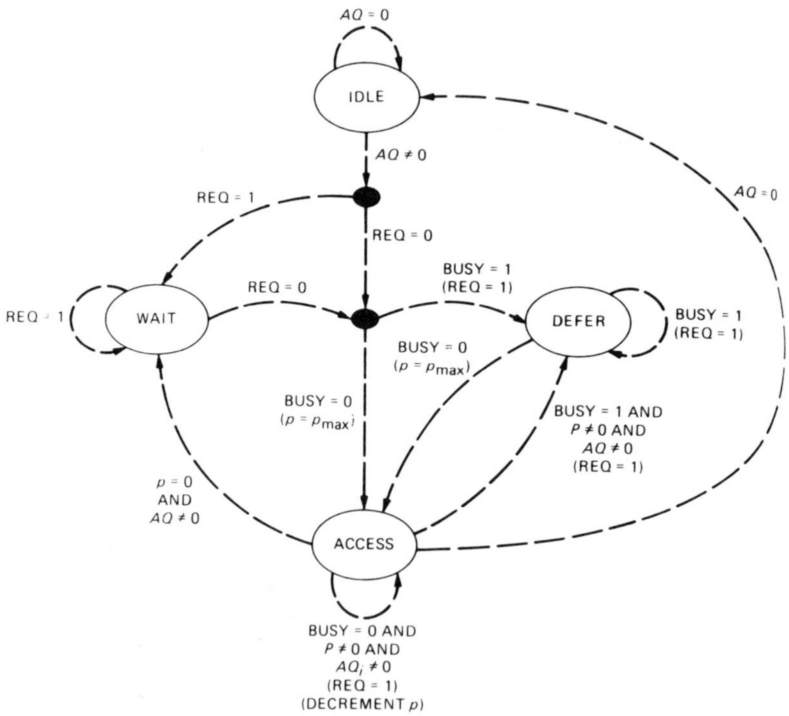

Fig. 8—State transition diagram of a distributed version of the Fasnet protocol in which unique functions in the head and end stations are not employed. Statements in parentheses are actions executed on making the transition. The filled circles denote intermediate states.

last frame before the next START = 1 will have REQ = 0 and the estimate of cycle length would remain unchanged.

Of the three methods for reducing intercycle gap, the former two are distributed while the latter requires additional intelligence in the end station.

VII. TRAFFIC CONTROL

Different types of traffic have very different transmission requirements. Voice traffic is an interesting example of where an initial restriction of access (blocking) is preferable to losing voice packets. Real-time traffic such as voice and video requires a guaranteed maximum delay if information is not to be lost. In contrast, some types of computer traffic are rather tolerant of delay. It is not our aim to develop here a comprehensive algorithm for handling different traffic types; it would depend closely on the particular environment and the mix of traffic. However, we would like to indicate the mechanisms that Fasnet can support to control integration of traffic.

We can identify four different types of control mechanisms.

(*i*) Selection of traffic—The ability to individually control a class of traffic.

(*ii*) Request for access—The ability to communicate that service is required.

(*iii*) Blocking of traffic—The ability to prevent traffic of a specific class from gaining access to the system.

(*iv*) Continuation of service—The ability to delay traffic for later transmission. We will take these four control mechanisms in turn and consider how they may be implemented in Fasnet.

(*i*) Selection—We can borrow from the strategy used by Frata et al.[15] and Ulug et al.[6] START, instead of being a single bit, can be expanded to a multibit word. A START code can then be allocated to each class of traffic. A station would then be permitted to transmit only if the START corresponded to the class of traffic it is waiting to transmit. A class of traffic could denote a traffic type, as well as a priority. The term subcycle will be used to denote the period from a START of one class to the next occurring START. Cycle will be reserved for the period between two STARTs of the same class.

(*ii*) Request—The amount of request information can vary from knowing exactly which station wants to send what traffic in one extreme to knowing just that a station somewhere wants to send some type of traffic in the other extreme. A compromise would be to provide a request word adjacent to the END field in each packet on the return line. Each bit in the request word would denote a class of traffic. This information would enable the head station to determine that one or more stations required service of a particular type without indicating the extent of the demand. Information about demand for service would most likely be used to adapt the control strategy in the case where there was a change in the balance between traffic types.

(*iii*) Blocking—The channel capacity allocated to a class can be controlled by the time allocated to that class as suggested in Frata et al.[15] By issuing a different START after a given period, further traffic of the original class type wishing to transmit would be denied access. Blocking is typically used to handle overflow of the type of traffic that generates information periodically, such as real-time voice and video, and synchronous data traffic. In this instance, it is appropriate to speak of connections between source and destination which implies guaranteed access once a connection has been allocated.

To discuss the allocation of connections in a blocking traffic class requires that the previous definitions of WAIT state and ACCESS state be generalized:

WAIT—A station is waiting for permission to seek access to the line.

ACCESS—A station has a connection.

Stations that already have access take the first free slot available to them after the appropriate START. Allocation of freed up slots on a reasonably equitable basis would proceed as follows. Stations would be aware of say n slots becoming free from the position of the END bit on the return line. Note, no END is issued if the class is full. Stations in the DEFER state would be permitted to compete for the n empty slots at the end of the subcycle. This will favor stations close to the head end. However, a large degree of fairness is achieved by permitting stations to switch from the WAIT to DEFER state only when two consecutive ENDS are encountered for that subcycle. This will only occur when all traffic currently in DEFER state has been granted access. At this point, traffic in the WAIT state would switch to DEFER and then vie for empty slots as they become available. This strategy is related to the snapshot algorithm.[16]

(*iv*) Continuation—In contrast to blocking, continuation requires that traffic not able to access the link in the previous cycle be served before any new traffic is accommodated. This may be achieved in the following manner. Assume that the class type is non-blocking. If the head station should issue a new START before all traffic of the class has been served, the end station will not detect the end of the cycle and hence will not issue END = 1. The absence of an END = 1 would indicate to the head station that the p registers of the stations in that class should not be reset on the next cycle (i.e., the stations would not switch from WAIT to DEFER).[15] Thus, in the following cycle, remaining traffic would be served. For centralized control, the START for this traffic type could contain an additional bit to indicate whether the previous cycle is being continued for deferring traffic or a fresh cycle is being started for new traffic. For distributed control, each station could keep track of the sequence of STARTs and ENDs.

It is important that the control strategy be adaptive to changing traffic conditions. We expect that the traffic mix will change relatively slowly—over a period of seconds rather than ms. Thus, it would be feasible to have the adaptation achieved by a server process.

The control algorithm could be implemented as completely distributed, completely centralized, or somewhere in between. Economics and reliability will dictate, to a large extent, where the control should be placed. Nevertheless, a hybrid strategy would seem more in the spirit of the current design. For example, selection is probably best achieved by having the head station transmit the appropriate START code (centralized, but perhaps assumable), whereas traffic assignment and continuation is probably best achieved by having each station read and operate on the END field (distributed).

VIII. TOPOLOGY

8.1 Introduction

A population of stations may be connected together by either a single link (Fig. 9a) or by several interconnected links (Fig. 9b). The best topology will depend upon physical distribution, traffic patterns, and the particular performance measures that one seeks to optimize. We will not consider the general problem, but restrict ourselves to the linear interconnection of links forming closed loops. Fasnets may be connected as shown in Fig. 10. Packets in Fasnet 1 destined for Fasnet 2 are addressed to station S_N. Station S_N transmits the packets to station S_1 which puts them on Fasnet 2. Similarly, for packets from Fasnet 2 destined for Fasnet 1. To provide reliability against single-station failures, interconnection stations would be provided in pairs. Thus, a similar connection would be provided between S_{N-1} and S_2. A detailed procedure may be specified whereby control of the interconnection passes from the $S_N - S_1$ connection to the $S_{N-1} - S_2$ connection in case of failure of the former. In principle, the secondary connection monitors both Fasnets and assumes the interconnection function after a suitable time-out period in event of failure. Provision can also be made for the primary connection to periodically check that the secondary connection is operational.

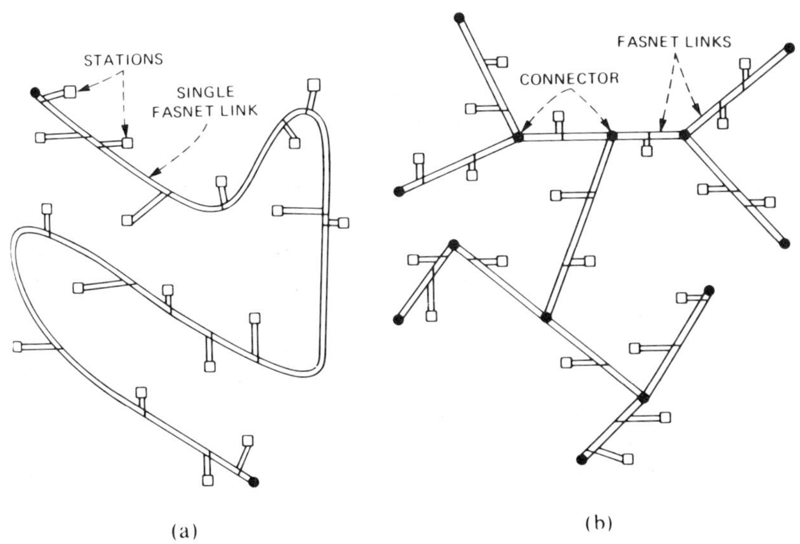

Fig. 9—(a) A cluster of stations being served by a single Fasnet link. (b) The same station population being served by several interconnected links.

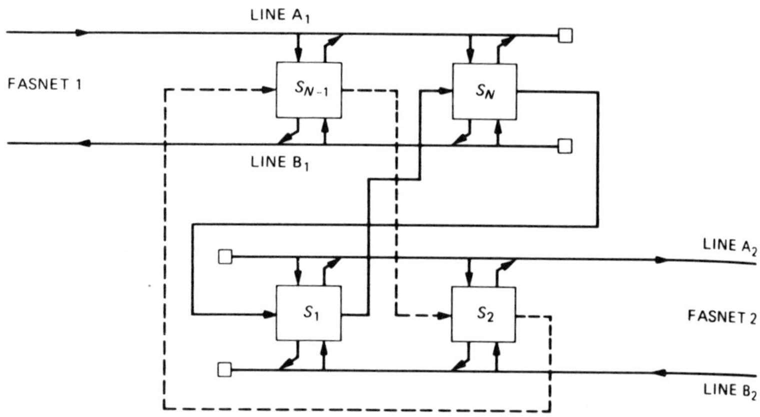

Fig. 10—Structure of a connector used to interconnect two Fasnet links.

Interconnection of Fasnets as shown in Fig. 10 permits traffic to pass from one link to another with a minimum of delay. Since the connection is to the first station on the link, the incoming packet can utilize the next occurring slot. Because of differences in frame timing between two links, it may be necessary to buffer a maximum of one complete packet; this amount of buffering is normally provided in a station interface. In general, interconnection of more than two Fasnets will require larger buffers to be employed to handle the condition where traffic arrives simultaneously for one Fasnet from connecting Fasnets.

8.2 Traffic localization

If stations on a Fasnet have traffic destined only for stations in the immediate vicinity, then total utilization can be significantly improved by dividing the single link into separate links that are connected. Only traffic that has not reached its destination link is allowed to cross the connector.

Consider two Fasnet links connected to form a ring with an inner and outer loop as shown in Fig. 11. The two connectors transfer traffic from one link to the other.

Assume that the ring has a length of unity and that the first link has a length l where $l \leq ½$. For a given packet, let d be the distance of the destination station from the source station. This distance is measured along the ring with the anticlockwise direction as positive, the clockwise direction as negative, and the source station as the origin. Assume that d is a random variable with a uniform distribution over $[-½, ½]$. When $d \epsilon [-½, 0]$, the source station accesses the outer loop; when $d \epsilon [0, ½]$, the source station accesses the inner loop. Thus, the source station selects the shortest distance to the destination along the ring.

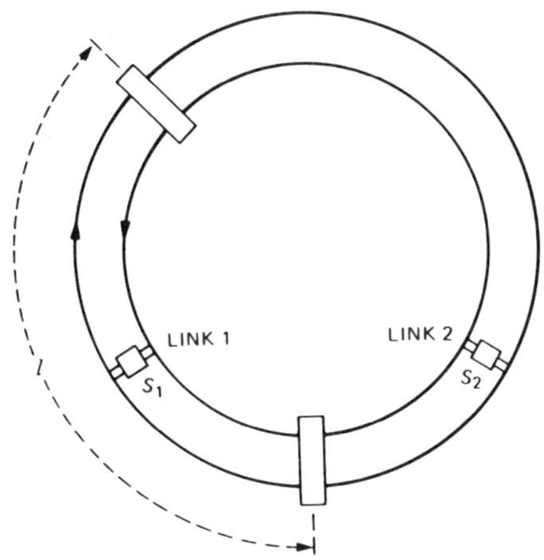

Fig. 11—Two Fasnet links interconnected to form a link using two connectors.

Depending on l and d, the traffic to any destination may have to traverse zero, one, or two connectors before it can be removed.

We first wish to determine the probabilities of the above events when the stations are distributed uniformly on the ring. For station S_j selected at random on link $j\,(j = 1, 2)$, let

p_j = probability that the destination is on the inner loop, and the same link, and the traffic traverses zero connectors,

q_j = probability that the destination is on the inner loop, but the other link, and the traffic traverses one connector,

r_j = probability that the destination is on the inner loop, and the same link, but the traffic traverses two connectors ($r_j \neq 0$ only if the length of the link is $>\frac{1}{2}$ and the shortest path between two stations at the ends of the link is through the other link).

It can easily be shown that

$$\begin{aligned} p_1 &= \frac{l}{2} & p_2 &= \frac{1}{2}\left(\frac{3}{4} - l\right)\frac{1}{1-l} \\ q_1 &= \frac{1-l}{2} & q_2 &= \frac{1}{2}\,l \\ r_1 &= 0 & r_2 &= \frac{1}{2}\,\frac{\left(\frac{1}{2} - l\right)^2}{1-l}. \\ &\left(\text{as } l < \frac{1}{2}\right) \end{aligned} \tag{3}$$

Note that $p_j + q_j + r_j = ½$, $(j = 1, 2)$ as any station accesses the inner loop with probability ½. Now suppose that there are N_1 active stations on link l and N_2 active stations on link 2. If each active station requires unit capacity, then the average traffic $T_i (i = 1, 2)$ in units of capacity on the inner loop in links 1 and 2 is given by

$$\begin{aligned} T_1 &= N_1(p_1 + q_1 + r_1) + N_2(q_2 + r_2) + N_1 r_1 \\ T_2 &= N_2(p_2 + q_2 + r_2) + N_1(q_1 + r_1) + N_2 r_2 \end{aligned} \tag{4}$$

because the traffic on a line in any link is the sum of three components: (*i*) all the traffic generated in that link, (*ii*) that fraction of the traffic generated in the other link that traverses this link, and (*iii*) that fraction of the traffic generated in this link that transverses the other link and then returns to the first link. If N_1 and N_2 are very large, then by the law of large numbers we have

$$\begin{aligned} \text{traffic on link 1} &= T_1 \le C \\ \text{traffic on link 2} &= T_2 \le C, \end{aligned} \tag{5}$$

where C is the line capacity.

It can be shown that $N_1 + N_2$ is maximum when $l = ½$ and

$$N_1 + N_2 = \frac{8}{3} C.$$

If no connectors were used, we have a single Fasnet link (it can no longer be a closed ring) for which

$$N_1 + N_2 = 2C.$$

Hence the gain $G = (8C/3)/2C = 4/3$. Thus, we are able to obtain a 33 percent increase in the effective network capacity even for uniformly distributed traffic by having two diametrically located connectors.*

We now extend the above analysis to the case of K connectors C_1, C_2 $\cdots$ C_K, which are located symmetrically around the ring as shown in Fig. 12.

It can be shown by similar means that

$$G = \frac{4}{1 + \dfrac{4}{K}}. \tag{6}$$

This is plotted in Fig. 13.

We then extend the same analysis to the case with K symmetric connectors C_1, C_2 $\cdots$ C_K, but with an arbitrary traffic distribution symmetric about the source station and extending from $-½$ to $+½$

* While this analysis does not pertain to a specific access protocol, the effective gain can be closely realized by the Fasnet protocol and the connector structure of Fig. 10.

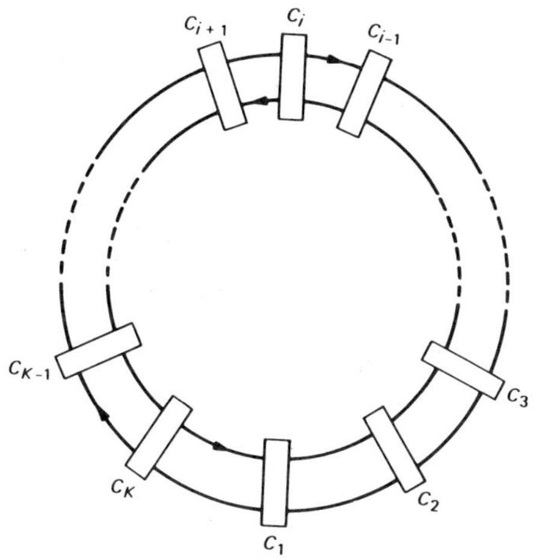

Fig. 12—A Fasnet link with multiple connectors.

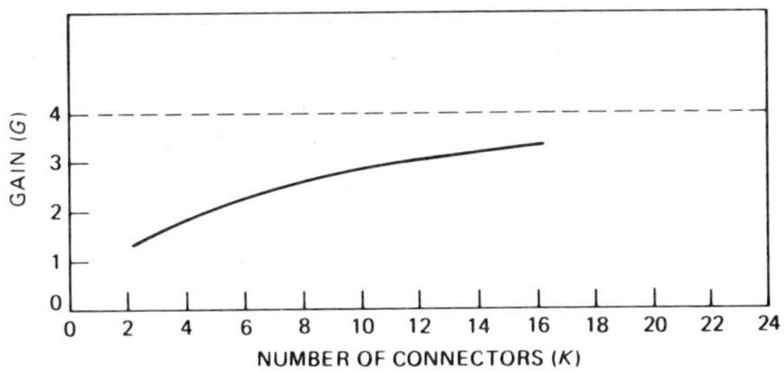

Fig. 13—Plot of the gain in traffic handling capacity of interconnected links relative to a single link as a function of K, the number of connectors.

along the ring. This preserves the shortest path routing, as well as sharing the load equally between the two loops.

It can be shown by similar means that for large K

$$G \approx \frac{K}{1 + KD} = \frac{1}{\frac{1}{K} + D}, \tag{7}$$

where D is the expected value of $|d|$, the absolute value of the source-destination distance. The approximation becomes exact if the distribution is uniform or if $K \to \infty$. Note that for the uniform distribution, $D = 1/4$ and

$$G = \frac{1}{\frac{1}{K} + \frac{1}{4}} = \frac{4}{1 + 4/K}$$

as before. For a given traffic distribution, (7) is a good design formula for how capacity costs can be reduced at the cost of extra connectors. However, this trade-off is useful only if accurate estimates of capacity and connector costs are available.

The above analysis highlights certain interesting features. For the uniform traffic distribution, it is seen that the gain, G, does not increase uniformly with K. This is fairly intuitive. Since the traffic is uniform, an extra connector, when K is large, removes the traffic over only a short link and results in only a marginal increase in the gain. On the other hand, as the traffic distribution becomes more localized (i.e., $D \rightarrow 0$)G increases uniformly with K. This is again fairly obvious, as with a high degree of localization the traffic on each link is almost independent of the traffic on the other links.

We have considered here some configurations of interconnected links. There are interesting graph theoretic questions relating to reliability. For example, given a graph like Fig. 9b, what is the minimum number of additional links and their position so that full connectivity is still maintained if any link is cut at a single point? Development of realistic models of the physical traffic and cost structures of local environments still remains. There is a paucity of statistics (except for Ref. 2) on the local network parameters. Future operational local networks will hopefully furnish statistics on which to build more accurate models.

IX. CONCLUSION

The physical configuration of Fasnet consists of two communication lines passing each station. One line carries traffic in one direction, while the other line carries traffic in the opposite direction. Thus, this configuration carries twice the traffic of a previous system in which the two lines were connected at one end so that traffic was written on the outbound line and read from the inbound line. Each station makes two connections to each line, a nondirectional read tap and a directional write tap. Reliability of the physical medium is high because it contains no active electronics. The access protocol is partly centralized in that bit synchronization, framing, and start-of-cycle are provided by the end stations; however, these functions would be assumable by any station upon failure of an end station.

The access protocol is as follows: Upon reading a start-of-cycle, a station may transmit a prespecified number of packets in the first available empty slots. When all stations have transmitted their packets, a signal is sent on the return line to inform the head station to start a new cycle. The efficiency of Fasnet increases as the length of a cycle increases; cycle length depends upon the length of a packet, the number of active stations, and the number of packets, p_{max}, that each station is permitted to send in a cycle. By adaptively changing p_{max}, efficiency can be maintained at a high level even for a small number of active stations. A number of techniques for further improving efficiency are suggested. A trade-off is necessary between increasing the complexity of the protocol, on one hand, and the resulting small improvements in efficiency on the other.

Bit synchronization of the stations is achieved through adding an out-of-band pilot tone, while framing is achieved through a periodically inserted code word. A three-level bipolar line code is preferred.

The potential of Fasnet for operation at high transmission rates makes it attractive as a conduit for the various types of traffic that may flow in a business environment. Mechanisms have been proposed to implement blocking, delaying, and request-for-service operations that are needed if mixed traffic is to be handled efficiently within a single medium. These operations can be implemented centrally or they can be distributed. The low-level access operations are best distributed while the more complex operations are best centralized.

Fasnets may be interconnected to increase the load that may be carried or to improve reliability. Investigation has been restricted to the connection of Fasnets to form a ring. As the number of segments in the ring increases, the throughput first increases rapidly (assuming uniformly distributed traffic). After about five segments, the increase is very small. Exploration of other topologies presents a challenge.

X. ACKNOWLEDGMENT

We thank B. Gopinath and N. F. Maxemchuk of Bell Laboratories and Z. L. Budrikis of the University of Western Australia for helpful discussions during various stages of this work.

REFERENCES

1. R. M. Metcalfe and D. R. Boggs, "Ethernet: Distributed Packet Switching for Local Computer Networks," Commun. of the ACM *19*, 7 (July 1976), pp. 395–404.
2. J. F. Shoch and J. A. Hupp, "Measured Performance of an Ethernet Local Network," Presented at Local Area Communications Network Symposium, Boston (May 1979), pp. 113–34.
3. The Ethernet, A Local Area Network, Data Link Layer and Physical Layer Specifications, DEC, INTEL, XEROX, Version 1.0 (September 30, 1980).
4. F. A. Tobagi, "Multiaccess Protocols in Packet Communication Systems," IEEE Trans. Commun. *COM-28*, 4 (April 1980), pp. 468–88.
5. J. O. Limb, "High Speed Operation of Broadcast Local Networks," Proc. ICC, June 1982.
6. M. E. Ulug, G. M. White, and W. L. Adams, "Bidirectional Token Flow System, Proc. Seventh Data Commun. Symp. (October 1981), pp. 149–55.
7. M. V. Wilkes and D. J. Wheeler, "The Cambridge Digital Communication Ring," Proc. Local Area Commun. Network Symp. (May 1979), pp. 47–61.
8. C. S. Skrzypczak and W. E. Falconer, "Bell System Planning of ISDN," Proc. ICC, June 1981, pp. 19.6.1–6.
9. N. F. Maxemchuk and H. A. Wilder, "Virtual Editing: I. The Concept," Office Information Systems Workshop, St. Maximim, France, October 1981.

10. W. Pferd, L. A. Peralta, and F. X. Prendergast, "Inactive Graphics Teleconferencing," Computer *12* (November 1979), pp. 62–71.
11. W. A. Rheinfelder, *CATV System Engineering*, Third Ed., Blue Ridge Summit, Pa.: TAB Books, 1972.
12. J. O. Limb, "Fasnet: Proposal for a High Speed Local Network," Office Information Systems Workshop, St. Maximim, France, October 1981.
13. IEEE 802 Standards Committee, Interim Report.
14. F. D. Waldhauer, "A 2-Level, 274 Mb/s Regenerative Repeater for T4M," Proc. of the ICC 1975, *3*, pp. 48-13–17.
15. L. Frata, F. Borgonovo, and F. A. Tobagi, "The Express-Net: A Local Area Communication Network Integrating Voice and Data," in *Performance of Data Communication Systems*, G. Pujolle, Ed., Amsterdam: North Holland, 1981, pp. 77–88.
16. R. C. Chen, "Bus Communication Systems," Report No. PB 235-897, Carnegie-Mellon University, January 1974.

Integrated Voice/Data Services on Fasnet

By J. W. MARK* and J. O. LIMB†

(Manuscript received July 8, 1983)

Fasnet is a high-speed local area network with a pair of unidirectional communication channels. While individual stations control their own access to the medium, global scheduling of the medium for various types of traffic is controlled centrally. Centralized scheduling permits a slot to be endowed with a type, so it is feasible to integrate synchronous and asynchronous traffic services on Fasnet. In this paper we describe a policy for admitting new voice calls into the system and an approximate analysis of the queueing behaviour of data packets in an integrated voice and data services environment. Queueing behaviour can also be studied by means of simulation, which is, however, an expensive undertaking, particularly at high transmission rates, for example, at 100 Mb/s. The approximate method described in this paper allows the performance evaluation for high-speed local area networks.

I. INTRODUCTION

Packet-switching technology has enhanced the potential for complete integration of different types of traffic on a common transmission facility, more particularly within a local environment where signal propagation delays are relatively small. Over the past decade a number of papers have been published on the integration of voice and data.[1-8]

We will address the problem of integrated voice and data services using a specific local area network (LAN), Fasnet,[9] as the backbone support. The techniques explored and evaluated here would carry over

* University of Waterloo, Waterloo, Ontario, Canada. † AT&T Bell Laboratories; present affiliation Bell Communications Research, Inc.

to other related systems (e.g., certain implicit token-passing systems and some ring networks) with small modifications.

Fasnet is a dual-channel, packet-switched LAN. Channel time is partitioned into contiguous slots, each of a duration long enough to transmit one packet of data plus the necessary overhead to allow for access control, addressing, error checking, etc. A voice source, engaged in a conversation, may be silent (either listening to the other party in the conversation or in a short pause between speech spurts) or active (producing a speech spurt). An active voice source is partitioned into segments, each of length T seconds. Each segment of voice is mapped into one information packet. The length of an information packet is thus determined by the code rate of the incoming voice source and the transmission rate of Fasnet. To maintain voice continuity, it is necessary to transmit one voice packet of the same source every T seconds. From the LAN's point of view, to maintain an approximate continuous service, it must offer service to voice once every T seconds. The initial portion of the T-second interval will be devoted to serving voice calls. The interval T between consecutive starts of voice service will be referred to as a service cycle. The service cycle period can be expressed in terms of the number of slots, and it is conceivable that the number of voice sources connected to Fasnet is larger than the number of slots in one service cycle. Obviously, the system cannot accommodate more simultaneously "active" voice source than the number of slots available in one service cycle without degrading the signal in some way; some kind of admission control needs to be exercised.

In this paper we assume that a station may be handling voice, data, or both. If we consider voice only, a station that is not in a voice session will be referred to as *idle*. A *nonidle* station producing a speech spurt (talk spurt) will be referred to as *active*; it will be referred to as *silent* if its voice session is in a silent condition.

From an operational point of view, calls already in progress, whether they be in a talk spurt or in a silent interval, must be guaranteed channel time when needed. Since data can be buffered and voice needs continuity, voice will be given a higher priority than data. New calls that are not admitted will be blocked. In the event that a subscriber is blocked, an appropriate tone (short busy) is returned to the subscriber and he/she would try again at a later time. (Of course a satisfactory system design would provide sufficient channel capacity that such an event would occur with no more than a prespecified probability.) On the other hand, unused voice capacity is available to data traffic until there is an increase in voice traffic. In this paper we propose an admission policy for new calls under the constraint that ongoing sessions are guaranteed channel access and that the average number of active voice calls is controlled to within a small neighbourhood of the target operating value, N_t. (Due to statistical variations in the number of active voice calls and the fact that ongoing voice sessions cannot be dropped, the mean number of active voice calls can only be controlled to within a neighbourhood of N_t.) We evaluate the performance of data under this admission policy. An alternative strategy has been explored in Ref. 10, in which data traffic is guaranteed a minimum capacity. Voice stations returning from the silent state to the active state are not guaranteed a voice slot, and clipping of a speech spurt can occur as experienced in an overloaded Time Assignment Speech Interpolation (TASI) system.[11] The maximum voice load is thus determined by the amount of clipping that may be tolerated. The study showed that, in the distributed environment of the LAN, performance approached very closely that of an ideal TASI system.

The priority admission mechanism and the data queueing behaviour in Fasnet are similar to those in Welnet.[8] The difference lies in the media access control, the network architecture, and the method of switching from one traffic type to another. In Fasnet, a "dead time", referred to as a sojourn time, usually accompanies switching from one type of service to another. In Welnet, provided the packet transmission time is greater than the propagation delay, there is no dead time due to switching of traffic type. Otherwise, the service disciplines of Fasnet and Welnet are similar.

Allowing voice sessions to capture the channel when needed will cause the service rate for data to fall below the data arrival rate during some epochs when the system is heavily loaded with voice and data. When such an event takes place, the traffic intensity, defined as the ratio of data arrival rate to service rate, will temporarily exceed unity. In this mode of operation, the aggregate data queue will experience a sustained growth with a nonstationary distribution, and the queueing process will be temporarily unstable. An approximate analysis method, based on a periodic switching single-server model and the fluid approximation method,[12] was developed in Ref. 8 to analyze the data queueing behaviour for integrated voice and data services on Welnet. The approximate analysis method involves a decomposition of the queueing process into stationary and nonstationary queues and a determination of the mean queue contribution from each of the operational modes. With a suitable modification to account for the sojourn time, the approximate analysis method developed in Ref. 8 is used to analyze the data queueing behaviour for integrated voice and data services on Fasnet.

This paper is organized as follows: Section II briefly describes the Fasnet architecture, which includes a description of the frame format of a Fasnet slot and the concept of a voice cycle. Section III discusses the evolution of the state equation for voice, which leads to the proposal of an ad hoc admission policy for new calls in Section IV. Fasnet provides centralized control of classes of traffic, where a class may be a traffic type, a priority, or some combination. However, access by individual stations is a distributed function, being determined by the station itself. It is important to assure that the centralized control is implemented in such a way that service to each traffic type remains fair. (We define a fair system as one in which all active stations have an equal chance to access the common channel.) In Section V we evaluate the performance of data for a "fair" Fasnet, under the new voice admission policy introduced in Section IV and the constraint that ongoing voice calls are guaranteed service.

II. FASNET ARCHITECTURE

Fasnet uses two channels with unidirectional signal propagation in each.[9] The first station (the head station in Fig. 1a) in each channel exercises centralized access control. In the event the first station malfunctions, the second station senses the lack of timing information and assumes the role of the first station. Since there is complete symmetry in the two channels, we will consider the information flow in one of them only.

Each channel has M taps, each of which supports a signal processor that handles both transmit and receive functions. The processor will be referred to as a station. Each station will support a number of user machines or processes. The first station permits access to each class by issuing a slot of a given type. A new class of traffic may be admitted

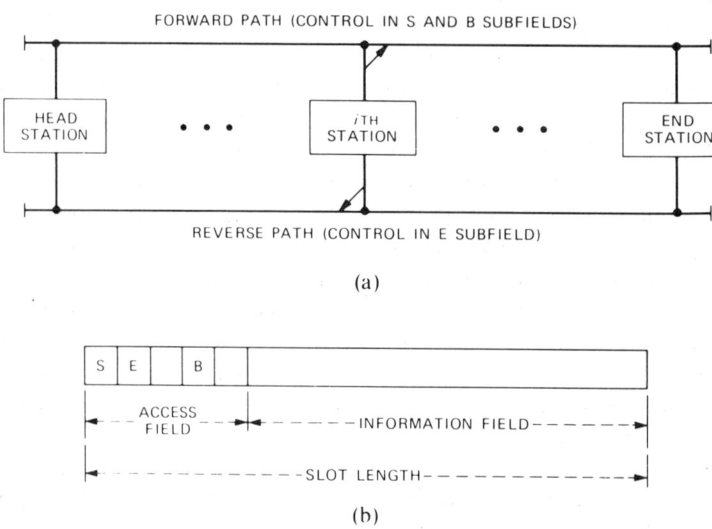

Fig. 1—Fasnet: (a) topology; (b) frame format.

by sending a request to the first station, which, in turn, upgrades its traffic classification to accommodate the new class.

2.1 Fasnet frame format

Channel time is divided into slots, each of a duration sufficiently long to transmit an information packet plus control, addressing, and cyclic redundancy checking information. A Fasnet slot comprises an access field and an information field in a manner depicted in Fig. 1. The access field is endowed with an S (Start), an E (End), and a B (Busy) subfield. The S subfield contains a code that designates the start of a sequence of slots of a particular traffic type. The B subfield is a single-bit subfield that indicates whether the information field is full or empty. Individual stations have the capability to read and write into the B bit. Guard times on either side of the B subfield allow for the read and write deliberations. As shown in Fig. 1, the contents of the information field are under the control of the individual stations. A slot is initiated with an empty information field and the content of the B subfield is a "0". If B = "1", the slot is busy; if B = "0", the slot is free. Stations can only write into free slots of the correct type.

2.2 Channel occupancy

Let r in b/s be the code rate of a voice source, R in b/s be the network transmission rate, and H be the number of overhead bits. H includes the access field, the source/destination addresses, control, and Cyclic Redundancy Check (CRC). Then, the slot length in bits is given by

$$F = H + rT \text{ bits/slot.} \tag{1}$$

The number of bits transmitted by the system in T seconds is RT, and the number of bits in an information packet is rT. The voice cycle length, L, in slots, is then

$$L = \frac{RT}{H + rT} \text{ slots.} \tag{2}$$

Let $C = 1$ be the normalized channel capacity, i.e., the channel can be utilized to transmit at most one packet per slot. Since a slot is longer than the time necessary to transmit one information packet, the maximum channel occupany, $U_{\max}$, is

$$U_{\max} \equiv \frac{rT}{H + rT} \text{ slots.} \tag{3}$$

The capacity required to coordinate packet transmission, C_s, is

$$C_s \equiv C - U_{\max}. \tag{4}$$

2.3 *Code designations of priority scheduling*

A voice cycle repeats every L slots. All admitted voice sessions are guaranteed one slot per voice cycle. It is expected that the number of slots taken by voice calls during any voice cycle will be less than L slots so that there will be slots available for data transmissions on a per-cycle basis. The concept of a voice cycle is depicted in Fig. 2, with a single slot expanded to emphasize the characteristics of the access field. The access field of each slot has three subfields, namely, S, E, and B. As indicated in Fig. 1a, the first (or head) station issues slots in the forward path, and the last (or end) station echoes acknowledgment in the reverse path. The S subfield contains a code word that designates the traffic type for that slot. The B subfield contains a token bit to indicate whether the slot is full or empty. The code word in the S subfield is set by the head station when the slot is issued. The B bit is to be set by the transmitting station. The end station observes whether a slot, after traversing the entire channel, is full or empty. If the slot is empty, then every station with the designated type of traffic has transmitted the permitted number of packets. The status of an empty slot is indicated (by the end station) in the E subfield. The occurrence of an empty slot signals the end of one type of service and the start of a new service type. The dead time associated with the circulation of an empty slot, as indicated to the head station by the marked E subfield, has been referred to earlier as the sojourn time.

The S subfield, which takes on one of the code words H, G, V, D, and C, determines the mode of the slot. To ensure "fairness", admission

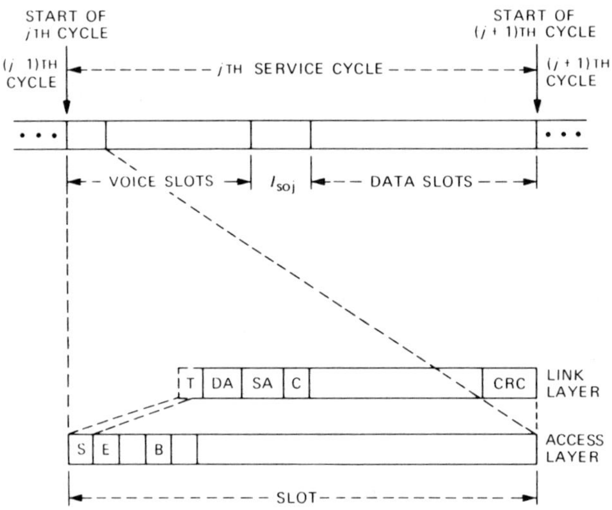

Fig. 2—Service cycle.

of new voice is carried out in two stages. Upon arrival, a new call enters the *wait* state. When the new voice in the wait state sees an H in the S subfield, it transits to the *hold* state and ultimately transmits when a visiting slot is free. The station model for new voice admission is depicted in Fig. 3. The code words H, G, V, D, and C have the following connotation:

H is a type designation that allows new voice in the wait state to transit to the hold state and to start competing for a slot

G allows new voice in the hold state to start competing for a slot

V allows old "active" voice to start competing for a slot

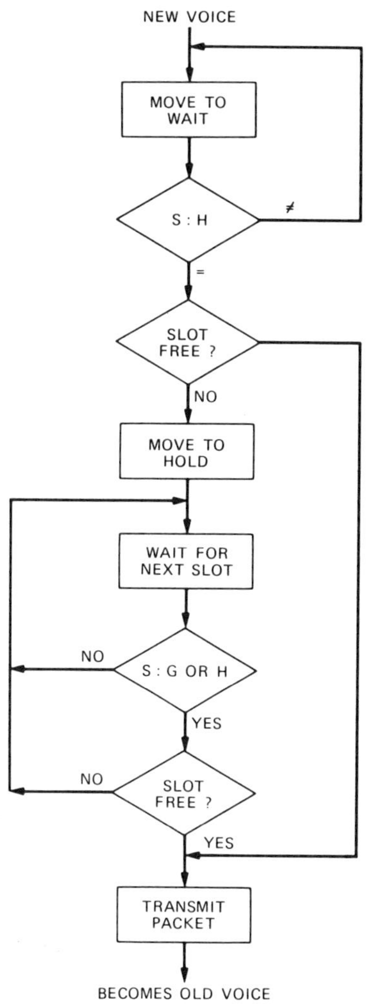

Fig. 3—New voice admission algorithm.

D allows "active" data stations to transit from the wait state to the hold state and to start competing for a slot

C allows active data stations in the hold state to start competing for a slot.

A data subcycle differs from a voice subcycle. Since a data subcycle may be interrupted by a voice subcycle, a data subcycle may span more than one service cycle. The access control code, D, starts a new data subcycle after all data stations in the wait state have been served so as to maintain fairness amongst the data stations. There are basically three types of services, namely, new voice calls, old voice calls, and data. Both H and G admit new voice calls, and both D and C allow data stations to transmit. However, each H (D) slot is issued at most once per voice (data) subcycle. That is, an H slot is followed by a string of G slots and a D slot is followed by a string of C slots. Fig. 4 shows the flowchart for the issuance of slots by the head station using feedback information, denoted by an asterisk, that arrives from the end station, after one channel propagation delay. The quantity Ω_j, determined in Section IV, is the number of new voice calls to be admitted in the jth admission cycle.

III. EVOLUTION OF STATE EQUATION

Consider just the voice subcycle. A station can be either in a voice session or idle. A voice session can be in a talk spurt or silent. Figure 5 shows the state diagram of a voice source.

Figure 5 depicts the state transitions at the start of the jth service cycle [or at the completion of the $(j-1)$th cycle]. The symbols used have the following notations:

n_{j-1} denotes the number of *active* voice stations in the $(j-1)$th cycle

d_{j-1} denotes the number of stations moving to the *silent* state after the completion of the $(j-1)$th cycle

D_{j-1} denotes the number of stations terminating their voice sessions at the completion of the $(j-1)$th cycle

b_j denotes the number of stations moving from the silent to the active state during the $(j-1)$th cycle

B_j denotes the number of stations with new calls that are admitted in the jth voice cycle

G_j denotes the number of new calls arriving in the jth voice cycle.

Then, the number of calls not accepted in the jth voice cycle is $G - B_j$. The state equation at the beginning of the jth voice cycle is

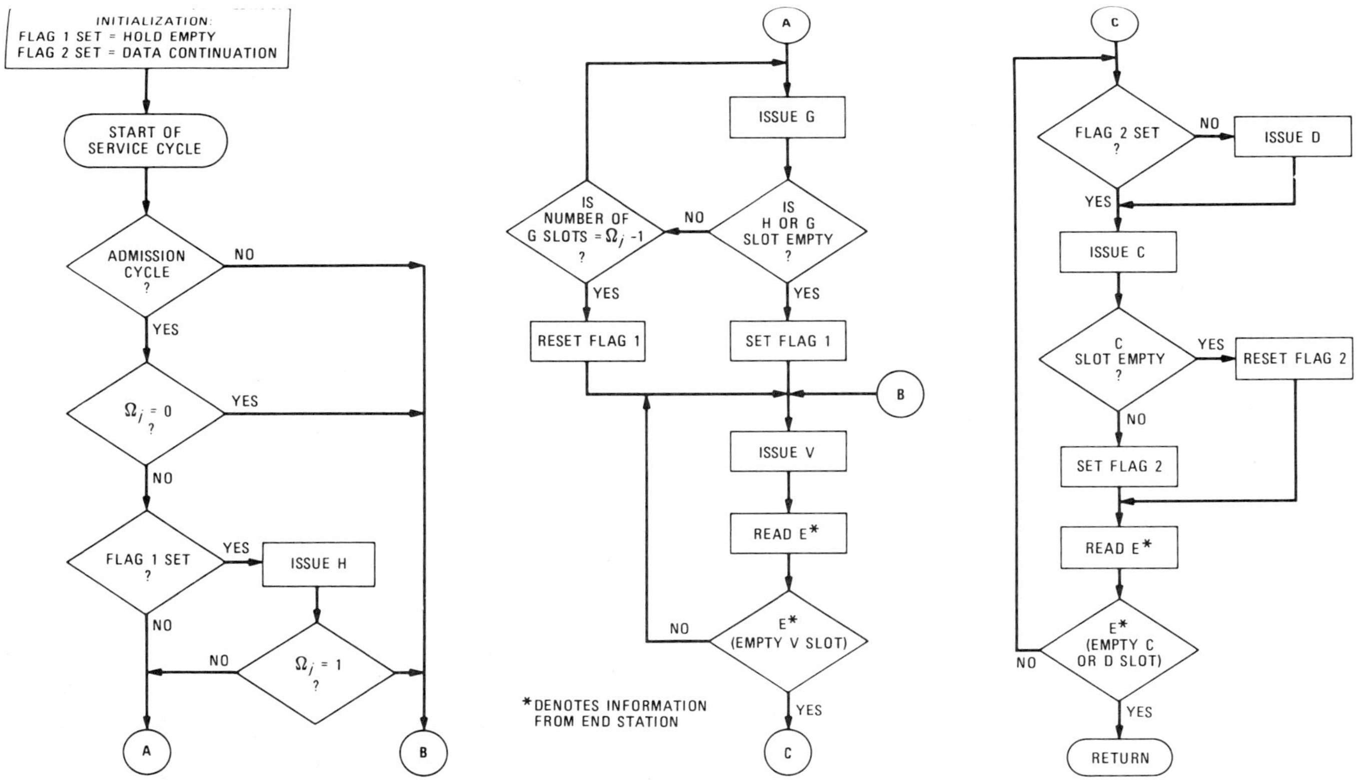

Fig. 4—Flow chart for slot issuance by head station with feedback from end station.

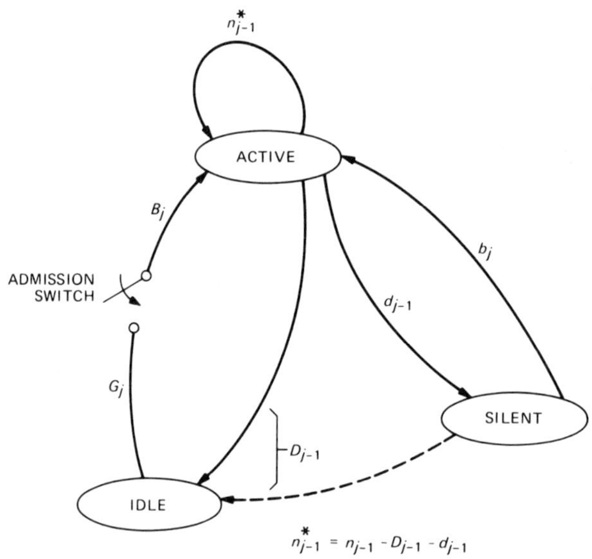

Fig. 5—State transition diagram for voice at start of jth service cycle, assuming deaths occur while in active state.

$$n_j = n_{j-1} - d_{j-1} - D_{j-1} + b_j + B_j. \qquad (5)$$

Under the access control mechanism described in Section 2.1, n_{j-1} is the only known quantity in the above equation. (It is possible to obtain information about D_{j-1} and d_{j-1} by the incorporation of a more complex access control mechanism.) We will describe, in the next section, an admission policy for new calls under the Fasnet structure described in Section II and the state development in eq. (5).

IV. NEW VOICE ADMISSION CONTROL

It is desirable to maintain an equilibrium operating state such that the mean number of active voice sessions (in talk spurt) will be $\bar{n} \approx$ N_t, where N_t is the target number of active voice sessions. Let N be the total number of voice sessions to be supported and p be the fraction of voice sessions that are active. Then, $\bar{n} = pN$. In this paper, we let $p = 0.4$, i.e., 40 percent of the time a (nonidle) voice source is in talk spurt.

Suppose we call each run of identically designated slots, i.e., H, G, V, D, and C, a sequence denoted by S_x, $x \in \{H, G, V, D, C\}$. The termination of a sequence is detected by the last station whenever it sees the first empty slot. Before a new sequence of slots can be initiated, a sojourn time, I_{soj}, elapses. I_{soj} equals one slot time plus one round trip propagation delay. (The round trip propagation delay equals

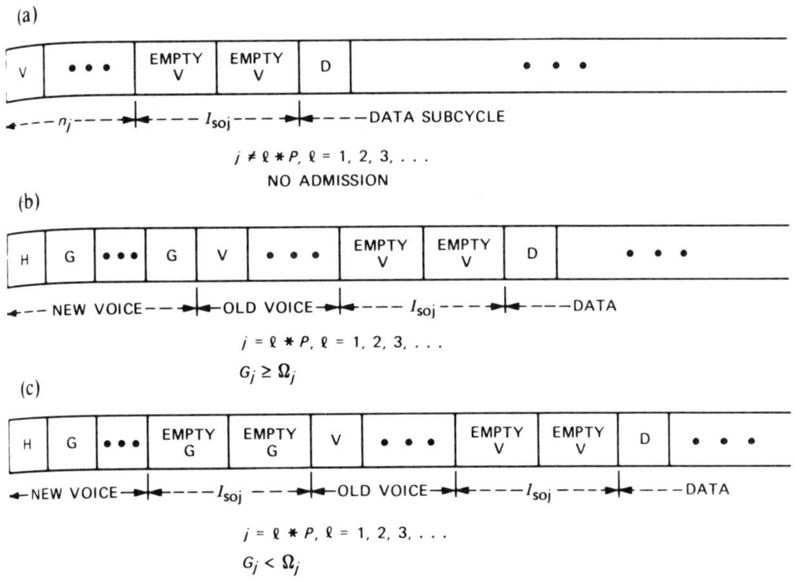

Fig. 6—Concept of sojourn time in jth cycle.

$2 \cdot \lceil \delta \rceil$ where δ, measured in slots, is the propagation delay in one direction, if the starts occur at the same instant in each channel, and the symbol $\lceil \cdot \rceil$ denotes the integer greater than or equal to its argument.)

At the start of a new service cycle, the first station issues a sequence, S_G, of Ω_j H- or G-designated slots (see Fig. 4). The number actually admitted is

$$B_j = \begin{cases} G_j & \text{if} \quad G_j < \Omega_j \\ \Omega_j & \text{if} \quad G_j \geq \Omega_j. \end{cases} \tag{6}$$

The format of a voice cycle is shown in Fig. 6. A sequence S_V will follow S_G. If $G_j \geq \Omega_j$, there will be no sojourn time separating S_G and S_V. Otherwise, there will be a sojourn time $\leq I_{soj}$ slots separating S_G and S_V. Also note from Fig. 6 that there is at least one sojourn time between an S_V and an S_C sequence within a service cycle, where S_C denotes a sequence of D or C starts.

Admission of new voice calls is controlled by the design parameter N_t. Since a cycle time is quite short, it is only necessary to admit new voice calls once every P cycles. Let Δ_j be the accumulated number of calls that left the system since the last admission cycle. Then Δ_j may be estimated by

$$\hat{\Delta}_j = \sum_{i=j-P+1}^{j} \frac{n_i}{p} \frac{T}{T_h}, \tag{7}$$

where T is the cycle length in seconds, T_h is the holding time of a voice session, and p is the probability that a voice session is in talk spurt. The value n_j/p is an estimate of the total number of voice sessions in the system in the jth cycle. On the assumption that, at steady state, the net movement from silent to talk spurt and vice versa is zero, the maximum number of active voice in the jth admission cycle can be approximated as

$$n_j = n_{j-P} + B_j - \hat{\Delta}_j. \tag{8}$$

The objective is to drive the system to operate at a point such that the average number of active voice approximately equals the design parameter N_t. We propose to compute a running average $\bar{n}_j$ by the following recursive formula:

$$\bar{n}_j = \gamma \bar{n}_{j-1} + (1 - \gamma) n_j, \tag{9}$$

where γ, in the range $0 < \gamma < 1$, is a constant factor that determines the relative weighting of the last running average and the current number of active voice. The constant γ also fades the memory of the recursive process and hence affects the size of the moving average window. To have a relatively large window, γ should only be slightly less than unity, e.g., $\gamma = 0.98$. Thus, during start-up, $\bar{n}_j$ will be small. By the same token, the number admitted is expected to be considerably larger than the number that leaves the system. Then, during start-up, the number of new voice already admitted should play an important role in controlling the admission process. When the system attains an equilibrium condition, the running average will approximately equal N_t and should assume primary control of the admission process. That is, at equilibrium, the number admitted should balance the number that left the system.

Let c_j be the parameter that accounts for the difference between the number admitted and the number that left. The number admitted in the past admission cycles should only be remembered until the system attains steady state, i.e., until $\bar{n}_j$ attains a value close to N_t. It is proposed to compute the control variable c_j by a recursive equation of the form

$$c_j = \eta c_{j-1} + B_j - \hat{\Delta}_j, \tag{10}$$

where $0 < \eta < 1$ is chosen to fade the memory at an appropriate rate. η should be related to the time constant with which the system attains steady state from start-up. In fact, the combination of (9) and (10) forms a robust iterative algorithm so that an optimum choice of η is not critically important. The strategy is to admit new voice calls whenever $\bar{n}_j + c_j < N_t$. The number to be admitted in the jth cycle is governed by the equation

$$\Omega_j = \max(0, N_t - \bar{n}_j - c_j). \tag{11}$$

We have run numerous tests of the admission control algorithm described in this section. In all cases, the system quickly locks onto a small neighbourhood of the target operating point, N_t. It is noted that admission control takes effect only when the total number of active voice exceeds N_t. If the total number of voice calls is fewer than N, so that the total number of active voice in the system is fewer than N_t, no admission control would be needed, and all calls would be admitted into the system.

V. QUEUEING ANALYSIS OF DATA PACKETS

Although the topology and the media access protocol of Fasnet[9] differ from those of Welnet,[8] the operational features of the two systems are similar. That is, each of these two systems has a deterministic media access protocol, which allows for priority scheduling and a bounded packet delivery time. With appropriate modifications, the periodic switching model and the mean value approximation method used in Ref. 8 to analyze the data buffer queueing behaviour are applicable here. The queueing analysis described below assumes that the procedure for admitting new voice calls is able to maintain

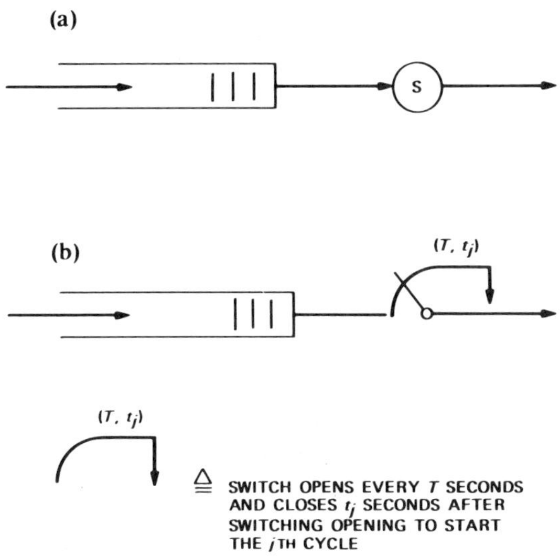

Fig. 7—(a) Single-server model. (b) Periodic switching model.

the voice service operating at the target value N_t, i.e., the mean number of active voice in the system approximately equals N_t.

The periodic switching model shown in Fig. 7 opens every T seconds and closes t_j seconds after the opening of the switch at the start of the jth service cycle. Here, t_j is the time needed to serve n_j active voice plus the total (i.e., including also those sojourns occurring in the data subcycles) sojourn time incurred in the jth service cycle. We have hitherto used n_j to represent the number of active voice calls. Since each active voice occupies one slot per cycle, in what follows, n_j will also denote the number of slots occupied by voice in the jth cycle.

Let m_j denote the number of sojourns in the jth cycle. We then have

$$t_j = \frac{(n_j + m_j I_{\text{soj}})T}{L}, \text{ in seconds,} \tag{12}$$

where T, L, and I_{soj} are as previously defined. The time available for servicing data packets in the jth cycle is then $T - t_j$ seconds or $(L - n_j - m_j I_{\text{soj}})$ slots. Since the transmission of each packet requires one slot time $(L - n_j - m_j I_{\text{soj}})$, data packets can be transmitted in the jth cycle. We define

$$\mu_k = \frac{L - (n_j = k) - m_j\, I_{\text{soj}}}{T} \text{ packets/s} \tag{13}$$

to be the service rate available to data in the jth cycle. The mean service rate is then

$$\mu_d = \frac{L - \bar{n} - \bar{m} I_{\text{soj}}}{T} \text{ packets/s.} \tag{14}$$

Let λ_d be the aggregate data arrival rate in packet/s. The data utilization, ρ_d, is defined by

$$\rho_d = \frac{\lambda_d}{\mu_d} = \frac{\lambda_d T}{L - \bar{n} - \bar{m} I_{\text{soj}}}. \tag{15}$$

For stable operation, it is required that $\rho_d < 1$. For a given total number of voice calls in the system, N, there is a small probability (caused by the statistical fluctuation in the number of active voice calls in the jth cycle) that, in a certain jth cycle, $\lambda_j > [(L - (n_j = k) - m_j I_{\text{soj}})]/T$ and the system is said to experience temporary overload, where λ_j is the number of arrivals in the jth cycle. Under the fluid approximation model to be discussed below, we approximate λ_j by λ_d, the mean arrival rate, and define the data utilization in the jth cycle by

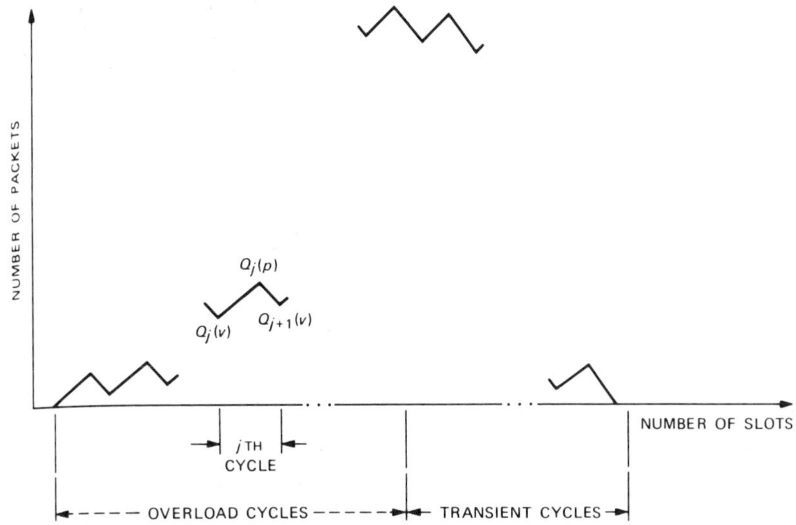

Fig. 8—Fluid approximation model for the jth cycle.

$$\rho_k = \frac{\lambda_d T}{L - (n_j = k) - m_j I_{soj}}. \tag{16}$$

Depending on the value of k, ρ_k can exceed unity.

The mean value, or fluid, approximation method[12] is depicted in Fig. 8, where $N_c(v)$ is the critical number of active voice at which the service available to data in the fluid approximation model equals the number of data packets that have arrived in the same cycle. The jth cycle is on the verge of overload when the combination of voice, sojourn time, and data is such that the queue at the end is the same as at the beginning of the jth cycle, i.e., $Q_{j+1} = Q_j$. At this operating point, the number of active voice is $N_c(v)$. If $n_j > N_c(v)$, there is a net growth in the queue and $Q_{j+1} > Q_j$. The queueing process temporarily becomes unstable and the system experiences overload. Since the number of voice in talk spurt obeys a binomial distribution and since the mean time between transitions from silent to talk spurt, and vice versa, is long compared to a cycle time (e.g., 1.50 seconds compared to 10 ms), once the system enters an overload state, there is every likelihood that it will stay in the overload state for many consecutive service cycles.

To analyze the queueing behaviour, we separate the service cycles into two sets, a stable set in which the queueing process is stationary and an overload set in which the queueing process is nonstationary. In addition, during the interval when the system is servicing voice only, new data arrivals will have to be scheduled for transmission. Let

Ω_s and Ω_o be the sets of stable and overload states, respectively. The average queue in the system is given by the sum of the contributions from the stable and overload states and the temporary queueing due to scheduling:

$$\bar{Q} = \bar{Q}(s) + \bar{Q}(o) + \bar{Q}_{sch}, \tag{17a}$$

where

$$\bar{Q}(s) = E[Q \mid k \in \Omega_s] \cdot \Pr(k \in \Omega_s) \tag{17b}$$

is the contribution from the stable states,

$$\bar{Q}(o) = E[Q \mid k \in \Omega_o] \cdot \Pr(k \in \Omega_o) \tag{17c}$$

is the contribution from the overload state, and $\bar{Q}_{sch}$ is mean queue due to scheduling of the data packets during a service cycle.

An approximate queueing analysis method, based on a decomposition of stable and overload operational modes, was developed in Ref. 8 for the queueing analysis of integrated voice/data services on Welnet. Functionally, the media access protocols of Fasnet and Welnet differ only in that Fasnet may incur a sojourn time whenever a change in the type of traffic to be served takes place. With the incorporation of the sojourn time into the analysis model, the method developed in Ref. 8 can be used to analyze the data queueing behaviour on Fasnet.

When the number of active data stations is small (at light loads), there will be many data subcycles within one service cycle and consecutive data subcycles will be separated by a sojourn time. While the existence of sojourn time represents a loss of channel capacity, not every occurrence of a sojourn affects the data queueing process. The ultimate limit occurs when every data station has packets to be transmitted (as is likely to happen in an overload state). We define the mean number of sojourn times per service cycle, which represents a loss in channel capacity, by

$$\bar{m} = \frac{L - \bar{n}(o) - \bar{m} \cdot I_{soj}}{M \cdot P_{max}} + 1, \tag{18a}$$

where the 1 in the right-hand side of (18a) represents the sojourn between the end of the voice subcycle and the beginning of the data subcycle, $\bar{n}(o)$ is the mean number of active voice in the overload state, M is the number of data stations, and $P_{max} \geq 1$ is the maximum number of packets that an active data station is permitted to send per data subcycle. A value of $P_{max} > 1$ is beneficial for small M, so that the mean number of sojourn times in the overload state is a minimum. For $M > L - N_c(v)$, a value of $P_{max} = 1$ suffices. The expression for $\bar{m}$ can be rearranged to yield

$$\bar{m} = \frac{L - \bar{n}(o) + M \cdot P_{\max}}{M \cdot P_{\max} + I_{\text{soj}}}. \tag{18b}$$

It can be shown that the mean number of voice while in the overload state, $\bar{n}(o)$, is given by Ref. 8:

$$\bar{n}(o) = \frac{\sum_{k=N_c(v)+1}^{N} (n_j = k) \cdot \Pr(n_j = k)}{\sum_{k=N_c(v)+1}^{N} \Pr(n_j = k)}. \tag{19}$$

From Fig. 8, we deduce that

$$\lambda_d T = L - N_c(v) - \bar{m} I_{\text{soj}}. \tag{20}$$

That is, when the number of active voice $n_j = N_c(v)$, the number of packets arrived in the jth cycle equals the number of packets served in the same cycle. The above equation can be rearranged to yield

$$N_c(v) = L - \lambda_d T - \bar{m} I_{\text{soj}}.$$

Substituting $\lambda_d T$ by $\rho_d(L - \bar{n} - \bar{m} I_{\text{soj}})$ from (20) and rearranging, we have

$$N_c(v) = L(1 - \rho_d) + \rho_d \cdot \bar{n} - \bar{m} \cdot I_{\text{soj}}(1 - \rho_d). \tag{21}$$

The parameters $\bar{m}$ and $N_c(v)$ are related by a set of transcendental equations. $\bar{m}$ and $N_c(v)$ can be computed by the following algorithm:

1. Substitute $N_c(v)$ *for* $\bar{n}(o)$ in (18b) and then combine the result with (21) to obtain

$$N_c(v) = \frac{[L(1 - \rho_d) + \bar{n}\rho_d](M \cdot P_{\max} + I_{\text{soj}}) - I_{\text{soj}}(1 - \rho_d)(L + M \cdot P_{\max})}{I_{\text{soj}}(1 - \rho_d)(L + M \cdot P_{\max})}.$$

2. Calculate $N_c(v)$.
3. Using this value of $N_c(v)$, calculate $\bar{n}(o)$ by eq. (19).
4. Calculate $\bar{m}$ using (18b).
5. Recalculate $N_c(v)$ using (21).

The approximate analysis model developed in Ref. 8 is based on the fact that the number of active voice obeys a binomial distribution and that the system (for voice) is at equilibrium when the number of active voice equals the mean, $\bar{n} = pN$, where p is the probability that a voice call is in talk spurt and N is the total number of voice calls in the system. When the number of active voice calls deviates from $\bar{n}$, there is a tendency for the system to move towards the equilibrium point.

Define d_k to be the net drift toward the equilibrium point. Then, at any operating point $n_j = k$, the net drift, d_k, is given by Ref. 8:

$$d_k = \alpha \cdot T \cdot (n_j = k) - \beta \cdot T \cdot (N - (n_j = k)) \,|\, k \in \Omega_o \quad \text{slots/cycle}, \tag{22}$$

where α^{-1} and β^{-1} are, respectively, the mean holding times of a talk spurt and a silent event and T, as previously defined, is the cycle length in seconds. With probability Pr[overload] = $\Pr[k \in \Omega_o] = \Pr[n_j > N_c(v)]$, the system is in the overload state. Once it enters the overload state, the mean time that the system spends in the overload state is a function of the net drift d_k given by (22). The mean and second moment of the duration that the system spends in the overload state is given by Ref. 8:

$$\tau_o = \frac{\displaystyle\sum_{k=N_c(v)+1}^{N} \frac{(N - (n_j = k))\Pr(n_j = k)}{d_k}}{\displaystyle\sum_{k=N_c(v)+1}^{N} (N - (n_j = k))\Pr(n_j = k)}$$

$$= \frac{\displaystyle\sum_{k=N_c(v)+1}^{N} \frac{(N - (n_j = k))\Pr(n_j = k)}{\alpha \cdot T \cdot (n_j = k) - \beta \cdot T \cdot (N - (n_j = k))}}{\displaystyle\sum_{k=N_c(v)+1}^{N} (N - (n_j = k))\Pr(n_j = k)}, \tag{23}$$

and

$$s_o = \frac{\displaystyle\sum_{k=N_c(v)+1}^{N} \frac{(N - (n_j = k))\Pr(n_j = k)}{[\alpha \cdot T \cdot (n_k = k) - \beta \cdot T \cdot (N - (n_j = k))]^2}}{\displaystyle\sum_{k=N_c(v)+1}^{N} (N - (n_j = k))\Pr(n_j = k)}. \tag{24}$$

The coefficient of variation of the duration in the overload state is then given by

$$C_o^2 \equiv \frac{s_o}{[\tau_o]^2} - 1. \tag{25}$$

It turns out that $C_o \ll 1$. This is a consequence of the fact that the increase in the number of active voice, which drives the data queueing process into an overload condition, is a slowly fluctuating process. Thus, the second moment of the overload duration has negligible effect on the queueing behaviour when the system is in the overload state.

As observed in Ref. 8 and depicted in Fig. 9, a transient state always follows an overload state. The chain of service cycles depicted in Fig. 9 will be referred to as an "overload/transient" chain. The contribution to the average queue in the system from an overload/transient chain is given by the queue in a typical overload/transient chain multiplied

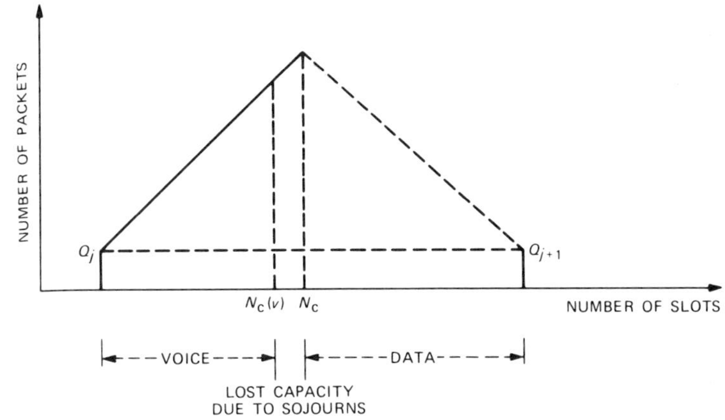

Fig. 9—Chaining of overload/transient cycles.

by the probability of the occurrence of an overload/transient chain. The mean duration of the overload portion of the chain depicted in Fig. 9 is τ_o. The queue growth in the overload state is given by the difference of the queue size at the ends of consecutive service cycles, i.e., $\Delta Q_j = q_{j+1}(v) - Q_j(v)$. The average queue growth per cycle will be denoted by $E[\Delta Q]$. The average queue growth in an overload state is then given by an accumulation of the growth per cycle over the overload duration, i.e.,

$$\sum_{i=1}^{\lceil \tau_o \rceil} iE[\Delta Q],$$

where $\lceil x \rceil$ denotes the integer part of $x + 0.5$, plus $\lambda_d N_c(v)P/2R$, which is the average growth in the cycle when the system enters the transient state. The queue shrinkage will be slightly faster than the queue growth so that the mean duration in the transient state is nominally shorter than τ_o. This is due to the fact that the number of active voice while the system is in the overload state is larger than that when the system is in the transient state. The average number of slots available for data per overload cycle is $L - \bar{n}(o) - \bar{m} \cdot I_{soj}$; that per transient cycle is $L - \bar{n}(t) - \bar{m} \cdot I_{soj}$, where $\bar{n}(o)$ and $\bar{n}(t)$ are the mean number of active voice in the overload and transient states, respectively. $\bar{n}(o)$ is given by (19), and $\bar{n}(t) \le N_c(v)$. For convenience, we approximate $\bar{n}(t)$ by $N_c(v)$ and define

$$h = \frac{L - \bar{n}(o) - \bar{m} \cdot I_{soj}}{L - N_c(v) - \bar{m} \cdot I_{soj}} \le 1 \tag{26}$$

to be the fractional contribution from the transient portion to the queue size in an overload/transient chain. Then, the number of packets in an average overload/transient chain is given by Ref. 8:

$$E[Q \mid k \in \Omega_o] = (1 + h)\left\{\frac{\lceil \tau_o \rceil \cdot (\lceil \tau_o \rceil + 1)}{2} E[\Delta Q] + \frac{\lambda_d \cdot \bar{n}(o) \cdot P}{2R}\right\}, \tag{27}$$

where

$$E[\Delta Q] = \lambda_d T - \{L - \bar{n}(o) - \bar{m} \cdot I_{soj}\} \tag{28}$$

is the average queue growth per cycle when the system is in an overload state. The average queue contribution from the overload/transient chain is then given by

$$\bar{Q}(o) = E[Q \mid k \in \Omega_o] \cdot \Pr[k \in \Omega_o]. \tag{29}$$

The stationary contribution can be written as

$$\bar{Q}(s) = \sum_{k=0}^{N_c(v)-1} E[Q \mid n_j = k] \Pr(n_j = k), \tag{30}$$

where $E[Q \mid n_j = k]$ is the mean number of packets in the system given the number of voice in the jth cycle is less than $N_c(v)$ and, for Poisson arrivals, is given by the M/G/1 formula:

$$E[Q \mid n_j = k] = \rho_k + \frac{\rho_k^2(1 + C_d^2)}{2(1 - \rho_k)}. \tag{31}$$

ρ_k is the utilization defined by (16) and C_d^2 is the squared coefficient of variation of service time. Since voice changes slowly compared to a cycle time, $C_d^2 \ll 1$.

The Pollaczek-Khinchin M/G/1 formula describes the queueing effect at the imbedded points of the service process which, in our case, correspond to the ends of service cycles. During the initial part of the service cycle when the system is serving voice stations, the data stations experience a temporary queueing. The average temporary queue per cycle corresponds to a delay during which a typical data packet is scheduled for channel access. Depending on the state of the system, the number served in a given cycle may be fewer than the number that has arrived during the same cycle. The queueing of those new arrivals that are not served in the same cycle is accounted for by the M/G/1 formula. To compute the queueing due to scheduling, it is only necessary to consider new arrivals when the system is in the stable mode. Consider the fluid approximation model depicted in Fig. 10. The total number of packets that need to be queued temporarily

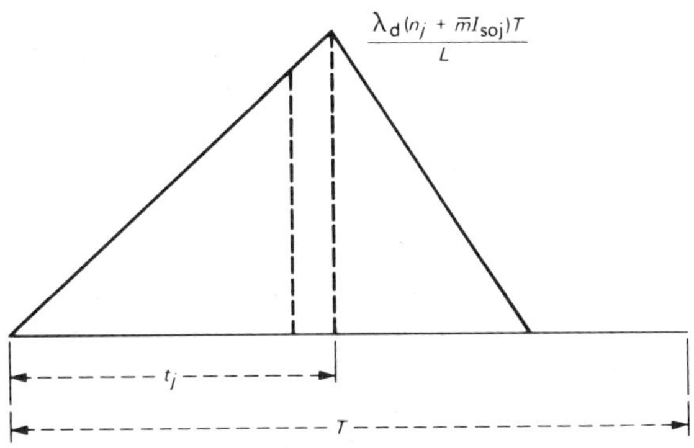

Fig. 10—Fluid approximation model for the evaluation of mean queue due to scheduling.

in the jth cycle is given by the area under the triangle; the average number of packets in the temporary queue is given by the area divided by the base. Thus, the average number of packets in the temporary queue due to scheduling during one cycle is given by

$$\bar{Q}_{\rm sch} = \frac{\lambda_d T}{2L} \sum_{k=0}^{N_c(v)} [(n_j = k) + \bar{m} \cdot I_{\rm soj}] \cdot \Pr(n_j = k). \tag{32}$$

The mean number of packets in the system is then given, as indicated in (17a), by the sum of $\bar{Q}(o)$, $\bar{Q}(s)$, and $\bar{Q}_{\rm sch}$, which are prescribed respectively by (29), (30), and (32).

VI. NUMERICAL RESULTS

Recognizing the difficulty with an exact analysis of the data queueing behaviour in an integrated voice and data service environment, we have introduced an approximate analysis in Section V. Another approach to evaluating the system performance is by means of computer simulation or emulation. Emulation portrays all the steps involved in the actual system. The drawback in an emulation process is that it is extremely time-consuming and that it is not feasible to emulate a high-rate system. We compare the approximate analysis results with those obtained by means of computer emulation. Based on the closeness of the approximate analysis results to those obtained from emulation on a 10-Mb/s Fasnet, we conjecture that the performance results obtained for a high-rate Fasnet (e.g., 100 Mb/s) using the approximate analysis method are reasonable portrayals of system performance.

6.1 Simulation

The simulation consisted of an emulation at the packet level. We assume the system operates at 10 Mb/s. The length of the line was assumed to be 2.5 km, resulting in normalized propagation and signal processing delay for a single line of 0.2 slot. Data and voice stations were randomly positioned along the line. We assumed that 142 voice station pairs were in conversation with an additional 20 active data stations. The other parameters used in the simulation are r = 64 kb/s, T = 10 ms, and P = 700 b/packet.

An ongoing voice conversation is modeled as a two-state Markov process. It spends 40 percent of the time in the active state and 60 percent in the silent state (activity factor = 0.4). The average talk spurt duration is 1.5 seconds and the average silent interval is 2.25 seconds, both with an exponential length distribution. All data stations generated traffic at the same mean rate and each station transmits a maximum of three packets from its queue during one data subcycle ($P_{max} = 3$). Data packets arrive at each station according to a Poisson distribution.

Each simulation run was for 500,000 packets, corresponding to a duration of 35 seconds. The simulation started with voice conversations assigned to talk and silent stations in the same proportion as the steady-state distribution. To ensure that the steady state was reached before data were recorded, statistics were not gathered until after the first 20,000 packets.

Performance is measured in terms of the average number of packets in the system. In a given service cycle, there is temporary queueing when the system is serving the voice stations. The queue at the end of each cycle is smaller than it is during the cycle. The simulation results shown in Fig. 11 reflect the effects of temporary queueing due to scheduling during one service cycle.

For each data point shown in Fig. 11, the results for between five and twenty independent runs were combined. At lower data rates, the variance is small and few runs were required to obtain accurate measures of the mean data queue. At higher data rates, the variance was significantly larger. The simulation did not guarantee that the average amount of voice traffic would be equal to the value used to calculate the theoretical results. This affects the measurement of the average queue size for data since the channel capacity available for data depends directly on the capacity taken by the voice stations. To improve the estimate of the mean queue for data, the average number of active voice conversations (which was measured for each run) was regressed against mean queue and, from this regression, the value at 56.8 (0.4 × 142) active calls (the value used in the calculations) was estimated. An example of this is shown in Fig. 12, where the experi-

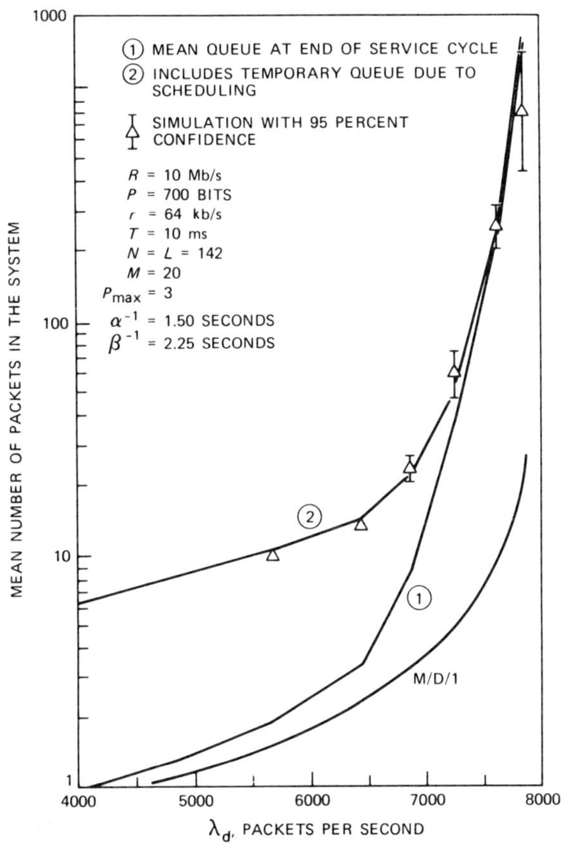

Fig. 11—Data queueing characteristics.

mental points, regression line, and 95-percent confidence interval are shown.

6.2 *Calculation*

The same parameters used in the simulation were employed to calculate the average queue contributions from the stable and overload modes and channel access scheduling, using the equations developed in Section V. The calculated results are tabulated in Table I. For comparison purposes, the mean number of packets in the system is plotted in Fig. 11 for the cases with and without the mean scheduling queue per cycle. Inspection of Fig. 11 shows that the calculated results closely approximate those obtained from simulation.

The data arrival rate in packets/slot is expressible as $\lambda_d T/L$. Then, by Little's formula,[13] the mean packet delay is given by

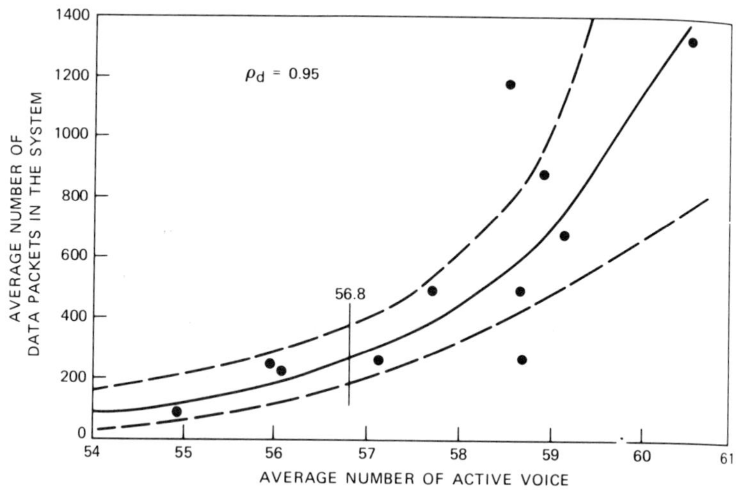

Fig. 12—Regression analysis of mean data queue vs. mean number of active voice with 95-percent confidence.

Table I—Mean number of packets in the system as a function of the arrival rate λ_d in packets/s*

ρ_d	λ_d	$N_c(v)$	$\bar{m}$	Pr[overload]	$\bar{Q}_{sch}$	$\bar{Q}(s)$	$\bar{Q}(o)$	$\bar{Q}$
0.50	4091.80	97.72	1.68	2.6449e-12	5.19	1.01	—	6.21
0.60	4891.12	89.43	1.82	1.6032e-08	6.86	1.38	—	8.24
0.70	5691.21	81.19	1.95	1.4707e-05	8.78	1.96	—	10.73
0.80	6483.12	73.01	2.08	2.3096e-03	11.02	3.25	0.23	14.51
0.85	6877.15	68.94	2.15	2.3249e-02	12.01	4.71	4.31	21.03
0.90	7269.90	64.88	2.21	9.4180e-02	12.16	6.86	30.56	49.58
0.95	7895.67	60.83	2.28	2.6206e-01	10.54	8.64	228.68	247.86
0.98	7895.67	58.41	2.32	3.8351e-01	8.02	6.30	1512.02	1526.34

* R = 10 Mb/s, r = 64 kb/s, H = 60 bits, P = 700 b/packet, α^{-1} = 1.50 seconds, β^{-1} = 2.25 seconds, L = 142 slots, $N = L$, M = 20, P_{max} = 3, I_{soj} = 2.

$$\bar{d} = \bar{Q}L/\lambda_d T.$$

The analysis and simulation results of the delay-throughput characteristics are shown in Fig. 13. These curves are an alternative presentation of the results shown in Fig. 11.

To demonstrate the effects on data queueing when the number of voice calls is reduced, we compare in Fig. 14 the cases for N = 128, 136, and 142. It is observed that at a relatively high data rate, the average queue decreases significantly when the number of voice calls is reduced by approximately 5 percent, i.e., from 142 to 136. The improvement in queueing comes from the fact that a reduction in the number of voice calls is accompanied by a relatively large reduction in the probability that the system is in the overload state.

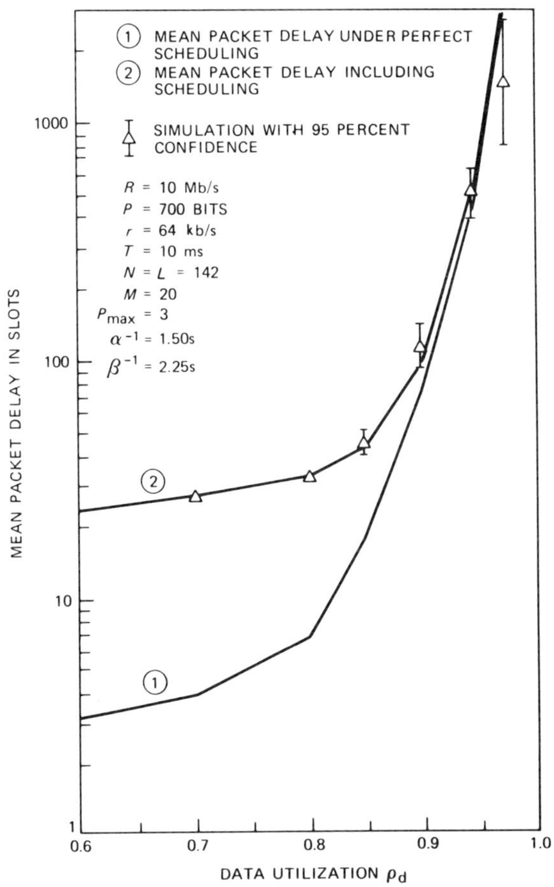

Fig. 13—Mean data packet delay characteristics.

As the voice arrival rate r decreases or the network transmission rate increases, the number of slots in a cycle increases. For example, with $R = 100$ Mb/s, $r = 64$ kb/s, $P = 700$ b/packet, and $T = 0.01$ second, the number of slots per cycle is approximately 1408. For $r = 32$ kb/s and $P = 380$ b/packet ($H = 60$ bits in each case), the number of slots per cycle is approximately 2624. Letting the number of voice calls $N = L$, the average number of data packets in the system as a function of the traffic intensity ρ_d is plotted in Fig. 15. It is noted that the mean contribution to the queue due to scheduling is the same for the two cases, since the packet length for the $r = 64$ kb/s case is 1.8421 times longer than that of the $r = 32$ kb/s case. The longer cycle length (due to a shorter packet length) case exhibits better long-term queueing performance (excluding scheduling queueing). Consider for

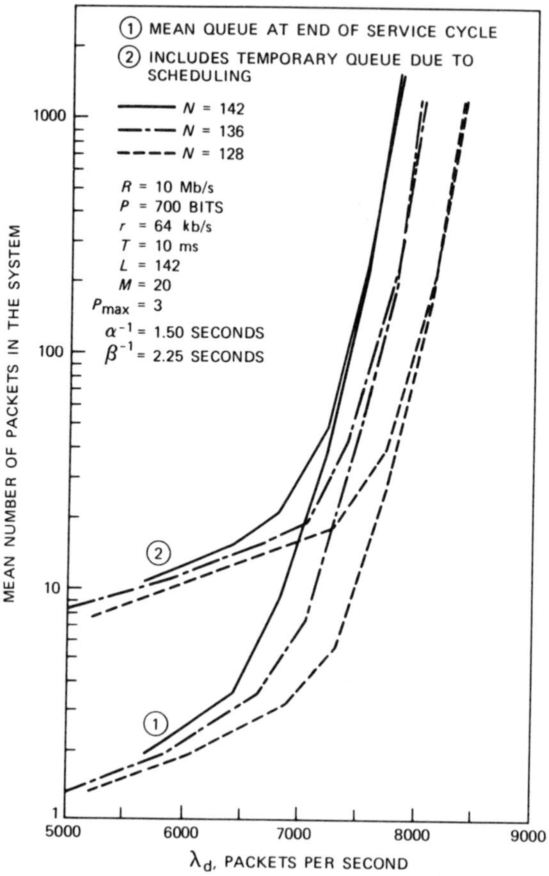

Fig. 14—Effects on queueing with a decrease in the total number of voice stations.

example, at $\rho_d = 0.95$, the average number in the system (excluding scheduling queueing) is 18.21 packets for the $r = 64$ kb/s case and 12.72 packets for $r = 32$ kb/s. Since the relative packet length ratio is 700/380, the effective mean queue for the 64 kb/s case is 2.64 times that of the shorter packet case. Thus, the mean queue for the case where $P = 700$ bits is equivalent to 33.58 380-bit packets at $\rho_d = 0.95$ compared to 12.72 380-bit packets for the 32 kb/s case at the same utilization.

For comparison purposes, the M/D/1 curve is also plotted in Figs. 11 and 15. The M/D/1 result is based strictly on a stationary distribution, i.e., ignoring the statistical variations associated with the speech spurts. At light loads where the system always operates in a stable mode, the M/D/1 represents a reasonable lower bound. At heavy loads where the system experiences a reasonable percentage of

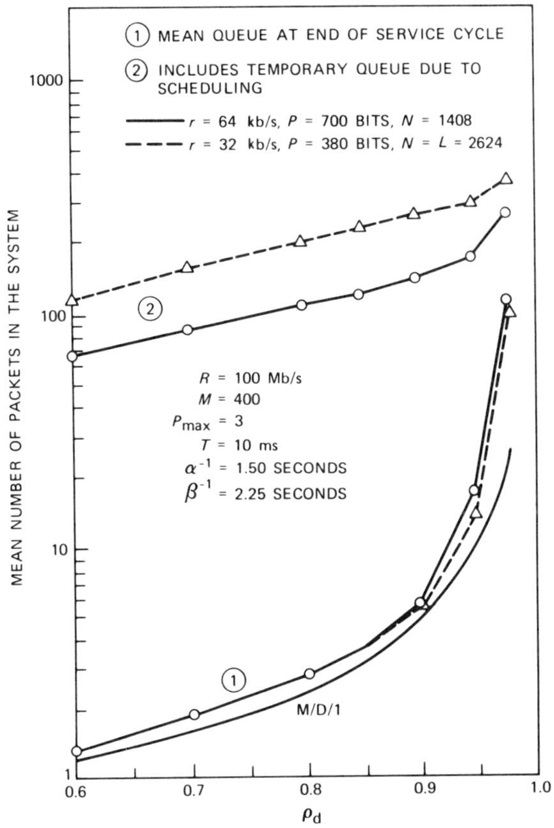

Fig. 15—Queueing characteristics for different packet lengths.

overload, the M/D/1 curve serves no useful comparison (see Fig. 11). Of significance are the numerical results shown in Fig. 15, in which the perfect scheduling Fasnet results deviate from the M/D/1 curve by almost a constant amount up to 95-percent data utilization. This is a consequence of the fact that, at the higher speed ($R = 100$ Mb/s), overload does not become significant until the data utilization approaches unity. The discrepancy between the M/D/1 and the perfect scheduling Fasnet curves can be attributed to the presence of sojourn time in the Fasnet access discipline.

VII. DISCUSSIONS AND CONCLUSIONS

In previous work, Limb and Flamm[10] described a scheme for accommodating voice and data that differs from the present method in that calls returning from idle may get clipped. New calls are accepted until an unacceptable level of clipping is approached, whereupon they are blocked. Unused voice capacity may be utilized by data, although a sojourn time is required to switch between the voice subcycle and the data subcycle unless the voice subcycle is full.

A direct comparison between the two schemes is difficult because of the different way in which the voice traffic is treated—the new calls admission policy and the clipping of talk spurts in one case and the absence of clipping in the other. However, a couple of observations can be made: Both systems have comparable performance since they both utilize the available slots efficiently when traffic is heavy; clipping may be undesirable in certain instances, for example, where a number of Fasnet links are connected in tandem. A speech spurt could be clipped more than once, leading to a noticeable degradation. In this instance, the present method would be preferable.

The global access control that is enabled by the endowment of each slot with a type designation renders Fasnet particularly suitable for integrated services. We have described the ramifications for integrated voice and data services on Fasnet and a policy for admitting new voice into the system. Once admitted, a voice station is guaranteed service when needed. Voice stations are served on a periodic basis, so that each voice source sees almost synchronous service, i.e., the service is synchronous within the statistical variations dictated by the binomial distribution of the silent and talk spurt intervals in a voice process. The mean interpacket delay can be expected to be negligibly small. Note that the service discipline described will not clip any voice packets from a talk spurt.

Since data traffic can be buffered, it is not necessary to maintain continuous service for data. During intervals when the system is servicing voice stations, data stations are temporarily locked out and have to be buffered. Depending on the data input load relative to the service time available for data during one service cycle, the temporary queues may be cleared entirely at the end of the service cycle, or some residual packets will carry over from one cycle to the next. If the latter phenomenon occurs over many consecutive cycles, the system will sustain a net growth in its queues. Under this condition the system experiences overloading, and the data utilization during these cycles will exceed unity for a short period and the data queues will experience significant growth.

The conventional approach to analyzing the queueing behaviour is the generating function analysis method. However, the size of the state space in a practical network renders the generating function method impractical. An approximate analysis method, based on a merge of a periodic service single-server model and a fluid approximation method, has been developed in Ref. 8 to analyze the queueing behaviour for integrated voice and data services on Welnet, a local area network in which the service discipline is functionally similar to that of Fasnet. With a suitable modification to account for the sojourn time that accompanies each switch in the type of service by the Fasnet access control, the approximate analysis method introduced in Ref. 8 was used to analyze the data buffer queueing behaviour for integrated voice and data services on Fasnet.

System performance may also be evaluated by means of computer simulation. Simulation is time-consuming and it is impractical to simulate high transmission rate systems. In this study, we simulated a 10-Mb/s Fasnet and measured performance in terms of the mean queue in the system as a function of the data input load. The analytic results closely approximate those obtained from simulation. It is conjectured that the analysis method yields reasonably accurate results. Analytical results for a 100-Mb/s Fasnet have been presented. An interesting observation is that the larger bandwidth has the effect of reducing the system overload probability.

In this paper we used an approximate model to analyze the network performance. We verified the reasonableness of the approximate model. An alternative approach is to approximate the correlated voice source by a memoryless process, e.g., a birth-and-death process. The validity of such an approximation needs to be shown. An analysis of the backlog in the buffers for a birth-and-death process is discussed in Ref. 14.

REFERENCES

1. K. Kummerle, "Multiplex Performance of Integrated Line and Packet-Switched Traffic," Proc. 2nd Int. Conf. Comp. Commun., Stockholm, Sweden, August 1974, pp. 517–23.
2. M. J. Fischer and T. C. Harris, "A Model for Evaluating the Performance of an Integrated Circuit-and Packet-Switched Multiplex Structure," IEEE Trans. Commun., *COM-24* (February 1976), pp. 195–202.
3. C. J. Weinstein, M. K. Malpass, and M. J. Fischer, "Data Traffic Performance of an Integrated Circuit-and Packet-Switched Multiplex Structure," IEEE Trans. Commun., *COM-28* (June 1980), pp. 873–7.
4. B. Maglaris and M. Schwartz, "Performance Evaluation of a Variable Frame Multiplexer for Integrated Switched Networks," IEEE Trans. Commun., *COM-29* (June 1981), pp. 800–7.
5. I. Gitman, W-N. Hsieh, and B. J. Occhiogross, "Analysis and Design of Hybrid Switching Networks," IEEE Trans. Commun., *COM-29* (September 1981), pp. 1290–1300.
6. B. R. Iyer and J. B. Sinclair, "Analysis of Data Delays in an Integrated Data-Voice Link," Technical Report 8204, Electrical Engineering Department, Rice University, Houston, Texas, April 1982.
7. A. Leon-Garcia, R. H. Kwong, and G. F. Williams, "Performance Evaluation Methods for an Integrated Voice/Data Link," IEEE Trans. Commun., *COM-30* (August 1982), pp. 1848–58.
8. J. W. Mark, "Priority Scheduling for Integrated Voice/Data Services in Local Area Networks," CCNG Report E-111, Computer Communication Networks Group, University of Waterloo, Waterloo, Ont., Canada, March 1983.
9. J. O. Limb and C. Flores, "Description of Fasnet—A Unidirectional Local Area Communications Network," B.S.T.J., *61,* No. 7, Part 1 (September 1982), pp. 1413–40.

10. J. O. Limb and L. E. Flamm, "A Distributed Local Area Network Packet Protocol for Combined Voice and Data Transmission," GLOBECOM '82 Conf. Rec., November 29–December 2, 1982, Miami, Florida, pp. B2.5.1–6.
11. C. J. Weinstein, "Fractional Speech Loss and Talker Activity Model for TASI Packet-Switched Speech," IEEE Trans. Commun., *COM-26* No. 8 (August 1978), pp. 1253–7.
12. G. F. Newell, *Applications of Queueing Theory*, London: Chapman and Hall Ltd., 1972, Chapter 2.
13. J. D. C. Little, "A Proof of the Queueing Formula $L = \lambda W$," Oper. Res., *9* (May 1961), pp. 383–7.
14. S. Halfin, "The Backlog of Data in Buffers with Variable Input and Output Rates," Proc. of the Second International Symposium on Performance of Computer-Communications Systems, Zurich, March 21–23, 1984.

AUTHORS

John O. Limb, B.E.E., 1963, Ph.D. (Electrical Engineering), 1967, University of Western Australia, Nedlands; Research Laboratory, Australian Post Office, 1966–1967; AT&T Bell Laboratories, 1967–1983. Present affiliation Bell Communications Research, Inc. At AT&T Bell Laboratories, Mr. Limb worked for a number of years on the coding of color and monochrome picture signals to reduce channel capacity requirements, and has published widely in this area. He is now manager of the Communication Research Department in Bell Communications Research, Inc. His current research interest is wideband, local area networks. Fellow, Institute of Electrical and Electronics Engineers; member, Association for Computing Machinery, Association for Research in Vision and Opthalmology, Optical Society of America.

Jon W. Mark, B.A. Sc., 1962, University of Toronto; M.Eng., 1968, Ph.D., 1970, McMaster University; Canadian Westinghouse, 1962–1968; IBM Thomas Watson Research Center, 1976–1977; AT&T Bell Laboratories, 1982-1983; University of Waterloo, 1970—. At the IBM Research Center Mr. Mark was a visiting scientist. At AT&T Bell Laboratories, he was a consultant to the Computer Technology Research Laboratory. He worked for a number of years in the areas of radar and sonar signal processing, equalization, data compression, and spread spectrum communication. At present, Mr. Mark is a Professor of Electrical Engineering at the University of Waterloo. His current research interests are network modeling and performance evaluation, satellite networks, and local area networks. Senior Member, IEEE.

Regular Mesh Topologies in Local and Metropolitan Area Networks

By N. F. MAXEMCHUK*

(Manuscript received January 14, 1985)

The throughput per user in loop and bus configured local area networks decreases linearly with the number of users. These networks cannot be extended to a metropolitan area with many users. A class of mesh networks is described that increases the throughput of conventional local area networks by decreasing the fraction of the network capacity needed to transmit information between a source and a destination. These networks have multiple paths that increase the reliability of the networks, and have point-to-point links that can cover a metropolitan area. In general, mesh networks require complex store-and-forward nodes that also route messages, control the flow of data entering the network, resequence packets at the destination, and recover packets with errors. However, there are characteristics of the local or metropolitan area that allow these functions to be simplified. As a result of these simplifications, loop access protocols are extended to mesh networks and the need to store and forward data is eliminated. A file transfer protocol that does not require packet resequencing is described. Three mesh networks are studied, and the desirable characteristics of networks are determined. One network, the Manhattan street network, has many of the desirable characteristics.

I. INTRODUCTION

Loop topologies[1] and random access strategies[2] were first applied to local data networks in the late 1960's. In that era,

- Low-bit-rate terminals were connected to large central computers,
- Computers and terminals were shared by a few computer experts,

* AT&T Bell Laboratories.

- Large-scale integrated circuits did not exist, and
- High-bit-rate transmission facilities were not readily available across public right of ways.

As a result, these networks trade reliability, total throughput, and the distance the network can span[3] for simple access and transmission strategies. Today, for comparison,

- Simple terminals are evolving into personal computers with bit mapped, rather than character, displays,
- Computer usage is becoming universal,
- Very-Large-Scale Integration (VLSI) is becoming commonplace, and
- The increased deployment of optical fibers and CATV systems makes it possible to obtain high-bit-rate communications over wider areas.

Personal computers use larger bandwidths than simple terminals to communicate with centralized support facilities and distribute processing. The increasing use of these devices and the increased distances that high-bit-rate networks can span increase the throughput required of the interconnecting network. In loop and bus systems the total throughput is constant. The average capacity available to each user decreases linearly as the number of users increases. Therefore, to support more users with greater individual requirements, alternative topologies must be considered. The complexity of the devices being connected to networks and advances in VLSI make more complex network interfaces feasible. This increases the class of networks and access strategies that can be considered.

A large number of users, dispersed over a large area, can be accommodated by interconnecting conventional local area networks with gateways. Schlatter and Massey have analyzed this type of network.[4] Their system consists of loops interconnected by switching elements, as proposed by Pierce.[5] Messages use a smaller fraction of the total network capacity than they would if the system were a single loop. Therefore, the maximum throughput of the system increases. Users who communicate the most often are placed on the same loop, which minimizes the interference between subgroups of users. The main disadvantage with this approach is that the gateways are different from the access units and are complex store-and-forward elements.

Yemini[6] and Saadawi and Schwartz[7] are investigating a tree topology. In this network, users are at the leaves of the tree and the nodes of the tree are switching points. Depending on the location of the destination, the switches direct messages toward the root of the tree or toward the leaves. In Yemini's system, the switches establish separate broadcast networks, and in Saadawi's, the switches store packets until the desired path is available. In these systems, messages only use a portion of the network capacity. Locating users who communicate frequently—who are near one another in the tree hierarchy—minimizes the interference between subgroups of users. The advantage of this approach over gateways is that there is only one type of element in the network. The disadvantage is that the network either stores and forwards packets or retains the distance constraints of broadcast networks.

To a certain extent these alternatives remove the throughput constraints of loop and bus systems. However, they still have single points of failure. To make networks more reliable, there must be multiple paths between each source and destination. By adding paths appropriately, the average and maximum distance between nodes decreases, messages use a smaller fraction of the network bandwidth, and the throughput increases. Multiple paths also make it possible to avoid heavily used segments of the network to equalize the load. Mesh networks, like loop networks, have point-to-point communication channels between nodes. This results in less expensive line drivers and receivers than multidrop broadcast systems and is compatible with current optical fiber transmission capabilities.

In general, mesh-configured networks and some of the local network alternatives require complex store-and-forward nodes. A queue of messages is maintained because packets arriving on several of the incoming links may be destined for the same outgoing link. In addition, store-and-forward networks must do routing, flow control, packet resequencing, and error control. Long-distance networks, such as the ARPA network,[8] perform these functions, but their interfaces are more complex than personal computers. Therefore, these networks are not a reasonable interconnection alternative for personal computers. There are, however, characteristics of the local or metropolitan area environment that make simpler mesh networks possible.

Local or metropolitan area networks differ from general long-distance networks in that the

- Physical location of the nodes does not dictate the topology of the network to as great an extent,
- Error rates are much lower, and
- Communications lines are less expensive.

In the local environment, it is not always necessary to connect the closest nodes together. Occasionally, connecting nodes that are further apart can make the topology of the network regular, and simplify tasks such as routing. The lower error rates make it more likely that a packet will traverse the entire network without error. Therefore, error control protocols can operate on an end-to-end, rather than on a link-by-link basis, and networks can have unidirectional links. When messages are not stored at the intermediate nodes for possible retransmission, the class of possible access and transmission strategies increases. In general, there is also a trade-off between the system complexity and communications efficiency. The cost of communications lines, and the additional access strategies and network options that are possible, result in a different network solution in local networks than in long-distance networks. The result is that local networks are significantly less complex.

There is a description in Section III of several regular networks with simple routing strategies. End-to-end error control protocols make it possible to extend the slotted system and register-insertion techniques developed for loop systems to mesh networks. This is shown in Section IV. With these techniques, flow control on mesh networks can be done by throttling the sources, as on loop networks. A trade-off exists between the buffering in these systems and the efficiency with which the communication lines are used. One attempt to take advantage of this trade-off is the Floodnet system.[9] There is a description in Section IV of how this trade-off is applied to mesh networks with slotted system and register-insertion interfaces. The result is that for certain topologies mesh networks without buffering are reasonable. Finally, there is a description in Section V of several file-transfer protocols that do not resequence packets.

II. EXAMPLE

Before discussing the implementation of mesh networks, we show that these networks provide a potential to increase the throughput of conventional local area networks. In this example, two-connected networks, with as few as 64 nodes, increase the throughput of bus configured networks by a factor of 20 to 30. This comparison assumes that the same rate communication lines are used in both the mesh and random access networks. A factor of two increase in throughput is obtained because there is twice as much capacity emanating from each node. However, the major portion of the increase occurs because messages in the mesh network use only a fraction of the total network capacity. Greater increases are obtained in larger networks.

Two traffic distributions are considered, a uniform distribution and a skewed distribution. In the uniform case, each node sends an equal amount of data to each of the other nodes. The skewed distribution corresponds to what might occur in a network of personal computers and file servers. The network is divided into communities of interest, each consisting of a file server and seven personal computers. A personal computer directs 80 percent of its traffic to its own file server and 20 percent to the other file servers. The computer receives an equal amount of traffic from the file servers.

For each traffic distribution, the throughput for six network topologies is investigated. The first two networks are the conventional broadcast bus and loop configured networks. The throughput of the bus network is calculated assuming that the link utilization can approach one, and is an upper bound on the achievable throughput. In the loop network, the packets only use the links between the source and destination. In this network, and in the remaining networks, the throughput is determined by increasing the traffic levels from the sources until the utilization on any link equals one. The remaining four networks are two-connected networks with two links arriving at, and two links emanating from, each node. The first of these networks is a conventional bidirectional loop. For the skewed distribution, the file server is in the middle of the seven personal computers it is servicing. The next two networks are regular arrays called the modified shuffle exchange and the Manhattan street network. These networks are described in Section III. The Manhattan street networks with 16, 32, 48, and 64 nodes are 4×4, 8×4, 8×6, and 8×8 arrays, respectively. For the skewed distribution, the seven personal computers and the file server in a community of interest are arranged in a 4×2 array on the network. This is shown for the 16-node network

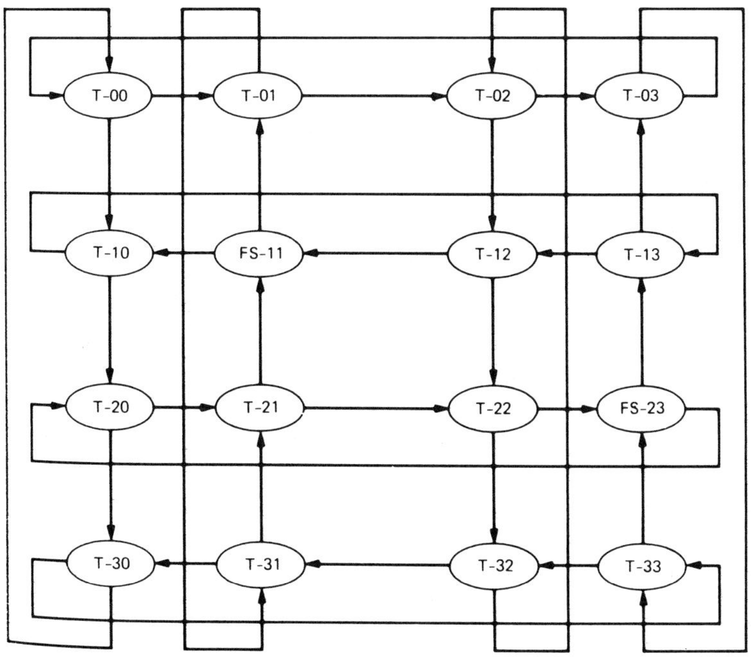

Fig. 1—A 16-node Manhattan street network with two communities of seven personal computers (T) and a file server (FS).

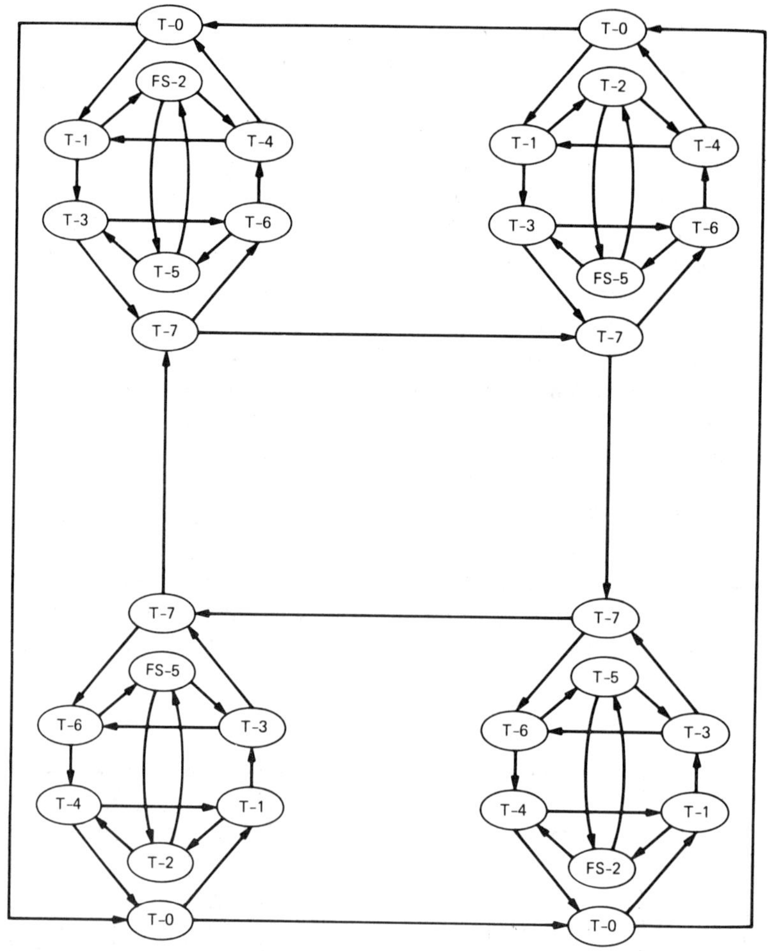

Fig. 2—A 32-node hierarchical network with four communities of seven personal computers (T) and a file server (FS) interconnected by shuffle-exchange networks. The four shuffle-exchange networks are connected by a bidirectional loop.

in Fig. 1. The final network is a hierarchical shuffle exchange, consisting of shuffle-exchange networks with the eight devices in a community of interest, interconnected by a bidirectional loop. Figure 2 shows a 32-node, hierarchical network consisting of four 8-node shuffle-exchange networks.

Traffic on the two-connected networks is placed on the shortest path between the source and destination. If there are several paths of equal length between a source and a destination, the path with the smallest flow is selected. Traffic with the shortest distance between a source and destination is assigned to the network first. Once a source destination requirement is assigned a path, the path is not changed if

Table I—The average megabits per user in networks with 10-Mb/s channels and the improvement over a broadcast network

Network	16 Nodes		32 Nodes		48 Nodes		64 Nodes	
	Mb/Usr	Imprv.	Mb/Usr	Imprv.	Mb/Usr	Imprv.	Mb/Usr	Imprv.
			Uniform Requirements					
BDCST	0.63	—	0.31	—	0.21	—	0.16	—
Loop	1.25	2.00	0.63	2.00	0.42	2.00	0.31	2.00
BDL	4.69	7.50	2.42	7.75	1.63	7.83	1.23	7.87
S-X	5.77	9.23	4.03	12.88			3.09	19.76
MSN	6.52	10.43	4.56	14.59	4.31	20.70	3.94	25.20
HS-X	4.69	7.50	1.96	6.28	1.41	6.75	1.13	7.20
			Skewed Requirements					
BDCST	0.36	—	0.18	—	0.12	—	0.09	—
Loop	0.71	2.00	0.36	2.00	0.24	2.00	0.18	2.00
BDL	2.63	7.37	2.08	11.67	1.80	15.11	1.59	17.82
S-X	1.61	4.52	1.01	5.68			0.81	9.10
MSN	2.63	7.37	2.17	12.17	2.12	17.80	2.12	23.76
HS-X	2.78	7.78	2.78	15.56	2.66	22.34	2.63	29.47

BDCST = Broadcast, BDL = Bidirectional, MSN = Manhattan street exchange, HS-X = Hierarchical street exchange, S-X = street exchange.

a link on the path becomes saturated, and the requirements are not split if two equally good paths exist. This procedure does not lead to the optimum throughput, but gives a reasonably good idea of what can be achieved.

The results of this investigation are presented in Table I. For each network, the average bit rate a user obtains in a network with 10-megabit-per-second transmission links, and the improvement this represents over a broadcast network, is presented. For the conventional broadcast and loop network, the fraction of the capacity a user obtains decreases linearly with the number of users, as expected. The loop system provides about twice as much throughput per user as the broadcast network because, on the average, a packet transmitted on this network uses only half of the network capacity. The two-connected networks obtain a factor of two increase in throughput because there is twice as much capacity emanating from each node, and an additional increase because the networks use a smaller fraction of the network capacity to transfer a packet between the sources and destinations.

The bidirectional loop, the Manhattan street network, and the hierarchical shuffle exchange respond well to the skewed requirements. These networks are capable of allowing complete connectivity while preventing users in different communities of interest from interfering with one another. This characteristic is extremely important in designing large networks. The shuffle-exchange and Manhattan street networks also respond well to a large group of users with uniform transmission requirements. This occurs because the average distance between users does not increase as rapidly in these networks as in the other networks. The average distance between users in the Manhattan street network is greater than that in the shuffle-exchange network; however, the throughput of the Manhattan street network is greater. The Manhattan street network can support a larger throughput because there are more equal-length shortest paths, and bottlenecks can be avoided.

III. TOPOLOGY

In this section, three two-connected networks are described, the bidirectional loop, the modified shuffle exchange, and the Manhattan street network. These networks have two independent paths between any node, and they can survive a single loop or node failure. While these networks are not optimal, they show what measures can be used to compare topologies, and what network characteristics are desirable. Lower bounds on two measures, the average and maximum shortest path between nodes, are derived and compared with these three topologies.

3.1 Bidirectional loops

In a bidirectional loop with N nodes, labeled 0 to $N - 1$, node i is connected to nodes $(i - 1) \bmod N$, and $(i + 1) \bmod N$. This is the only two-connected network with bidirectional paths between all of the nodes. If the transmission protocols require a response each time a packet of data is transferred between two intermediate nodes on a path, this is the only possible two-connected network. This network was initially considered as a mechanism to make loop networks more reliable.

This network has many of the topological advantages of loop systems. It

- Is defined for any number of nodes,
- Makes geographical sense,
- Has a simple rule for expanding the network by one node at a time, and
- Has a simple routing rule.

When a node is added to the network, the two existing nodes closest to this node are disconnected from one another and connected to the new node. Even if the network covers a large geographical area, there are not many long wires. Shortest-path routing in this network is straightforward. The nodes in the network are sequentially numbered from 0 to $N - 1$. The distance from a source node s to a destination node d is $(d - s) \bmod N$ on the incremental path and $(s - d) \bmod N$ on the decremental path. At the source, the shorter of these two paths is selected. Once a packet in this system starts on a path, it remains on that path. Therefore, a complete routing decision is made at the source, and the system is implemented as two separate loop systems.

A disadvantage of this network is that the node addresses and the value of N changes whenever a node is added to the system. Either an addressing and routing scheme that does not use this information must be found, or this information must be distributed each time the network is changed. Another disadvantage of this network is that the throughput is not as great as that in the shuffle-exchange and Manhattan street networks.

3.2 Modified shuffle exchange

The modified shuffle-exchange network is based on the shuffle-exchange multistage switch. The network is defined for N nodes, where N is constrained to be a power of two. Node i is connected to nodes 2^*i mod N and $(2^*i + 1)$ mod N. This results in self-loops at nodes 0 and $N - 1$, which are not used to transmit packets. They also make the network less reliable in that a single link failure can disconnect a node from the network. In the modified network, the self-loops are removed and nodes 0 and $N - 1$ are connected to one another, as shown in Fig. 3. When the shuffle-exchange network is part of a hierarchical structure, as in the previous section, the self-loops are replaced by connections to the higher-level network.

Routing in this network is straightforward. Initially, ignore the two paths that were added to the modified network. Represent the address of node i by $M = \log_2 N$ bits, and label the paths to nodes 2^*i mod N and node $(2^*i + 1)$ mod N as 0 and 1, respectively. When a packet is transmitted from node i, the address of the new node has the low-order $M - 1$ bits of node i's address in the high-order $M - 1$ bits. The low-order bit of the new address is 0 or 1, depending on the path selected. To find the shortest path between a source and destination, match as many of the high-order bits of the destination address with the low-order bits of the source address as possible. To get to the destination, shift the low-order bits of the destination address that are not included in this match into the address and determine the path that must be selected.

For instance, assume that the source address is 11011 and the destination address is 11001. The first two bits of the destination address match the last two bits of the source address. The bits 001 must be shifted into the address to get to the destination, and the distance to the destination is 3. To get to the destination, first path 0 is taken to node 10110, then path 0 is taken to node 01100, and finally path 1 is taken to node 11001.

If none of the high-order bits of the destination match the low-order bits of the source, then the distance to the destination is $\log_2 N$ steps.

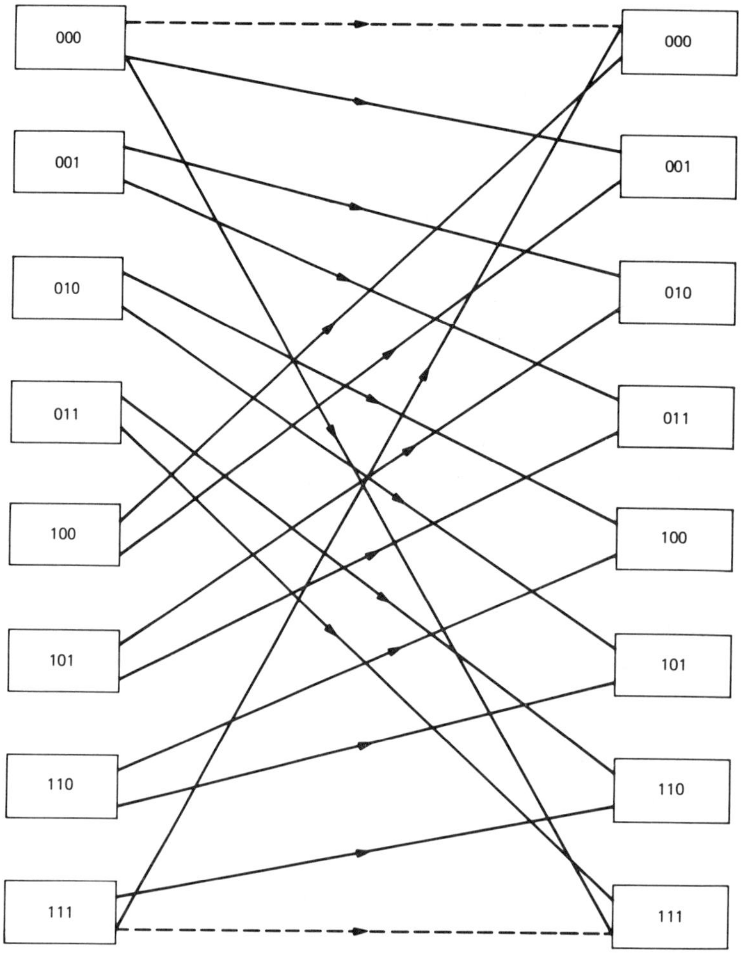

Fig. 3—An 8-node modified shuffle-exchange network.

This is the maximum distance between any source and any destination. In Section 3.4, this will be shown to be the minimum maximum distance between nodes for a two-connected network of this size.

The two paths that are added to the modified network make it possible to go from the all-one address to the all-zero address, and vice versa, in a single step. This shortens the average distance between nodes. These paths cannot change the maximum distance between nodes since this is already a minimum. The effect of these paths on the distance between nodes is shown in Table II. The network S-X is the shuffle exchange with two self loops and MS-X is the modified network. It is evident that the additional paths do not provide a great decrease in the average minimum distance between nodes. They should

Table II—The average and maximum shortest distance between nodes in the shuffle-exchange network and the modified shuffle-exchange network

Nodes	Average S-X	Average MS-X	Maximum
4	1.50	1.33	2
8	2.11	1.96	3
16	2.83	2.73	4
32	3.65	3.58	5
64	4.53	4.49	6
128	5.46	5.44	7
256	6.42	6.40	8

be included to allow an alternate path when failures occur, but unless a simple routing rule is found to use them under normal operation, they should not be used.

There are several problems with this type of a network. The first problem is that the physical layout of this network does not make sense geographically. If half of the nodes in the network are in one area and half of the nodes are in another remote area, then half of the connections must be between the two remote areas. Therefore, the network can only be used in a small area where the length of the interconnections do not make a difference. Shuffle-exchange networks in physically disjoint areas must be interconnected by a hierarchical network that can make sense geographically, as in the example in Section II. The second problem is that the network is only defined if the number of nodes equals 2^i. At present, no way has been found to add one node at a time—changing a small number of connections—and move from a network with 2^i nodes to a network with 2^{i+1} nodes. The third problem is that the alternate paths between a source and a destination are not good. If the preferred path is blocked or inoperable, the alternate paths are much longer.

3.3 *Manhattan street network*

The Manhattan street network is based on a grid of alternatingly directed streets and avenues, as shown in Fig. 1. The nodes exist on the corner of a street and an avenue. The rationale for this type of a network is that routing from a particular street and avenue to a destination should be straightforward. As in a city with this layout, any destination street and avenue can be found without asking directions, even when some roads are blocked. In addition, it should be possible to lay out the network to make sense geographically.

The principal difference between a grid connecting corners with streets and a grid connecting nodes with wires is that the physical constraints associated with a two-dimensional surface can be violated more easily with wires. For instance, in the example in Section II, the file servers and terminals forming a community of interest are in the same neighborhood. Assume that the file servers in this system are in the same room and that the personal computers in the same community of interest are in the same physical area. By connecting the file server to the region of the network with the terminals, rather than basing the connections strictly on the physical location of devices, the file server appears to be in the same neighborhood as the terminals. This reduces the interference between terminals in different communities of interest.

The difference in physical constraints also allows the extremes of the grid to be connected. These connections form the grid on the surface of a torus instead of a flat surface. The advantage of this cyclic surface is that there are no corners. Therefore, the maximum distance from a source to a destination is not the distance between two corners of the grid, but the distance between the center and one of the corners. The graph can also be flipped so that the links leaving the center node are always pointed in the same direction. This allows the same routing decision function to be used at every node.

Consider a network with r rows and c columns. The current node has coordinates (i_s, j_s), and the destination node has coordinates (i_d, j_d). The current node is considered to be at location (0, 0), and the relative location of the destination (i, j), is expressed as

$$i = \left\{[1 - 2(j_s \bmod 2)](i_d - i_s) + \frac{r}{2} - 1\right\} \bmod r - \left(\frac{r}{2} - 1\right)$$

$$j = \left\{[1 - 2^*(i_s \bmod 2)](j_d - j_s) + \frac{c}{2} - 1\right\} \bmod c - \left(\frac{c}{2} - 1\right).$$

The current node is now in the center of the network. The value of i is between $-(r/2 - 1)$ and $r/2$, and j is between $-(c/2 - 1)$ and $c/2$. The factors $1 - 2^*(j_s \bmod 2)$ and $1 - 2^*(i_s \bmod 2)$ guarantee that the links leaving the current node point toward increasing i and j. The routing decision now depends only on the relative location of the destination and not on the current node.

The routing preference from the central node to outlying nodes for a 12 × 12, 12 × 14, and 14 × 14 Manhattan street network is shown in Fig. 4. In this network, the two links emanating from the central node are directed upwards and to the right. The routing preference is the shortest distance from the central node to the destination when the link to the right is taken, minus the shortest path to the destination

	-5	-4	-3	-2	-1	0	1	2	3	4	5	6
6	0	4	0	4	0	4	0	0	0	0	0	0
5	4	4	4	4	4	4	0	0	0	0	0	0
4	0	4	4	4	4	4	4	0	0	0	0	0
3	4	4	4	4	4	4	0	0	0	0	0	0
2	0	4	4	4	4	4	4	0	0	0	0	0
1	4	4	4	4	4	4	0	-4	0	-4	0	0
0	0	0	4	0	4	0	-4	-4	-4	-4	-4	-4
-1	0	0	0	0	0	-4	-4	-4	-4	-4	-4	0
-2	0	0	0	0	0	0	-4	-4	-4	-4	-4	-4
-3	0	0	0	0	0	-4	-4	-4	-4	-4	-4	0
-4	0	0	0	0	0	0	-4	-4	-4	-4	-4	-4
-5	0	0	0	0	0	0	-4	0	-4	0	-4	0

	-6	-5	-4	-3	-2	-1	0	1	2	3	4	5	6	7
6	2	0	4	0	4	0	4	0	0	0	0	0	0	0
5	2	4	4	4	4	4	4	0	0	0	0	0	0	2
4	2	2	4	4	4	4	4	4	0	0	0	0	0	0
3	2	4	4	4	4	4	4	0	0	0	0	0	0	2
2	2	2	4	4	4	4	4	4	0	0	0	0	0	0
1	2	4	4	4	4	4	4	0	-4	0	-4	0	-2	2
0	-2	2	0	4	0	4	0	-4	-4	-4	-4	-4	-4	-2
-1	0	0	0	0	0	0	-4	-4	-4	-4	-4	-4	-2	-2
-2	-2	0	0	0	0	0	0	-4	-4	-4	-4	-4	-4	-2
-3	0	0	0	0	0	0	-4	-4	-4	-4	-4	-4	-2	-2
-4	-2	0	0	0	0	0	0	-4	-4	-4	-4	-4	-4	-2
-5	0	0	0	0	0	0	0	-4	0	-4	0	-4	0	-2

	-6	-5	-4	-3	-2	-1	0	1	2	3	4	5	6	7
7	2	2	2	2	2	2	2	-2	0	-2	0	-2	0	0
6	2	2	4	2	4	2	4	2	0	0	0	0	0	0
5	2	4	4	4	4	4	4	0	0	0	0	0	0	2
4	2	2	4	4	4	4	4	4	0	0	0	0	0	0
3	2	4	4	4	4	4	4	0	0	0	0	0	0	2
2	2	2	4	4	4	4	4	4	0	0	0	0	0	0
1	2	4	4	4	4	4	4	0	-4	0	-4	0	-2	2
0	-2	2	0	4	0	4	0	-4	-4	-4	-4	-4	-4	-2
-1	0	0	0	0	0	0	-4	-4	-4	-4	-4	-4	-2	-2
-2	-2	0	0	0	0	0	0	-4	-4	-4	-4	-4	-4	-2
-3	0	0	0	0	0	0	-4	-4	-4	-4	-4	-4	-2	-2
-4	-2	0	0	0	0	0	0	-4	-4	-4	-4	-4	-4	-2
-5	0	0	0	0	0	0	-2	-4	-2	-4	-2	-4	-2	-2
-6	0	0	2	0	2	0	2	-2	-2	-2	-2	-2	-2	-2

Fig. 4—Routing preference in a 12 × 12, 12 × 14, and 14 × 14 Manhattan street network.

when the upwards-directed link is taken. Therefore, a negative number implies that the right link leads to the shortest path to the destination, and a positive number implies that the upwards link yields the shortest path to the destination. The magnitude of the number shows how much longer the distance to the destination would be if a packet were forced to take a less desirable path. A zero implies that the distance to the destination is the same along either path. The figures show that to get to half of the nodes either path can be taken, to get to a quarter of the nodes the left path should be taken, and to get to the other quarter of the nodes the right path should be taken. The figures also

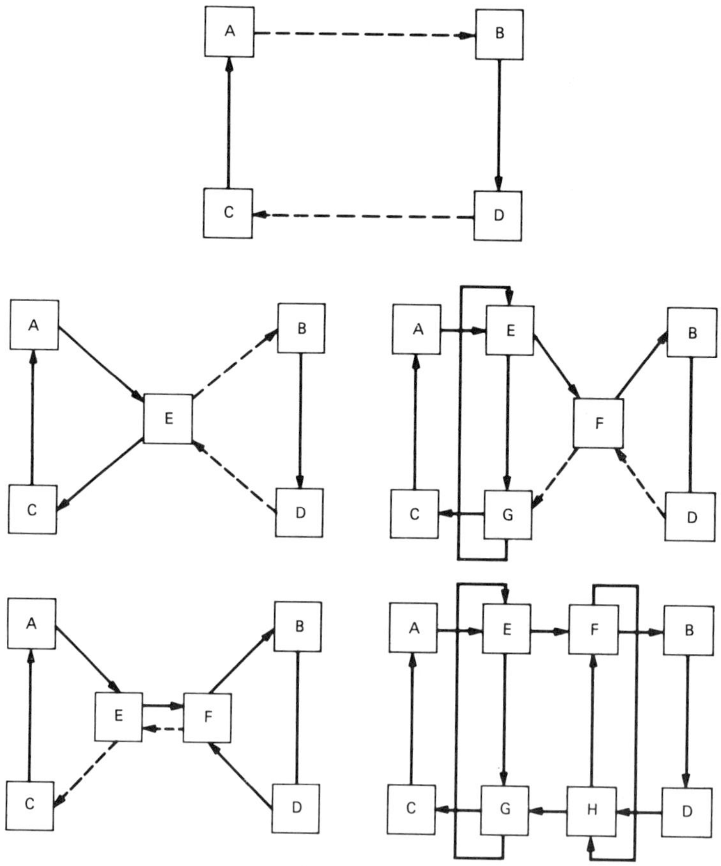

Fig. 5—Adding nodes E, F, G, and H one at a time to the basic rectangular structure consisting of nodes A, B, C, and D in a Manhattan street network.

show that if a packet is forced to take the wrong path, the increase in path length to the destination is never more than four.

One problem with the shuffle-exchange network is the difficulty in changing the number of nodes in the network. Figures 5 and 6 show how nodes may be added to the Manhattan street network. Figure 5 shows how two columns are added to the basic square structure within the Manhattan street network. The dotted lines show the links that will be broken when the next node is added. Figure 6 shows how the procedure is continued to add nodes to partially full columns. Each time a new node is added, two links are broken and connected to the new node. This is no greater than the number of links that must be broken in the bidirectional loop. Eventually this procedure leads to a network with two additional rows or columns, and the pattern of alternatingly directed rows and columns is preserved.

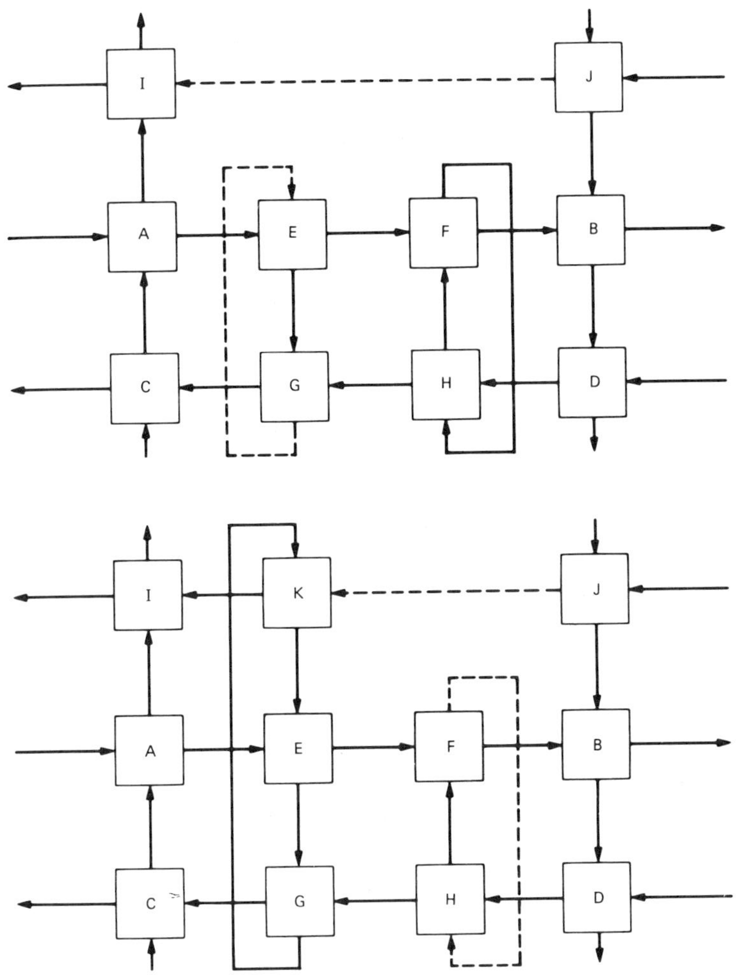

Fig. 6—Adding a node K to a partially full column in a Manhattan street network.

When adding a new node both the physical position of the node and the topology of the network must be considered. It is desirable to connect the node to the nearest existing nodes, but it is also desirable to start as few new rows or columns as possible, and to keep the number of rows and columns equal. When rows and columns are kept approximately equal, the average and maximum shortest paths between nodes increase as shown in Table III.

In the shuffle-exchange network it is occasionally better to establish a hierarchy of networks rather than make a single network larger. Hierarchical structures are also useful in Manhattan street networks. They are used to

Table III—Distances between nodes in $2i$ by $2j$ Manhattan street networks

Nodes	Rows	Columns	Shortest Paths Between Nodes	
			Average	Maximum
4	2	2	1.33	2
8	2	4	2.00	3
16	4	4	2.93	5
24	4	6	3.30	5
36	6	6	3.71	6
48	6	8	4.34	7
64	8	8	5.02	9
80	8	10	5.42	9
100	10	10	5.84	10
120	10	12	6.42	11
144	12	12	7.02	13
168	12	14	7.45	13
196	14	14	7.89	14
224	14	16	8.45	15
256	16	16	9.02	17

- Decrease the number of paths between physically distant sections of the network,
- Eliminate long paths between communities of interest, and
- Prevent traffic between communities of interest from affecting communications in other communities of interest.

The two-connected strategy can be maintained, as in Fig. 7. However, this will make routing more complex. An alternative is to connect one or more of the nodes in a local area to a higher-level network, as shown in Fig. 8. By using this approach, routing decisions in a local area are not affected by network changes in other areas, and addresses in different local areas are assigned independently. A hierarchical addressing and routing structure, similar to that used in the telephone system, can be used. For example, the address within the local area corresponds to a phone number, and the address of the local network on the higher-level network corresponds to the area code. When sending a packet within the local area an area code is not required.

3.4 An optimal two-connected network

Certain characteristics of the "best" networks are difficult to quantify. For instance, it should be possible to add nodes without making major reconfigurations, create geographically dispersed networks without adding excessive numbers of long links, and establish communities of interest. Other characteristics, such as the average and maximum number of links between nodes, can be compared and bounded.

Consider the class of two-connected networks. From a particular node, at most two nodes can be reached in one step, four additional

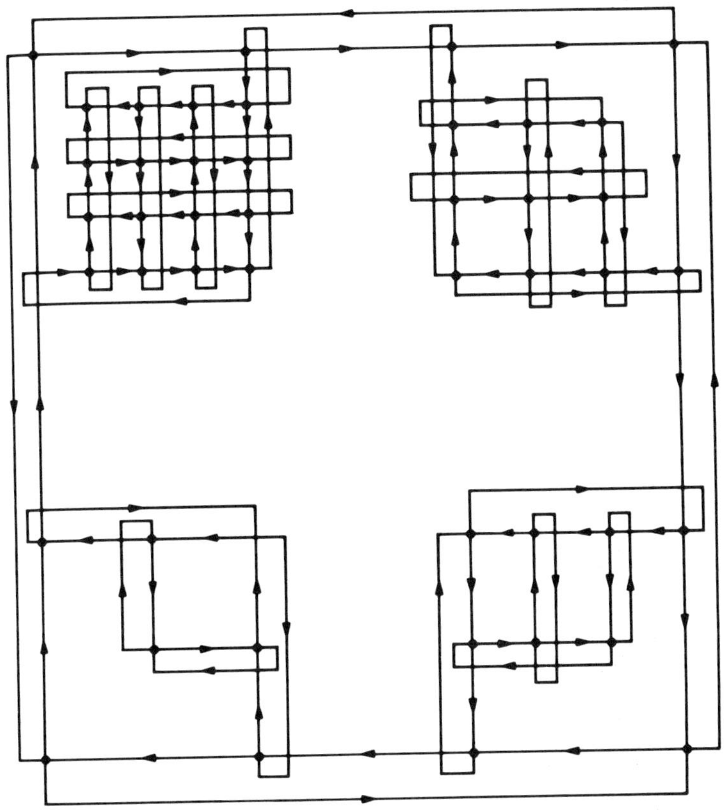

Fig. 7—A hierarchical Manhattan street network in which all of the nodes are two-connected.

nodes in two steps, and so on. The destination nodes form a binary tree. If, at any level in the tree, a destination node recurs, the number of new nodes that can be reached in future levels is reduced by the descendants of that node. Therefore, the maximum number of nodes that can be reached in m steps is

$$\sum_{i=1}^{i=m} 2^i = 2^{m+1} - 2.$$

If every node in a two-connected network can reach this number of new nodes in each m steps, then the network has the smallest average and maximum distance between nodes. In general, networks with these characteristics do not exist. However, this is a lower bound on these distance characteristics.

In the shuffle-exchange network with 2^j nodes, each node must reach $2^j - 1$ nodes, and the maximum minimum distance between

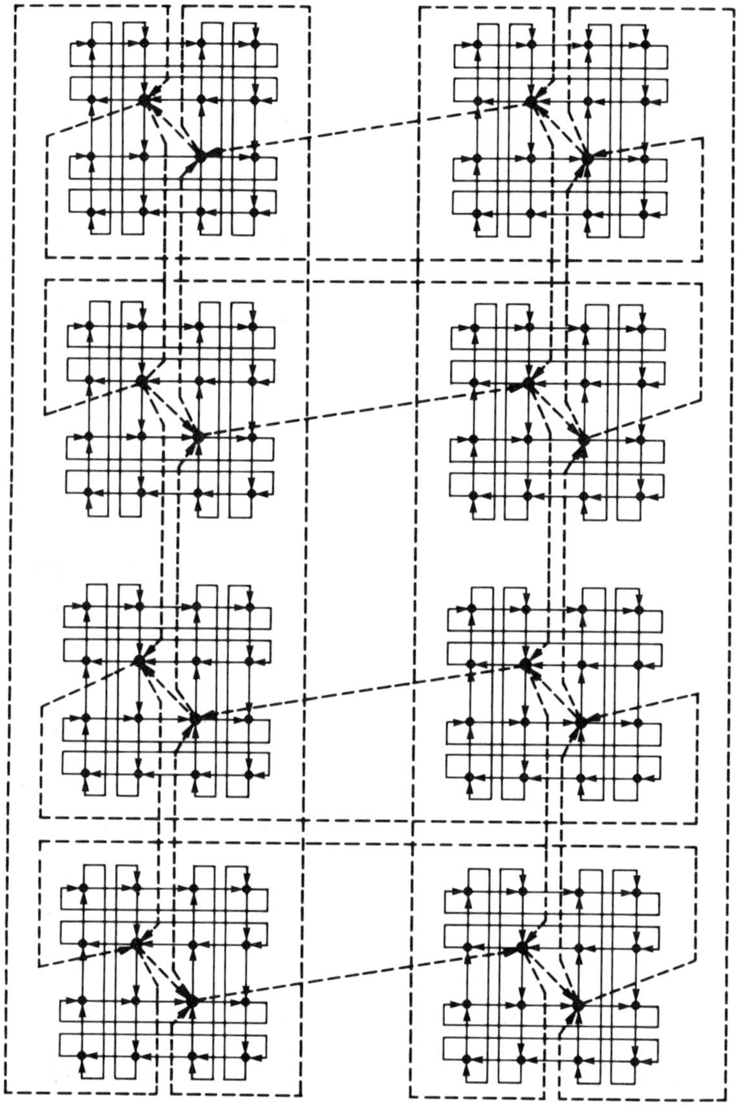

Fig. 8—A hierarchical Manhattan street network in which the nodes connected to the hierarchical network are four-connected.

nodes is j. The number of nodes that are reached in $j-1$ steps in the optimum network is 2^j-2. Therefore, if 2^j-1 nodes must be reached on the optimum network, the minimum distance to the furthest node is j, and the largest minimum distance between nodes in the shuffle-exchange network is less than or equal to that in any network with 2^j nodes.

A comparison of the average and maximum distance between nodes

Table IV—A comparison of the distances between nodes in several networks

Net	Distance	16	64	256
		Number of Nodes		
Opt	Avg	2.53	4.19	6.06
	Max	4	6	8
S-X	Avg	2.73	4.49	6.40
	Max	4	6	8
MSN	Avg	2.93	5.02	9.02
	Max	5	9	17
BDL	Avg	4.26	16.25	64.25
	Max	8	32	128

for the optimum, shuffle-exchange, Manhattan street network, and the bidirectional loop is shown in Table IV. In both the optimum network and the shuffle-exchange network, the maximum distance between nodes varies as the log of the number of nodes. In the Manhattan street network, the maximum distance between nodes varies as the square root of the number of nodes, and, in the bidirectional loop, this distance varies linearly with the number of nodes. The same relationship is also noted between the number of nodes in these networks and the average distance between nodes. The average distance between nodes shows what fraction of the network resources is used to transfer a packet and provides an indication of the relative throughput of the networks. Although, as shown in Section II, there are other factors that also affect the throughput.

IV. IMPLEMENTATION

In a mesh network, as in a loop network, the communications lines are point-to-point links with a single transmitter and a single receiver. Transmission on these links is much simpler than in a broadcast network with a shared communication channel. The access protocols are simpler because there is only one source, and it is not necessary to multiplex users on the communication channel or resolve collisons. The receiver is simpler because the distance between the source and destination is constant, and the signal strength does not change by a large amount from packet to packet. Regenerating the signal to eliminate distance constraints is simpler because signals only propagate in one direction. Timing recovery is simpler because the source can transmit continuously and bit synchronization does not have to be reestablished at the beginning of each packet. In addition, the communication line does not have many taps and is compatible with the current generation of fiber-optic equipment.

In a two-connected network there are two links and a local source inputting data to a node, and two links and a local sink removing data from the node. Occasionally, multiple inputs try to transmit data to the same outgoing link. One way to resolve this problem is to queue packets waiting for a link. The network now assumes the complexity of a store-and-forward network. Not only must potentially large packet queues be maintained, but adaptive-routing, flow-control, deadlock-avoidance, and packet-resequencing issues must be addressed.

In this section, the slotted-system and register-insertion techniques, developed for loop communication systems, will be extended to mesh networks with equal in and out degrees. The general strategy guarantees that every packet arriving on an incoming link, and not destined for the node, will be transmitted on one of the outgoing links. Therefore, it is not necessary to maintain a packet queue for the links emanating from the node. The requirement that packets passing through the node take one of the outgoing paths results in longer paths when the shortest path to the destination is busy. However, it is possible to design networks to reduce the effects of incorrect paths. For instance, in the Manhattan street network the path to only half of the destinations is increased if a packet is forced to take one path rather than the other. In addition, if a packet is forced to take a less desirable path, the distance to the destination is increased by at most four.

The storage between the local source and the network is also limited. Packets from the local source are only transmitted when one of the outgoing links is not being used by an incoming link. It is assumed that either the local source can be throttled when the network is busy, or that the source provides data at a low rate relative to the network transmission rate. In the latter case, when a packet is lost, it must be recovered by a higher-level protocol. If the network delivers packets faster than the local sink can accept them, packets are either transmitted on one of the outgoing links or discarded. In the former case, the network is used for storage. Since new packets cannot enter the network when it is recirculating old packets, this transmission strategy acts as a flow-control mechanism. The assumptions on the local source and sink are implicit in all loop-configured systems without infinite storage.

The packets of data in a slotted system are fixed size. A node continuously transmits bits on each of the links emanating from the node, and periodically transmits a start-of-slot indication. The start-of-slot indication is followed by a packet of data or an empty slot. In the interval between the start-of-slot transmissions, at most one packet of data is received on each of the incoming links. The packets that are received between start-of-slot transmissions are forwarded after the start of slot is transmitted. These packets are switched to one of the outgoing links or the local sink before data from the local

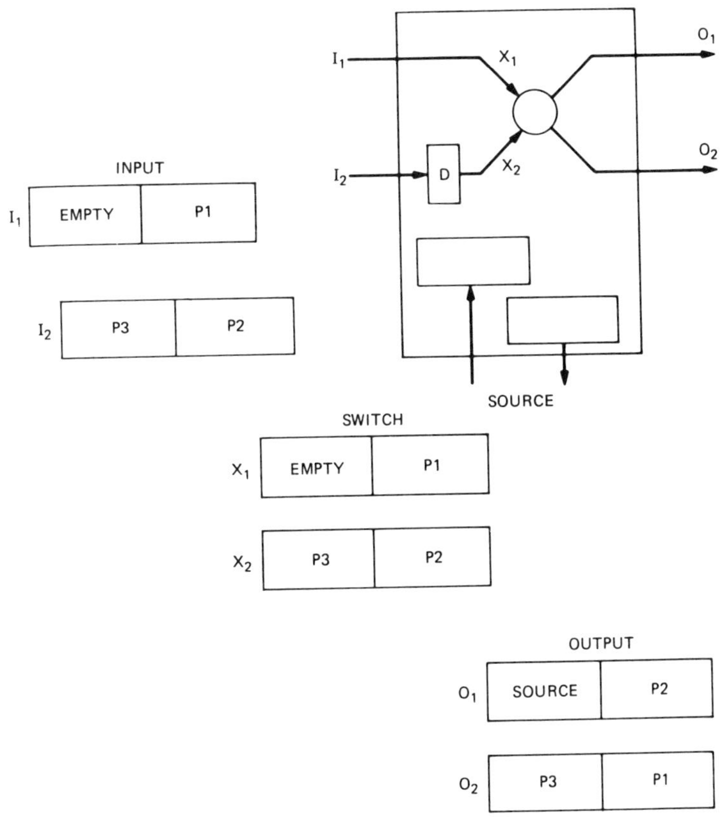

Fig. 9—Extension of slotted-loop systems to a mesh network.

source are given access to the slot. Since there are the same number of links arriving and leaving from each node, and the local source can be throttled, a queue of packets will not accumulate. The operation of a slotted system without a packet queue is shown in Fig. 9.

The interface for a register-insertion loop is shown in Fig. 10. Packets in this system are variable in size, but constrained to be less than the storage register W_l. The local source is only allowed to transmit when register W_l is empty. Since a packet from the local source is less than W_l, all data received from the loop while the local source is transmitting can be stored in W_l. When register W_l is not empty, bits from this register are transmitted on the loop. Therefore, the length of this register remains the same when bits are being received, and decreases when bits are not received. As long as this register is not empty, gaps between arriving packets are removed.

The register-insertion technique can be applied to a mesh network in which the in degree and out degree of the nodes are the same by

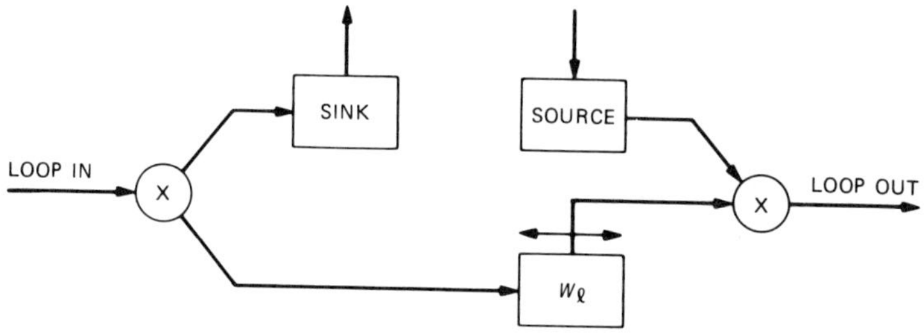

Fig. 10—A register-insertion access unit in a loop system.

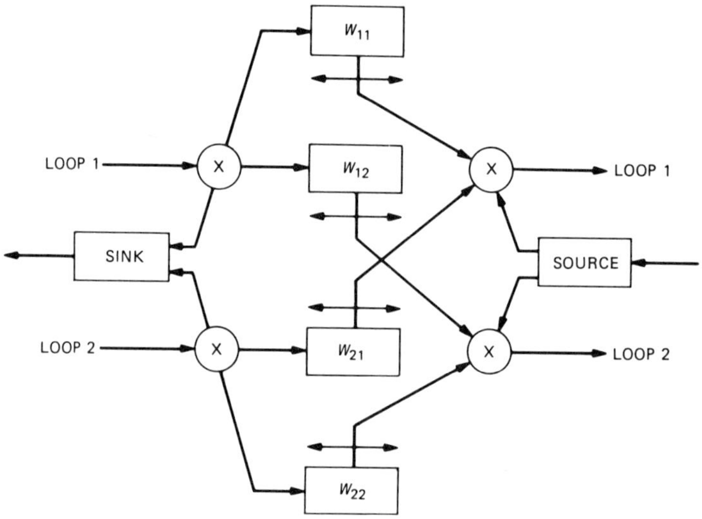

Fig. 11—Extension of register-insertion systems to a mesh network.

making the node appear as if several loops are passing through it. This is shown in Fig. 11. Registers W_{11} and W_{22} correspond to register W_l for loops 1 and 2, respectively. In addition to the local sink, register W_{12} appears to be a sink for loop 1. And, in addition to the local source, register W_{12} appears to be a source of data for loop 2. Therefore, register W_{12} allows messages on loop 1 to transfer to loop 2. As in a loop system, if buffer W_{12} is full the packet must continue around loop 1. Buffer W_{21} serves the same purpose for packets crossing from loop 2 to loop 1.

The register-insertion technique allows variable-length packets. When the system is busy, each node eliminates the null space between incoming packets to efficiently use the outgoing links. The slotted-system technique uses fixed-size packets. If there is less than a packet of data, the packet is partially empty. Therefore, the register-insertion technique uses the channel more efficiently. In a slotted system, the packets from the incoming links are aligned at the switching point. A packet only traverses a longer path if more than one incoming packet requires the same outgoing channel. In a register-insertion system, a packet can only transfer from one loop to the other if the crossover buffer is empty. It is possible that both packets on the through loops would rather be on the other loop, but cannot cross over because the buffers are full. Therefore, the register-insertion technique uses individual links more efficiently, but the slotted system takes a shorter path between the source and destination. The technique that provides the greater throughput depends on both the message-length distribution and the network topology.

A small amount of buffering can be included in either the slotted or register-insertion system to reduce the probability of a packet taking a longer path. In the slotted system, fixed-size packet buffers are inserted at the output channels. The probability that a packet must take a longer path is the probability that two arriving packets must take the same path and the buffer for that path is full. Without buffering, one packet must take the longer path whenever two arriving packets want to take the same path. In the register-insertion system, the additional storage is inserted at the crossover point. A packet cannot cross over if, when it is received, there are fewer bits available in the crossover buffer than there are in a maximum-size packet. In the original system, a packet cannot cross over if, when it arrives, the crossover buffer is not empty. Decreasing the probability that a packet takes a longer path decreases the fraction of the network resources that a packet uses, and increases the throughput of the system. The trade-off between buffering and system throughput remains to be investigated.

V. FILE TRANSFER

A file transfer consists of several packets being transmitted from a source to the same destination. In a system in which packets do not take the same path, it is possible that packets are not received in the same order that they are transmitted. Packets may be resequenced at the receiver, however, it is preferable to avoid this task.

One possible solution to this problem is to transmit one packet at a time and wait for an acknowledgment. This reduces the file-transfer rate. However, because of the small delays at each node, this solution is not as bad in mesh networks as it is in store-and-forward networks. For instance, in a slotted system the average delay per node is half a slot time, and the average round trip delay equals the average number of hops between nodes, $\bar{L}$, times the slot time. Therefore, there is an average of $\bar{L}$ slots between each packet in the file, and the file-transfer rate equals the channel rate divided by $\bar{L} + 1$. Higher file-transfer rates can be achieved by end-to-end protocols that take advantage of the delay characteristics of the system, or node protocols that take advantage of specific hardware structure.

Because of the small amount of delay at each node, it is unlikely that packets that take routes that have approximately the same length will arrive out of sequence. This probability can be reduced by allowing a small number of slots between packets in the same file transfer. A simple file-transfer protocol, which takes advantage of this characteristic, operates like the window protocols of store-and-forward networks and the go-back-N protocols of satellite systems. This protocol labels packets in a file transfer with a sequence number and a retry number. At the beginning of a file transfer, the transmitter and receiver start with the same sequence and retry number. The sequence number is the order of the packets. The receiver

- Increments its retry number when a packet with the expected retry number and a larger sequence number is received,
- Sends a positive acknowledgment if a packet has a sequence less than or equal to the expected number,
- Sends a negative acknowledgment with its retry number and expected sequence number if the packet has a larger sequence number than it expects, and
- Commits a packet if it has the expected sequence number.

The transmitter

- Stops saving a packet for retransmission when it receives a positive acknowledgment for a packet with a sequence number greater than or equal to the expected number,
- Adopts a new retry number and starts retransmitting from a negatively acknowledged sequence number if a negative acknowledgment with a larger retry number is received, and
- Periodically retransmits the last packet in a file transfer until it receives an acknowledgment.

The transmitter initially transmits packets in the file transfer in every available slot. However, when it receives negative acknowledgments, it increases the number of slots between subsequent packets.

Since this protocol only accepts packets in the correct order, packets do not have to be resequenced at the receiver. In a mesh network, several packets can be in transit between the source and destination. If the receiver misses a packet, it must send a negative acknowledgment for every packet with a larger sequence number than expected to be certain that the trasmitter receives the negative acknowledgment. The retry number is included so that the transmitter only backs up and starts retransmitting when it receives the first negative acknowledgment to an outstanding packet.

The transmitter adaptively changes the number of slots between packets according to network load and the rate of the receiver. When the network is lightly loaded, all packets follow the best path to the receiver, and arrive in sequence. If the receiver can accept packets at this rate, there are no negative acknowledgments, except for infrequent transmission errors, and the file-transfer rate equals the channel rate. When the network is heavily loaded, the packets follow different paths, are received out of sequence, and the file-transfer rate decreases. If the receiver cannot accept packets as fast as the transmitter can deliver them, the buffer in the interface unit is full when the packets arrive. In the systems described, these packets are directed to one of the output links at the node, and recirculate in the network until the buffer is available. These packets arrive out of sequence, negative acknowledgments are transmitted, and the transmitter slows down.

Additional improvements are possible by taking the structure of the nodes into account. In a slotted system, subsequent packets in a file transfer can be marked. At a node, packets in a file transfer, which follow immediately behind one another, can be directed along the same path. The end-to-end protocol can be implemented with empty slots only occurring when the source cannot deliver packets quickly enough, or when the channel at the source node is busy with traffic passing through the node. This will improve the file-transfer rate on moderately used channels. If this modification is used, the transmitter and receiver must negotiate the number of packets in a continuous sequence to be certain that the receiver can accept them at the channel rate.

In a register-insertion system, if the loop paths define a Hamiltonian circuit, packets in a file transfer can be constrained to follow these loops. Since a packet can never be denied access to this loop, all packets follow the same path and will be received in sequence. This allows file transfers to occur at the channel rate without resequencing, but requires file transfers to use a larger fraction of the network resources.

In both the register-insertion and slotted systems, it is possible to use a higher-level protocol to set up a limited number of virtual circuits along which file transfers can occur efficiently at the channel rate without resequencing. In the slotted system, the higher-level protocol is used to assign an input at a node to an output. When a file-transfer packet arrives at a node input, it is given first priority to the assigned output. Therefore, all packets in the file transfer follow the same path and do not have to be resequenced. The function of the higher-level protocol is to make the assigned paths for a file transfer use as few links as possible. The problem with this approach is that the preferred paths must be established at the beginning of each file transfer and disabled at the end of the transfer. In addition, a file transfer may be temporarily blocked by previously assigned paths, creating a need for a scheduler. In the register-insertion system, the preferred paths are established as the paths through the node. This has the same problems as the slotted system.

VI. CONCLUSION

Mesh networks increase the throughput of conventional local area networks by decreasing the fraction of the network capacity needed to transmit information between a source and a destination. These networks have multiple paths between each source and destination, thus increasing the reliability of local networks. The networks consist of point-to-point links, and can be extended to cover a metropolitan, rather than a local, area.

In general, mesh networks require complex store-and-forward nodes that also route messages, control the flow of data entering the network, resequence packets at the destination, and recover packets with errors. There are characteristics of the local or metropolitan area that allow these functions to be simplified. In the local environment, regular network topologies can be selected in which routing is straightforward. The lower error rates make it reasonable to recover errors on an end-to-end basis. This allows loop-access protocols to be extended to mesh networks, eliminating the need for buffering and additional flow-control protocols. Extensions for the slotted system and register-insertion techniques used in loop systems have been shown. Buffering can be included in these systems to improve the channel utilization; however, channels are less expensive in the local environment. The small node delays in these systems also make it reasonable to implement file-transfer protocols that do not require packet resequencing.

Three mesh networks have been studied, and the desirable characteristics of networks have been determined. Networks should have regular structures with simple routing rules, and should not have single points of failure. By minimizing the average and maximum distance between nodes, the fraction of the network resources used to transmit a packet decreases, and the throughput increases. This can be done by packing topologies with these characteristics noted, and by locating communities of terminals that communicate frequently in the same area of the network. Equal-length alternate paths between sources and destinations reduce the probability of bottlenecks and the need for buffering within a node. Networks will change, and it must be possible to add or delete nodes without changing a large number of connections. If the network covers a large area, it must be possible to limit the connections between nodes in different areas. Of the networks studied, the Manhattan street network has all of these characteristics.

REFERENCES

1. E. H. Steward, "A Loop Transmission System," Conf. Rec. Intern. Conf. Commun., San Francisco, June 1970, pp. 36-1–9.
2. N. Abramson, "The ALOHA-System—Another Alternative for Computer Communications," University of Hawaii Tech. Rep. B70-1, April 1970, AD707853.
3. R. M. Metcalf and D. R. Boggs, "Ethernet: Distributed Packet Switching for Local Computer Networks," Commun. ACM, *19* (July 1976), pp. 395–404.
4. M. Schlatter and J. L. Massey, "Capacity of Interconnection Ring Communication Systems With Unique Loop-Free Routing," IEEE Trans Inform. Theory, *IT-29,* No. 5 (September 1983), pp. 774–8.
5. J. R. Pierce, "How Far Can Loops Go," IEEE Trans. Commun., *COM-20,* No. 3 (June 1972), pp. 527–30.
6. Y. Yemini, "Tinkernet: Or Is There Life Between LANs and PBXs," Proc. ICC'83, *3,* Boston, June 1984, pp. 1501–5.
7. T. N. Saadawi and M. Schwartz, "Distributed Switching for Data Transmission Over Two-Way CATV," Proc. ICC'84, Amsterdam, May 1984, pp. 1409–13.
8. D. C. Walden, "Experiences in Building, Operating and Using the ARPA Network," Second USA-Japan Computer Conference, Tokyo, August 1975.
9. C. Petitpierre, "Meshed Local Computer Networks," IEEE Commun. Mag., *22,* No. 4 (August 1984), pp. 36–40.

AUTHOR

Nicholas F. Maxemchuk, B.S.E.E., 1968, City College of New York; M.S.E.E., 1970, Ph.D., 1975, University of Pennsylvania; RCA David Sarnoff Research Center 1968–1976; AT&T Bell Laboratories, 1976—. Mr. Maxemchuk is presently Head of the Distributed Systems Research Department. Since joining AT&T Bell Laboratories, he has done research on computer-communication networks, virtual and speech editing, and picture processing. From 1980 to the present, he has been on the adjunct faculty of the University of Pennsylvania, where he teaches a course on computer communications networks. From 1980 to 1985, he was the Associate Editor, then the Editor of Data Communications for the IEEE Transactions on Communications. Member, Eta Kappa Nu, Tau Beta Pi.

Information Outlet, ULAN versus ISDN

Edouard Y. Rocher

This article discusses two key concepts: the information outlet (a universal interconnect standard) and a ULAN (universal local area network) designed to provide a universal, low cost, wideband, block-switched information transport service

With the emergence of Local Area Networks, the 1985–1995 decade will witness fundamental changes in information transport technology at the local level. A century ago [1][1], similar changes occurred in the electric power industry, the "battle of the currents" took place between the DC proponents, Thomas A. Edison and General Electric Co. and the AC supporters, George Westinghouse and Nikola Tesla. In order to avoid a repetition of such a lengthy (at least half a century) battle, two key concepts are presented and discussed in this article:

- The *Information Outlet,* the analog of the power outlet, a universal interconnect standard, with all its potential benefits to the information industry (end-users, carriers, vendors).
- A *Universal Local Area Network (ULAN)* designed to provide rapidly, through the Information Outlet, a universal, low cost, wideband (typically 50 Mb/s), block-switched information transport service.

A hierarchy of *active* bidirectional fiber optic token rings will first be deployed in the business environment as private networks which then will be progressively interconnected within a metropolitan area. Finally, ULAN services will be extended to the residence market. As a model, a system supporting one million subscribers with a three-level hierarchy and a maximum of 100 nodes per ring is discussed. TV quality video services will not be supported by ULAN; they will be provided by CATV networks.

High degree of standardization, support of a wide range of new applications, low initial investments, and low incremental costs are expected to be the main driving forces for a rapid deployment of ULAN's.

In order to avoid a costly transition from the circuit switched (the analog of DC), centralized approach followed by ISDN to ULAN's block-switched[2] (the analog of AC), decentralized approach, a solution is proposed to accelerate and coordinate a very complex, but necessary, standardization process.

Introduction to ISDN and ULAN

With AT&T's divestiture of its local operating companies [2], a competitive environment has been created to take advantage of the new technologies in order to deploy

---

This article is an expansion of a communication made at FOC/LAN 85 in September 1985 in San Francisco, CA and of presentations made in November 1986 at the San Diego, CA meeting of IEEE 802.

[1]Clearly, the issues raised in this article have become a primary concern to all the different players in the information industry. There is abundant and rapidly growing literature on most issues. Every effort has been made here for selecting the latest and the most relevant references on each subject. The references can be used as pointers.

[2]For legal and regulatory reasons, it is essential to make the technical distinction between packet and block switching. The distinction is discussed in the section titled "The ULAN Design." In this article, the term "block switching" is used to describe the switching technique used in LANs. There are also good arguments to adopt the reverse terminology. See also [13].

innovative information transport systems. The current long-range plan of the telephone industry is to evolve the existing copper plant, a multi-billion dollars investment[3], to provide medium speed (for example, 64 kb/s, circuit switched) data communications to all telecommunications users in large business, campus, small business, and finally into the home. The general strategy is to first replace the analog telephone switching systems by digital switches (central exchange or PBX) and then to digitalize the telephone sets and concurrently the local loop. Optical technology will then progressively displace copper in the local plant, starting from the central exchange toward the subscriber, and accordingly, it will help create new services and hold down the cost of basic services. With this approach data communication services will be integrated with the conventional voice communication services. The objective is to deploy progressively an Integrated Services Digital Network (ISDN) [3]. The ISDN concept is an international one [4]; it has been the strategy of the telecommunications industry for more than a decade; it is currently under study by the Consultative Committee on International Telephony and Telegraphy (CCITT) [5]. In this article, ISDN refers to the abundantly documented developments aimed at using the existing copper plant to provide a medium speed digital service (for example, 2B + D = 144 kb/s) through a copper-based information outlet. When referring to ISDN long-range plans for a fiber-based information outlet, the expressions "second generation ISDN" or "B-ISDN" (Broadband ISDN) will be used. With the current evolutionary plans, it will take a long time for a B-ISDN to happen. The industry need for a wideband (multi-megabit per second) Information Outlet is discussed in the section titled "Needs and Problems: The Information Outlet."

ULAN will be a new long-range competitive approach to ISDN first and second generations local information *transport* services. ULAN is aimed at a more rapid and more efficient deployment of fiber in metropolitan areas through fundamental changes in transport network architecture. The strategy is to start from scratch, designing a Universal Local Area Network (ULAN) by exploiting an optimum combination of two mature technologies, VLSI [6] and fiber optics [7], and by coordinating and focusing current efforts in the development of token rings [8]. The principle and the architecture of the ULAN approach are described in the sections titled "System Requirements" and "The ULAN Design." The goals are to transport digital information (data, voice, image, low definition video) with more flexibility, higher performance and at lower costs than will be possible with ISDN by upgrading the copper-based telephone network [9], and to provide the new wideband services much faster than could be achieved with the second generation ISDN. ULAN will bypass the copper plant. In metropolitan areas, ULAN will be deployed following the backbone strategy of CATV networks. A more practical description of ULAN capabilities, with some comparison with ISDN's, is made in "Comparison Between ULAN and ISDN," where three somewhat arbitrary examples are taken: support of one million telephone subscribers in a metropolitan area; integrated voice, DP, image traffic through an Information Outlet; ULAN's support of 100,000 simultaneous players in an interactive TV game. Critical standardization issues raised by ULAN are discussed in "Need for a High Level Interconnection Standard." Finally, the section titled "ULAN Project" proposes an approach to accelerate the necessary standardization process and to coordinate the required developments.

It is expected that until the end of the century, ULAN will compete with and progressively replace the ISDN local transport services. During this period of transition, ISDN and ULAN will be interconnected by gateways, with ULAN supporting an increasing number of wideband applications that ISDN will not be able to support efficiently, due to the limitations of an architecture initially constrained by the capacity of the local copper plant.

ISDN and ULAN differ in several aspects:

- ISDN's architecture is optimized for voice; ULAN's architecture will be optimized for data.
- ISDN is developed mainly by the conventional telephone industry; ULAN will be developed mainly by the data processing industry.
- Both ISDN and ULAN want to provide the "Information Outlet": ISDN's is based on copper; ULAN's will be based on fiber.
- ISDN will progressively deploy fiber from the central office toward the subscriber; ULAN will start deployment of fiber at the subscriber level in a multiplicity of LAN's which later on will be interconnected.
- ISDN's second generation (fiber-based) information outlet will support digital TV video broadcasting and switching; ULAN will not support TV quality full-motion video. It will use conventional analog CATV cable-casting services.
- ISDN and B-ISDN will lead to centralized systems, and will tend to cluster network resources; ULAN will be fully distributed, and is expected to create a more competitive environment for equipment and for application services.

In summary, by the year 2000, 90 percent of the digital equipment will attach to their local networks at 64 kb/s (circuit switched) if the ISDN strategy prevails [10], and at 50 Mb/s (block switched) if the ULAN strategy prevails: ISDN is evolution; ULAN, in comparison, will be a revolution.

Needs and Problems: The "Information Outlet"

By developing and applying new technologies, human society keeps evolving toward greater efficiency. Today, the information society is emerging. It is aiming at a more efficient utilization of information in all of its operations: office automation; plant automation; checkless, cashless, paperless society; home interactive applications, etc. The goal is to open up the wide market of

[3]The local loop plant investment of the former Bell operating companies is about $45 billion, or almost half of their total assets.

information services to every individual and to every human-controlled activity.

Technology is applied to progress in two complementary fields: "information handling" (collection, presentation; storage, retrieval; treatment) and "information transport" (transmission; switching). The last three decades have witnessed remarkable advances in information handling through steady improvements in VLSI, magnetic technology, and so forth. Fiber optics, combined with VLSI, is going to play a key role in the next decades to bring similar advances to information transport. For long haul networks, single-mode fiber will soon be used to transmit at very high data rates (for example, 1.7 Gb/s[4] to 20 Gb/s[5]) over very long unrepeated distances [7]. Shortly the available transmission capacity of the long haul network will far exceed the national needs, unless the local bottleneck created by the embedded copper plant is by-passed by new fiber based wideband networks, which can be deployed *quickly,* that is, at low costs and with low initial investments.

Designing a new local information transport network taking advantage of fiber optics is a very complex task from several points of view: business, regulatory [12], legal [13], technical, social, etc. Independently, the design has to respond to the needs of the end users. These needs are difficult to express because they are based on innovative applications not yet developed (for example, based on high definition image, graphics, distributed databases), or on a new social environment not yet created because of a lack of technology support (for example, work from home).

A practical definition of the critical need characterizing information transport can be gained by an analogy. Let us reflect on the convenience with which we use the electric power generated by large power plants. This convenience is due to the existence of a *universal power outlet.* A sophisticated electric transport network terminates at a multiplicity of identical power outlets, whose universality has three complementary aspects:

1) A well-defined interconnection and service standard (117V, 60Hz);
2) Accordingly, all equipment is designed to that standard and, once a power outlet has been installed, it can provide power to a wide variety of electric equipment (lighting, home appliances, tools, and so forth);
3) The power outlets are ubiquitous and can be found in any environment (large or small business, government office, residence).

TABLE I
COMPARISON OF THE STANDARDIZATION OF THE POWER AND INFORMATION OUTLETS

Power Outlet	Information Outlet
Cu/Al	Fiber optics/Cu
2-3 wires, flat/cylindrical prongs	1-2 fiber + 0-1 copper pair connectors
AC/DC	block/circuit switching
117/220V, etc.	64 kb/s - 50 Mb/s, etc.
60/50 Hz	BSC/HDLC/X25s
5A-20A, etc.	block size/holding time

By analogy, for user convenience and to stimulate competition in equipment and in integrated services, a new sophisticated information transport network should have:

- As a design goal, a universal, ubiquitous *Information Outlet,* a technical challenge;
- As a market objective, the widest and quickest *penetration* possible, a legal and a regulatory challenge;
- As a standard objective, a unique and complete set of interconnect *specifications,* a national and international standardization challenge.

The power/information outlet analogy can be pushed further to analyze the main issues: technical complexity of the standard; difficulty for the industry to work jointly toward a new standard. Table I proposes a correspondence between characteristics of the power outlet and of the Information Outlet. In both cases, for equipment to be able to connect with the network requires compliance with all the parameters which define the outlet standard. Two groups of parameters are shown in Table I. The first three parameters concern hardware; they raise similar and relatively simple questions for both types of outlets. The last four parameters are more complex to standardize because they have profound system implications. The AC/DC versus block/circuit switching analogy correspond to the key choice between LAN and PBX, and more generally between ULAN and ISDN. The voltage versus bit rate analogy shows a network parameter which is easy to change as long as the next parameter, the frequency or the link protocol is not changed. Finally, the amperage versus block size/holding time analogy is an example of a parameter which is defined by its maximum value to protect the network, and which restricts utilization of the outlet by certain equipment.

Generally, the problems raised by the Information Outlet are far more complex than for the power outlet, but also the number of available solutions is greater, which tends to create a Tower of Babel[6] environment in the standardization process. The business challenge is to show, using the analogy with the electric utility, that, in the long range, an Information Outlet would be of great

[4]1985 announcement by AT&T of the F-1.6G plans for 1.7 Gb/s links for the 1987-1988 time frame [11]; 1986 field trials by British Telecom of an experimental system operating at 2.5 Gb/s, and by Nippon Telegraph and Telephone Corp. of an experimental system at 4.8 Gb/s.

[5]By wavelength multiplexing of ten 2 Gb/s channels on one fiber, in 1985, AT&T Bell Laboratories have demonstrated feasibility of transmitting 20 Gb/s on one single mode fiber over a distance of 50 km without repeater.

[6]Note that, this time, it is not brick and mortar technologies which are involved, but highly sophisticated technologies to turn sand into ultrapure silicon and glass.

value to the end users as well as to the equipment and service industry. There is more value in having a unique, ubiquitous and universal[7] approach than in having a multiplicity of incompatible offerings tailored to each particular situation. *What would have happened if each electric power company had its own socket and frequency voltage?*

Another dimension to the difficulty of designing the network which will support the Information Outlet is the rapid evolution of the equipment market. Figure 1 compares the evolution of the numbers of telephone sets, television sets and data terminals in the United States from 1880 to 2000. It can be expected that by the year 2000, there will be between 100 and 200 million microprocessor based devices attached to networks. The logarithmic vertical scale used in Fig. 1 hides the problems of market penetration. Figure 2 shows the same curves with a linear scale for the 1970-2000 time frame. It indicates that:

- *Now* is the time to address the critical issue of interconnect standardization in order to obtain information transport compatibility and,
- Despite the fact that initially the new wideband networks will be deployed in the business environment, the design has to be optimized for the residence market, which is more cost sensitive. As soon as the home computer market develops, it will rapidly require the same wideband capabilities as the business environment. The availability of low cost wideband communication to the residence market will signal the birth of the Information Society.

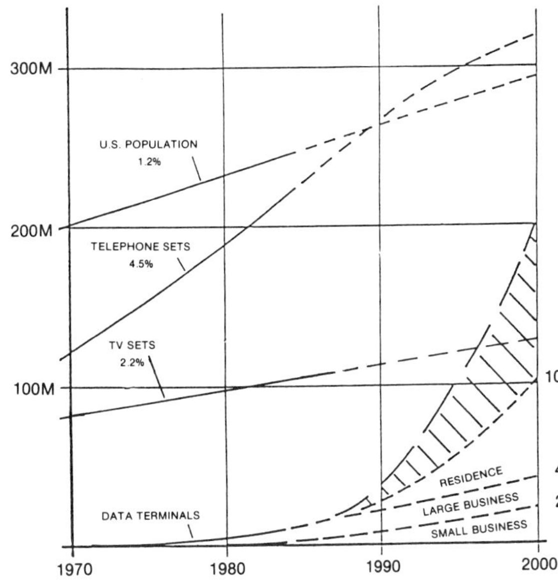

Fig. 2. A scenario for data terminals market penetration (percent figures indicate annual growth rates).

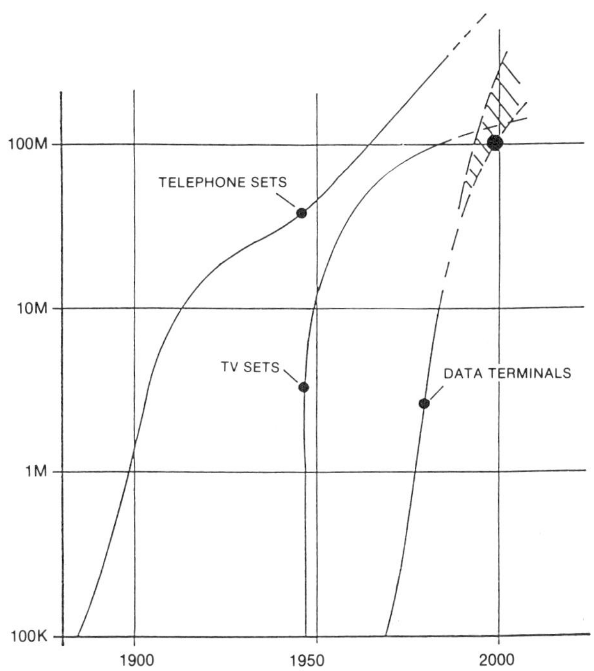

Fig. 1. Number of telephones, TV sets and data terminals in the United States 1880-2000 (Sources: FCC Common Carrier Statistics. Statistical Abstract of the United States. Data extrapolated beyond 1985 in order to indicate trends).

System Requirements

The new technologies of the last decade, VLSI and fiber optics, create the opportunity to combine the functional advantages of a centralized system (privacy, network control, and so on) with the deployment advantages of a decentralized system (low initial investment, expandability, reliability, serviceability). The ULAN goal is to deploy the new wideband networks rapidly, and consequently to choose the most cost-effective design to respond to the users' needs. For this reason, ULAN will not support present and future television quality video through the Information Outlet. This is the main difference with the service plans currently considered for a second generation ISDN.

Providing full connectivity two-way television quality video requires a dedicated wideband link (70 to 140 Mb/s circuit switched) for each user and thus a network with a star topology. Several broadband (B-ISDN) research projects have implemented star networks based on fiber optics.[8] The corresponding investment is high (estimated to be at least $3000 per subscriber) due to the complexity of the cable plant and also due to the cost of the

[7] Universal is used here, in opposition to "multiversal," which would correspond to a transport network capable of supporting a multiplicity of standards [14] in an "open system." A multiversal power utility would offer a multiplicity of services: 117 V, 60 Hz; 220 V, 50 Hz; 50 V, DC, etc. and charge the end users according to the costs of providing each of these services. A multiversal television network would broadcast in NTSC, PAL, and SECAM to support all vendors' equipment.

[8] Japan 1972-1979, Hi-OVIS; Canada 1977-1979, Yorkville; Germany 1979-1985, BIGFON, Hannover, Berlin, and so forth; France 1980-1984, Biarritz; and so on.

broadband circuit switching equipment. In order to match the fiber bandwidth, the capacity of the switches will have to be at least three orders of magnitude higher than current voice digital switches. When compared with the investments in a conventional cablecast network (the CATV industry invests between $300 and $400 per subscriber), the fiber star network investment is too high[9] to be justified by the incremental value of the two main high quality two-way video applications made possible by full circuit-switched connectivity [15]: video-on-demand, which will have to compete with the rapidly expanding video-cassette and video-recorder markets; videotelephony, which still has to prove market acceptance and can be implemented with much lower bandwidths. Finally, the CATV technology has a substantial advantage over ISDN for broadcast TV quality or high definition television, it is an analog technology which is signal-compatible with the more convenient but spectrum limited off-the-air broadcasting technology. Although B-ISDN could make use of analog TV signals [16] to provide CATV services in a metropolitan area, instead of the presently contemplated digital techniques, it would then lose most of the advantages of integration of the information transport services.

The ULAN strategy for television quality video in the office and to the home is to use the CATV technology (coaxial cable or fiber [16, 17] in a physically separate video broadcast network with limited video return capacity. The two networks, ULAN and CATV, can be installed simultaneously; they can be logically integrated, for example, for television interactive applications. An example is discussed at the end of "Comparison Between ULAN and ISDN." Concerning video and image communications, the Information Outlet will support full connectivity for high resolution and, later, high definition image (such as, catalog shopping) and compressed full motion video for special applications (for example, video conferencing, videophone).

Broadband LAN's (Sytek, Wangnet, and so forth) represent an intermediary solution: they use the analog bandwidth available in two-way CATV networks (450 MHz) to transmit several frequency division multiplexed wideband digital channels (typically 1 to 6 Mb/s in a conventional 6 MHz channel). General cablecast television channels, a few two-way video channels and high-speed data channels are then available by tapping the cable; with the proper access method the digital bandwidth can be shared among a large number of devices. The broadband approach is not acceptable for the long-range objectives of ULAN:

1) Broadband LAN's are hybrid analog/digital systems, with the relevant *cost* penalty and performance limitations;
2) The topology is not adapted to large digital networks: the tree structure is ideal for broadcasting signals from the head end toward a multiplicity of users, but it presents a *worst case topology* for the return channels from the users toward the head end;
3) For a large network (metropolitan), the tree topology presents an *availability* exposure, which is not acceptable for data or voice applications; it combines the weaknesses of a centralized system (vulnerability of the head-end), and of some distributed systems (single physical path between two users);
4) The architecture cannot be extended to very high performances *(Gb/s)* for future integration of main-frames[10] into the network;
5) Finally, broadband LAN's are based on information broadcasting and thus present a *privacy* exposure.

Privacy and security are two complementary, important attributes of the information networks of the future. An example of privacy violation would be for an individual to monitor with whom and when his next door or next office neighbor communicates, such as, bank, lawyer, doctor. An example of security violation would be to access and eventually modify someone else's bank account. *Privacy* mainly relates to the design of the transport network, the bottom-up approach to network design, *security* mainly to the design of the systems network, the top-down approach to network design [18]. The privacy exposure of broadband networks is not specific to this technology; in fact, it is common to the four LAN media access control standards (802.3, 4, 5, 6) [19] being worked on within the IEEE 802 committee [20], as well as to all Metropolitan Area Networks (MAN) proposed thus far [21]. The four IEEE standards are based on *logically passive* network: all the intelligence (network control) is in the attached devices, which *must* all have access to all the information circulating on the bus or ring, a privacy exposure. Passive LAN do *not* respond to the functional requirements of the ISDN/ULAN market.

The objective of ULAN is to take advantage of two mature technologies, VLSI and fiber optics, to deploy block-switched decentralized local transport networks, functionally equivalent to the current circuit-switched centralized star networks. The star topology has been traditionally supported by the telecommunications industry, because it makes it easier to control: access to the network (privacy, billing); traffic (load control, routing); maintenance (network monitoring, diagnostics, re-routing, and so forth). A decentralized local transport network is easier to deploy:

- It requires lower initial investments;
- It is easier to expand;
- It is less vulnerable to catastrophic failures (fire, earthquake).

[9]As discussed in the section titled "Need for a High Level Interconnection Standard," one of the objectives of the first phase of the ULAN Project will be to make a comparative economic analysis of ULAN and of ISDN first and second generation deployment philosophies using the same cost parameters.

[10]The total channel capacity of a large mainframe is close to (100 channels) × (3 megabytes per second) = 2.4 Gb/s. In actuality, the average utilization of a channel is generally less than 10 percent.

TABLE II
COMPARISON OF ULAN AND ISDN SERVICES

	ULAN	ISDN
Basic Services:		
• Information Outlet	• 50* Mb/s (unique)	• 8, 16, 64 kb/s, T1, T2, etc.
• Hierarchy	• 500* Mb/s, 5* Gb/s	• 500* Mb/s, 5* Gb/s
• Traffic	• Bursty	• Continuous
Target Services:		
• Voice	• Yes (Packetized)	• Yes (64 kb/s)
• DP Equipment	• Integrated	• Transparent
• High Definition Image	• Yes (50 Mb/s)	• Yes (64 kb/s)
• Low Definition Videophone	• Yes (Packetized)	• Yes (64* kb/s or T1)
• Videoconferencing	• Yes (Packetized)	• Yes (T1*)
• Broadcast TV	• No (Use CATV)	• Yes** (fiber 100* Mb/s)
• TV on Demand	• No	• Yes** (fiber 100* Mb/s)
• TV Quality Videophone	• No	• ?

*Typical
**B-ISDN (fiber local loop)

A blocked switched local transport network is the best design for a distributed system:

- It integrates network control and transport functions;
- It integrates processor controlled equipment into the network;
- It makes more efficient use of the transmission media.

A general comparison of resulting ULAN and ISDN services is given in Table II.

The ULAN Design

The ULAN design is based on a hierarchy of *active* bidirectional token rings with a global integrated block switching protocol:

- Block switching, as discussed in the next section, is well adapted to the bursty traffic generated by a wide range of new applications [8, 22]:
- Token passing gives fast and deterministic access to the full bandwidth of the shared media:
- Token passing is not limited in performances (for example, bandwidth × distance product limitation for IEEE 802.3 or Ethernet's CSMA/CD):
- Token passing is the most efficient access method [22a], for a distributed system:
- Token passing can be generalized to a hierarchical metropolitan area [23].

The objective is to deploy a highly available wideband information transport network with a level of investment comparable to CATV investment ($300 to $400 per subscriber). The design will be optimized:

- To reduce hardware costs by reducing component counts;
- To reduce maintenance costs by real time self-diagnostics and automatic reconfiguration, and by systems diagnostics; and,
- To reduce installation costs by organized sharing of transmission facilities.

Reduction of component counts is illustrated in Fig. 3, which shows how to pass from a star to a bidirectional ring topology. A conventional central switch (SW) and its controller (C) are represented in the center, with a line terminal module (LT in CCITT terms [24]) controlling the link AB which terminates at a network termination unit (NT) for connection of the user equipment. Details show two wall-mounted NT's with the information outlet jack. The concept is to *decentralize the functions performed by the switch* in the NT and into distributed controllers. For this purpose, the LT end A of the link (generally two copper pairs or two fibers) is directly connected to another NT (A′) to form a ring. With the new structure the amount of wiring is dramatically reduced and there is no need for a central switch. The

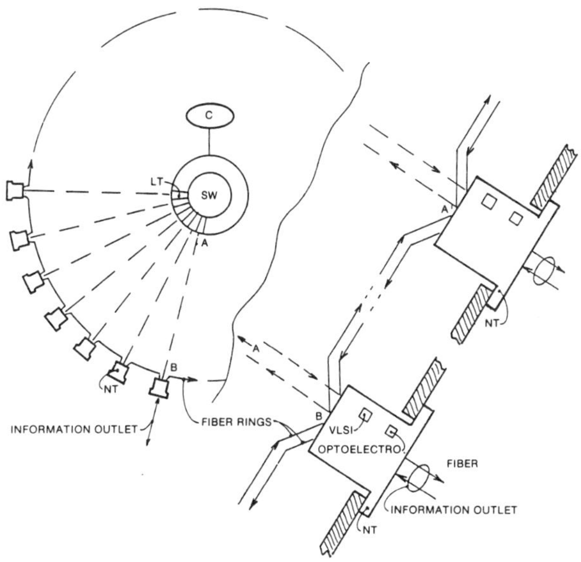

Fig. 3. First cost advantages of active *ring/star, reduced* component counts: *no switch, less wiring, all controllers connected to the network by a single port, lower initial investment.*

network control functions are performed by controllers attached to the ring, at low cost, because with a single port they can monitor all the traffic and access all NT's. The important implementation savings are realized at the cost of average performances: since all attached devices share the same transmission medium, the full bandwidth of the link can be allocated to an Information Outlet only for a short period of time. The ULAN design will use fiber optics for all the links, so the link data rate will be limited at the low end by the performances of low cost VLSI technology (probably in the 50 Mb/s to 100 Mb/s range). As discussed in the next section, the traffic handling characteristics of the network are well matched to the bursty traffic generated by the equipment and the applications, and also, to the intermittent traffic generated by voice.

When a device X has data to send to a device Y, it requests connection (if not already connected) from the active network controller. The controller establishes the connection between X and Y by setting the proper parameters in their respective NT's. Once connected, the device X passes its message through the Information Outlet to its NT, which assembles a block of data with Y (and X) NT address. The NT address is a geographical address.[11] The NT waits (milliseconds)[12] for a token to grant access to the link and then it transmits the block. The block is passed from NT to NT, which act as repeaters and switches; when Y's NT recognizes its address, it delivers the data block to Y, thus completing the block switching function.[13] The control of the circulation of the block is implemented in a VLSI chip inside the NT; the presence of intelligence in the NT is the important feature of the ULAN *active* ring design. In current ring implementations (for example, IEEE 802.5 [19c] or IBM token-passing LAN [8]), which are passive, there is no VLSI chip in the NT; all the intelligence is in the attached device, or in a node controller, which has to listen to *all* the information circulating on the medium, for the system to work. With ULAN, the NT is an integral part of the transport network; it insures *privacy* by forwarding to the attached device only those data blocks which are addressed to this particular NT.

Like in the other block-switched LAN's, IEEE 802.3-5, the stations X and Y have immediate[12] access to, and use of, the *full* bandwidth of the LAN when, and only when, they need it during a given session. The implications concerning traffic, response time and transport network processing load, are briefly discussed with three examples in the section titled "Comparison Between ULAN and ISDN." Once connected, X and Y do not have to request access and bandwidth from the controller for each transaction, like they would in a fast connect/disconnect circuit-switched environment (PBX, ISDN). Immediate access to the full bandwidth differentiates the token rings from the slotted rings (IEEE 802.6, Cambridge ring, NEC LOOP 6830, FDDI-II, and all rings interconnecting digital PBX's) which guarantee and set aside a fraction of the LAN bandwidth by pre-assigning time slots for the duration of a connection (permanent virtual circuits, the equivalent of DC). With a block switched system (the equivalent of AC), a session between X and Y can last for hours with only a few transactions per minute; due to the fast access and the wide bandwidth, the transport network does not affect the response time, the average load created by the XY communications is very low.

Block switching is well adapted to the highly bursty nature of the traffic generated by data, graphics and image applications. The technique can also be used for more continuous sources of information, like voice and video, as illustrated in the next section. When applied to packetized voice, block switching permits relaxing the synchronization requirements on the transport network; when applied to video-conferencing or videophone applications, availability of a variable bandwidth will lead to the development of compression algorithms more efficient than the current algorithms, which have to use a constant bandwidth (such as, a circuit-switched T.1 1.536 MB/s ISDN service) despite wide fluctuations in the amount of new information generated by the source.

Sophisticated connection control, switching, network management and maintenance functions can be implemented at very low incremental cost in each NT of an active ring, once have been resolved the power distribution problems. In the current LAN standards efforts (IEEE 802, FDDI), the station management functions are implemented (mostly in software) in the terminal, for example: allocation of isochronous channel in a slotted ring (802.6 or FDDI-II); claim token MAC frame or beacon MAC frame in IEEE 802.5 token ring; control of configuration management, fault isolation and recovery in FDDI. With an active NT, these transport network functions will be implemented very efficiently in the NT and in distributed network controllers, which will be an integral part of the transport network. Also changing configuration can be done electronically, without requiring an optical switch (FDDI, 802.6), with all the advantages in reliability, speed, and relaxation of the transmitter/receiver specifications.

---

[11] ULAN proposes to use an addressing scheme based on a hexadecimal extension of the present postal zip codes (an international trend). Another addressing philosophy is proposed in conjunction with the IEEE 802 standard, whereby each device has its own identity number. This number is used as an address on any LAN to which the device is connected. To understand the implications of this last addressing scheme, consider what would happen if one were to use Social Security numbers as mailing addresses: the task of the Post Office would be unmanageable.

[12] It is supposed that at regular intervals (for example, 5 ms at the lowest level), a new polling cycle is created. During a cycle, all stations have the opportunity to transmit at least once for sending control or synchronous information (such as packetized voice). Then, "immediate" means within a cycle, or 5 ms.

[13] It is hereby proposed to reserve the term "packet switching" to the technique used for communications between the upper peer layers of the architecture. Packet switching can then be implemented on a circuit switched (ISDN) or a block switched transport network (ULAN). The expression "block switching" was used in the late 1960s [25] to distinguish this switching technique from the traditional circuit and line switching. The natural combination of packet switching with block switching corresponds to an integration of the network control and transport functions.

In the same way, critical parameters can be monitored by an active network at very low additional costs, for example: NT power supply voltage, temperature, humidity, as well as node-to-node transmission delays, and so forth, thus enhancing the maintainability and serviceability of the network.

Reduction of maintenance costs comes from several features, among which are the reduction of the number of part numbers resulting from standardization and simplification of the design. This advantage would also be partially shared with a second generation ISDN. An important property, which is specific to ULAN, comes from the structure of the ring. In the preceding comparison with the star network, since each link AB had one transmit fiber and one receive fiber, the resulting decentralized network is in fact composed of two rings, each with its own direction of propagation. Distribution networks with a bidirectional ring topology have been used for several decades for electric power distribution, as early as 1963 for data communication networks [26] and are considered for fiber optics CATV franchise interconnect [16]. This bidirectional structure greatly improves network availability [27]. For example, it took Mexican Light and Power Co. only 72 hours to restore service to its 3.2 million customers, after the earthquake of September 1985. The transmission network has the topology of a ring surrounding Mexico City [28]. Two weeks later, the telephone networks were operating at less than 20 percent for international calls, because of the disruption of an important tandem switch. As shown in Fig. 4, when a link or a NT fails, the adjacent NTs reconfigurate the network into a single ring without interruption[12] of service and, at the same time, the controller can diagnose the type and the location of the failure for prompt repair. Two simultaneous failures on a ring are necessary to isolate a group of users, since there are two possible paths between NT's on the same bidirectional ring.

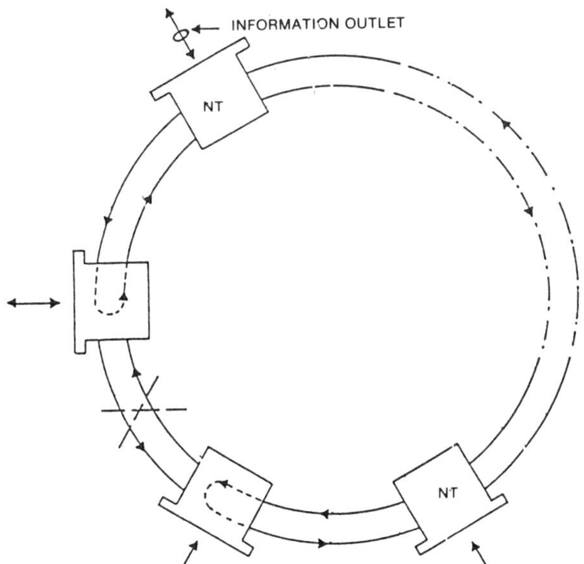

Fig. 4. Second cost advantage of active *ring/star, reduced* maintenance *cost: network diagnostics in each NT (sensors); automatic reconfiguration in case of failure (high mean time between failures); precise failure localization (low mean time to repair).*

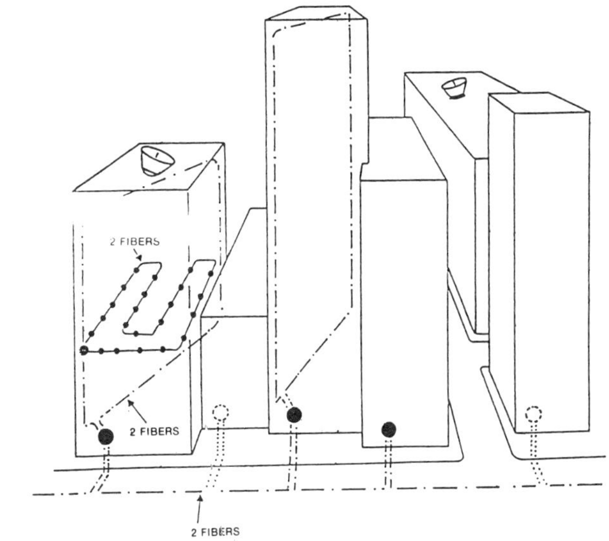

Fig. 5. Third cost advantage of active *ring/star, reduced* installation *cost: once the backbone cable has been installed, human intervention is required only where the new installation must be made.*

In practice, the number of NT's on a ring is limited (typically 100) for several reasons: availability, powering, traffic, performances, costs, privacy. A large network is made of a hierarchy of rings interconnected by bridges or by gateways which can act as stand-alone controllers and isolate a private network from a metropolitan network. The data rates at the different levels of the hierarchy will be determined by the peak traffic forecasts. Hypothetical examples are discussed in the next section.

Reduction of installation cost is due to a combination of the above features. Figure 5 illustrates how a distributed metropolitan network can be deployed. A user would wire one floor with a two-fiber cable going from Information Outlet to Information Outlet. Once the system is running, new Information Outlets can be inserted at low cost because, thanks to the high functionality of the NT, the only place where human intervention is required, is where the new NT must be installed. There is no need to pull new cable nor to install new hardware at the switch: place the new NT, cut the existing cable (the rings will reconfigurate into a single ring), attach connectors and connect the new NT (the ring will reconfigure back into bidirectional rings). The same technique can be used to interconnect one or two bridges on each floor to a two-fiber building ring, which in turn can be connected to the metropolitan network deployed under the same architecture with two fibers running from building to building.

In the same way, at the floor level, the equipment of the computer room of a particular user can be inserted in the local distribution loop and be interconnected

TABLE III
TECHNICAL COMPARISON OF ULAN AND ISDN LOCAL PLANT

	ULAN	ISDN
Architecture:		
• Switching	• Distributed (block)	• Centralized (circuit)
• Control	• Distributed	• Centralized
• Topology	• Bidirectional active ring	• Star/tree
• Alternate Path in Local Loop	• Yes	• No
• Hierarchy	• Yes	• Yes
Network Functions:		
• Network Access Control	• Yes in NT and Controllers	• Yes in CO or PBX
• Network Maintenance	• Yes in NT and Controllers	• Yes in CO or PBX
• Network Recovery	• Yes in NT and Controllers	• Yes in CO or PBX
• DP Host Integration	• Yes	• No
Network Deployment:		
• Business	• Progressive interconnection of clusters	• From central office toward user equipment
• Residence	• Fiber network deployed with CATV philosophy	• Existing copper plant (fiber in year 2000+)

internally[14] by higher speed rings [29], which under the ULAN architecture will conform to the same internal or external protocols as the low end rings (60Hz analogy) but will run at higher data rates (such as 500 Mb/s and 5 Gb/s, by analogy with the 13,800 V and 765,000 V of the electric distribution network). The ULAN design has to respond to two sets of requirements: at the low end, serving ultimately the residential market and, at the high end, integrating main-frames and large data-processing resources into the network. The ULAN protocols will be selected for maximum simplicity, in order to get low cost at the low end and high performances at the high end.

In all cases, thanks to its design features and its decentralized architecture, ULAN can be deployed on a large scale with a great flexibility, rapidly, and with low initial investments: stand-alone ULAN's will first be deployed simultaneously in the business environment, where a high density of workstations exist, requiring wideband for sharing of resources, and where the large DP centers are located, then these independent networks will be progressively interconnected. There will be no *need for a large central switch* with a bulky distribution frame, and there will be no need to *pre-install fiber optics cables with one pair pre-assigned to each of the potential future users.* The ULAN design takes full advantage of the very wide bandwidth of the fiber optics medium to reduce network deployment costs, and of VLSI to control shared access to that bandwidth in an efficient and convenient way, as depicted by three examples in the next section.

Comparison Between ULAN and ISDN

Both ULAN and ISDN want to define an Information Outlet. ISDN current objective is to provide a medium speed digital service through the existing copper plant, and later on to evolve the same architecture to support fiber optics bandwidths. ULAN objective is to define rapidly a fiber based Information Outlet to respond to the market demand for wideband bursty traffic. The comparison between ULAN's block switched services and ISDN's circuit switched services was summarized in Table II. A general comparison of the technical characteristics of the corresponding local plants is given in Table III. There are three main differences: 1) ULAN architecture has the potential for higher reliability, availability and serviceability because switching and control are fully distributed and the topology features an alternate path in the local loop; 2) ULAN is designed to handle bursty traffic and integrates DP host and equipment into the network; 3) ULAN will be deployed as clusters of independent networks, which will later on be progressively interconnected.

In order to illustrate *ULAN integration of voice into data,* and ULAN efficient support of the bursty traffic generated by new applications, three somewhat arbitrary examples are proposed. Table IV shows the voice traffic load on a metropolitan area network supporting one million subscribers, with a three-level hierarchy of rings counting one hundred nodes each. With block switching, bandwidth is used only when needed, that is, for voice, only when a subscriber talks; the resulting 60 percent average bandwidth requirement reduction has been included in the load estimates. Even with the unrealistically high voice traffic assumed (20 percent of the subscribers simultaneously engaged in telephone conversations, that is, an average busy hour local loop traffic of 7.2 CCS), the voice load at each level of the system is still low.

Clearly the economic justification for the deployment of ULAN cannot come from the conventional telephone applications alone. In order to displace and replace the existing local plant, computer related applications have to be developed which cannot be supported by ISDN currently planned copper-based Information Outlet. Two such examples are given in Tables V and VI: one

[14]The processor or equipment might be directly connected to the bidirectional ring, which will then replace the current I/O channels. (See [10])

Table IV
Example of Voice Support by ULAN: 1 Million Subscribers in a Metropolitan Area

Nominal Bit Rates	Level 1 50 Mb/s × 2	Level 2 500 Mb/s × 2	Level 3 5 Gb/s × 2
Number of users in subsystem:	100	10,000	1,000,000
Number of simultaneous conversations in subsystem[1]	10	1000	100,000
Total voice load in subsystem[2] (Mb/s):	0.25 × 2	20 × 2[3]	600 × 2[4]
Percentage of systems bandwidth used by voice:	0.5 percent	4 percent	12 percent

Assumptions (somewhat arbitrary):
[1] Conversation between two subscribers
[2] (64 kb/s per channel) × (40 percent talking time), overhead neglected, both rings active
[3] 80 percent of level 1 traffic transiting through level 2
[4] 30 percent of level 2 traffic transiting through level 3

Table V
Comparison of Traffic Patterns: Voice, Data Processing, Image

	Voice	DP	Image
Application:	Conversation	Program	High Definition
Amount of Data:	25 kb/s[1] × Duration	64 KBytes	1 MByte
ULAN Level 1 Transfer Time:	Duration	Less than 0.1 Second	Less than 1 Second
ISDN Transfer Time:[2]	Duration	8 Seconds	2 Minutes
Request Repetitive Rate:	Continuous	1 Every 10 Minutes	1 Every 10 Seconds
Relative Load on ULAN:	100	3	320
Percent of ULAN IO Capacity:[3]	0.05 percent	0.002 percent	1.6 percent

Assumptions:
[1] (64 kb/s) × (40 percent talking time)
[2] With a 64 kb/s B channel
[3] Averaged over a long period of time

compares wideband applications as viewed by one end user; the other shows the advantages of ULAN in a highly interactive environment.

Table V illustrates the *bursty traffic patterns* generated by data processing (DP) and image applications, compares them with the continuous traffic pattern generated by voice and finally compares how ISDN and ULAN support both. For the DP application, it is supposed that a 64 kBytes program is called by a user who will run it for 10 minutes. For the image application, a non-compressed high definition image is supposed to be called by a user on the average every 10 seconds (such as, catalog shopping). The table shows that:

Table VI
Example of Interactive Application: 100,000 Simultaneous Players in a TV Game

- PHASE 0: Sign up (5 minutes)
  - — ULAN controllers bind Players/TV game organizer
  - — Processing average: 350 calls per second
- PHASE 1: Question asked on TV screen
- PHASE 2: Player response
  - —All players respond (5 Bytes within 1 sec)
  - —IO assembles 25 B block to game organizer

—Traffic load:	level 1	level 2	level 3
kb/s:	2	200	20,000
Level utilization (up)	0.002%	0.02%	0.2%
Level utilization (down)	20%	2%	0.2%

- PHASE 3: Game organizer response, and so forth

1) The load generated by bursty applications can be much smaller or much larger than the average load generated by voice.[15] The ULAN design is well adapted to handle bursty traffic and to integrate voice into data.
2) Fast transfer times (implying short response times) are the main advantage of a wideband service. ULAN sub-second transfer times will be unnoticed by the user in the DP and image applications. First generation ISDN transfer times will be marginally acceptable for the DP application, but for the image application the two-minute transfer time will be totally unacceptable.

The last example shows the efficiency and short response times which can be expected from a distributed block switched transport network serving a large number of users. The scenario (Table VI) is an *interactive TV game* with 100,000 simultaneous players in the metropolitan area environment used in the first example. In a sign-up phase the game is introduced on TV and the

[15] The very low utilizations indicated on the last line of Tables IV and V show that the challenge is not so much to develop sophisticated models comparing LAN access methods, but rather to understand the applications which will generate the system's traffic. It also supports ULAN's philosophy of selecting protocols "for maximum simplicity, in order to get low cost at the low end and high performances at the high end."

conditions for participation are explained. Players request access to the computer of the game organizer. The (distributed) network controllers establish the necessary connections and eventually provide a billing service to the game organizer. If the sign-up phase lasts for 5 minutes, the network controllers have to process on the average 350 calls per second, that is, an average of 3.5 calls per second each, if the controllers are located in the level 2/3 bridges. After that, the role of the controllers during the game is eventually limited to traffic monitoring for billing purposes, and later on to disconnection when the game is over. As soon as the first question has been asked on the TV screen, a one second burst of responses is forwarded to the game organizer. With five byte responses, resulting in 25 byte blocks assembled by the IO, the traffic load at levels 1, 2, and 3 is less than .2 percent for all rings except for the level 1 and 2 rings to which the game processor(s) is connected. On these two rings the utilization is 2 percent at level 2, and eventually 20 percent at level 1. Similar traffic patterns are created when the game processor(s) responds to each player. Very short response times can be obtained for each question/answer/response cycles for several reasons: the connections have been established during the sign-up phase; all the messages are block multiplexed and enter/exit the game processor(s) through a single port, in an orderly fashion; the game processor can be implemented with a coordinated array of processors which are individually assigned to a particular set of players at connection time for the duration of the game.

Need for a High Level Interconnection Standard

To get from the concepts of an Information Outlet and behind it a decentralized fiber optics network, to the reality of rapidly deploying such a network requires an unusual planning effort. Recently Kenneth H. Olsen, President of Digital Equipment Corporation, predicted that "The most important contribution of computer applications is going to be networking. It is the way we are going to tie everything together . . . To be successful we first of all must know what we want to do, then we have to be ready to have a discipline to make sure we do it correctly and we do it accurately." He said he believed that making a large network work requires a level of discipline that few organizations of mankind have ever had to face.

End users justify their data processing and telecommunication systems by the number of applications that their systems can support. With the rapid cost/performances improvements driven by the evolution of the technology, there is a growing open variety of applications which can be developed to improve upon the efficiency of society. In order to discipline the implementation of these applications, the International Standards Organization (ISO) [30] has defined a layered network architecture model,[16] for Open System Interconnection (OSI) [31]. Seven sequential functional layers have been

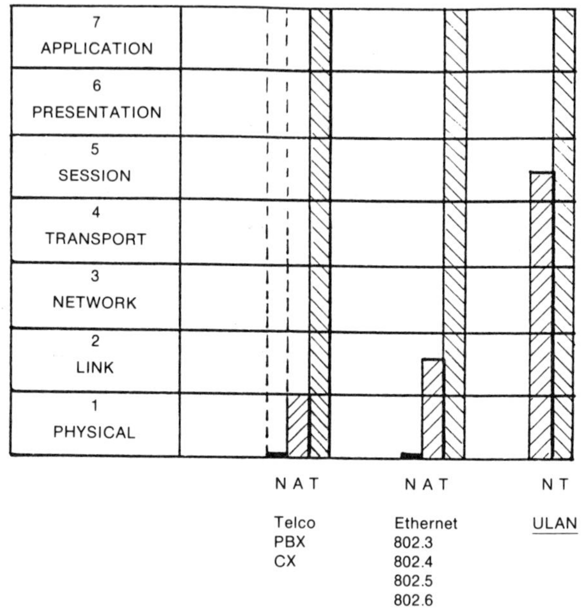

Fig. 6. ISO seven layer OSI architecture and repartition of functions in network (N), adaptor (A) and terminal (T) for three environments.

defined with the applications at the top and the transmission medium at the bottom (Fig. 6). The problem of "tying everything together" can be partitioned in two:

1) System architecture, where one proceeds from the applications down, and,
2) Transport architecture, where one proceeds from the physical layer up.

Where the two approaches meet defines the functions respectively performed by the transport and the system networks, and consequently, the standard associated with the Information Outlet (117V, 60Hz analogy). Information network optimization requires that some of the functions traditionally performed by the system migrate into the transport network. Consequently, the functionality and the complexity of the network/equipment interface (Information Outlet) has to increase.

Openness is the main requirement for the top layers, due to the open multiplicity of applications.[15] The still growing diversity of data applications and the wide variety of possible implementations have created deep incompatibilities between systems: operating systems, data base structure, character sets, formats, and so on. In the ULAN time frame, most of these incompatibilities in the higher levels will find a solution, thanks to two new technological achievements, which will improve the mobility and transportability of software:

1) Very low cost storage (64 kBytes to 1 MByte chips [6])
2) Wideband data communications (LAN's: multi-Mb/s at very low cost)

By contrast with the requirement for openness and variety in the upper layers, that is, in the system

[16]The layering of the OSI model can be viewed as an architecture for network architecture. In a similar way, Maxwell's equations encompass both DC and AC.

architecture, usage convenience and flexibility require *uniqueness* and *standardization* in the lower layers in order to achieve full connectivity in the transport architecture. By prompting the concept of Information Outlet, one must obtain *universality*[7] from the bottom up in the transport network.

Figure 6 shows a functional comparison of three typical distribution networks: the telephone network, Ethernet or any current IEEE 802.X LAN's, and ULAN. In each case, the functions performed by the network (N), the adaptor (A) and the attached terminal equipment (T) are shown. In all cases, the terminal has to support the functions of the seven layers, since it is always connected to the network or an adaptor (such as, modem) by a physical connection. In some instances, the adaptor is built in the terminal. When the degree of functional integration of the terminal equipment into the transport network increases, the highest layer of interaction of the equipment with the network raises. The figure shows the functional evolution of the network/equipment interface. With the current analog telephone network, once the proper switching equipment has been activated and a connection has been established (dotted lines), the network is fully transparent and appears to the modem as an ubiquitous analog information transport medium (solid bar). The DCE/DTE (Data Circuit-terminating Equipment/Data Terminal Equipment) interface covers the mechanical, electrical, functional, and procedural characteristics of layer 1, which are necessary to exchange bit streams across the interface. With the LAN's currently developed within the IEEE 802.3-6 committees, the network is passive, the block or circuit switching function is implemented in the equipment or in an adaptor controlled by the equipment. Again, the network is a transparent medium for the equipment (solid bar); the link protocols are designed for direct terminal-to-terminal communications.

As shown in Figure 6, with the ULAN approach, the terminal equipment interacts with the network directly for all the functions of the lower layers up to the middle of the Session layer (level 5);[17] for the upper layers, the terminal interacts directly and transparently with the equipment it is connected to, as usual. For example, with the ULAN block-switched local transport sub-system, voice and video applications will be handled by layers 6 and 7 of the OSI model; layers 1 to 5 will handle digital blocks of data, image, voice or video information packets in a transparent and open fashion as described in the preceding section.

Table VII compares ULAN objectives with other standardization effort currently under way for fiber based LAN's. From a design point of view, the differences between an active and a passive transport network are fundamental. For example, for an active network, two kinds of protocols have to be specified: internal protocols for NT-to-NT communications; external protocols for equipment-to-network connection. The external protocols correspond to the Information Outlet standard. From a VLSI design point of view, native support (within the equipment) of a passive network by an equipment requires only one chip (or a set of chips). For an active network, like ULAN, a similar chip is required for native support of the Information Outlet standard (external protocol) by the equipment, and inside ULAN's NT, one has to design the counterpart of this chip, plus the duplicated token control part (internal protocol), plus all the necessary network control functions. The design of an active network is far more complex than the design of a passive network.

[17]This approach might seem to deviate from the OSI model. In fact, it only deviates from the usual concept of a functionally transparent transport network; it exemplifies the integration of the equipment into an active network.

TABLE VII
COMPARISON OF ULAN WITH CURRENT FIBER-BASED LAN's

	ULAN	802.X	FDDI
Network Functions:			
• Mode	• Active (NT)	• Passive (all)	• Passive
• Access, Maintenance, Recovery Control	• Network (NT, Controller)	• DP Equipment	• DP Equipment
• OSI Level of Equipment/ Network Interaction	• 5	• 0 (medium)	• 0 (medium)
Services:			
• Basic data rate	50* Mb/s	• 1-4-10 Mb/s	• 100 Mb/s
• Hierarchy	Yes (500* Mb/s, 5* Gb/s)	• No	• No
• Basic Switching	• Block	• 802.3-4-5 Block • 802.6 Circuit	• Block • FDDI-II (Block + Circuit)
• Access Method	• Token	• 802.3 CSMA/CD • 802.4-5 Token • 802.6 Slot	• Token • FDDI-II (Slot)
Deployment			
• Campus	• Yes	• Yes	• Yes
• Metropolitan	• Yes	• No (802.6-?)	• No
• Number of Users	• 10^6*	• 10^3*	• 10^3*

*Typical

ULAN Project

A consensus is building up indicating that ULAN should and could happen: the needs are clear; the technologies are here; the know-how and the know-why are at hand. In summary, ULAN would result from the token-ring LAN's currently being proposed if:

- The adaptor (NT) were made logically active, that is, more intelligent (using VLSI);
- Optical fiber were used throughout;
- Internal standards were defined at the adaptor-adaptor interface;
- The same protocols were architected for several levels of performances;
- Information outlet specifications were established at the adaptor-terminal interface.

Nevertheless, ULAN and a fiber based Information Outlet might never[18] happen, because in every domain very complex issues have to be resolved. More specifically:

- *Technical:* feasibility of an active NT has to be proved; the concept of transparent transport network has to evolve, functions have to migrate from the equipment into the transport network; new applications of low-cost wideband communications have to be created (image, mobility of software, etc.), which have deep implications.
- *Market:* a critical mass must be gained to guarantee the rapid market penetration which is required to justify the necessary changes; the strategic value of the local loop, and the services it can provide, was one of the main technical reasons for the divestiture; ISDN has accumulated a considerable momentum; the challenge is not to compete with the second generation ISDN, but with the current plans for a copper-based information outlet.
- *Legal:* a well-coordinated cooperation of the vendors will be necessary to implement the ULAN concepts; complex legal restrictions will have to be levied to insure that a joint effort involving the major manufacturers fosters, rather than restricts, competition in information equipment and services.
- *Regulatory:* by intending to by-pass the embedded copper plant, ULAN proposes a phasing out of the existing telephone plant, which is more rapid than planned by ISDN. The implications have to be understood and the interests of the rate payers have to be protected.
- *Standard:* the Information Outlet will require precise *specifications* for the functions of the first five layers of the ISO model (fibers, connectors, modulation, bit-rates, addressing, error control, routing, connect/disconnect, synchronize, bind/unbind); the most complex part of the standards will be engraved in silicon rather than written in blue or red books; standardization of the Information Outlet will require coordination[19] of the efforts and support of a multiplicity of national and international[20] organizations.

In order to avoid a repeat of a costly and lengthy "battle of the currents," the information industry (users, carriers and vendors) needs to act now. Two steps are proposed: the first one mainly to accelerate the process; the second one mainly to reduce and share the costs.

1) *Creation of a new project,* the ULAN Project, which will act as a catalyst and will activate a reaction from the industry. The Project will:
   - Quantify, justify, and specify the needs of the end users[15] and of the LEC's for an Information Outlet and for the rapid deployment[9] of a Universal Local Area Network;
   - Prove feasibility of the ULAN concepts, by planning, developing and testing prototypes of network equipment, terminal equipment and wideband services;
   - Propose the specifications of the Information Outlet and specify the architecture of local information transport networks.
2) *Creation of a joint venture,* which will serve as a focal point to concentrate the efforts of the industry. The joint venture will:
   - Manage the ULAN Project;
   - Coordinate development among vendors;
   - Interface with national and international standards organizations.

Acknowledgments

The author is grateful to several sets of editors and reviewers, anonymous IEEE colleagues, for their encouragement and suggestions to bring this article to its present form.

References

[1] Charles R. Wright: "Taming the wild west with power: a tale of the world's first AC plants," *The Institute,* vol. 9, no. 12 p. 11, Dec. 1985.

[2a] John Sodolski and Thomas Plevyak, Guest Editors, "Divestiture: Two Years Later," *IEEE Communications Magazine,* vol. 23, no. 12 Dec. 1985.

[2b] Edward A. Torerro, Issue Editor, "Hello again. The future of telecommunications," *IEEE Spectrum,* pp. 43–114, Nov. 1985.

[3a] J. J. Cochrane, W. E. Falconer, V. S. Mummert, and W. E. Strich, "Latest network trends," *IEEE Communications Magazine,* vol. 23, no. 10, pp. 17–31, Oct. 1985.

[3b] Richard P. Skillen, Guest Editor, "Integrated services digital networks," *IEEE Communications Magazine,* vol. 24, no. 3, March 1986.

[3c] M. Decina, W. S. Gifford, R. Potter, and A. A. Robrock, Guest Editors, "Integrated services digital network. Recommendations and field trials—I," *IEEE Jour. on Selected Areas in Communications,* vol. SAC-4, no. 3, May 1986.

[18] Or in the year 2010, i.e., half a century after the data processing industry reached the 1 billion dollars mark.

[19] Even within a single standard committee, for example, IEEE 802, functionally equivalent but actually incompatible solutions are developed (for example, 802.3, 4, 5, 6).

[20] It is important to avoid a repeat of the failure of the international standard organizations to standardize the power outlet.

[3d] Same Guest Editors as [3c], "ISDN: technology and implementation—II," *IEEE Jour. on Selected Areas in Communications,* vol. SAC-4, no. 8, Nov. 1986.

[3e] Dana Blankenhorn, "Waiting for ISDN," *Datamation,* pp. 83-88, July 1, 1986.

[4a] Two sets of seven articles each, on ISDN field trials in Canada, England, France, Germany, Italy, Japan and the United States in Ref. [3b] and [3c].

[4b] T. Irmer, "Our Task and Approach for the Development of an Advanced Information Society" in "Worldwide Trends Towards the ISDN—Facts and Trends," NTT International Symposium, pp. 39-47, Tokyo, Feb. 28, 1983.

[5] Eckart Hummel, "The CCITT—the streamlining of modern telecommunications developments in a worldwide forum," *IEEE Communications Magazine,* vol. 23, no. 1, pp. 8-11, Jan. 1985. Special issue on telecommunications standards.

[6] K. A. Pickar and J. D. Meindl, Guest Editors, "Integrated circuit technologies of the future," *Proc. of the IEEE,* pp. 1605-1798, Dec. 1986.

[7a] Paul S. Henry, "Introduction to lightwave transmission," *IEEE Communications Magazine,* vol. 23, no. 5, pp. 12-16, May 1985. Special issue on fiber optics.

[7b] Herwig Kogelnik, "High-speed lightwave transmission in optical fibers," *Science,* vol. 228, no. 4703, pp. 1043-1048, May 31, 1985.

[7c] Ian Garrett, "Towards the fundamental limits of optical fiber communications," *Jour. Lightwave Technology,* vol. LT-1, no. 1, pp. 131-138, March 1983.

[7d] J. Salz, "Modulation and detection for coherent lightwave communications," *IEEE Communications Magazine,* vol. 24, no. 6, pp. 38-49, June 1986.

[8] M. Willett, Guest Editor, "Token Ring Local Area Networks," *IEEE Network,* Premier Issue, Jan. 1987. Five papers on *passive* token rings.

[9a] Ian M. Ross, "Two years into the future—a view from AT&T Bell Laboratories," Ref. [2a], pp. 64-66.

[9b] Paul Wallich and Glenn Zorpette, "The innovation revolution awaits," Ref. [2b], pp. 104-110.

[10] Nguyen Q. Duc and Eng K. Chew, "ISDN protocol architecture," *IEEE Communications Magazine,* vol. 23, no. 3, pp. 15-22, March 1985. Special issue on telecommunications protocols.

[11a] Trudy E. Bell, "Technology '86 communications," *IEEE Spectrum,* pp. 49-52, Jan. 1986. Special issue on Technology '86.

[11b] D. A. Fishman, et al. "1.7 Gb/s lightwave transmission field experiment," *OFC '86,* PDP-11, Feb. 1986.

[12a] Mark S. Fowler, "The FCC and the aftermath of divestiture," Ref. [2a], pp. 28-31.

[12b] David B. Newman, Jr., Guest Editor, "Telecommunications regulation," *IEEE Communications Magazine,* vol. 25, no. 1, Jan. 1987.

[13] "Interview with Professor William Baxter," Ref. [2a], pp. 25-26.

[14] A. G. Fraser, "Towards a universal data transport system," *IEEE Jour. on Selected Area in Communications,* vol. SAC-1, no. 5, pp. 803-815, Nov. 1983. Special issue on LAN.

[15a] Pierre-Andre Probst, "Services and transmission in broadband integrated services local digital network," *Proc. FOC/LAN 85,* pp. 260-265, San Francisco, Sept. 1985.

[15b] Horst Ohnsorge, "Introduction and overview of broadband communication systems," *IEEE Jour. on Selected Areas in Communications,* vol. SAC-4, no. 4, July 1986. Special issue on B-ISDN.

[15c] David J. Harrold, Robert D. Strock, "The Broadband Universal Telecommunications Network," in [12b] pp. 69-79.

[16] Edouard Y. Rocher, "Applications of monomode fiber to local networks," *IEEE Jour. on Selected Areas in Communications,* vol. SAC-3, no. 6, pp. 897-907, Nov. 1985. Special issue on fiber optics for local communications.

[17a] John J. Prisco and Robert J. Hoss, "Fiber optics regional area networks," *IEEE Communications Magazine,* vol. 23, no. 11, pp. 26-39, Nov. 1985.

[17b] C. W. Lundgren and P. S. Venkatesan, "Applications of video on fiber cable," *IEEE Communications Magazine,* vol. 24, no. 5, pp. 33-49, May 1986.

[18] Dennis K. Branstad, Guest Editor, "Computer security," *IEEE Communications Magazine,* vol. 23, no. 7, pp. 10-55 July 1985 and Ibid., vol. 23, no. 9, pp. 22-46, Sept. 1985.

[19] See IEEE publications:

[19a] Carrier Sense Multiple Access with Collision Detection (CSMA/CD), ANSI/IEEE Std. 802.3-1985 (ISO/DIS 8802/3).

[19b] Token-Passing Bus Access Method and Physical Layer Specifications, ANSI/IEEE Std. 802.4-1985 (ISO/DIS 8802/4).

[19c] Token Ring Access Method and Physical Layer Specifications, IEEE Std. 802.5-1985 (ISO/DP 8802/5).

[19d] Metropolitan Area Network (MAN), IEEE Std. 802.6, (slotted ring), under preparation.

[20] E. J. Cohen and William B. Wilkens, *"The IEEE role in telecommunications standards," IEEE Communications Magazine,* vol. 23, no. 1, pp 31-33, Jan. 1985.

[21] Robert W. Klessig, "Overview of metropolitan area networks," *IEEE Communications Magazine,* vol. 24, no. 1, pp. 9-15, Jan. 1986.

[22a] Werner Bux, F. H. Closs, K. Kuemmerle, H. J. Keller, and H. R. Mueller, "Architecture and design of a reliable token-ring network," *IEEE Jour. on Selected Areas in Communications,* vol. SAC-1, no. 5, pp. 756-765, Nov. 1983.

[22b] A. Leon-Garcia, Guest Editor, "Network performance evaluation" with six papers on "Polling and token passing systems," *IEEE Jour. on Selected Areas in Communications,* vol. SAC-4, no. 6, pp. 775-878, Sept. 1986.

[23] Werner Bux and Davide Grillo, "Flow control in local-area networks of interconnected token rings," *IEEE Trans. Communications,* vol. COM-33, no. 10, pp. 1058-1065, Oct. 1985.

[24] CCITT Recommendation I.411: "ISDN user-network interfaces, interfaces structures and access capabilities." CCITT—red book, Geneva 1985.

[25] J. R. Pierce, C. H. Coker, and W. J. Kropfl, "An experiment in addressed block data transmission around a loop," *IEEE Int. Conv. Rec.,* pp. 222-223, March 1971.

[26] H. K. M. Grosser and F. J. Schramel, "Data transmission and switching equipment for the seat reservation of United Airlines," *Philips Telecommunication Rev.,* vol. 24, no. 1, pp. 13-24, Feb. 1963.

[27] C. S. Raghavendra and J. A. Silvester, "Double loop network architectures—a performance study," *IEEE Trans. Communications,* vol. COM-33, no. 2, pp. 185-187, Feb. 1985.

[28] Jorge Gutierrez Vera, "Back on line in 72 hours," *IEEE Spectrum,* pp. 41-46, June 1986.

[29a] John D. Crow, "Computer applications for fiber optics," *IEEE Communications Magazine,* vol. 23, no. 2, pp. 16-20, Feb. 1985.

[29b] ANSI X3T9.5 "Draft proposed American National Standard Institute Fiber Distributed Data Interface token

ring media access control," Document ANSI X3T9.5/83-16, Update Mar. 1, 1985.

[29c] Floyd E. Ross, "FDDI—a tutorial," *IEEE Communications Magazine,* vol. 24, no. 5, pp. 10-17, May 1986.

[29d] W. E. Burr, "The FDDI Optical Data Link," Ibid. pp. 18-23.

[30] Edward Lohse, "The role of the ISO telecommunications and information systems standardization," Ibid. Ref. [5], pp. 18-24, Jan. 1985.

[31a] John Voelcker, "Helping computers communicate," *IEEE Spectrum,* pp. 61-70, March 1986.

[31b] J. D. Day and H. Zimmerman, "The OSI reference model of architecture for Open Systems Interconnection," *IEEE Trans. on Communications,* vol. COM.28, no. 4, pp. 425-432, April 1980.

[31c] Hubert Zimmerman, "OSI reference model—the ISO model of architecture for open systems interconnection," *IEEE Trans. on Communications,* vol. COM.28, no. 4, pp. 425-432, April 1980.

Edouard Y. Rocher received a MS/Ph.D. degree from Ecole Centrale de Paris in 1956 and a Ph.D. in Solid-State Physics from Darmstadt Technische Hochschule in 1963.

He started his career in 1960 with the development laboratories of IBM-France, where he worked on PBX solid-state crosspoints. In 1968, he transferred to IBM's Thomas J. Watson Research Center, Yorktown Heights, NY, where he worked on LSI, and later on link protocols (SDLC). For loop and ring, he contributed the token passing protocol and the error recovery procedure. From 1972 to 1974, he was a member of an IBM corporate task force on local area networks. In 1980, he joined AT&T, Business Marketing, Basking Ridge, NJ, which he represented at the IEEE 802 LAN Committee. In 1983, with Aetna Telecommunications Consultants, Centerville, MA, he led a project in which single-mode optical fibers were used for the first time in a private campus network (AEtna Life and Casualty, Hartford). He is currently an independent consultant, specializing in LAN and fiber optics at P.O. Box 643, East Sandwich, MA 02537.

He has published papers and holds patents on solid-state physics, electroplating of magnetic alloys, space-division telephone switching, error control codes, link protocols, fiber-optics, LAN and in theoretical physics.

Dr. Rocher is a senior member of IEEE and a member of the American Physical Society, the American Association for the Advancement of Science, and the Association for Computing Machinery. ■

OPTIMAL LOOP TOPOLOGIES FOR DISTRIBUTED SYSTEMS*

C. S. Raghavendra and M. Gerla

UCLA Computer Science Department
University of California
Los Angeles, CA 90024

ABSTRACT

Double loop network architectures offer higher performance and reliability than single loop networks. In this paper, a double loop network which is optimal among all double loops is described. This network topology consists of a loop in the forward direction connecting all the neighboring nodes, and a backward loop connecting nodes that are separated by a distance $\lfloor\sqrt{N}\rfloor$, where N is the number of nodes in the network. We show that this network is optimal in terms of hop distance between nodes, delay, throughput, and terminal reliability. The paper includes derivation of closed form expressions for the maximum and average hop distance between nodes, number of distinct routes between two farthest nodes, and throughput. The effect of node and link failures on network performance is also considered.

1. INTRODUCTION

Local area networks have been extensively used in recent years to support distributed processing. One of the key design issues is the identification of a network topology which is best suited for a variety of different applications. Loop or ring architectures are quite popular, since they require simple control software and provide high data transfer rates. Many loop network architectures have been described in the literature, some using centralized control, others using distributed control [FARM 69, PIER 72, OHLI 77]. In this paper, we consider a loop operated in the "check and forward" mode [GRNA 80]. Referring to Figure 4, the destination address of each received message is checked to determine the output channel (either to the local host, or to a neighboring node) onto which the message should be forwarded. If the channel is free, the message is forwarded "on the fly", without buffering. Otherwise, the message is stored in a delay insertion register, waiting for the channel to become free.

A major disadvantage of a unidirectional **single** loop system is vulnerability, since any link or interface failure partitions the network. Another disadvantage is the fact that some messages must travel along the entire loop before reaching their destinations. This affects the delay performance.

In order to improve **reliability** and **performance** of loop networks, double loop architectures have been proposed. Liu proposed a Distributed Double Loop Computer Network (DDLCN) [LIU 78, WOLF 79a], which is designed as a fault-tolerant distributed system, and claimed to have achieved better reliability as well as better communications performance than any other loops. More recently, another double loop network, called daisy-chain loop, was proposed by Grnarov et. al. [GRNA 80]. This network was shown to be superior to DDLCN, in terms of reliability and communications performance in both fault free and fault mode of operation [GRNA 80].

In this paper, we present a generalized double loop network, which is optimal among all double loop topologies. The DDLCN and daisy-chain become special cases of the proposed generalized network. We present the analysis of the optimal network and derive closed form expressions for maximum number of hops, average number of hops, and throughput. We also study the effects of node and link failures and show the improvements obtained with respect to the daisy-chain loop.

2. OPTIMAL DOUBLE LOOP NETWORK

The optimal double loop consists of a forward loop connecting all the neighboring nodes and a backward loop whose interconnection pattern depends on N, the number of nodes in the network. As shown later, for optimum performance and reliability the skip distance of the backward loop is $\lfloor\sqrt{N}\rfloor$, where skip distance is the hop distance (measured on the forward loop) between two nodes which are adjacent in the backward loop. Except for the fact that skip distance is a function of N, all other features are similar to those of the daisy-chain loop [GRNA 80]. The proposed topology for a 15-node network is shown in Figure 1, where the skip distance in the backward loop is 3. For comparison

* This research is supported in part by the ONR contract N00014-79-C0866
(Research in Distributed Processing)

purposes, a 15-node daisy-chain loop network is shown in Figure 2, and a 15-node DDLCN is shown in Figure 3.

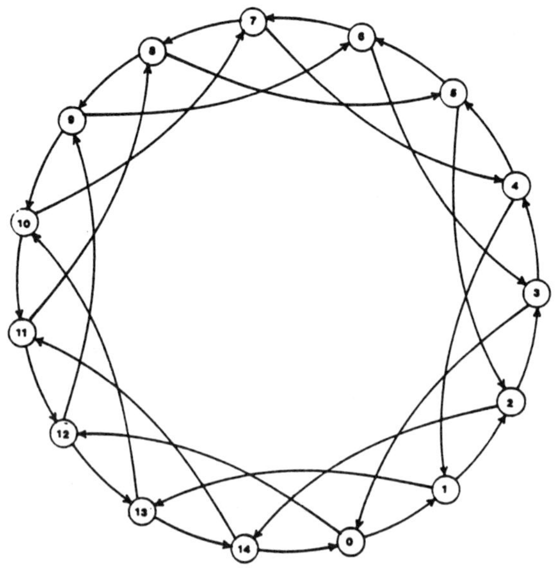

Figure 1. A 15-node Optimal Loop Topology

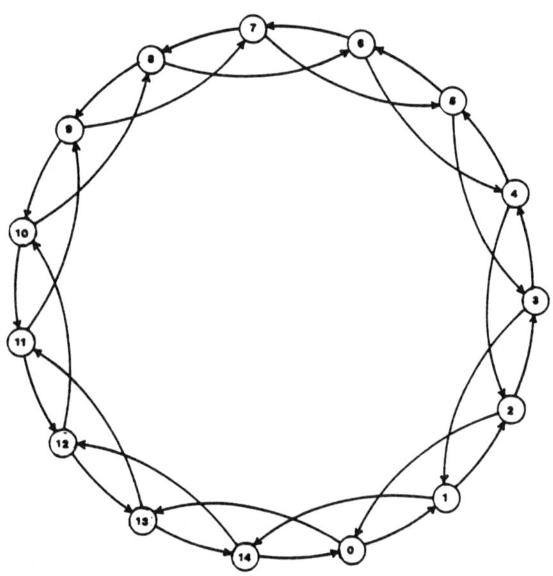

Figure 2. A 15-node Daisy-Chain Loop Network

In the optimal loop network, both forward and backward links are active, and there exist several paths from a source to a destination. The network can tolerate several link and nodal interface failures before becoming partitioned. Intuitively, delay and reliability are improved since the skipping of several nodes creates "short cuts", and also provides more alternate paths. We will prove this more rigorously later in the paper.

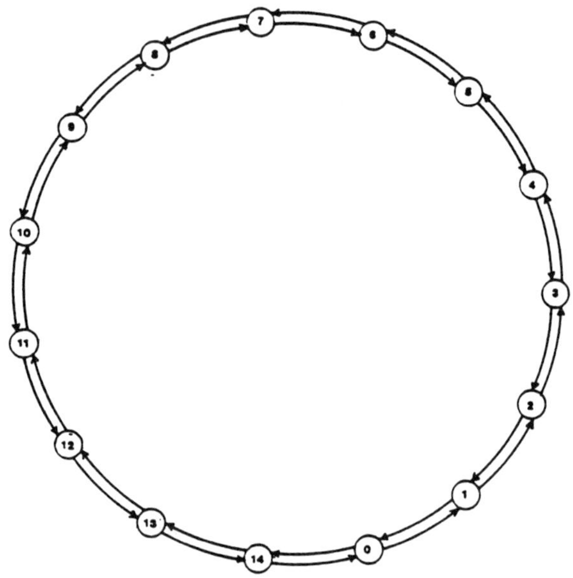

Figure 3.

A 15-node Distributed Double Loop Computer Network

The functional organization of a loop interface is shown in Figure 4. An interface receives messages from the host it is connected to. It also receives messages relayed by the neighboring interfaces. Conflicts of the simultaneous arrivals of messages from the three streams are resolved in the same way as in DDLCN [WOLF 79b], i.e., by delaying incoming relayed messages in variable-length shift registers located in the loop interface. To achieve better reliability, a loop interface is split into two identical modules which have separate control and separate line driver/receiver (D/R). Both modules share transmitter (T) and receiver (R) for communications with the host (H) connected to this interface.

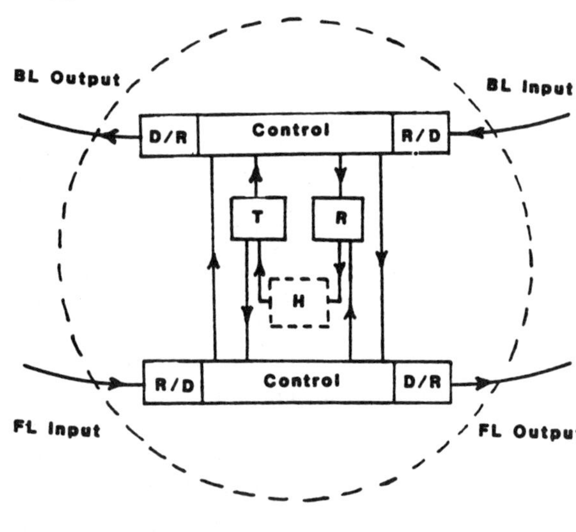

Figure 4. Functional Organization of a Loop Interface

In the message format shown in Figure 5, the destination address field is used in selecting the path. If the destination address matches with the interface address, the message is removed from the loop and routed to the host. Otherwise, the message is forwarded one "short" step (on the forward loop) or, when appropriate, one "giant" step (on the backward loop). Messages circulate on the loops until they reach their intended destinations. An adaptive routing algorithm can be used to route messages when there are failures in the network.

Flag	Destination address	Origin address	Control	Data	CRC	Flag

Figure 5. A Message Format

3. MAXIMUM AND AVERAGE NUMBER OF HOPS

Recall that both forward and backward loop are unidirectional, with the forward loop connecting adjacent nodes and the backward loop connecting nodes at hop (i.e., skip) distance $h > 1$. The shortest route between a source and a destination may therefore involve a combination of forward and backward hops. For each node, there is some other node which is diametrically opposite, i.e., it is the farthest away in the hop distance sense. Indeed, in designing the backward loop topology, the skip distance h will be selected such that the diameter d, i.e., the maximum over all node pair distances is minimized. Clearly, the hop distance from a node to all other nodes will vary from 1 to d. In the following we derive expressions for maximum and average number of hops and find their optimum values.

Let N be the number of nodes in the network;
h be the skip distance in the backward loop;
and d be the diameter.

We want to choose h so as to minimize,
d = diameter = max d_{ij} over all i,j.
where d_{ij} is the shortest hop distance from node i to node j.

The nodes in the network are numbered from 0 to N-1. To get the shortest path between i and j we use backward hops as long as it is advantageous. For two diametrically opposite nodes, backward hops are advantageous as long as d is greater than (N-bh), where b is the number of backward hops used. In addition to the backward hops, h-1 forward hops will be required to reach the farthest node. The relationship between maximum number of hops d and skip distance h is,

$$d = N - (d - h + 1)h + (h - 1)$$

$$d = N - dh + h^2 - 1$$

$$d = \left\lfloor \frac{N}{(h+1)} \right\rfloor + (h - 1)$$

For d to be minimum, the optimum value of h is $\lfloor \sqrt{N} \rfloor$.

If h = 1, we get Liu's DDLCN of diameter d = $\lfloor N/2 \rfloor$;
If h = 2, we get daisy-chain of diameter d = $\left\lfloor \frac{N}{3} \right\rfloor + 1$.

The average number of hops is defined as,

$$\frac{1}{N}\sum_{i=0}^{N-1}\left[\frac{1}{(N-1)}\sum_{j=0}^{N-1} d_{ij}\right]$$

where $d_{ii} = 0$.
As the network is symmetric, the term in parentheses is the same for all nodes i, thus:

$$\text{Avg. no. of hops} = \frac{1}{(N-1)}\sum_{j=0}^{N-1} d_{ij} \quad \text{for an arbitrary i.}$$

Thus we need to calculate the sum $S = \sum_{j=0}^{N-1} d_{ij}$ of the distances to all other nodes in the network. Note that we always consider the shortest distance between any node pair. Recall that the maximum number of backward hops used in any route is,

$$b = \left\lfloor \frac{N}{(h+1)} \right\rfloor$$

Using this result we have,

$$S = [1 + 2 + 3 + \ldots + (N - bh - 1)] + [1 + 2 + \ldots + h] + [2 + 3 + \ldots + (h + 1)] \ldots + [b + (b + 1) + \ldots + (b + h - 1)]$$

Each term in the sum corresponds to a route with 0, 1, 2, ..., b number of backward "giant" hops (i.e., steps on a backward loop) respectively.

Manipulating the previous expression, we obtain:

$$S = [1 + 2 + 3 + \ldots + (N - bh - 1)] + h(1 + 2 + \ldots + b) + b(1 + 2 + \ldots + h-1)$$

which simplifies to,

$$S = \frac{(N - bh)(N - bh - 1)}{2} + \frac{bh}{2}(b + h)$$

The average number of hops is given by S/(N-1), i.e.,

$$AVG = \frac{1}{(N-1)}\left[\frac{(N - bh)(N - bh - 1)}{2} + bh\frac{(b + h)}{2}\right]$$

Substituting the value of b,

$$AVG = \frac{1}{2(N-1)}\left[\left(N - \left\lfloor\frac{N}{h+1}\right\rfloor h\right)\left(N - \left\lfloor\frac{N}{h+1}\right\rfloor h - 1\right) + \left\lfloor\frac{N}{h+1}\right\rfloor h\left(\left\lfloor\frac{N}{h+1}\right\rfloor + h\right)\right]$$

For the sake of comparison, we report DDLCN and daisy-chain results. These are:

For DDLCN, AVG = $\frac{1}{(N-1)}\left\lfloor\frac{N^2}{4}\right\rfloor$

For daisy-chain,

$$AVG = \frac{1}{2(N-1)}\left[\left(N-\left\lfloor\frac{N}{3}\right\rfloor 2\right)\left(N-\left\lfloor\frac{N}{3}\right\rfloor 2-1\right) + \left\lfloor\frac{N}{3}\right\rfloor 2\left(\left\lfloor\frac{N}{3}\right\rfloor + 2\right)\right]$$

Next, we want to minimize the average number of hops, i.e, we want to minimize the following expression:

$$A = \left(N-\left\lfloor\frac{N}{h+1}\right\rfloor h\right)\left(N-\left\lfloor\frac{N}{h+1}\right\rfloor h-1\right) + \left\lfloor\frac{N}{h+1}\right\rfloor h\left(\left\lfloor\frac{N}{h+1}\right\rfloor + h\right)$$

For optimality,

$$h+1 = \frac{N}{(h+1)}$$

That is, $h = \lfloor\sqrt{N}\rfloor$

Note that the same optimum value of $h = \lfloor\sqrt{N}\rfloor$ minimizes both maximum number of hops and average number of hops. In Table 1 we summerize maximum and average hop results for DDLCN, daisy-chain, and optimal loop for typical values of N.

N	DDLCN Max	DDLCN Avg	Daisy-Chain Max	Daisy-Chain Avg	Optimal Loop Max	Optimal Loop Avg
10	5	2.80	3	2.33	3	2.33
15	7	4.00	6	3.21	5	3.00
30	15	7.76	11	5.69	9	4.65
50	25	12.75	17	9.00	12	6.14
100	50	25.25	34	17.33	18	9.09

Table 1. Maximum and average hop distances for three different loop architectures.

4. THROUGHPUT ANALYSIS

In this section we perform the throughput analysis to determine the total message traffic that the network can handle before saturation occurs. When link utilization reaches unity, the network is saturated. The total message arrival rate at which this occurs is the maximum traffic the network can handle, and is defined to be the throughput. For the analysis we make the following assumptions [KLEI 76]:

1) Poisson message generation with total aggregate rate γ, (msg/sec)
2) Uniform traffic pattern, and
3) Saturated throughput conditions, i.e., message traffic on each link equals link capacity.

We compute the maximum traffic in a way similar to Wolf's [WOLF 79b] relating each link utilization to the total message arrival rate and setting the maximal utilization to one.

We assume that shortest routes only are used for message transmissions. If there is more than one shortest route, traffic will be **equally** divided among the alternate routes. The total traffic on a forward link may be different from the total traffic on a backward link. As the average number of hops AVG is computed by considering the shortest routes from a node to all other nodes, the total network traffic caused by distributing γ/N units of traffic from one source to all other destinations is:

$$T_f + T_b = AVG^{*}\gamma/N$$

where T_f is the contribution to total traffic on a forward link and T_b is the contribution to total traffic on a backward link. By using suitable policies for routing traffic when there is more than one shortest route between a node pair, it is possible to balance the flow on forward and backward links. That is,

$$T_f = T_b = \frac{1}{2}AVG^{*}\gamma/N$$

When all sources are active, saturation is reached if:

$$(T_f + T_b)N = 2N\mu$$

where μ is the link capacity (msg/sec)

Substituting for T_f and T_b we obtain:

$$\gamma_{SAT}{}^{*}\,AVG = 2\mu N$$

or:

$$\gamma_{SAT} = \frac{2\mu N}{AVG}$$

Finally, substituting for AVG we obtain:

$$\gamma_{SAT} = \frac{2(N)(N-1)\,\mu}{\frac{1}{2}\left[\left(N-\left\lfloor\frac{N}{h+1}\right\rfloor h\right)\left(N-\left\lfloor\frac{N}{h+1}\right\rfloor h-1\right) + \left\lfloor\frac{N}{h+1}\right\rfloor h\left(\left\lfloor\frac{N}{h+1}\right\rfloor + h\right)\right]}$$

γ_{SAT} is maximum if average number of hops is minimum. Therefore, γ_{SAT} is also maximized when $h = \lfloor\sqrt{N}\rfloor$

We may compare this result with DDLCN and daisy-chain results.

For DDLCN, $\gamma_{SAT} = \frac{2(N)(N-1)}{\left\lfloor\frac{N^2}{4}\right\rfloor}\mu$

For daisy-chain,

$$\gamma_{SAT} = \frac{4(N)(N-1)\,\mu}{\left[\left(N-\left\lfloor\frac{N}{3}\right\rfloor 2\right)\left(N-\left\lfloor\frac{N}{3}\right\rfloor 2-1\right) + \left\lfloor\frac{N}{3}\right\rfloor 2\left(\left\lfloor\frac{N}{3}\right\rfloor + 2\right)\right]}$$

Since γ_{SAT} is proportional to the inverse of AVG, and AVG is minimized by our topology, the throughput is also optimal for our topology.

As a special case, when N is the square of an integer, we get the following closed form expressions for maximum number of hops, average number of hops, total flow, and throughput.

$$\text{Diameter d} = 2(\sqrt{N} - 1)$$

$$\text{Avg. no. of hops} = \frac{N}{(\sqrt{N} + 1)}$$

$$\gamma_{SAT} = \frac{2(N-1)}{(\sqrt{N} - 1)} \mu$$

For example, when N = 64 we have,

Diameter d = 14
Avg. no. of hops = 64/9
$\gamma_{SAT} = 18\ \mu$

When there is a link failure, the maximum and average number of hops will increase. However, the performance of the optimal loop can be shown to be still superior to that of DDLCN or daisy-chain loop in similar failure conditions.

5. FAULT TOLERANCE

The optimal double loop architecture provides good protection against node and link failures. As many alternate routes exist between any node pair, paths can be selected adaptively in case of failures. The effect of a node failure is to increase the maximum number of hops for some nodes and increase the average number of hops slightly. Figure 6 shows the effect of node #0 failure on optimal loop and daisy-chain loop, and Figure 7 shows the effect of link 0→1 failure on optimal loop and daisy-chain loop.

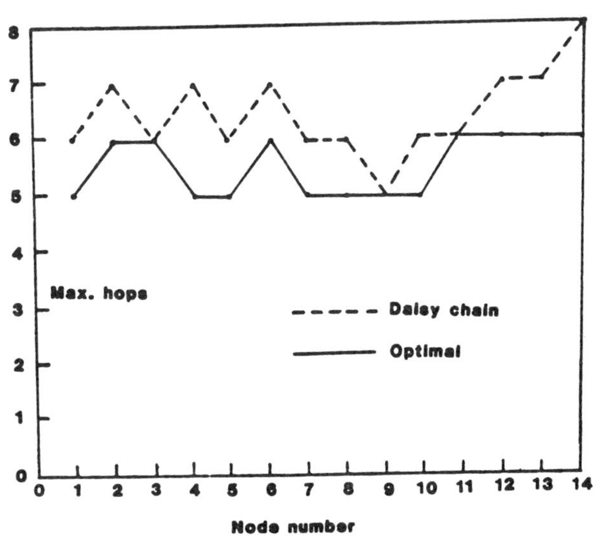

Figure 6.
Effect of a Node Failure on Maximum No. of Hops

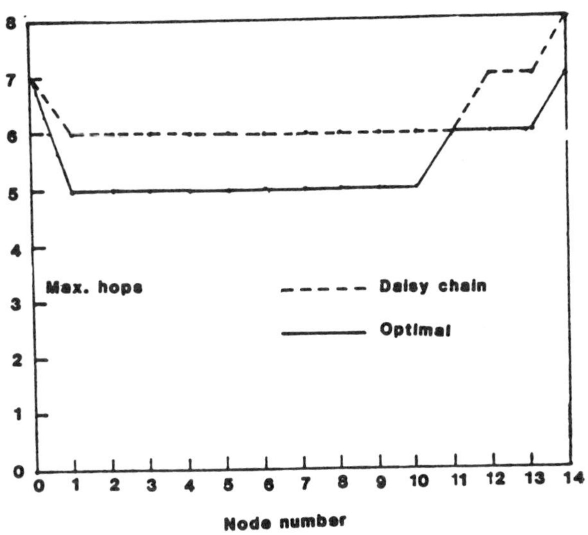

Figure 7.
Effect of a Link Failure on Maximum No. of Hops

A useful measure of fault-tolerance is to compute the terminal pair reliabilities. We consider two farthest nodes for terminal reliability analysis. In the optimal double loop network there exist many alternate routes between any pair of nodes, and therefore we expect high terminal pair reliability. For the two farthest nodes, one can intuitively see that the following properties hold:

Terminal reliability α No. of alternate routes; and

Terminal reliability α $\frac{1}{\text{diameter}}$

We have already shown that the diameter is minimized when $h = \lfloor\sqrt{N}\rfloor$. Also, for this value of h, the number of alternate routes between the two farthest nodes will be higher in the optimal loop than in the daisy-chain loop, as shown in the following derivation. Therefore, we can conclude that the optimal loop network gives higher terminal pair reliability than the daisy-chain loop.

Now we compute the number of alternate routes between any two farthest nodes in the network. There is one path using only forward links, which may not be a shortest path. There are several additional paths which use backward links and forward links. All these routes are shortest routes between that node pair. Each of these routes consists of $b = \left\lfloor \frac{N}{h+1} \right\rfloor$ backward hops and (h-1) forward hops. We get different routes depending on the relative positions of "giant" steps and forward hops along the route. Therefore, the number of alternate routes between the two farthest nodes is the number of combinations of b out of (b+h-1) objects, that is:

$$S = R - 1 = \binom{b+h-1}{b} = \binom{b+h-1}{h-1}$$

The expression for the number of distinct routes is given by,

$$R = 1 + \binom{b + h - 1}{h - 1}$$

For Liu's loop R = 2;

For Daisy-chain $R = 1 + \binom{b+1}{1} = 2 + \left\lfloor \frac{N}{3} \right\rfloor$;

For the optimal loop, with $h = \lfloor \sqrt{N} \rfloor$,

$$R = 1 + \binom{\left\lfloor \frac{N}{h+1} \right\rfloor + h - 1}{h - 1}$$

For example, a 15-node network has h=3, b=3, and the number of distinct routes between any two farthest nodes is, $1 + \binom{5}{2} = 11$.

If we maximize R with respect to h, we find that the optimum value of h is $\lceil \sqrt{N} \rceil$. However, for most values of N, $h = \lfloor \sqrt{N} \rfloor$ also gives the same maximum value for R. Therefore for most values of N, and for N the square of an integer, the optimal loop will also maximize the number of distinct routes between any two farthest nodes.

6. CONCLUSIONS

We have presented an optimal double loop topology for locally distributed networks. This network is shown to provide the optimum performance with respect to maximum number of hops, average number of hops, throughput rate and terminal reliability. In particular, better performance is achieved with respect to previously proposed topologies such as DDLCN and daisy-chain loop. The cost of this topology in length of loop cable is $N(\lfloor \sqrt{N} \rfloor + 1)$, as compared with 2N for DDLCN and 3N for daisy-chain. Nodal interface costs are the same for all these architectures. One should realize that in a local network cable cost represents only a small fraction of total system cost. Therefore it appears that the additional cable cost is well compensated for by the performance benefits of the optimal loop topology.

This type of interconnection scheme among host processors may find applications in tightly coupled multiprocessor system. A separate host processor can perform the reconfiguration in case of failures. The reconfiguration of interconnections under failures, and addition of new nodes can be done easily if the interfaces are brought to one place. Interestingly, if all the links are bidirectional, the optimal double loop interconnection is precisely the same as the Illiac IV structure.

ACKNOWLEDGMENTS

The authors wish to thank Prof. Avizienis for his encouragement and support during this research and Ms. Rita Hall for her help in the preparation of the final manuscript.

REFERENCES

[BARN 68] G. H. Barnes et. al., "The Illiac IV Computer", IEEE Transactions on Computers, C-17, No. 8, August 1968, pp 746-757.

[FARM 69] D. Farmer, E. Newhall, "An Experimental Distributed Switching System to Handle Bursty Computer Traffic", Proc. ACM Symposium on Problems in the Optimizations of Data Commns., October 1969, pp 1-23.

[GRNA 79] A. Grnarov, L. Kleinrock, M. Gerla, "A New Algorithm for Network Reliability Computation", Proc. Computer Networking Symposium, NBS, Gaithersberg, December 1979, pp 17-20.

[GRNA 80] A. Grnarov, L. Kleinrock, M. Gerla, "A Highly Reliable Distributed Loop Network Architecture", Proc. 1980 Int. Symp. Fault-Tolerant Computing, Kyoto, October 1980, pp 319-324.

[KLEI 76] L. Kleinrock, Queueing Systems Vol. II : Computer Applications, Wiley Interscience, New York, 1976.

[LIU 78] M. T. Liu, "Distributed Loop Computer Networks", Advances in Computers, Vol. 17, Academic Press, 1978, pp 163-221.

[OHLI 77] Y. Oh, M. T. Liu, "Interface Design for Distributed Control Loop Networks", Proc. 1977 National Telecommunications conference, Los Angeles, December 1977, pp 31:4(1)-31:4(6).

[PIER 72] J. Pierce, "Network for Block Switches of Data", Bell Syst. Tech. Jour., Vol. 51, No. 6, July/August 1972, pp 1133-1145.

[WOLF79a] J. Wolf, M. T. Liu, B. Weide, D. Tsay, "Design of a Distributed Fault-Tolerant Loop Network", The 9th Annual Int. Symp. Fault-Tolerant Computing, Madison, June 1979, pp 17-24.

[WOLF79b] J. Wolf, B. Weide, M. T. Liu, "Analysis and Simulation of the Distributed Double Loop Computer Networks", Proc. Computer Networking Symposium, NBS, Gaithersberg, December 1979, pp 82-89.

Chapter 4
Broadband ISDN

The ISDN standard discussed in Chapter 1 has the "narrowband" of $2B + D$. Services that require wider bandwidth than the narrowband ISDN capabilities are currently handled by private point-to-point or circuit-switched networks. The broadband ISDN (B-ISDN) attempts to provide wide bandwidth services by way of standardized switched connectivities in long-haul networks. A considerable amount of effort is spent at the CCITT to define standards for interfaces, services, and management. Spears's article summarizes a set of services intended for B-ISDN networks. Perhaps the most distinct from earlier networks is the availability of video services in the form of switched TV and HDTV (high-definition television).

In a 1988 CCITT meeting, a packet-oriented transfer mode, called ATM (asynchronous transfer mode), was selected as the universal transfer mode of all B-ISDN services. In ATM, information is divided into small, fixed-size blocks called cells. These cells are multiplexed and switched over virtual circuits based on their header contents. Details of cell sizes and fields are still to be worked out. Principles and structures of ATM are described in Händel's article. The next article by Lee discusses the related issue of an integrated transport technique for the B-ISDN.

Transmission of ATM cells is to be achieved by SONET (synchronous optical network). SONET represents the standard interface for optical transmission signals and the standard frame format of multiplexing ATM cells. Ballart and Ching's article provides the details of SONET interface as well as the history of its development.

The next four articles offer alternative viewpoints on B-ISDN. Turner's integrated packet switching approach is often referred to as the fast packet switching (FPS) technique. It shares many similarities with the ATM, although FPS is an adaptation of the pure packet switching concept to a high-speed environment. Turner's FPS is closely related to several high-speed switching experiments, based on Banyan or copy networks [White *et al.*, 1987; Coudreuse *et al.*, 1988]. Turner correctly and eloquently points out that high-speed networks need to employ an efficient routing technique, such as the self-routing capability, and to disperse conventional network functions toward the edges of the network so that node complexity is minimized.

While Turner advocates a statistical packet switching approach, Gechter and O'Reilly raise concerns about the ATM. Specifically, they question whether the ATM is suitable for maintaining the desired grade of service of real-time traffic, which may be a valid concern if the network is dominated by synchronous traffic. As pointed out, the ATM may then reduce to an inefficient and expensive form of circuit switching.

Finally, Parekh compares fixed- and variable-length cell formats in the ATM. The article points out that fixed-length cells generally require less memory space at the expense of longer delays. However, implementation of the variable-length format incurs more overhead to detect cell boundaries. In the same CCITT meeting at which a fixed-length cell format was approved, a compromise proposal was put forth to provide a support for variable-length packets, but in *multiples* of fixed-length cells. The proposal may be considered a realistic compromise by reducing the overhead associated with variable-length packets while providing higher throughput for high volume data such as video.

Broadband ISDN Switching Capabilities from a Services Perspective

DAN R. SPEARS, MEMBER, IEEE

Abstract—**Broadband switching and transmission technologies are maturing rapidly, promising to make broadband ISDN (B-ISDN) possible and potentially economical in the near future. The diversity of possible B-ISDN services provides much impetus for the implementation of B-ISDN, but complicates the task of specifying network switching requirements. This paper examines a few representative services in order to generate a set of basis vectors in capability space. New services are then projected onto this basis; if a new service cannot be constructed from the basis vectors, other vectors are added to the basis set. B-ISDN architects can then specify switching requirements by selecting those capability basis vectors that are required for the services most in demand.**

For full motion entertainment video services, circuit switching capability at about 150 Mbit/s may be required, while selective broadcast and remote switching capabilities may reduce the cost of these services. For subvideo-rate services, new transfer mode (NTM) capability may be required to allow users the flexibility to allocate access channels dynamically on a per-call basis. Multiple connection (e.g., voice and video) and multiple party call capabilities may be required for many B-ISDN services, implying a need to extend the ISDN signaling protocol. The capability for the user to request network code conversion between different coding formats (e.g., NTSC or HDTV), and network compression of signals to reduce the cost of interoffice facilities, may also be required. Finally, fast call setup capability and high calling volume capabilities may be required, implying a need to distribute call processing functions over multiple processors in the central office.

I. Introduction

BROADBAND switching and transmission technologies are maturing rapidly. The cost of single-mode fiber, which offers virtually unlimited bandwidth, dropped from \$5/m in 1982 to about 45¢/m in 1984, with goals of 12¢/m by 1987 and 4¢/m by 1993 [1]. A 40-input 40-output analog video switch occupied a volume of 8 ft^3 three years ago; a 64-input 64-output digital video switch can now be placed on a single card measuring 4 in^2 [2]. A few years ago, packet switches could switch less than 500 packets/s with a network transit delay of 100–300 ms; researchers are now building a single-shelf Batcher-banyan packet switching fabric that can switch millions of packets/s with an input-to-output delay of less than 1 ms [3]. These and other advances offer the possibility of a practical and economical broadband ISDN (B-ISDN) in the near future [4]–[6].

The driving forces for B-ISDN are varied. To researchers, it represents a new challenge to create technology that is faster, less expensive, and more flexible. To operating companies, it represents new information transport capabilities and new revenue opportunities. Most importantly, to end users, many of whom may not even know the difference between optical fiber and copper wire, it represents new services. And it is their willingness to pay for these new services that will ultimately determine the success or failure of B-ISDN.

B-ISDN, as shown in Fig. 1, brings the vision of many new service possibilities. Imagine a bedridden grandmother using videotelephone to be present at a distant first birthday party of a grandchild. A student gets help with a physics homework problem by calling a video library and watching the professor work similar problems. A transferee does preliminary house-hunting by viewing a videotaped walk-through of selected homes. A researcher quickly accesses current technical articles, viewing text, graphics, and pictures on a high resolution terminal. A couple comes home from a busy day at the office and relaxes by selecting a movie from a video-on-demand service or browsing through a home video shopping service [7].

Business communications may be completely changed by desktop teleconferencing over B-ISDN with each user's terminal displaying windows containing full-motion images of the other parties in the conference. Viewgraphs, sketches, photographs, and text files are exchanged and either displayed or printed. Other parties are added to the conference as needed for consultation, and subconferences are created for private conversations.

The diversity of these services provides much impetus for building a B-ISDN network, but also complicates the task of specifying network switching requirements. Networks in the past often focused on a specific service, such as speech communications, and the needs of that service determined the network requirements, such as bandwidth, signal-to-noise ratio, call setup time, and blocking objectives. B-ISDN, on the other hand, will support many widely varying services, each having different network requirements. Rather than studying all possible B-ISDN services, an alternative is to examine a few representative services in order to generate a set of basis vectors in capability space. New services can then be projected onto this set of basis vectors, as shown in Fig. 2; if the new service cannot be constructed from the basis vectors, other vectors are added to the basis set. B-ISDN architects can then specify switching requirements by identifying those

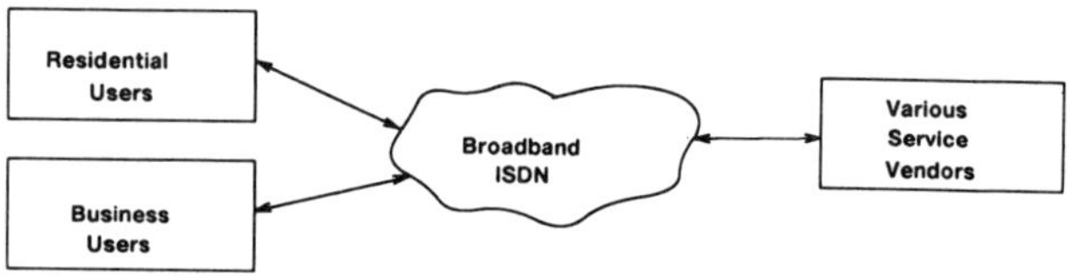

Fig. 1. Broadband ISDN.

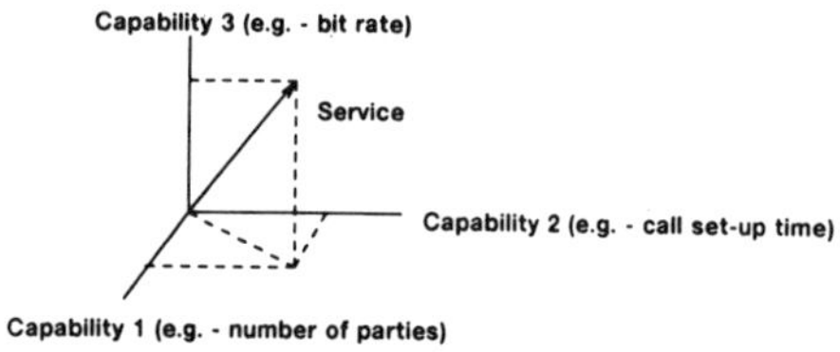

Fig. 2. Projection of a service onto switching capabilities.

capabilities that are required for the services expected to be most in demand.

This paper examines several B-ISDN services and identifies the switching capabilities required for those services. The mappings between services and switching capabilities are summarized at the end of the paper.

II. Full-Motion Video Services

A. Switched Access Television Service

Switched access television service, the delivery of entertainment video to residential customers over switched facilities, is a good candidate for a representative B-ISDN application. From a marketing standpoint, it is attractive because it taps a large existing market and offers several advantages over current delivery alternatives. These advantages include access to a wider variety of programming, elimination of the need for scrambling equipment, opportunities for creative pricing strategies such as pay-per-view, and higher quality video such as high definition television (HDTV). From a capability standpoint, it allows us to identify several basis vectors that are also required for many other broadband applications.

The most obvious capability requirement for this application is a *high bandwidth downstream digital channel.* Straightforward digitization of an NTSC quality video signal requires about 86–115 Mbit/s; using differential pulse code modulation (DPCM) this rate can be reduced to the 44.7 Mbit/s DS-3 rate [8]. Extended quality video (EQTV) can be coded at 216–243 Mbit/s [9], or reduced to 100–150 Mbit/s using diagonal filtering and decimation [10] or DPCM and variable length coding [11]. A full bandwidth HDTV signal occupies 1.2 Gbit/s; however, compression techniques may reduce this to 200–300 Mbit/s [6]. Therefore, a video transmission rate of about 150 Mbit/s would provide for near-term transmission of NTSC or EQTV signals, with HDTV being transported using two channels. In the long term as coding technology advances, HDTV may potentially be transported on a single 150 Mbit/s channel.

The transmission requirements between the user and the switch in order to handle switched access television are determined by the bit rate requirements for each video channel and by the number of video channels needed simultaneously. For example, a user requiring three simultaneous 150 Mbit/s video channels will need a 450 Mbit/s transmission stream for switched access television.

Video switching is a second capability required for switched access television. The switching of 150 Mbit/s channels exceeds the capabilities of current implementations of time slot interchangers (TSI) and packet switches. A promising alternative for video switching is *space division circuit switching* using ECL or CMOS technology [12]–[15]. These IC technologies and new packaging methods now make it possible to create a compact 64 × 64 switching subsystem using a 4 × 4 matrix of 16 × 16 CMOS crosspoint chips on a single card measuring 4 in^2 [2].

Many users in a central office may be simultaneously accessing the same video source. To reduce interoffice transmission costs, the switching fabric will need a *selective broadcast* capability. For example, assume that 50 users associated with the same central office access the same video source simultaneously. Only one incoming video trunk is needed to serve those users if the switching fabric can selectively broadcast the same information to all 50 users, as shown in Fig. 3. Furthermore, if the switching fabric has a *remote switching* capability, also shown in Fig. 3, the length of fiber dedicated to the user is shortened since the source channel can be selectively broadcast at the remote node and use shared facilities between the central office and the remote node.

Fast call setup capability is required for switched access television. Using infrared remote control units, today's television viewers are accustomed to a channel change response time of less than 1 s. B-ISDN must be able to match this response time if switched access television is to be successful. Closely associated with the call setup response time is the rate at which call requests must be handled. In the voice network, the calling rate per line is about 2–3 calls/h, giving a call processing load of 20 000–30 000 calls/h for a 10 000 line office. With switched access television every channel change results in a signaling request, so when programs end at $\frac{1}{2}$-h intervals, peak calling rates of 2–3 calls/min can be expected, giving a peak call processing load of 600 000–18 000 000 calls/h for a 10 000 line office. Therefore, call processing in a B-ISDN must have the capability to handle a *high calling volume*. For fast call setup and high calling volume capabilities, it is likely that call processing will need to be distributed over multiple processors in the central office.

B. Video-on-Demand and Videotelephone Services

Video-on-demand and videotelephone services have similar capability requirements. Like switched access television service, they both require full motion video transmission channels; however, videotelephone requires both an upstream and a downstream video channel, while video-on-demand requires only a downstream video chan-

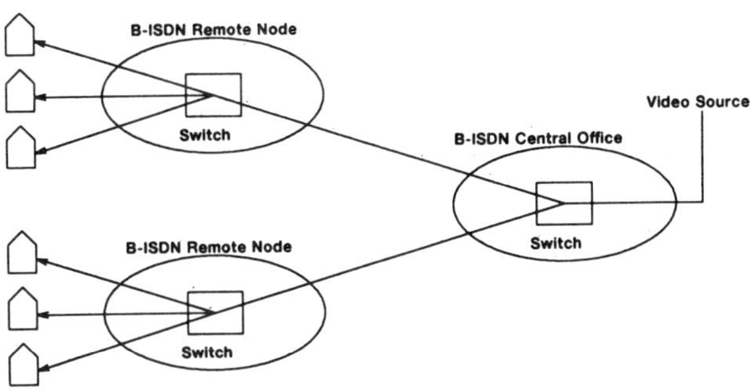

Fig. 3. Selective broadcast capability of switching fabric and remote switching capability.

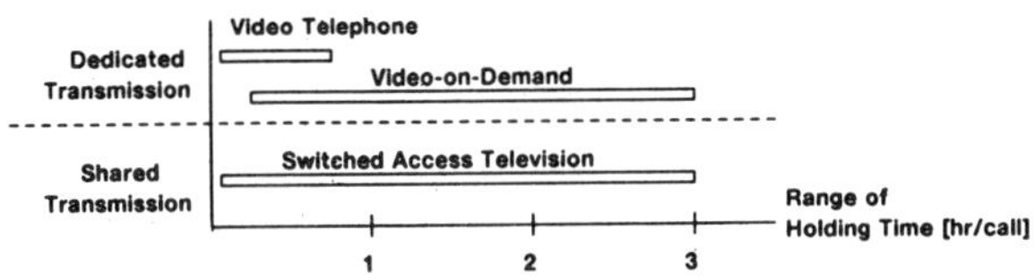

Fig. 4. Comparison of holding time range for full-motion video services.

nel. Both services have lower calling rates and less stringent setup response time requirements than switched access television, so they place fewer demands on call processing.

These services are nonbroadcast in nature, so neither benefits from selective broadcast or remote switching capability. However, for the same reason, they place a higher demand on interoffice trunking resources. The extent of this demand depends on the calling rate and the holding time. For videotelephone, the holding time may be similar to that of today's voice traffic (1–45 min/call), but for video-on-demand the holding time will be much longer (15 min/call–3 h/call), as shown in Fig. 4. When B-ISDN is first introduced, the calling rate for these services is expected to be lower than that of today's voice traffic (2-3 call/h during the busy hour) due to the limited number of users that will initially have video cameras and the limited number of video-on-demand vendors available initially.

Compressing interoffice video signals using a *code conversion* capability in the central office will reduce the transmission component of the cost of these services. The compressed video signal can be converted back to a 150 Mbit/s signal and circuit switched at intermediate switching offices; alternatively, it may be more economical to packet switch the compressed video signal at intermediate switching offices, depending on the amount of the compression.

III. Subvideo-Rate Services

The previous section discussed services that require channels operating at full motion video rates. This section discusses the switching capabilities required for services that operate at information transfer rates less than 150 Mbit/s.

A. Audio Communications

Speech communications will be as important in B-ISDN as in the current telephone network. The speech coding technique most often used today is called pulse code modulation (PCM). After band-limiting the speech signal to less than 4 kHz and sampling at a rate of 8000 samples/s, each sample is nonuniformly quantized in order to provide a near-constant signal-to-noise ratio over the full dynamic range. Two encoding laws, commonly referred to as A-law and μ-law, are recommended for converting each sample into an 8-bit code [16], resulting in an information transfer rate of 64 kbit/s.

Speech compression algorithms have been proposed for decreasing the bit rate requirements. Adaptive differential pulse code modulation (ADPCM) [17] is a code conversion technique for converting a 64 kbit/s A-law or μ-law PCM signal to a 32 kbit/s signal. Other algorithms have been proposed for compressing speech to 9.6 kbit/s and below, but with some degradation of the speech quality [18].

Higher quality speech coding algorithms have also been proposed. This is especially important in business teleconferencing services where users say that improving the audio quality is more important than adding video capability. Standards already exist for the digital coding of 7 and 15 kHz speech [19], [20], and involve sampling the speech signal at 16 and 32 kHz, respectively. The encoding process uses uniformly quantized 14-bit per sample PCM techniques with 11-segment 14- to 11-bit instantaneous A-law companding or 5-range 14- to 10-bit near-instantaneous companding. After including parity bits for error protection, the information transfer rates required are 192 and 384 kbit/s, respectively.

In order for B-ISDN to handle these different speech coding algorithms, the switching fabric needs the capability for handling a *broad range of channel rates*, including 9.6, 32, 64, 192, and 384 kbit/s. In addition, the network needs the capability for *code conversion* in order to support communications between users with different terminals. The current telephone network already supports some code conversion for digital speech, but the coding algorithm of each user is determined by locality (μ-law in the United States and A-law in Europe) rather than by user preference. To support user selection on a per-call basis, the signaling protocol needs to contain a field allowing each party to specify the speech coding algorithm to be used.

B. Services Involving Still Picture Communications

The belief that a picture is worth a thousand words may cause service involving the transmission of still pictures to mushroom. Imagine a video merchandising service in which users flip through pictures of merchandise in much the same way that they now flip through catalog pages. Do-it-yourself enthusiasts get faucet and car repair information, sewing patterns, and deck and patio construction plans through a B-ISDN. High school students get pictures of animals in India or weapons used in the Civil War to include in a research project. Graduate students and university professors access technical journals with the ability to see figures as well as text. In each of these services, the pictures are viewed on a video display or printed locally.

The information transfer rate required for pictures depends on two factors—the rate at which pictures must be delivered and the information required per picture. Teleshoppers will need to scan rapidly flipping through pictures at about the same rate they flip through catalog pages—about 2 pictures/s. With some other services, lower rates may be acceptable.

The information required per picture depends on the resolution of the picture and the number of colors. Using straightforward coding, a 320 × 200 low resolution picture with 256 colors requires about 0.5 Mbit/picture, while a 640 × 480 medium resolution picture requires 2.6 Mbit/picture for 256 colors and 7.4 Mbit/picture for 256 levels each of red, green, and blue. A 1000 × 1000 picture requires 1 Mbit/picture for a bilevel picture, 8 Mbit/picture for 256 gray levels or 256 colors, and 24 Mbit/picture for 256 levels each of red, green, and blue. The information transfer rates range from 50 kbit/s for low resolution pictures delivered at a rate of 0.1 picture/s, to 5.2 Mbit/s for 640 × 480 pictures delivered at a rate of 2 pictures/s, to 48 Mbit/s for 1000 × 1000 pictures delivered at a rate of 2 pictures/s. Therefore, services involving picture transmission require the capability for handling a *broad range of channel sizes* ranging from 50 kbit/s to 48 Mbit/s.

C. Remote Disk Server

In the 1960's, time sharing operating systems made it possible for users to gain access to a share of a centralized processor's computing power from their desks, as shown in Fig. 5. In the 1970's, low-cost microprocessors made personal computers possible, and both processing and disk storage moved to the user's desk. Personal disk storage, however, meant that users had to be responsible for backing up their own data, and made it more difficult to share files. In the 1980's, local area networks (LAN's) increased data communications speeds over short distances, making it possible to centralize data storage while leaving the processors on the user's desk. This gave the user the best of both worlds, local processing and centralized file storage. In addition, remote login and remote execution capability blurred the distinction between local and remote processors, allowing any processor on the network to run tasks for any authorized user.

The use of B-ISDN for remote disk access removes the distance limitations of local area networks, allowing centralized file servers to be accessed from home or other work locations. To provide remote disk access service, a B-ISDN needs *fast call setup* and *high transfer rate* capabilities comparable to latency times and transfer rates of disk drives. For a floppy disk drive, the track-to-track seek time is about 6 ms, with a head settling time of 15 ms, and the transfer rate is about 250 kbit/s. For a fixed head disk drive, the average latency time is about 8 ms and the transfer rate is about 5 Mbit/s.

	USER'S DESK	COMMUNICATIONS	CENTRAL LOCATION
1960s	Terminal	low speed data	Processor and Disk
1970s	Processor and Disk		
1980s	Processor	LAN	Disk (nearby)
Future	Processor	Broadband ISDN	Disk (anywhere)

Fig. 5. Migration of processors and disks over time.

IV. Switching for Subvideo-Rate Services

A. Multiple Switching Fabrics

Many services discussed above require a *broad range of channel rates*. One way to handle these services is to use an access structure of the form

$$\sum_i n_i * r_i$$

where n_i is the number of channels operating at a rate r_i. At the customer's premises each channel is time-division-multiplexed onto a higher-bit-rate transmission facility. At the central office, the individual channels are demultiplexed and cross connected to the appropriate switching fabrics, one for each channel type, as shown in Fig. 6.

Historically, economic studies have not favored alternatives involving multiple switching fabrics. The high first cost of installing multiple switching fabrics is an initial penalty, while the increased operational costs are a recurring penalty. For example, if a user initially wants three 64 kbit/s channels, two 192 kbit/s channels, one 5 Mbit/s channel, and two 150 Mbit/s channels, seven cross connects are required. Then, whenever the user wants to change the number of channels of a particular type, another service order must be issued and a craftsperson must travel to the central office to change and test the line circuit cards and rearrange the cross connects. Larger distributing frames are needed in these central offices due to the additional cross connects required, increasing the length of jumpers and increasing crosstalk problems. Finally, costs increase for maintaining records of customer assignments to line circuits and switch ports.

B. Single Packet Switching Fabric with Dynamically Allocated Channels

Another alternative for handling subvideo-rate services is the use of high speed packet switching. CCITT Study Group XVIII is now studying a broadband packet technique called new transfer mode (NTM) or asynchronous transfer mode (ATM). ATM may afford a potential for more effective resource utilization and/or more flexibility

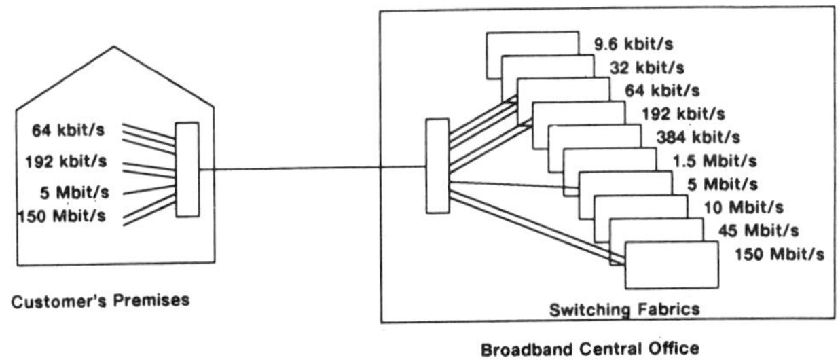

Fig. 6. Multiple switching fabrics in central office.

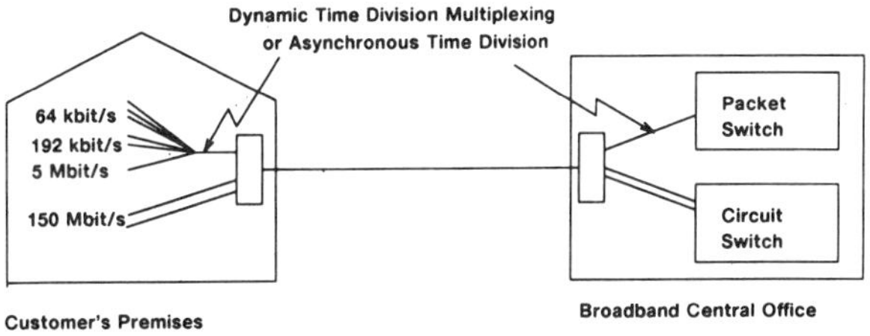

Fig. 7. Use of DTDM or ATD for bandwidth flexibility.

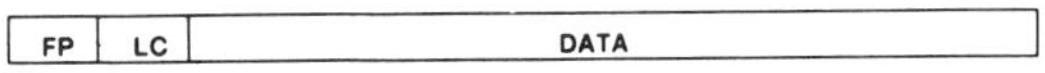

Fig. 8. Possible format for DTDM packets.

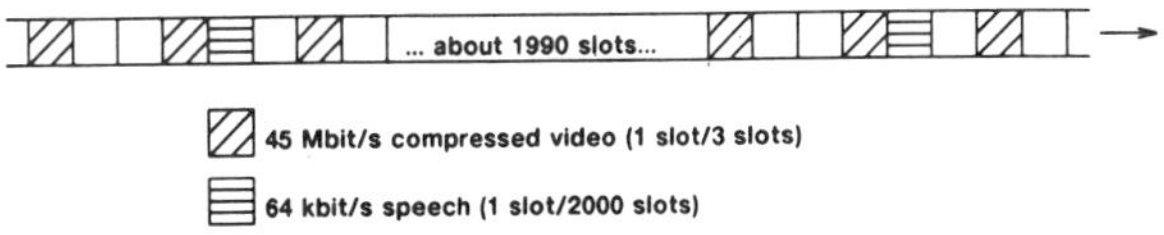

Fig. 9. Illustration of video and speech channel on DTDM interface.

to accommodate a wide variety of service needs [21]. Several similar techniques have been proposed for ATM. One technique, called asynchronous time division (ATD), merges the simplicity of circuit switching with the flexibility of packet switching by using fixed length packets with virtual circuit routing, but no flow control or error detection [22], [23]. A similar technique, called dynamic time division multiplexing (DTDM), subdivides a single high speed physical access channel into multiple virtual channels that are dynamically allocated and sized as needed on a per-call basis [24], [25]. Both techniques use a single physical access channel, shown in Fig. 7, connected to a packet switching fabric that routes each virtual channel to the proper destination. Since only one physical cross connect, one switching fabric, and one type of line interface circuit are needed to support a wide range of subvideo channel types and sizes, both initial costs and recurring costs are reduced.

DTDM and ATD combine the favorable characteristics of both packet switching and TDM switching. Historically, store-and-forward packet switching has been used to multiplex bursty communications onto a common transmission facility, with large buffers being placed at each switching node to levelize the burstiness. Packet switching has been ideal for computer-to-computer communications such as file transfers, but has been unsuitable for voice communications due to its limited throughput and its variable delays. TDM switching, which has a constant delay, has been the preferred switching technique for voice traffic. DTDM combines the explicit channel identification of packet switching with the framing pattern and control information of TDM switching to form a transmission format for slotted packets on synchronous facilities. Unlike older packet switching systems, DTDM is optimized for high throughput, low overhead, and low delay. For example, fixed size packets allow the use of a very simple framing mechanism, which is more suitable for the high transmission rates of B-ISDN than the flag sequence and zero insertion transparency mechanism used on the D-channel of narrowband ISDN [26].

As an example of DTDM, suppose that the physical channel operates at 150 Mbit/s and that each slot contains 260 octets[1] comprising a 2 octet framing pattern (FP), a 2 octet DTDM logical channel number (LC), and a 256 octet data field (DATA), as shown in Fig. 8.

About 70 000 slots/s are available, each with a duration of 14 μs. Speech transmission using full packets requires buffering 256 speech samples, which gives a fixed packetization delay of about 30 ms if 64 kbit/s PCM speech coding is used. A slot is required for this signal every 30 ms, or about 1 slot out of every 2000, as shown in Fig. 9. Doubling the packet rate to 64 packets/s reduces the fixed packetization delay to 15 ms and uses 1 slot out of every 1000. This can be achieved either by decreasing the information content of each packet to 128 octets or by using a higher quality speech coding algorithm that has a higher bit rate.

Low resolution picture transmission at a rate of 0.1 picture/s requires about 50 kbit/s (25 packets/s) or about 1 slot out of every 2800. Medium resolution pictures transmitted at a rate of 2 pictures/s use 1 slot out of every 27, and high resolution pictures transmitted at a rate of 2 pictures/s use 1 slot out of every 3. Full motion video might also be packet switched if desired. For example, an NTSC video signal compressed to a rate of 45 Mbit/s requires about 1 slot out of every 3.

The primary motivation for DTDM is flexibility. It can support speech communications at 9.6, 32, 64, 192, and 384 kbit/s, and any other rate that might be needed in the future. It can support picture communications from 50 kbit/s to 5 Mbit/s to 48 Mbit/s. It even has the potential to support compressed full motion video transmission, all using a single switching fabric.

Determination of the optimal physical transmission rate for DTDM is an important issue. The minimum transmission rate required is clearly dependent on the number and the size of the logical channels needed simultaneously, but other factors support using a rate that may far exceed this minimum. For example, suppose that a user needs three 150 Mbit/s circuit-switched channels for video use and also 10 Mbit/s for DTDM virtual channels. In order to multiplex these four physical channels onto a common

[1]The optimal slot size is a difficult problem involving tradeoffs between service flexibility, transmission efficiency, and equipment simplicity. This slot size has been chosen for illustrative purposes only.

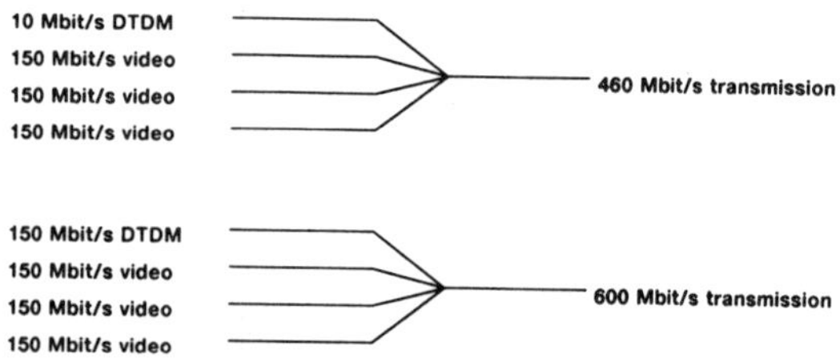

Fig. 10. Possible access structures for B-ISDN.

higher speed facility, a higher speed transmission rate of at least 460 Mbit/s is required, as shown in Fig. 10.

If the rate of the physical channel supporting DTDM virtual channels is chosen to be 150 Mbit/s instead of 10 Mbit/s, the rate of the higher speed channel increases to 600 Mbit/s, which still allows the same technology to be used for optical transmitters and receivers. More importantly, using a single common physical transmission rate for video circuit channels and for DTDM virtual channels simplifies time division multiplexer design and allows the same interface components to be used for both circuit and packet channels reducing development costs. The higher transmission rate also makes growth capacity available for handling future increases in DTDM traffic and reduces the transit delay through packet switches. Finally, if a user has a specific service needs that require DTDM virtual channels exceeding a total of 150 Mbit/s, using a common transmission rate allows circuit channels to be easily converted to packet usage.

C. DTDM Packet Switches and Multiplexers

High speed packet switches using a Batcher sorting network followed by a Banyan self-routing network are in the prototyping stage in research laboratories [3], [27]. These switches offer the potential of building a single-shelf packet switching fabric with 256 input ports and 256 output ports, each operating at 150 Mbit/s, giving a total throughput of more than 10 million packets/s and an input-to-output delay of less than 1 ms.

In order to build a packet switch capable of serving an office of 10 000 lines, one alternative is to use 40 256 × 256 packet switching modules as the first stage of a multistage packet switch, as shown in Fig. 11. Alternatively, if the average user requires only a small fraction of the 150 Mbit/s capacity of a DTDM facility, a packet multiplexer could be used as a front end for each port on the packet switch, as shown in Fig. 12. If the average bandwidth required per line is 1 Mbit/s (500 packets/s), then the total throughput required to handle all 10 000 users is 5 million packets/s. Therefore, by using packet multiplexers, a single packet switching module is sufficient to handle an entire 10 000 line office. Furthermore, the packet multiplexers can be placed at a remote switching node to reduce the cost of dedicated fiber.

One consideration with packet multiplexing is that the distribution of high bandwidth users over the multiplexers may vary with time. Even if the average bandwidth required per user is only 1 Mbit/s, several users assigned to the same packet multiplexer may place high bandwidth

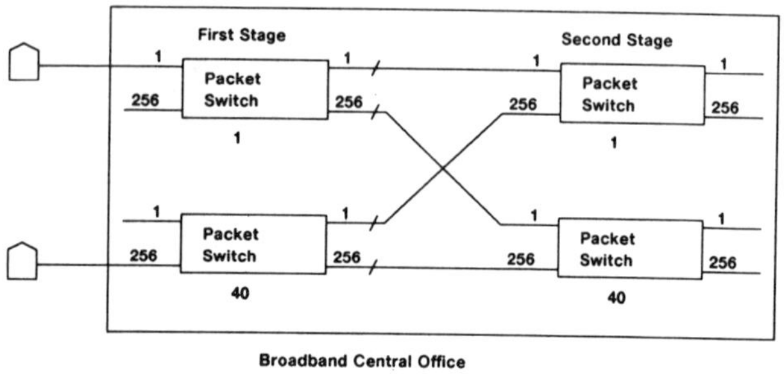

Fig. 11. A 10 000 line office with multiple packet switches, no remote switching, and no concentration.

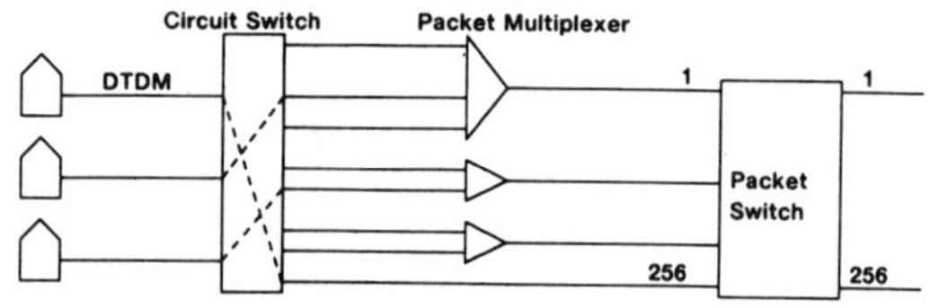

Fig. 12. A 10 000 line office using packet multiplexers.

calls simultaneously and exceed the output capacity of the packet multiplexer, resulting in the inability to complete one or more calls. If a common transmission rate is used for both circuit and packet channels, each user's packet channel can be routed through a circuit switch to gain access to the packet multiplexer, as shown in Fig. 12. Then, if a user requires a high bandwidth connection, the network can reroute that user's physical DTDM channel to a packet multiplexer with a lower concentration ratio instead of rejecting the call.

D. Bandwidth Management and Resources Allocation

The primary importance of DTDM is the ability to allocate bandwidth on demand. In conventional packet switching networks, it is not possible to allocate bandwidth; every user transmits information as desired and shares the available bandwidth. Whenever contention occurs at a concentration point, the excess packets are buffered (and therefore delayed) until they can be transmitted to the output port. So, excessive traffic generated by one user can cause another user's packets to be delayed or lost. With DTDM, the bandwidth is guaranteed so excessive traffic generated by one user cannot affect the communications of another user.

Bandwidth is allocated on the output ports of multiplexers and switches. For example, consider a Batcher–Banyan packet switch, which is nonblocking internally with contention occurring only at the output ports. One method for managing resource allocation is to limit the bandwidth reservation such that

$$\sum_i b_{ij} < f * C_j \qquad (f < 1)$$

where b_{ij} is the reserved bandwidth of the DTDM channel i on output port j, C_j is the capacity of output port j, and f is the output port fill factor determined by the packet loss rate delay requirements. Buffers are required in the switch to handle output port contention; the quantity required is determined by the packet loss rate requirements and the output fill factor.

One disadvantage of this algorithm for bandwidth allocation is that many services, such as remote disk servers and services involving still pictures, have very bursty transmissions—they require high bandwidth while actively transmitting, but are idle until the next burst. This means that the average load on an output port is low even when most of the bandwidth capacity is reserved, and users may be unwilling to pay for the reservation of bandwidth that is only required sporadically during a call. A second disadvantage of this algorithm is that different services have different packet loss rate requirements. For example, high quality speech transmission cannot afford lost packets because this reduces the speech quality, and using a reliable transmission protocol to recover from lost packets is unacceptable because it introduces a variable delay. For picture transmission and remote disk servers, on the other hand, a higher packet loss rate is acceptable, using a higher layer protocol to recover from lost packets.

The two disadvantages of a pure guaranteed-bandwidth allocation algorithm suggests the need for two *priority levels* in the DTDM protocol. Services requiring a low packet loss rate would use the higher priority, while services not affected by delay or packet loss could use the lower priority. The same network elements could be used to carry both types of traffic. In maintaining resource allocation information, separate sums would be kept at each output port for high and low priority bandwidth allocations. The high priority bandwidth sum would only be allowed to reach some fraction $f_h < 1$ of the output port bandwidth, while the low priority bandwidth could allow overbooking up to some factor $f_l > 1$ of the output port bandwidth. The maximum contention rate between high priority packets is determined by f_h; the packet loss rate for high priority packets is controlled by specifying f_h and the quantity of buffers. When a high priority packet contends with a low priority packet, the high priority packet passes through the switch first while the low priority packet is buffered. If a high priority packet contends with a low priority packet for a single available buffer, the high priority packet is buffered and the low priority packet is discarded. Use of such a priority scheme reduces costs by increasing the utilization of the output ports while still maintaining the low packet loss rates required for some services.

V. Multiple Connection and Multiple Party Services

Two capabilities required for many B-ISDN services are *multiple connection calls* and *multiple party calls*. For example, a simple videotelephone call consists of two connections—a video connection and a voice connection. The user needs to be able to set up both connections as a single call, but also be able to control each connection independently.

A more elaborate example that illustrates both multiple connection and multiple party capability is that of a real estate service, as shown in Fig. 13. A transferee calls a real estate vendor's database to search through a multiple listing of available real estate; this requires a data connection and possibly a voice connection if a real estate agent is desired. When the transferee requests more information on a few selected homes, the first vendor adds a second vendor to the call, who has the video information about these homes. The connections now include a data and a voice connection between the transferee and the first vendor, a data connection between the first and the second vendors, and a video connection from the second vendor to the transferee.

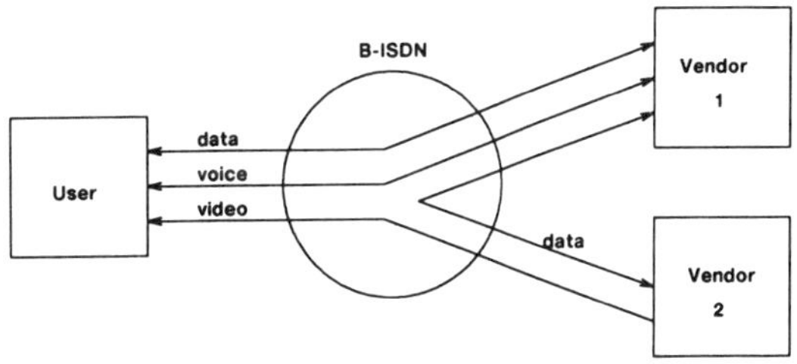

Fig. 13. Illustration of a multiple party, multiple connection call.

In order for B-ISDN to support multiple connection calls, the signaling protocol for ISDN (Q.931/I.451) [28] needs to be extended to include a *connection reference* element, which allows users to refer to specific connections in a call. Another signaling capability needed is support for multiple party calls. This is important for teleconferencing, and also for services such as the real estate service mentioned above. To handle multiple party calls, the signaling protocol needs to be extended to include a *party reference* element, which allows users to refer to specific parties in a call.

Signaling protocol extensions to handle multiple connection and multiple party calls require a new model for a call [29]. One proposed call model has been implemented in a new signaling protocol that is based on extensions to the Q.931 signaling protocol. In addition to multiple party and multiple connection capability, this model also allows each user to specify, on a per-connection basis, the type of coding used on his or her access interface. Therefore, if two users use different coding formats, the network can provide code conversion of compatibility. In addition, this model allows the user to specify the quality of service required for a connection. For example, if a user in New York with an HDTV video codec calls a user in San Francisco with an identical codec, they may not necessarily need HDTV quality for a simple video conference. By allowing the user to specify the quality of video desired, the network can compress the video to a lower rate in order to use a lower cost trunk. Therefore, users are not required to purchase multiple video codecs in order to communicate with different terminals or to take advantage of lower cost transmission.

VI. Summary and Conclusions

B-ISDN switching capability requirements must be driven by the service needs. However, the diversity of

CAPABILITY \ SERVICE	Switched access television	Video-on-demand	Videotelephone	Audio communications	Still picture services	Remote disk server	Desktop teleconferencing
150 Mbit/s video channel	x	x	x				x
Fast call setup	x					x	
High calling volume	x					x	
Broad range of channel rates		x		x	x	x	x
Selective broadcast	x						x
Remote switching	x						
Circuit switching for video	x	x	x				x
NTM packet switching for flexibility		x		x	x	x	x
Multiple connection calls		x					x
Multiple party calls			x	x			x
Code conversion for compatibility		x	x	x			x
Compression for trans. cost reduction		x	x				x
Two priority levels in packet switch				x	x	x	x

Fig. 14. Service-capability mapping matrix.

possible B-ISDN services complicates the specification of switching requirements. One approach is to select a few representative services in order to generate a set of basis vectors in capability space. As new services are imagined, they are projected onto this set; if the new service cannot be constructed from the basis vectors, then other capability vectors are added to the set. This can aid B-ISDN architects in identifying the switching requirements that are required for the services expected to be most in demand.

This paper has examined several services ranging from speech to full-motion video and has identified several capability requirements. A summary mapping each service to the associated capability requirements is shown in Fig. 14.

To offer video services, circuit switching at 150 Mbit/s is needed. Selective broadcast and remote switching capabilities, while not absolutely essential to broadcast video services, can reduce the cost of dedicated fiber and make these services more economically attractive.

The information transfer rates needed for various non-video B-ISDN services range from a few bit/s to several Mbit/s. A promising alternative for handling this range of rates without requiring the use of multiple switching fabrics is dynamic time division multiplexing (DTDM) packet switching, which offers users the flexibility to allocate access channels dynamically on a per-call basis. By combining the explicit channel identification of packet switching with the framing pattern and control information of TDM switching, DTDM forms a transmission format for slotted packets on synchronous facilities with the favorable characteristics of both TDM switching and packet switching. Flexibility is further increased by using a single common physical transmission rate (150 Mbit/s) for both video circuit channels and DTDM logical channels. This also allows the same interface circuitry to be used for both circuit and packet channels.

The model of a call needs to be extended to support multiple connections and multiple parties in the same call. In addition, the model of a call should be flexible enough to support code conversion for compatibility between terminals with different coding formats, and compression of signals to reduce the cost of interoffice facilities. The ISDN signaling protocol needs to be extended to support this new call model.

Some services require fast call setup or have high calling rates. These services may require that call processing in the central office be distributed over multiple processors to handle the load.

Acknowledgment

The following individuals provided valuable input to this paper: M. Beckner, H. Carabelli, N. Haller, J. Irven, L. Linnell, V. Massa, S. Minzer, J. Patterson, B. Schatz, P. Shumate, L. Smoot, P. White, and L. Wu.

References

[1] "Further gains for Japan predicted in telecom war," *Lightwave*, Sept. 1986.
[2] G. Hayward, L. Linnell, D. Mahoney, and L. Smoot, "A broadband ISDN local access system using emerging-technology components," in *Proc. ISS*, 1987.
[3] C. Day, J. Giacopelli, and J. Hickey, "Applications of self-routing switches to LATA fiber optic networks," in *Proc. ISS*, 1987.
[4] P. E. White, "The broadband ISDN—The next generation telecommunications network," in *Proc. ICC*, 1986.
[5] H. Ohnsorge, "Introduction and overview of broad-band communications systems," *IEEE J. Select. Areas Commun.*, vol. SAC-4, July 1986.
[6] L. Linnell, "A wideband local access system using emerging-technology components," *IEEE J. Select. Areas Commun.*, vol. SAC-4, July 1986.
[7] "Home shopping," *Business Week*, pp. 62–69, Dec. 15, 1986.
[8] J. B. O'Neal, "Predictive quantizing systems for the transmission of TV signals," *Bell Syst. Tech. J.*, vol. 45, no. 5, pp. 689–721, May–June 1986.
[9] "Encoding parameters of digital television for studios," CCIR Rec. 601 (Mod 1), Nov. 25, 1983.
[10] G. J. Tonge, "Sampling of television images," IBA (U.K.), Rep. 112/81, May 1981.
[11] Y. Ninomiya, "A single channel HDTV broadcast system—The muse," NHK (Japan), Lab. Note 304, Sept. 1984.
[12] H. Yamada, H. Kataoka, T. Sampei, and T. Yano, "High-speed digital switching technology using space-division-switching LSI's," *J. Select. Areas Commun.*, vol. SAC-4, July 1986.
[13] H. Reiner, "Integrated circuits for broadband communications systems," *J. Select. Areas Commun.*, vol. SAC-4, July 1986.
[14] B. Schaffer, "Switching in the broad-band ISDN," *J. Select. Areas Commun.*, vol. SAC-4, July 1986.
[15] K. Langer, G. Lukanek, J. Vathke, and G. Walf, "Broadband switching network and TV switching for 70 Mbps," in *Proc. 1986 Int. Zurich Sem. Digital Commun.*, 1986.
[16] CCITT Recommendation G.711, "Pulse code modulation (PCM) of voice frequencies," *CCITT Red Book*, ITU, Geneva, Switzerland, 1985, vol. III, fascicle 3.
[17] CCITT Recommendation G.72, "32 kbit/s adaptive differential pulse code modulation (ADPCM)," *CCITT Red Book*, ITU, Geneva, Switzerland, 1985, vol. III, fascicle 3.
[18] L. R. Rabiner and R. W. Schafer, *Digital Processing of Speech Signals.* Englewood Cliffs, NJ: Prentice-Hall, 1978.
[19] CCITT Recommendation J.41, "Characteristics of equipment for the coding of analogue high quality sound programme signals for transmission on 384 kbit/s channels," *CCITT Red Book*, ITU, Geneva, Switzerland, 1985, vol. III, fascicle 4.
[20] CCITT Recommendation J.42, "Characteristics of equipment for the coding of analogue medium quality sound programme signals for transmission on 384 kbit/s channels," *CCITT Red Book*, ITU, Geneva, Switzerland, 1985, vol. III, fascicle 4.

[21] CCITT SG XVIII Working Party XVIII/1, "Draft status report of the task group on ISDN—Broadband aspects," *CCITT Temporary Doc.*, 16 (P), Geneva, July 16–18, 1986.

[22] P. Gonet, P. Adam, and J. Coudreuse, "Asynchronous time division switching: The way to flexible broadband communication networks," in *Proc. 1986 Int. Zurich Sem. Digital Commun.*, Mar. 1986.

[23] J. LeBris and M. Servel, "Integrated wideband networks using asynchronous time division techniques," in *Proc. ICC*, 1986.

[24] M. W. Beckner, T. T. Lee, and S. E. Minzer, "A protocol and prototype for broadband subscriber access to ISDN's," in *Proc. ISS*, 1987.

[25] L. T. Wu, S. H. Lee, and T. L. Lee, "Dynamic TDM: A packet approach to broadband networking," in *Proc. ICC*, 1987.

[26] CCITT Recommendation I.441, "ISDN user-network interface data link layer specification," *CCITT Red Book*, ITU, Geneva, Switzerland, 1985, vol. III, fascicle III.5.

[27] J. Hui, "A broadband packet switch for multi-rate services," in *Proc. ICC*, 1987.

[28] CCITT Recommendation I.451, "ISDN user-network interface layer 3 specification," *CCITT Red Book*, ITU, Geneva, Switzerland, 1985, vol. III, fascicle 5.

[29] S. E. Minzer and D. R. Spears, "Principles for defining a signaling protocol for complex calls in an ISDN," in *Proc. ICC*, 1987.

Dan R. Spears (M'75) received the B.S. degree in physics from the University of Kentucky, Lexington, in 1971 and the M.S. degree in electrical engineering from the University of Louisville, Louisville, KY, in 1979. He is currently working toward the Ph.D. degree in electrical engineering at Rutgers University, New Brunswick, NJ.

He joined South Central Bell in 1973 and has worked as a Facilities Equipment Engineer and a Switching Systems Network Planner. In 1979 he began a rotational assignment with the Bell System Center for Technical Education as an Instructor/Developer for various network planning courses. In 1981 he transferred to the New Switching Systems Lab at Bell Laboratories to work on new packet switching systems and customer programmable switching software, where he patented a new procedure for distributing packets in a packet switching system. In 1983 he transferred to Bell Communications Research where he wrote the software for the Video Resource Manager (VRM), a flexible video services testbed, and codeveloped the signaling protocol for the Experimental Research Prototype (ERP), a broadband ISDN testbed. He is currently writing call processing software for the ERP prototype.

Evolution of ISDN Towards Broadband ISDN

Rainer Händel

Since about 1985, worldwide activities to evolve Integrated Services Digital Network (ISDN) into an optical-fiber-based universal broadband network have resulted in the first baseline documents, such as CCITT Recommendation I.121. The Asynchronous Transfer Mode has been established as a key element of Broadband ISDN (BISDN). The adoption of the G.707–709 recommendations on the new digital transmission hierarchy will also strongly influence further BISDN development. This article addresses network aspects of the forthcoming BISDN, including some special impacts of ATM on network scenarios and network evolution.

The ISDN Concept of the CCITT I Series Recommendation

In 1984, the Plenary Assembly of the CCITT (International Telegraph and Telephone Consultative Committee) adopted the I series recommendations dealing with ISDN matters. CCITT stated that "an ISDN is a network...that provides end-to-end digital connectivity to support a wide range of services, including voice and non-voice services, to which users have access by a limited set of standard multipurpose user-network interfaces" [1]. Such an ISDN standard interface was defined (and called "basic access"), comprising two 64-kb/s B channels and a 16-kb/s signaling D channel. Another type of interface, the "primary rate access," with a gross bit rate of about 1.5 Mb/s or 2 Mb/s, respectively, offers the flexibility to allocate high-speed H channels or mixtures of B and H channels (see Table I).

Desired Evolution of the ISDN

The need for services employing bit rates greater than 2 Mb/s was clearly seen when the I series recommendations were written. Thus in the 1984 version, other H channels and interface types above 2 Mb/s were declared for further study. To this aim, CCITT COM XVIII installed the "Task Group on Broadband Aspects of ISDN" at theirLondon interim meeting in January 1985. From the very beginning, the great majority of delegates within CCITT COM XVIII were of the opinion that "ISDN from its definition is a network that will, due to its evolution, allow total integration of broadband services" [2] (i.e., bearer services as well as teleservices, interactive services, and distribution services).

In comparison to several dedicated networks, service and network integration has major advantages in economic planning, development, implementation, operation, and maintenance [3]. While dedicated networks require several distinct costly subscriber access lines, the BISDN access can be based on a single optical fiber for each customer. Large-scale production of highly integrated system components of a unique BISDN will lead to cost-effective solutions.

In principle, BISDN should be suitable for business applications and residential subscribers. To meet the requirements of possible future broadband services, a concept is necessary that is highly flexible and can support high-speed data transmission as well as moving picture services such as video telephony or TV program distribution. The Task Group on Broadband ISDN and other standardization bodies, like T1 in the United States and CEPT in Europe, have been trying to find such flexible solutions, taking into account both assumed service needs and probable network evolution.

In the following, I would like to present the current state of the BISDN discussion, especially within CCITT, briefly addressing broadband services and then concentrating on the user-network interface, as its definition is of utmost importance with respect to the introduction of a worldwide unique BISDN. Moreover, network scenarios and evolution and interworking aspects of the arising broadband realizations will be addressed.

Overview of Current BISDN Concepts

General Aspects

As already stated, there are new service needs (some can be seen quite clearly now, and others are more elusive) that have to be catered to by BISDN and by assumed network evolution capabilities.

Future optical-fiber-based networks will be suitable for the transmission of channel bit rates greater than 100 Mb/s. Connections through the networks can be built up with conventionally circuit-switched paths or may be based on new high-speed packet transfer techniques such as the ATM (provided that the appropriate switching technologies will be available). Last but not least, the development and costs of broadband terminals will influence the conception and implementation of BISDN. For example, it is essential to know which bit rate will preferably be employed in a picture codec of a video telephony or video-conference terminal, because this defines the number of simultaneously usable video connections on a broadband subscriber loop of a fixed total bit rate.

Broadband Services

As ISDN is conceived to support all kinds of existing and future services, a close look at possible types of services was taken by CCITT COM XVIII. A variety of potential interactive and distribution broadband services was listed [4] [5]:

- Broadband video telephony/video-conference
- Video surveillance

Table I. ISDN CHANNELS AND INTERFACE STRUCTURES*

Channel	Bit Rate**	interface	Gross Bit Rate	Structure
B	64	basic access	192	2 B + D16
HO	384			23B + D64/30B + D64
H11	1536	primary rate	1544/2048	or
H12	1920	access		3HO + D64/5HO + D64
				or
				H11 /H12 + D64
				or
				....

*According to CCITT Rec. I412
**kb/s

- High-speed unrestricted digital information transmission
- High-speed file transfer/tele-action/telefax
- Video/document retrieval service
- TV distribution (with existing TV/extended/high definition quality).

The above list comprises "bearer services" (to be defined by OSI Layers 1 to 3 characteristics) and "teleservices" (full specification of Layers 1 to 7 required). To be able to define these broadband services, new attribute values to supplement CCITT Recommendation I.130 were proposed and agreed upon (e.g., "high quality video" as a new information transfer capability, with high quality denoting at least existing TV standards quality). The first examples of new broadband teleservices, high quality video telephony/video-conference, and TV distribution were preliminarily described by the attribute method. In particular, these services should offer an information transfer capability of high quality video plus 15-kHz stereophonic sound (plus user-to-user messages, if required), and the information transfer rate should be above 30 Mb/s.

The dynamic description of the above services has not yet been worked out. Other important points that need further study are:

- The interworking between video services with different qualities and bit rates
- The adaptation of existing ISDN D channel signaling procedures to also cover the needs of distribution services (e.g., TV program selection)
- The investigation of "connectionless" applications in a BISDN environment
- The description of cell-multiplexing-based ATM services

ATM Principles and Impact

ATM (Asynchronous Transfer Mode) is considered the ground on which BISDN is to be built [6–8]. In ATM, all information to be transferred is packed into fixed-size slots called "cells," which are identified and switched by means of a label in the header (see Figures 1 and 2). The term "asynchronous" refers to the fact that cells allocated to the same connection may exhibit an irregular recurrence pattern, as cells are filled according to the actual demand (see Figure 1); this aspect is well-known from the existing packet transfer modes.

ATM is a multiplexing and switching technique that is confined to Layer 1 and basic functions of Layer 2 of the OSI model. The so-called ATM layer of CCITT's BISDN protocol model (see Figure 3) [9] sits on top of the physical media dependent layer (that is responsible for the proper bit transmission). The ATM layer in Figure 3 is common to all services. Its functions are represented by the ATM cell header functions. Thus, the boundary between the cell header and the cell information field corresponds to the boundary between the ATM layer and the higher layers.

The ATM header contains the label called Virtual Channel Identifier (VCI) and an error detection field (see Figure 2). Error detection on the ATM level is confined to the header. A special VCI may be used to indicate unassigned cells or, as an alternative, a special header field could be used instead. Other ATM functions are not yet defined. For example, the need for quality-of-service indication, payload type (to be able to distinguish between user data cells and

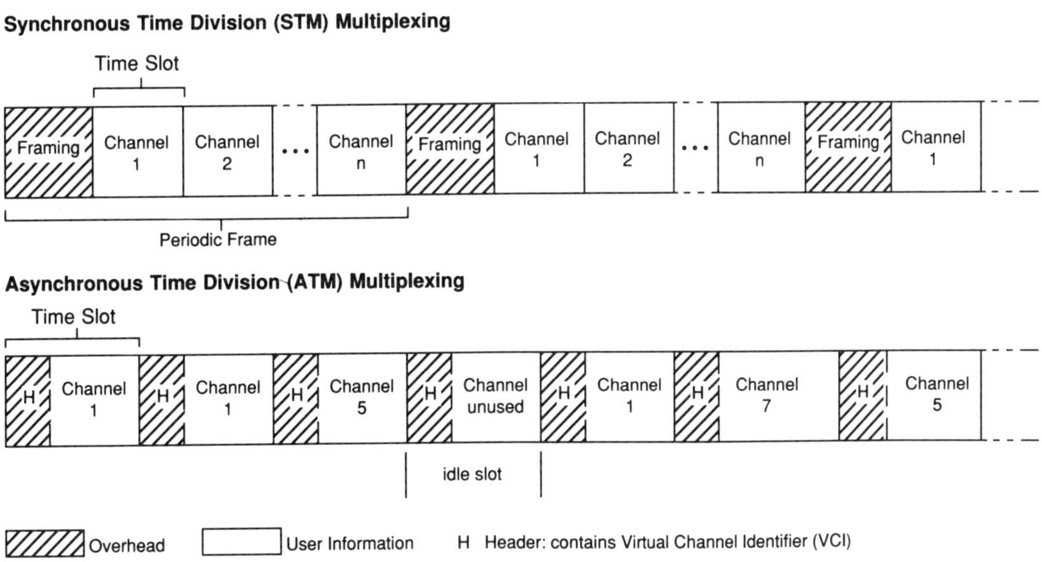

Fig. 1. STM and ATM principles.

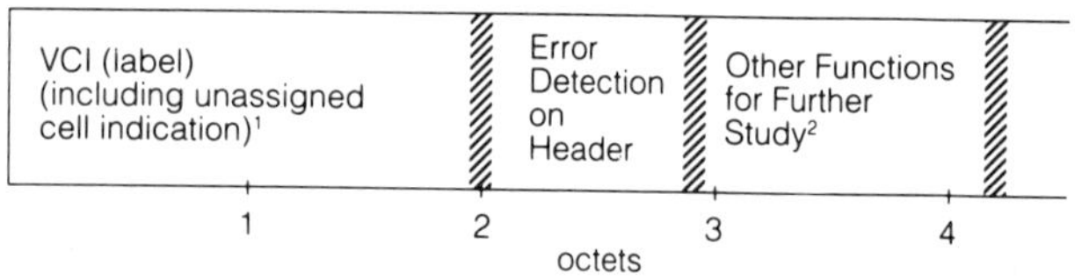

[1]Either explicit field, or indicated by special VCI
[2]e.g. error correction, quality-of-service indication (see text)

Fig. 2. ATM cell header function and fields.

OA & M cells), and others [6] is still under discussion. The ATM cell header size should therefore be at least 4 octets; it is felt that some spare capacity for further, up to now unidentified, ATM functions should be reserved.

The higher layers of the ATM protocol model are service-dependent. The adaptation layer allows higher layer information of either the user plane or the control plane to be mapped on to the ATM layer. Thus, the ATM layer can be the unique transfer basis for continuous bit-stream-oriented services as well as packet-type services.

ATM allows for bit rate allocation on demand, so the bit rate per connection can be chosen quite flexibly. In addition, the actual "channel mix" at the broadband interface can dynamically change over a vast range. The definition of high-speed channel bit rates is now, in contrast to the situation in an STM (Synchronous Transfer Mode) environment, a second-rank task.

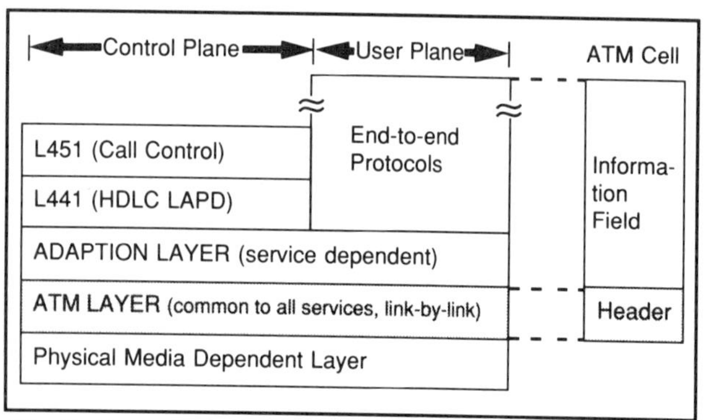

Fig. 3. Protocol model for ATM.

ATM in principle allows for:

- Labeled channels with any fixed bit rate in the range of some kb/s up to the total payload capacity of the interface
- Connections with varying bit rates (burst-type traffic)

Therefore, flexible definition of broadband channels (through-put) is possible, according to service needs.

ATM supports, for example, the following channels:

- H2 = 32.768 Mb/s, about 44 Mb/s
- H4 = 132–138.240 Mb/s

However, the ATM approach requires that many new problems be solved. For example, the impact of possible cell loss, cell delay, and cell jitter on the quality of all broadband and other services needs to be determined. One of these problems, the enhanced echo delay of ATM speech connections, will be briefly discussed in a later section.

Broadband Interfaces and User-Network Access Configurations

The required bit rates for typical broadband services and the number of connections to be offered simultaneously to a customer can give an indication of how broad the broadband user-network interface should be. Due to some uncertainties about service bit rates, simultaneous service mix, future improvement of video-coding, HDTV requirements, etc., people have decided to consider two interface candidates [6–8]):

- one, around 150 Mb/s
- the other, four times as much (i.e., around 600 Mb/s)

The 150 Mb/s interface (with symmetrical information transfer capability) is conceived for ATM use and effectively supports the flexible breakdown of the interface payload into several (labelled) channels on demand. This interface type is especially suitable for business customers who want to access a variety of mainly interactive services.

The other interface—obviously much more powerful in terms of usable bit rate—may be the adequate solution for certain needs, for example, simultaneously conveying several high quality TV programs to a residential customer. Therefore, it can be asymmetrical, with 150 Mb/s being a sufficient transfer capability in the direction of the network. The feasibility of a 600 Mb/s ATM interface is somewhat unclear (at least in the near future), so hybrid solutions (ATM/STM usage) or subdivision of the 600 Mb/s interface into smaller ATM modules (e.g., 150 Mb/s) are also under consideration.

It should be noted that there are some doubts about the suitability and economy of ATM as a universal transfer mode for all kinds of services. For this reason, the option of STM usage of broadband interfaces is being maintained within CCITT. This view has been especially expressed by Japan.

General Requirements of Broadband ISDN Interfaces

The discussion of broadband user-network accesses must consider the existing 64-kb/s ISDN recommendations as a basis. New ISDNs with broadband capabilities should, at least for a long interim period, support terminals already deployed and also the basic and primary rate interfaces of CCITT Recommendation I.412. Interworking with 64-kb/s services and with other non-ISDN networks (e.g., analog telephone networks and packet data networks) must be provided for.

Due to its inherent flexibility in terms of actual bandwidth consumption (according to individual needs of different services), the role of ATM for broadband applications has been recognized, almost unanimously, all over the world. CCITT Draft Recommendation I.121 ("Broadband Aspects of ISDN") states that the target network for BISDN is based on ATM.

On the other hand, the introduction of ATM networks will be a complex task that will take considerable time to be completed. For economic reasons, the existing networks will still be in use for some time. This is especially true for the 64-kb/s ISDNs that are just about to be implemented now. In addition to economical reasons, there may be technical obstacles to an immediate overall ATM deployment. Therefore, certain intermediate steps on the way to a universal ATM network will probably be implemented, and definition of broadband interfaces should take into account the required compatibility with assumed broadband network evolution scenarios.

The general requirements of BISDN interfaces can be summarized as follows:

- Support of existing terminals and interfaces (e.g. 64 kb/s-based ISDN, telephone networks, data networks)
- Suitability for the assumed broadband target network (ATM)
- Compatibility with evolution scenarios from interim solutions to a possible universal ATM network

User-Network Access Configurations

The general reference configuration for the broadband user-network access is shown in Figure 4. Several terminals are connected (via appropriate terminal interfaces at the S/S_B reference points) with the subscriber premises network, which accesses the local network itself via a standardized interface. The Broadband Network Termination (BNT) is the boundary between the local access network and the subscriber premises network. In Figure 4, an asymmetric interface with 150 Mb/s to the network and 600 Mb/s to the subscriber is drawn as an example.

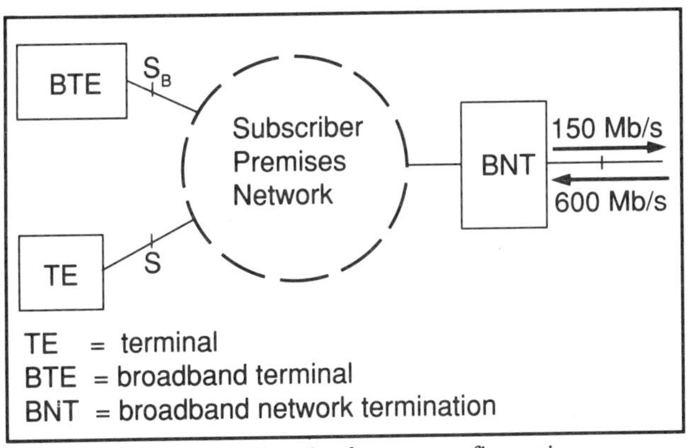

Fig. 4. General broadband reference configuration.

The subscriber premises networks may be, according to the customers' needs, either quite complex networks themselves (e.g., LANs or PABXs), with different topological structures (ring, bus, star, or mixture thereof) and different switching features, or may be rather simple installations. An example of the latter case, offering broadband interfaces and basic interfaces, is shown in Figure 5. The BNT provides the customer with a multiple broadband access (each of several star-shaped S_B interfaces allow the connection of a Broadband Terminal, or BTE), and, in addition, with several S interfaces, according to CCITT Recommendation I.430. These basic (2B + D16) interfaces are realized in the (passive) bus structure defined within the above-mentioned I.430 Recommendation. The provision of the basic interfaces by the BNT is optional; it will be required in those cases where the 64-kb/s ISDN customer installations are still to be used and BISDN accesses are implemented as an add-on communications means.

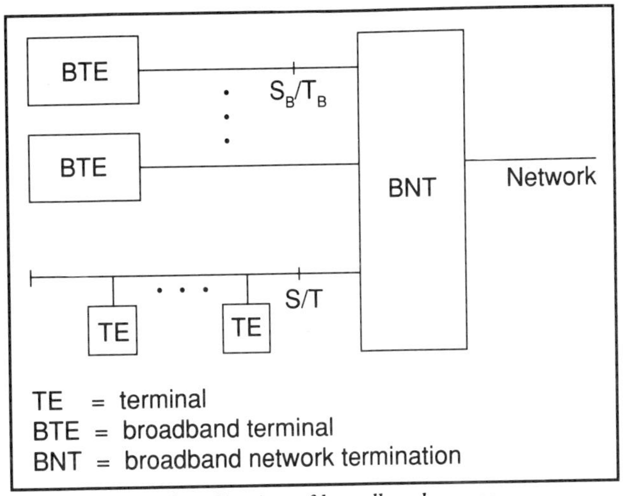

Fig. 5. Physical realization of broadband access.

In the case shown in Figure 5, the BNT could simply broadcast all downstream information to all BTEs and could statistically multiplex all upstream ATM cells, thus also getting rid of the access contention problem relating to the multiple broadband interfaces. If basic accesses are realized, the BNT will have to adapt these STM interfaces to ATM internally.

Some Special Interface Items

Transfer Mode, Frame Structure, Bit Rate

The 150 Mb/s interface is "full ATM," according to CCITT Recommendation I.121. Still under discussion are the method of framing and the exact bit rate. Figure 6 shows two different options for the 150 Mb/s interface: ATM with external frame (a good solution for a smooth evolution of BISDN and support of ATM/STM interworking) and purely cell-interleaved ATM (the optimum solution for a pure ATM target network). While in the latter case all overhead information is inserted into cells, in the former case an external frame has reserved overhead locations. In the case of pure cell multiplexing, regular insertion of framing cells is possible but not mandatory.

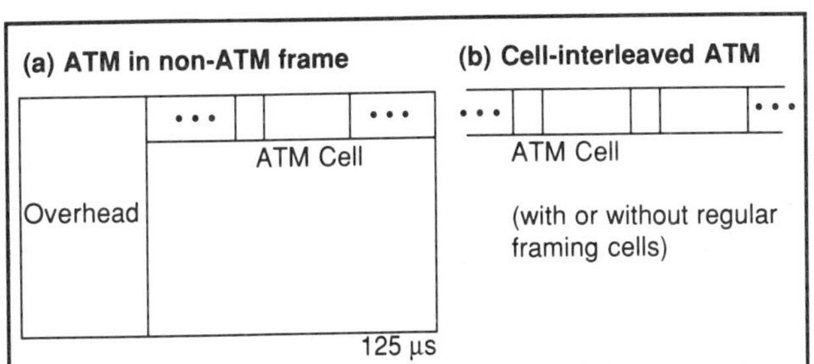

Fig. 6. Proposed frame structures for 150 Mb/s user-network interface.

If option (a) in Figure 6 is chosen, undoubtedly the 8-kHz, SONET-based solution with 155.52 Mb/s (now standardized in CCITT Recs. G.707, G.708, G.709) will be adopted. Forthcoming transmission systems will comply with this new synchronous hierarchy standard, and they "are expected to be deployed in the interoffice and access portions of some U.S. public networks prior to broadband aspects of ISDN being implemented. Compatibility with these systems would allow for lower costs to provide BISDN access which will require local access links and could require interoffice links to switches in central offices remote from the serving wire center" [10]. In this case, identity (in the physical layer) of the user-network and the network-node interface would yield the benefit of being able to use common equipment for transmission and network administration functions in the access and interoffice environments.

Interface option (a) of Figure 6 has some other advantages:

- The 155.52 Mb/s frame can easily be used as a building block for interfaces with higher bit rates ($n \times 155.52$ Mb/s, e.g. $n = 4$).
- The 8-kHz clock is directly provided and can be employed if required by LANs/MANs based on the emerging IEEE standard 802.6 [11].
- The same circuitry can be used on both sides of the BNT because no clock conversion and frame adaptation is necessary.

Other Layer 1 Aspects of the Interface

It is the goal of standardization to define a unique interface, or at least a very limited set of interfaces. Consequently, parameters such as transmission medium, coverable transmission length, etc. must be fixed.

CCITT Study Group XVIII felt that transmission lengths up to about 1 km in the subscriber premises networks should be catered to. This requirement could be met by employing optical fiber. Unfortunately, an optical fiber interface may be more difficult to install, maintain and operate, and is assumed to remain more expensive than electrical interfaces for a certain period of time. Therefore, as an alternative, an electrical interface can be conceived that may be extended in length by means of media adaptors that perform electrical/optical conversion, thus allowing for the insertion of a far-reaching optical transmission system.

CCITT stated, moreover, that no power supply across the user-network interface was foreseen; BNT and customer equipment should be power-fed from the mains (perhaps with additional battery back-up for emergency cases). To save electrical energy, a power-down state (requiring activation/deactivation procedures) would be advantageous.

Point-to-Point or Point-to-Multipoint Configuration?

In the 64 kb/s-based ISDN, the basic interface supports point-to-multipoint wiring configurations (passive bus, CCITT Recommendation I.430). An access contention resolution procedure is specified, which requires a D echo channel at the interface. This small additional capacity is always allocated within the transmission frame, whether or not it is actually used. With broadband interfaces of 150 Mb/s or even higher bit rates, physical point-to-multipoint implementations would require a considerable amount of extra bandwidth (that cannot be used by the terminals) in order to be able to guarantee undisturbed operation of several terminals that are all connected, with a shared medium, at varying local positions. This causes quite different transmission lengths and times.

Even if only a simple point-to-point interface is defined, which might be based on the same G.708 type framing at all reference points (S, T, etc.) in the network, simultaneous operation of several terminals is possible, e.g., via a LAN or a star installation (as depicted in Figure 5). Furthermore, a point-to-multipoint interface could be defined in addition to the standard interface.

Some Special Aspects of ATM Networks

Speech Connections in an ATM Environment

In an ATM speech connection, speech samples are collected until they fill an ATM cell. So, according to the chosen cell size, an additional transmission delay is encountered. This delay is 4 to 8 ms per packetization for cell information field sizes in the range of 32 to 64 octets. Depacketization also causes a delay of roughly 1 ms. The resulting total delay may exceed the tolerable limit, especially when there are several transitions from STM to ATM and vice versa within one speech connection. The main problem is not the one-way delay itself but the delayed echo at the sending side. Figure 7 gives an assessment of the ATM-related speech echo problem when an ATM area is inserted into an existing analog or digital telephone network. Depending on the chosen cell information field size (32 or 64 octets) and the type of connection, the employment of ATM more or less reduces the transmission length that can be covered without echo cancellation measures. Preferably no more than 1% objections to speech quality should be made in a network, according to CCITT Rec. G.131, but up to 10% objections may be tolerable.

To avoid extra costs for echo cancellers whenever possible, some requirements may be:

- Adjustment of the ATM cell information field size to the requirements of speech connections
- Avoidance of connections that cross more than a single ATM island
- Perhaps, during the introductory phase of ATM, sub-structuring cells to carry speech samples of more than one connection (composite cells), or the use of partly filled cells.

Support of Connectionless Protocols by ATM Networks

The ATM networks, as conceived by CCITT, will be based on a connection-oriented mode using a virtual channel that is identified in each cell header. On the other hand, connectionless mode operation would be more adequate for inter-LAN communications through ATM networks, since gateway functions could be simplified.

In [12] a user-network interface protocol for connectionless communications on ATM networks is introduced. It is proposed that the long connectionless data units (each in-

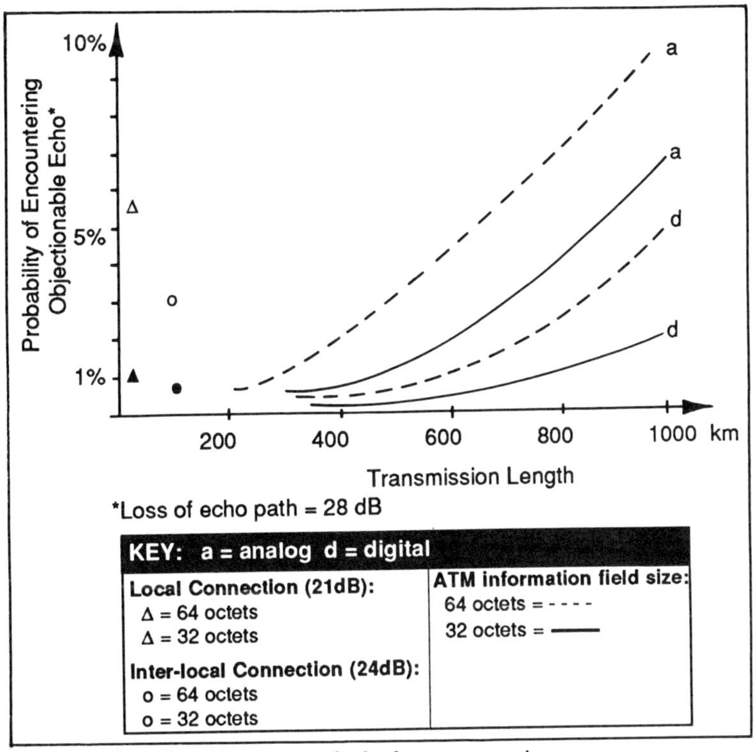

Fig. 7. Echo tolerance curves of telephone connections.

cluding destination and source addresses) be segmented into multiple cells by means of an adaptation header. The ATM cell header should not be burdened by additional, connectionless–mode-specific functions. ATM connection-oriented switches terminate lower layer protocols up to the ATM cell layer, and upper layer protocols are terminated by ATM connectionless switches in the network (Figure 8) [12].

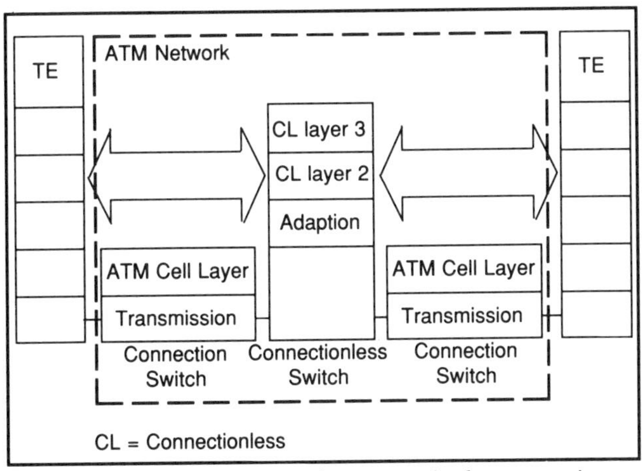

Fig. 8. Protocol termination for connectionless operation.

Future Development of Broadband ISDN

Standardization of Broadband ISDN

In the 1985–1988 study period of CCITT, the fundamental issues of BISDN were discussed and partly solved, yielding the new recommendations G.707–709 for the synchronous digital transmission hierarchy and I.121, which is a baseline document on broadband services, interfaces, and the ATM.

During the next study period (1989–1992), detailed specifications are expected; it is hoped that key parameters of BISDN, such as ATM cell functions and size and the user-network interface structure, will be agreed upon rather soon by virtue of CCITT's accelerated procedure.

Implementation of Broadband ISDNs

The implementation of ATM-based universal ISDNs will most probably be preceded by certain intermediate steps. One important task is to quickly provide customers who want to access broadband services with an integrated, flexible user-network interface that complies with the arising standards. Such an interface should have a bit rate of about 150 Mb/s and be based on ATM. The network should at least offer broadband transport capabilities, and access to and interworking with existing networks, such as telephone networks and 64 kb/s-based ISDNs (see Figure 9).

Starting from such a scenario, the integration of existing networks—including distribution networks—into an all-purpose broadband ISDN based on ATM switching and transmission can be accomplished gradually in a way that always takes into account the current technical development and economical assessment.

Acknowledgments

The author would like to thank his colleagues for all information received, especially Dr. A. Kammerl for the provision of his calculations of ATM speech echo delays.

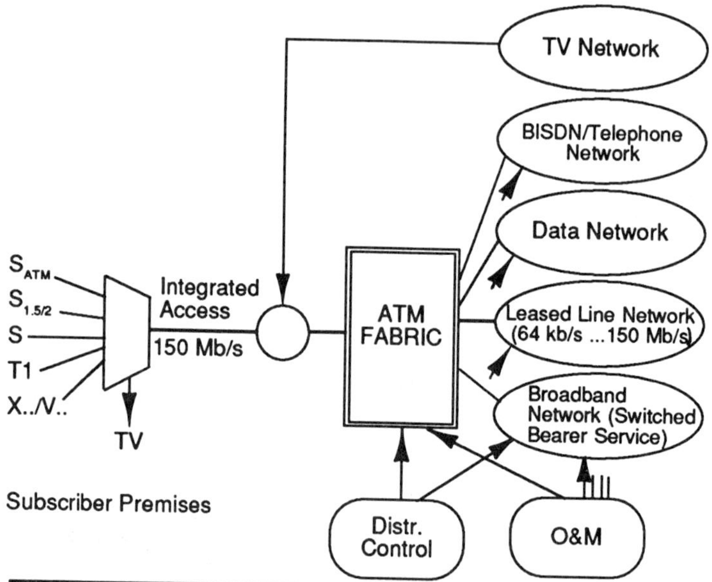

Fig. 9. Flexible integrated user access.

References

[1] CCITT COM XVIII-228-E, Mar. 1984.

[2] CCITT COM XVIII-R1-E, Feb. 1985.

[3] H. Armbruster and B. Schaffer, "Broadband ISDN: Realization Aspects of the Future," *Telecommunications Infrastructure*, Telecom Report 10, no. 5, pp. 206–216, 1986.

[4] H. J. Rothamel, "A Blueprint for Broadband ISDN Services," *Telephony*, International Edition, pp. 50–58, Nov. 17, 1986.

[5] H. Armbruster, "Applications of Future Broadband Services in the Office and Home," *IEEE J. on Sel. Areas in Commun.*, vol. 4, no. 4, pp. 429–437, 1986.

[6] CCITT Rec. I.121, Geneva, June 1988 (TD 140).

[7] T1S1.1/88-333: T1S1 Baseline Doc. "Broadband Aspects of ISDN," July 1988.

[8] Doc. T/NA5(88)/1: Report of the 10th Meeting of SWG NA5, The Hague, March 7–11,1988.

[9] V. Frantzen, "Trends in the Development of Public Telecommunication Networks," *Computer Networks and ISDN Systems*, vol. 14, pp. 339–358, 1987.

[10] CCITT COM XVIII D.1109, U.S. contribution to Hamburg Meeting, July 1987.

[11] Draft of Proposed IEEE Standard 802.6 Metropolitan Area Network (MAN) Distributed Queue Dual Bus, Physical Subsystem; Jan. 19, 1988.

[12] H. Ohmishi and T. Okada, "A User-Network Interface Protocol for Connectionless Communications on ATM Networks," Workshop on ATM, Geneva, June 1988.

Rainer Händel (born in 1950) received his physics diploma and doctorate from the University of Erlangen-Nürnberg (FRG). After finishing his studies he joined Siemens, Munich, in 1978. There he has been involved in the development of data and telephone switching systems. Since 1984 his responsibilities are in technical planning and standardization of broadband ISDN within the Central Laboratories of the Telecommunication Networks and Security Systems Group. He is a member of CCITT COM XVIII and is also active in CEPT NA5 and T1S1.

Dynamic TDM - An Integrated Transport Technique for Circuit and Packet Switched Traffic

Sang H. Lee

Bell Communications Research
Morristown, N.J. 07960

ABSTRACT

Digital transmission systems in the public switched network were developed mainly for the purpose of voice trunking, and have resulted in a hierarchical multiplexing structure. Although the hierarchical transmission plan is suitable for homogeneous voice traffic, its structure is not flexible enough to accommodate new and uncertain service demands in the future. Therefore, a packet mode of transport is favored for the fiber-based broadband network design due to its flexibility and simple interfacing structure. However, the question of how to integrate the existing transmission hierarchy into packet transport has not been resolved.

In this paper, an approach to achieving integrated transmission of packet and circuit traffic is proposed. Starting with the description of Dynamic Time Division Multiplexing (DTDM), integrated multiplexing techniques based on block level processing are described together with their design architecture. These multiplexing techniques allow the handling of both bursty traffic and continuous time-multiplexed signals through the same physical multiplexer, while maintaining a design complexity that is no greater than that of a conventional time division multiplexer. By incorporating these multiplexing techniques into the existing network, DTDM will provide a practical migration strategy from the existing predominantly circuit switched network to the future broadband packet switched network.

I. INTRODUCTION

Historically, the technology of digital transmission and digital switching in the public switched network has evolved independently in different time periods. In North America, the first digital transmission facilities were introduced in the early sixties to carry 24 analog voice channels using a single 1.544 Mb/s digital bit stream (DS-1 signal). The wide deployment of digital transmission facilities into the network required further multiplexing of several DS-1 signals into a single higher speed bit stream to efficiently utilize the transmission link. Since the frequencies of DS-1 signals are not synchronized with each other, a particular form of Time Division Multiplexing (TDM) technique, known as pulse stuffed multiplexing, was developed to satisfy the multiplexing need economically. When the network grew further, continuing efforts to effectively multiplex tributaries having different bit rates onto a common digital transmission facility resulted in the hierarchical multiplexing plan as illustrated in Figure 1. This multiplexing hierarchy provided a unified interface standard for digital transmission systems in the existing analog switching environment and economically satisfied network transport needs.

As digital time multiplexed switches (TMSs) were introduced into the network, the same digital transmission facilities were required to work with the digital switches through switch interfaces. Since the fundamental switching unit of a TMS is the 64 Kb/s time slot, the digital switch interface demultiplexes the incoming higher level hierarchical signal down to the basic digital trunk signal level (DS-1) and then re-multiplexes again according to the switch's internal format before switching occurs. This type of processing is necessary because the addition of framing and stuffing bits at each level of the multiplexing hierarchy makes it difficult to recognize the basic time slots directly from the high level hierarchical signal. As a result, this back-to-back multiplexing and demultiplexing makes the operation of switch interface quite complex.

In the recent years, the rapid deployment of digital switches into the public network has brought about the requirement of a tight network synchronization plan [1,2] to make sure that switching nodes and transmission links in the network are able to operate together at the same frequency. Network synchronization in principle can simplify the digital switch interface by making the basic channel observable in the higher rate bit streams. If the network is synchronized to the same frequency, then it is possible to multiplex these synchronized signals into the higher level of hierarchy without pulse stuffing. SYNTRAN (SYnchronous TRANsmission at DS-3) [3] is an example which can provide basic time slot observability directly from the 45 Mb/s signal. As optical transmission systems have become available since early eighties, the synchronous transmission technique has been further

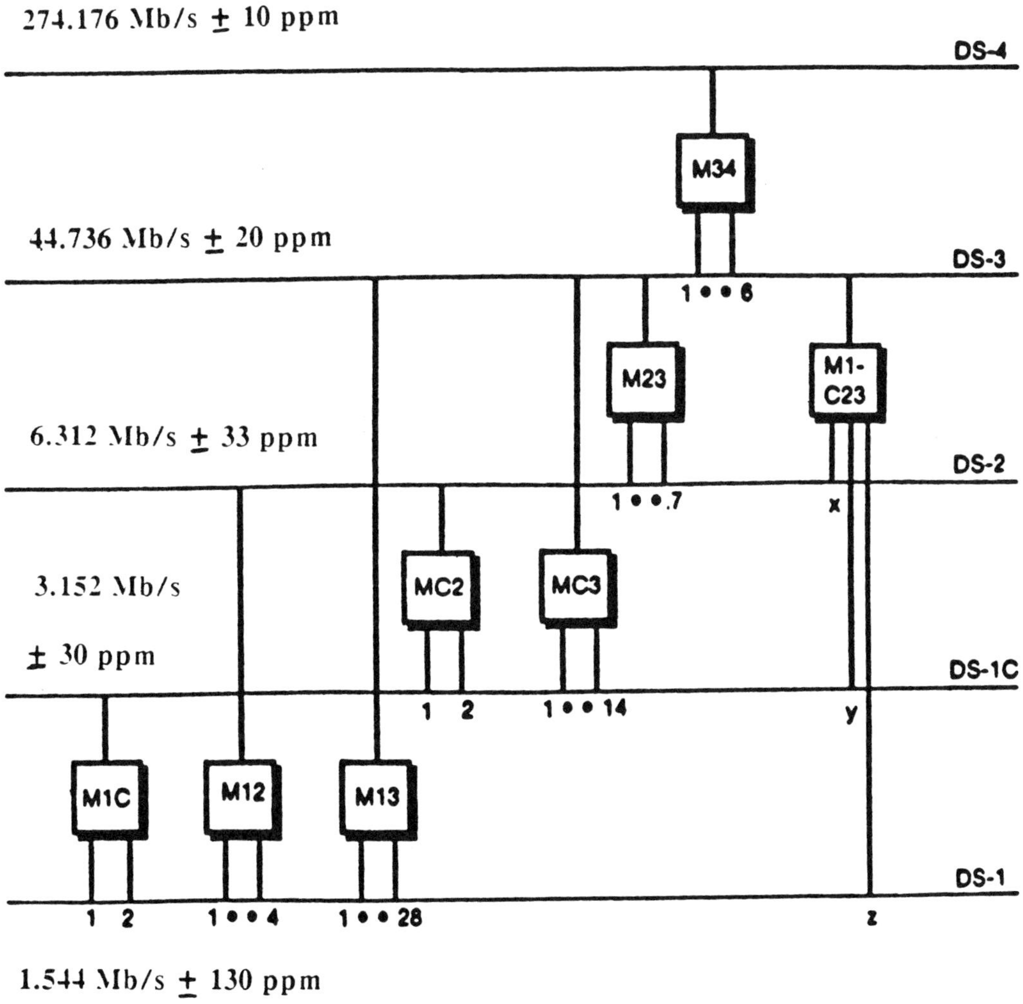

Figure 1 Digital Multiplexing Hierarchy (North America)

extended to bit rates beyond the existing multiplexing hierarchy. In particular, a new hierarchical transmission standard for fiber optic systems, called the Synchronous Optical NETwork (SONET) [4], has been proposed to provide universal compatibility in fiber optic transmission systems.

The major problem of the hierarchical transmission plan in the circuit switched network is the lack of flexibility. Since the network can only transmit a set of hierarchical level signals, every service must meet the stringent interface requirement of the specified bit rates instead of transmitting at its natural bit rate. For designing a future broadband network, it is essential to provide a flexible network transport technique which can handle vastly different types of traffic, ranging from low speed data and voice to full motion video. Therefore, packet-mode transport techniques are considered for future broadband network design due to their inherent bit rate flexibility.

Recently, a simplified packet transport technique based on a fixed length packet called asynchronous time division multiplexing (ATDM) [5,6] has been proposed to effectively carry both continuous and bursty traffic. In ATDM, the channel bandwidth is slotted and packets appear only on the slot boundaries. The slot timing is obtained from a special synchronization pattern inserted in unused time slots. Since the synchronization pattern appears only in unused time slots without periodicity, this approach is unstable for heavy traffic loading, particularly if we use packet transport to carry existing high speed hierarchical signals at close to 100 % loading.

In order to apply the packet transport technique successfully to the existing, predominantly circuit switched network, several important questions need to be resolved to achieve true "transport integration". First, there is the issue of integrating transmission hierarchy with packet transport techniques. Second, we need to identify an integrated multiplexing technique which is capable of handling both existing hierarchical traffic and packet traffic efficiently. More importantly, this integrated multiplexing technique should provide transparent transmission for the existing hierarchical signals while maintaining a design no more complex than that of a conventional TDM multiplexer. Third, similar to time slot observability in a digital circuit switched network, individual packets should be observable very easily at any transport signal level. Fourth, and finally, this multiplexing technique should not preclude any benefits that the establishment of a new fiber optic transmission hierarchy [4] can provide.

In this paper, extending the previous work on DTDM [7] we attempt to provide answers to the above questions. In section 2, starting with a brief description of DTDM we describe end-to-end multiplexing strategies in a DTDM network. By defining a packet identical to the conventional TDM frame, we can describe an integrated transport architecture that allows transmission at a flexible bit rate which is below a predetermined high backbone signal rate and provides simple transparent TDM beyond that rate. In section 3, we describe a bit rate flexible multiplexing technique, referred to here as DTDM multiplexing, capable of multiplexing both packet and circuit traffic through the same physical DTDM multiplexer to eliminate step-by-step hierarchical multiplexing, all the way up to the basic DTDM backbone signal rate. As an outgrowth of DTDM work, an alternative synchronous TDM multiplexing technique is described in section 4 for high speed point-to-point transmission between offices. We describe a simple scheme to achieve both tributary synchronization and frame alignment based on positive and negative block justification. In section 5, we briefly discuss the network synchronization issues in a broadband mixed circuit/packet switching environment and we show that it is not necessary to synchronize the entire network with DTDM transport. Finally, summary and concluding remarks are given in section 6.

II. DYNAMIC TDM

2.1 Definition

DTDM is an approach to transport both a conventional TDM hierarchical signal and packet traffic in a common integrated structure, while allowing complete bit rate flexibility. As illustrated in Figure 2, the transmission bit stream is divided into blocks like in a conventional hierarchical TDM frame. These bit stream blocks can also be viewed as a continuous chain of fixed length packets. We define this block as a fundamental unit of information transport and call it a DTDM block.

As shown in Figure 2, DTDM is a two level transport architecture that allows bit rate flexible multiplexing below a predetermined backbone rate, say 150 Mb/s, and provides simple transparent multiplexing beyond that rate. Unlike hierarchical TDM, the DTDM frame format is identical for any bit rate up to the backbone rate. Therefore, the same multiplexer can be used for a wide range of bit rates. Furthermore, straightforward block observability is preserved regardless of the previous multiplexing history.

2.2 DTDM Block Structure

Each block consists of two fixed-length fields as shown in Figure 3 ; the overhead field and the payload field. Determination of a specific block size, format and backbone bit rate is an application specific engineering decision and is not discussed in this paper. We focus

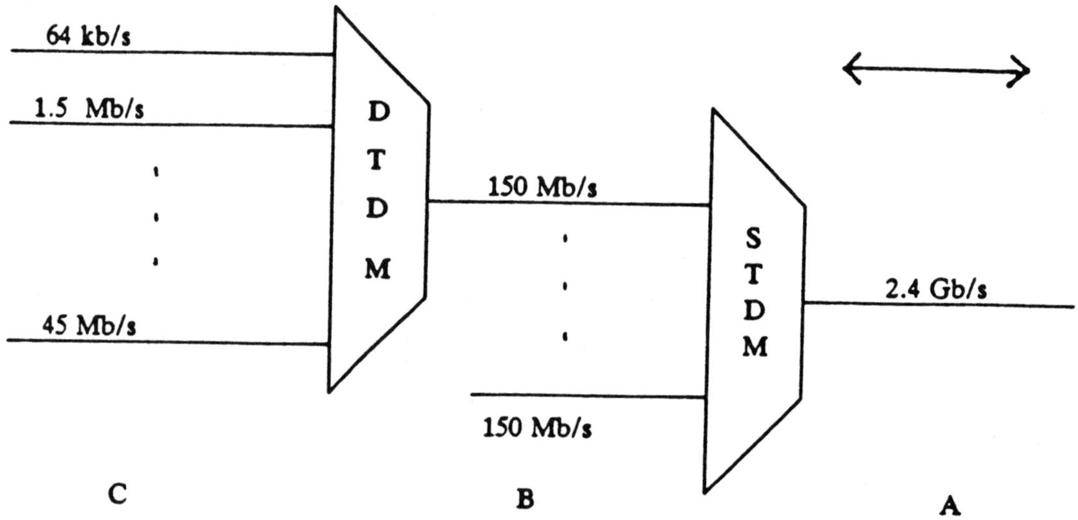

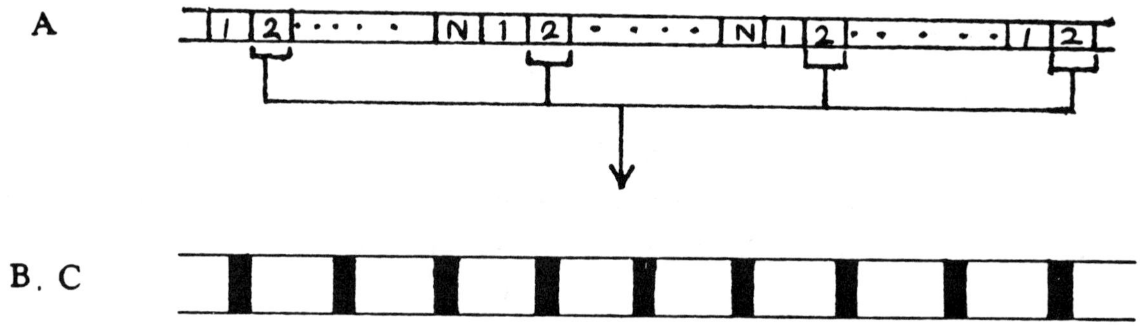

(Packet + Conventional TDM)

DTDM : Dynamic Time Division Multiplexing
STDM : Synchronous Time Division Multiplexing

Figure 2 Dynamic TDM - Two Level Transport Architecture

here on integrated multiplexing techniques which are not sensitive to a particular rate or format. In DTDM, for illustration of the basic concepts, one block has been chosen to consist of 130 bytes with a 10 byte overhead field [7], and the basic backbone rate is about 150 Mb/s, corresponding to the STS-3 bit rate in SONET [4].

The use of the overhead field in the DTDM block structure differs from both the existing hierarchical transmission and the fiber optic transmission system frame structure. In the existing hierarchical transmission scheme, bandwidth is permanently assigned for all overhead functions such as frame alignment, stuff control, error checking and performance monitoring. In the fiber optic transmission system, additional overhead bandwidth is reserved for maintenance and operation functions [4]. In DTDM transport, on the other hand, only the information which needs to appear in every block gets permanent bandwidth allocation in the overhead field. For other information such as performance monitoring, maintenance and operation, bandwidth is allocated dynamically within the payload whenever it is necessary. This simplifies the overhead access interface without requiring tight synchronization with the transport bit stream. As shown in Figure 3, the following information must be available from the overhead field of every DTDM block for proper multiplexing and demultiplexing: block alignment word (F) for block timing, empty/full (E/F) status of the block and signal identification (SID), for ownership of the frame (e.g., logical channel number). It is easy to show that with a 10-bit long F, maximum average reframe time is around 2 block periods. This means that the expected block loss will be less than 2 blocks on the average when we lose block timing.

2.3 Multiplexing Stages in a DTDM Network

As shown in Figure 4, several different multiplexing stages can be used to support end-to-end transport in a DTDM network. These include flexible subscriber access, network concentration in remote electronics (RE), and point-to-point transparent transmission between COs. The multiplexers shown in Figure 4 can be separated into two categories: the DTDM multiplexer for the access portion of the network, and the synchronous TDM (STDM) multiplexer for point-to-point inter-office transmission. The DTDM multiplexer combines vastly different types of traffic into the basic backbone rate of 150 Mb/s. It also provides a flexible single step multiplexing of existing hierarchical signals at various rates. The synchronous TDM multiplexer multiplexes several DTDM backbone rate signals into a high speed fiber optic transmission link for transparent point-to-point inter-office transmission. The asynchronous nature of multiplexing in the DTDM multiplexer allows the presence of empty blocks in the output bit stream. The deletion and addition of these empty blocks can be used for tributary synchronization in the STDM multiplexer in a manner

Example : 150 Mb/s

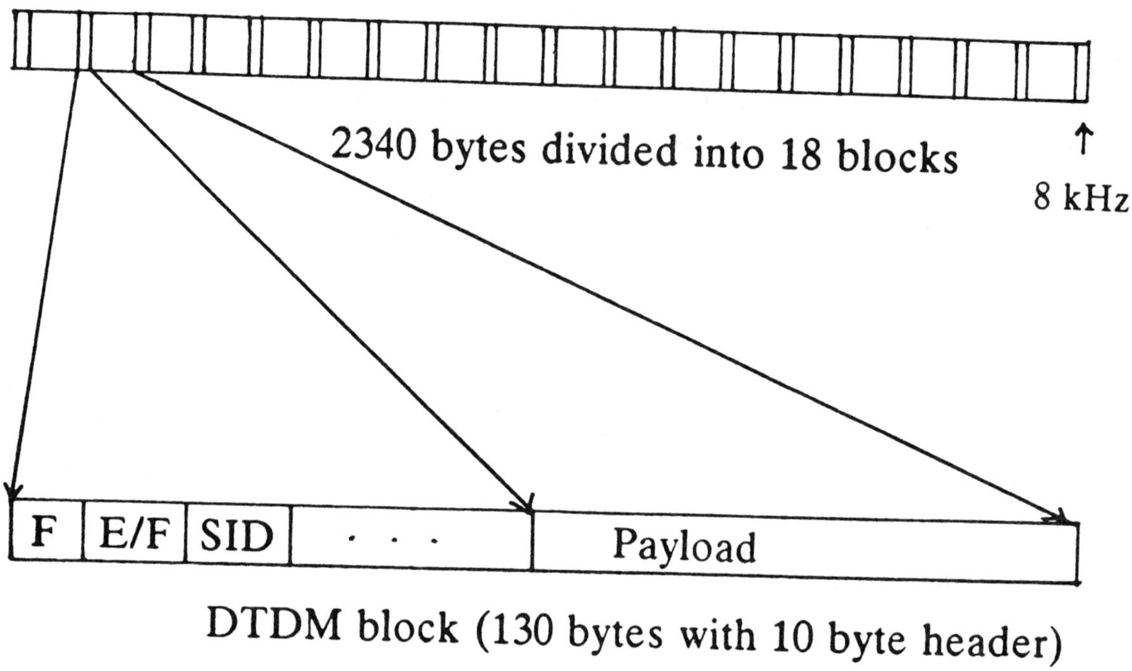

F : Block alignment word

E/F : Empty/Full status

SID : Signal ID

Figure 3 DTDM - Block Structure

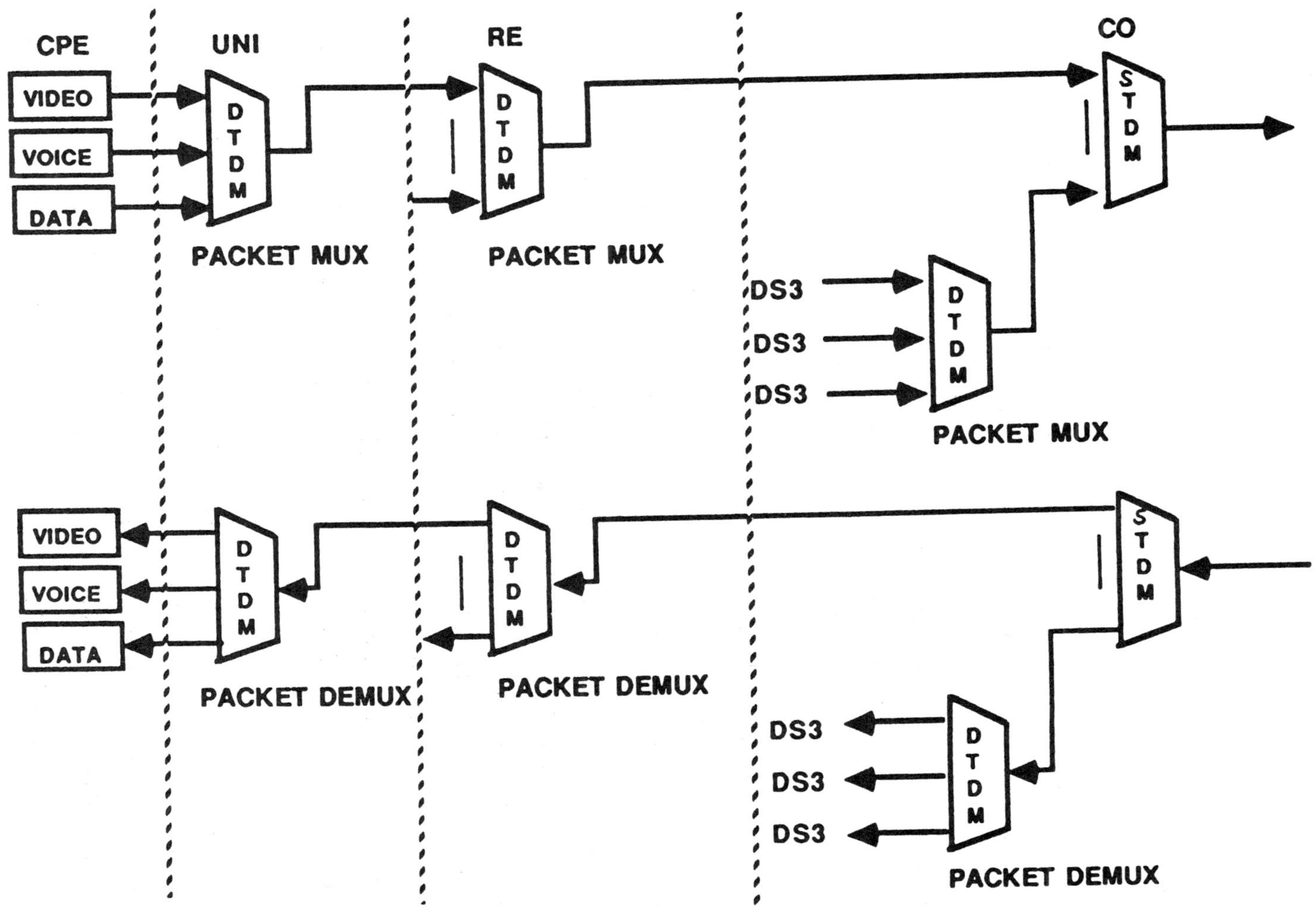

Figure 4 Multiplexing Stages in DTDM Network

similar to pulse stuffing in conventional TDM.

III. DTDM MULTIPLEXER/DEMULTIPLEXER

In this section, we describe a DTDM multiplexer design capable of multiplexing packet and circuit traffic through the same physical multiplexer. DTDM, as mentioned before, can also be interpreted as a flexible TDM technique based on block-level processing instead of bit- or byte-level processing. There are a number of design parameters which need to be specified for a DTDM multiplexer. Among these, input and output signal specifications and interleaving schemes are the key ones.

3.1 Input and Output Signal Specification

Figure 5 shows a general DTDM multiplexer having both continuous TDM signal and bursty packet input tributaries. Unlike the conventional TDM hierarchy, the bit rates of the input signals to the DTDM multiplexer can be arbitrary as long as the available payload of the output signal is large enough to carry all the incoming tributary traffic. For example, if each input signal is a lightly loaded packet traffic, then the output line bit rate can be less than the sum of the input tributary bit rates. On the other hand, if all tributary signals are continuous real time traffic, e.g., existing hierarchical signals, then the payload of the output signal should be larger than the sum of the incoming tributary bit rates to maintain transparency as in the conventional TDM multiplexer. As long as the above conditions are satisfied, the same DTDM multiplexer can be used as a statistical multiplexer, conventional TDM multiplexer or a mixture of both, depending on the characteristics of the input signals.

The framing format of the DTDM multiplexer can be generated at any place in the network depending on the application. DTDM blocks can be generated and occupied by a single service terminal such as a packet telephone or data terminal. In case of the existing hierarchical signal, the frame format is added at the DTDM multiplexer as shown in Figure 5 for DS-1 and DS-3 signals. Once the signals are converted to the DTDM format, no additional frame format is added for any subsequent DTDM multiplexing stages. Thus, low level signal observability is preserved at all points of the network, regardless of the multiplexing history.

DTDM achieves bit rate flexible multiplexing by stuffing and interleaving tributaries at a "block" level. This is very much like bit stuffing and interleaving in conventional TDM. There is one fundamental difference, however, which emanates from the statistical nature of multiplexing in DTDM. The output channel of the DTDM multiplexer is never 100 % occupied unless all the incoming tributaries and local multiplexers are synchronized to the

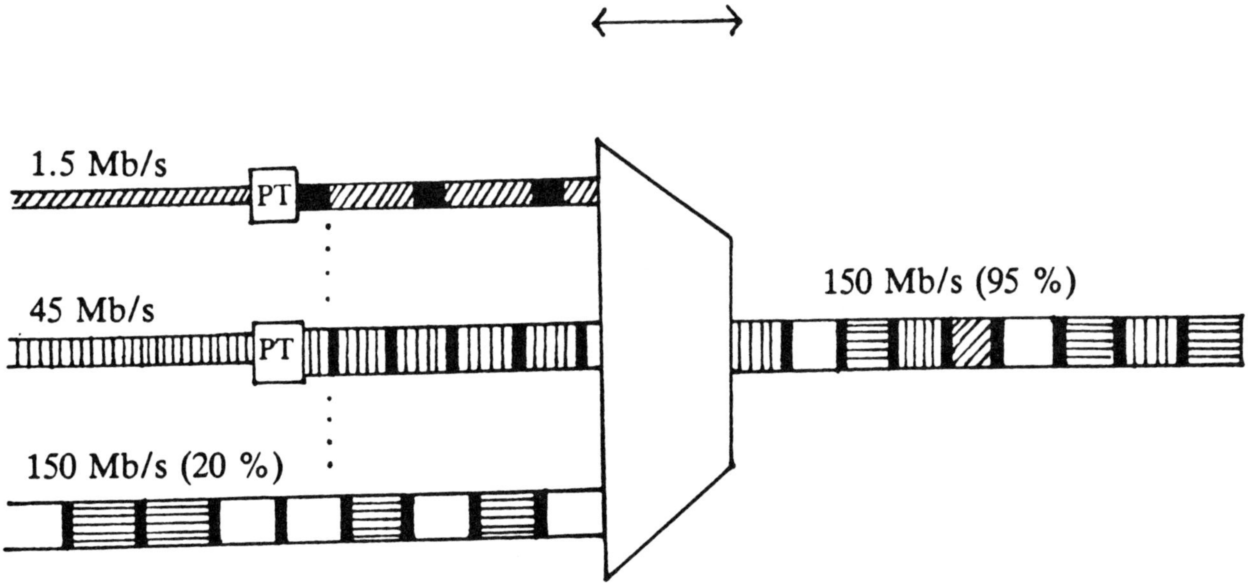

PT : Packetizer

Figure 5 DTDM Multiplexer / Demultiplexer

common clock. Another difference between the DTDM block stuffing and the conventional bit stuffing is that, unlike the latter, no prefixed positions are reserved for DTDM block stuffing. In other words, DTDM block stuffing is performed whenever it is necessary as long as the block boundaries are maintained. Therefore, the jitter caused by increasing the size of the stuffing unit can be absorbed by providing enough buffer with a digital phase locked loop (DPLL) at the receiving end [8].

3.2 Scheduling Scheme

The final step of multiplexing is the interleaving of tributary time slots into a high speed output signal. A scheduling scheme is required to properly assign output time slots to active input tributary time slots to resolve contention. The choice of a scheduling scheme depends on a tradeoff between functionality and design complexity. In conventional TDM, this is accomplished by a simple parallel to serial conversion of tributary bits after all tributaries are synchronized to the same frequency through pulse stuffing. However, in the DTDM multiplexer the same parallel-to-serial conversion is not applicable since the tributaries are not synchronized to a common frequency. Among possible scheduling alternatives, a working prototype operating at 150 Mb/s has been reported in [9] based on a simple first come first serve (FCFS) contention resolution circuitry for each incoming tributary. In this scheduling scheme equal priority is assigned to each incoming tributary, which could be a possible drawback. The assignment of equal priority to each incoming tributary makes it difficult to guarantee the transparency of continuous input tributaries due to bursty traffic on packet tributaries.

Figure 6 is a block diagram of another scheduling scheme based on a train structure which can guarantee the transparency of continuous input tributaries without requiring explicit contention-resolution circuitry. In this scheme, incoming tributaries are first converted to the DTDM block structure and stored in the first-in first-out (FIFO) buffer, waiting to be served onto the multiplexed line. For a packet traffic tributary, only the occupied blocks are stored in the buffer and the empty blocks are discarded. At the top of the multiplexer, a DTDM framing circuit generates a continuous chain of empty DTDM blocks at a constant frequency. For 150 Mb/s, it generates 144 K blocks per second with a 130-byte long DTDM block. This real time block stream is connected to each tributary buffer through the framing circuit which examines the occupancy of both block stream and tributary buffer. If the incoming block is empty and the tributary buffer has a block to send, then the framing circuit loads the block from the tributary buffer into the high speed block stream.

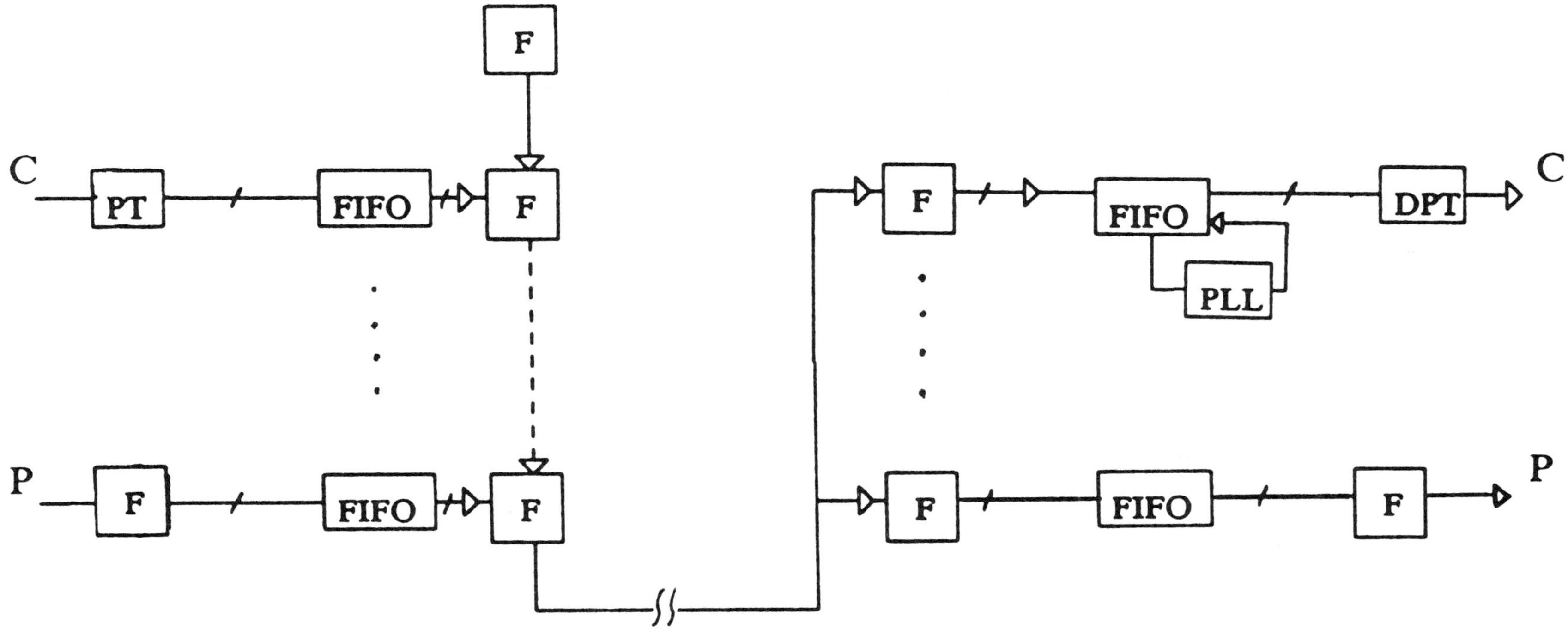

C : Circuit Tributary
P : Packet Tributary
PT : Packetizer
DPT : Depacketizer

F : Framing Circuit
FIFO : First-in First-out Buffer
PLL : (Digital) Phase Locked Loop

Figure 6 DTDM Multiplexer / Demultiplexer with "Train" Scheduling

At the receiving (demultiplexing) end, the incoming high speed bit stream is broadcast to the framing circuit in every tributary as in [9]. Only the matched blocks are selected and written into the corresponding tributary FIFO by the framing circuit and the reverse processing is performed. The above multiplexer can be built with one circuit board per tributary using only currently available technology components, and the framing circuit can be implemented in a single CMOS VLSI chip [10].

In this train structure, the contention problem is resolved automatically without requiring a separate controller. Also, by assigning real time continuous tributaries above the packet traffic tributaries, we can guarantee the transparency of the continuous tributaries independently of the packet traffic statistics. This is because the tributary closer to the top-most block generation framing circuitry gets the available output blocks first due to the inherent train structure. Since the bit rate of the input tributaries can be different, it allows one step multiplexing of different hierarchical level signals, while providing a design complexity much lower than that of the conventional pulse stuffing multiplexer. Therefore, the DTDM multiplexing technique provides an attractive replacement alternative to the conventional TDM multiplexer even for the existing hierarchical signal transmission.

3.3 Buffer Size Requirement for Periodic Tributary

Buffers are commonly used in TDM multiplexers to absorb the temporal variations of the incoming tributary time slots. In a statistical packet multiplexer, the size of the buffer is a function of buffer overflow probability and delay performance, and it depends on the underlying statistics of input tributaries. Performance analysis of a DTDM multiplexer based on a FCFS scheme can be found in [9] when all input tributaries have only packet traffic. Here, we are interested in determining the minimum buffer size required to multiplex circuit traffic through the DTDM multiplexer while preventing buffer overflow and underflow.

We make the following assumptions in determining the minimum buffer size for each input tributary:

- DTDM multiplexer has N input tributaries and one output channel; each input tributary bit rate is f_i b/s, i=1,2,..,N, and the output channel operates at f_0 b/s.

- At each tributary the incoming bit stream is first packetized according to the DTDM block structure and stored into a FIFO buffer with size k_n, n = 1, 2, ..., N.

- The length of the DTDM block is J bits with I bits of information payload.

- Each buffer sends a block ready signal to the server immediately after receiving a complete DTDM block.
- The server selects an active buffer from the input tributaries and loads one block in an empty DTDM block interval in the multiplexed output stream.

For given f_i, f_0, I and J, the constant block arrival rate of the input tributaries, λ_i and the constant block generation rate of the output channel, μ, can be expressed as

$$\lambda_i = \frac{f_i}{I}, \qquad i = 1, 2, \cdots, N \tag{1}$$

$$\mu = \frac{f_0}{J}. \tag{2}$$

To meet the transparency condition, μ should satisfy

$$\mu \geq \sum_{i=1}^{N} \lambda_i \tag{3}$$

where equality holds if and only if all λ_i's and μ derive timing from the same clock. If all arriving packets are served immediately after they arrive in the tributary buffer, then the waiting time at the tributary buffer is limited to 1 block time, i.e., $1 / \mu$ second, since the blocks are served only at the beginning of the block boundary. This is true only when each tributary packet arrives at the tributary buffer in disjoint output block intervals. If the different tributary packets arrive at the same output block interval, only one packet can be served immediately and others must wait in their respective buffers until their turn comes. Therefore, the buffer size requirement depends on the kind of scheduling scheme employed in the multiplexer as discussed below.

3.3.1 First Come First Serve (FCFS) Scheduling

Let us assume that all tributary packets arriving before time 0 are served before time $1/\mu$. Then the service time of the packets arriving at each tributary buffer in the time interval $[0\ ,\ 1/\mu)$ will start from the time interval $[1/\mu\ ,\ 2/\mu)$. Let the service time of the block from the $n-th$ tributary be $[t_n/\mu\ ,\ (t_n + 1)/\mu)$, i.e., the block from the $n-th$ tributary gets placed onto the multiplexed stream during the indicated interval. In the FCFS scheme t_n is equal to the number of blocks which have arrived earlier than the $n-th$ tributary block during the $[0\ ,\ 1/\mu)$ clock interval plus one for the service interval of the $n-th$ block itself. To prevent buffer overflow, the size of the $n-th$ tributary buffer must be greater than or equal to the number of blocks arriving at the $n-th$ tributary buffer in t_n/μ second. Thus, the buffer size, k_n can be expressed as

$$k_n \geq \left\lceil \frac{t_n}{\mu} \lambda_n \right\rceil \tag{4}$$

where $\lceil x \rceil = m$ if m-1 < x $\leq$ m, and m is a positive integer. Since the worst case for the FCFS scheme happens when $t_n = N$, the right hand side of Eq. (4) is bounded by

$$\left\lceil \frac{t_n}{\mu} \lambda_n \right\rceil \leq \left\lceil \frac{N}{\mu} \lambda_n \right\rceil$$

$$\leq N \tag{5}$$

From Eq. (4) and (5), we can easily see that if the size of the $n-th$ tributary buffer, k_n, is greater than or equal to N blocks then the buffer overflow can be prevented for any λ_n, n = 1, 2, ... , N, as long as Eq. (3) is satisfied. Therefore, the size of the buffer must be at least N blocks for each tributary to prevent block loss if we don't have any information about the exact value of λ_n, n = 1, 2, ... , N.

3.3.2 Train Scheduling

To preserve the transparency of the $n-th$ tributary signal, the first arriving block of the $n-th$ tributary should be served before the buffer with size k_n blocks overflows. To satisfy the above condition in the train structure, the available number of blocks in the high speed output channel during k_n/λ_n second must be greater than or equal to the total number of arriving blocks from all preceding i tributaries, i = 1, 2, ... ,$n-1$, within the same period k_n/λ_n second plus one for the $n-th$ tributary block itself. This can be expressed as

$$\frac{k_n}{\lambda_n} \mu \geq \sum_{i=1}^{n-1} \left\lceil \frac{k_n}{\lambda_n} \lambda_i \right\rceil + 1 , \qquad n = 1, 2, \cdots, N \tag{6}$$

Let

$$\left\lceil \frac{k_n}{\lambda_n} \lambda_i \right\rceil = \frac{k_n}{\lambda_n} \lambda_i + a_i , \qquad 0 \leq a_i < 1 \tag{7}$$

Using Eq. (7) in Eq. (6), we get

$$k_n \geq \frac{1 + \sum_{i=1}^{n-1} a_i}{\mu - \sum_{i=1}^{n-1} \lambda_i} \lambda_n \tag{8}$$

The right hand side of Eq. (8) is bounded by

$$\frac{1 + \sum_{i=1}^{n-1} a_i}{\mu - \sum_{i=1}^{n-1} \lambda_i} \lambda_n \leq \frac{\lambda_n}{\mu - \sum_{i=1}^{n-1} \lambda_i} n$$

$$\leq n \tag{9}$$

where $n = 1, 2, \ldots, N$. From Eq. (8) and (9) we can easily see that an n block buffer is sufficient for the n-th tributary, $n = 1, 2, \ldots, N$, for any λ_i, $i = 1, 2, \ldots, N$, to prevent buffer overflow. Unlike FCFS scheduling, the required buffer size is different in each tributary. As expected, the tributary closest to the top-most block generation framing circuitry requires the smallest buffer (1 block) while the farthest tributary requires the largest buffer (N blocks).

IV. SYNCHRONOUS TIME DIVISION MULTIPLEXER

Even if each individual subscriber traffic is bursty, the statistical effect of DTDM multiplexing (network concentration) could make the aggregated output traffic look almost like a smoothed stream. In other words, if the $150 Mb/s$ DTDM backbone rate is sufficiently utilized, then it might be better to maintain the transparency of that channel for further multiplexing to avoid additional queuing delays and possible block loss for packet traffic. Moreover, there could be many existing transmission hierarchical signals which are contained in the $150 Mb/s$ DTDM stream that is to be further multiplexed to a higher rate. Therefore, it appears necessary to design a simple transparent point-to-point multiplexer which can multiplex many $150 Mb/s$ DTDM backbone channels into a higher speed fiber optic transmission link. Today's fiber optic TDM transmission systems operate in the multi-gigabit range. Since bandwidth is limited by the speed limitations of electronic devices and not by the fiber itself, it is a common practice to perform the more complex electronic processing at the low-speed tributary rate and limit high speed circuitry only to the final stage of simple bit interleaving. Since the tributary backbone rate signals have the same nominal bit rate ($150Mb/s$), synchronous TDM is an attractive choice for this situation.

One of the major tasks in designing a TDM multiplexer is to synchronize the incoming tributary bit streams to a common bit rate prior to interleaving these synchronized tributary time slots to form a high speed serial bit stream. If the entire network is synchronized, then it is possible to multiplex these synchronized tributary signals by simple bit interleaving without additional overhead, such as framing and pulse stuffing. The frequency of the multiplexed bit stream would then become exactly an integer multiple of the common tributary frequency, thereby simplifying the clock distribution problem. However, each tributary still requires a slip buffer (typically 1-2 frame-length size buffer), instead of pulse stuffing circuitry, to effectively control faulty situations, such as clock failure and temperature variations in the transmission link. Unlike pulse stuffing, however, the slip buffer approach cannot guarantee transparency of the tributary signal. To overcome this problem, a positive and negative byte stuffing technique [11] has been proposed for the new fiber optic transmission standard (SONET) [4]. In this method, tributary synchronization can be achieved by adding or deleting a dummy byte in predetermined positions of the bit stream. Since the frequency of each tributary signal can be self-adjusted in either the positive or negative direction, it is possible to make the frequency of the multiplexed bit stream exactly an integer multiple of the nominal tributary frequency.

In the above STDM, the high speed frame format may not be available since bit or byte interleaving is performed with arbitrary and unknown frame displacement relationships between input tributaries. One way to overcome this problem is to align the embedded frame between the tributaries such that a unique framing pattern can be generated automatically with the interleaving process. Again, this process generally requires a frame buffer. In the synchronous circuit switched network, one frame is normally of 125 μs duration. This means that the number of bits in one frame increases as the bit rate increases. Therefore, to achieve the tributary frame alignment without buffering the entire frame, a movable frame boundary scheme along with the pointer method has been proposed in [11] and [12] independently. In this method, the transmission frame is decoupled from the source frame (information payload), and the distance between two frames is indicated by a pointer reserved in the transmission frame. By fixing the stuffing byte location to the beginning of the source frame, both tributary synchronization and the frame alignment problems are handled by updating the payload pointer in every multiplexing stage in the network. A simplified example is illustrated in Figure 7 for 2 to 1 multiplexing of $150Mb/s$ cases.

The tributary synchronization problem can be greatly simplified in the DTDM network by adding or deleting an empty DTDM block in each tributary whenever it is necessary. As shown in Figure 8, let us consider the simple case of multiplexing 2 plesiochronous 150*Mb/s* DTDM backbone signals (i.e., the nominal frequency of each tributary is 150*Mb/s* but differs by a small amount) into an exactly 300*Mb/s* signal without adding any new framing format. Incoming tributary block occupancy is examined by the same framing chip used in DTDM, and only the occupied blocks are written into the buffer. The block stored in the buffer is reloaded into a newly generated DTDM frame by the local multiplexer clock to align the frame between the tributaries.

If the frequency of the first tributary is less than the reading frequency of local multiplexer, then an empty block is added to the read out bit stream to prevent buffer underflow. As long as a block boundary can be maintained, addition of an empty block is an instantaneous operation. Therefore, any buffer size greater than one block length is sufficient. However, if the frequency of the tributary is larger than that of the local multiplexer read clock (like the second tributary of Figure 8), an empty block should be found and dropped before the buffer overflows. Therefore, the required number of block buffers is a function of frequency difference and the maximum separation between empty blocks in the bit stream. Let the maximum frequency difference between tributary writing clock and multiplexer reading clock be ϵ *ppm* (parts per million). Then we need an $\epsilon+1$ block-length buffer to accommodate one million block separation between empty blocks. For $\epsilon=10$, which can be easily achieved with a cheap crystal technology, a two block-length buffer can allow up to 99.999 % loading of the input tributary as long as we can find one empty block in every 10^5 blocks.

It is easy to see from section 3 that we can always control output loading of the DTDM multiplexer for continuous input traffic, by making sure that the information payload of the 150*Mb/s* output signal is larger than the sum of the incoming tributary bit rates. For example, 3 DS-3 (44.736 Mb/s) input signals to the DTDM multiplexer result in about 97.1 % loading of the 150 Mb/s channel. Again, note that it is possible to load the output channel 100 % if the network is synchronized to the common clock. For packet traffic it is not possible to load the output channel close to 100 % without excessive delay and packet loss. Therefore, a two block buffer would be sufficient to handle tributary synchronization in both cases.

Since each tributary contains a 2 block-length buffer, the size of the final interleaving unit can be either a bit, a byte or a block depending upon what the technology dictates. If one chooses the interleaving unit as a block, then STDM becomes the special case of a DTDM multiplexer and the multiplexed output bit stream has the same DTDM block

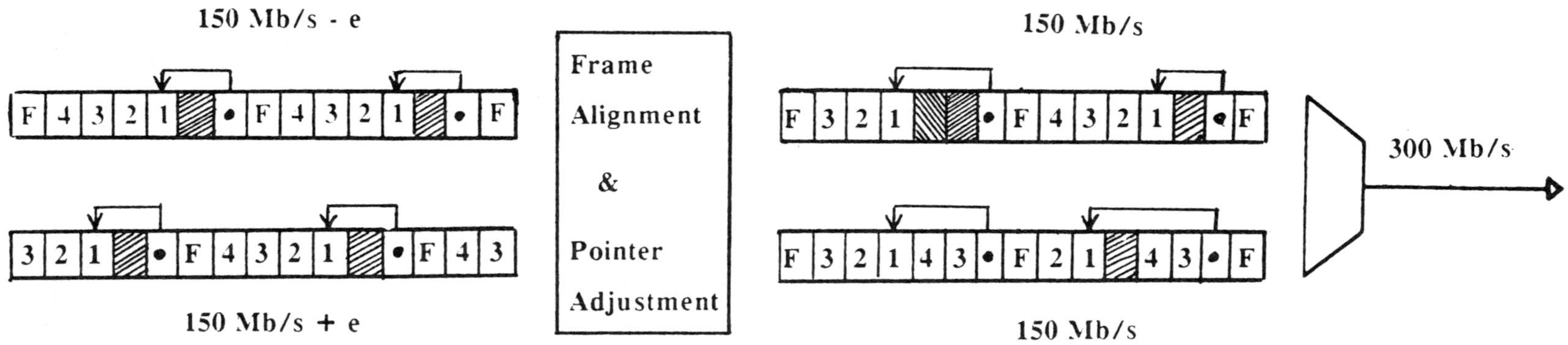

Figure 7 Synchronous TDM in SONET

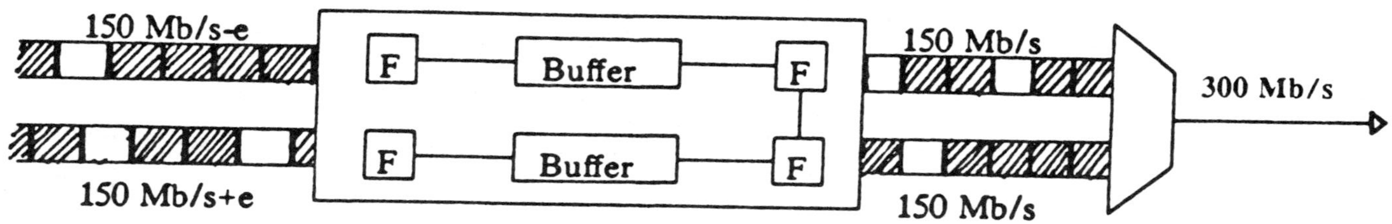

Figure 8 Synchronous TDM in DTDM Network

structure, thereby, allowing the flexible single transport architecture to grow as technology advances. With currently available high speed electronics, simple parallel to serial conversion in terms of bit or byte interleaving would be the most feasible choice for this final stage of high speed processing. Figure 9 shows a simplified block diagram of a bit interleaved STDM multiplexer and demultiplexer pair. Unlike the DTDM "train scheduling" multiplexer, a framing generation circuit produces empty blocks and broadcasts them to each tributary simultaneously to achieve frame (block) alignment between the tributaries. The demultiplexer has similar design structure to that described in [4, 13]. Implementation of this STDM multiplexer is simple, since no explicit stuffing circuitry is required. Moreover, by utilizing the byte-oriented DTDM block structure, most of the processing occurs at only a 20 M Hz clock rate with 8 bit parallel processing.

V. NETWORK SYNCHRONIZATION

The basic goal of network synchronization in a circuit switched network is to minimize the service impairment by controlling the slip rate [2]. The tolerable slip rates vary with the services provided. For the PCM voice signal, the audible effect of slips is an occasional "clicking" type of noise. If we assume that a duration of 5 minutes between audible clicks is acceptable, the tolerable slip rate would be 300 slips/hour, since only one slip in 25 results in an audible click [14]. Therefore, voice service is not the major driving force behind end-to-end network synchronization, although it is the dominant service. On the other hand, the effect of slips on voiceband data signals can be much more serious. A single slip can impair the operation of some voiceband modems for several seconds [15]. Therefore, the slip rate objective is set up as 1 slip per 5 hours in an end-to-end connection mainly based on performance considerations of the circuit switched voiceband data.

To satisfy the above tight slip rate objective, all clocks in the circuit switched network are classified into a hierarchy according to their accuracies, and are slaved to a master clock either directly or indirectly. Because the plan fully utilizes the existing clock sources and existing transmission facilities (T1) while not affecting the traffic carrying capacity in any way, it may be easily realizable. However, appropriate selection of both primary and back-up synchronization facilities in the current mesh-topology network while at the same time avoiding a timing loop is not an easy task [2]. Furthermore, tight synchronization is required all the way down to the circuit board level to ensure a synchronous network. Finally, performance tracking and bookkeeping of these synchronization facilities cause an additional overhead in maintenance, operations and associated administration.

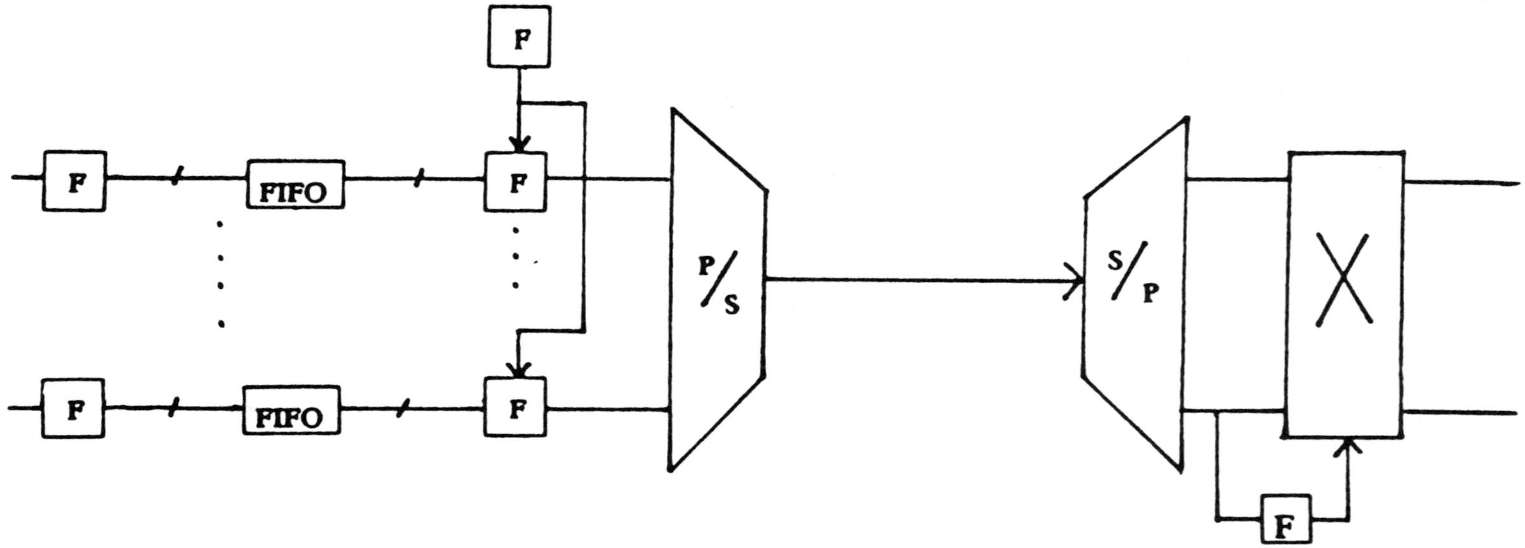

p/s : Parallel to Serial Conversion

Figure 9 STDM Multiplexer / Demultiplexer

In contrast to the above, network synchronization is not a necessary condition for a stand-alone packet switched network. Since packets are not occupying 100 % of the incoming bit stream, each local packet switching node can adapt to the frequency difference without losing information [16]. For the next 20 years, it is expected that the network will evolve from the existing predominantly circuit switched network to a broadband packet switched network. Since the broadband packet network is not being designed as a separate overlay network over the current circuit switched network, internetworking between the two networks would be inevitable during the transition period. For successful network evolution, the quality of the existing services should not be degraded for any signals transmitted over mixed packet and circuit connection [17]. For example, circuit switched data services still require network synchronization until they are converted to end-to-end packet switched connections. One natural question being raised is: What is the right level of network synchronism which should be present in a mixed packet and circuit switched network?

One answer to the above question is to tie the DTDM backbone signal with the circuit switched network synchronization plan by upgrading the timing distribution path from the copper based 1.5 Mb/s T1 facility to the fiber-based DTDM backbone transmission facility ($\geq$ 150 Mb/s). This will provide a time transparency for the existing circuit switched traffic by extracting the clock through the 150 Mb/s backbone rate signal. For packet switched traffic, flexible asynchronous access to the DTDM network is still preserved up to 150 Mb/s since network synchronization is imposed for the backbone rate and above. In the DTDM network, each local multiplexer can adjust for small frequency differences by adding or deleting empty DTDM blocks from the incoming tributaries. Therefore, associated network synchronization overhead is minimal since information transparency can be preserved even when operating off the local clock in case of network clock failure. This means that a plesiochronous (independent but nominally the same frequency) network would be sufficient for the future DTDM-based packet switched network while providing for a smooth transition from the present circuit switched network.

VI. SUMMARY and CONCLUSIONS

In this paper we have described an integrated transport technique suitable for both packet and circuit switched traffic. By defining a packet to be identical to the conventional transmission frame, we are able to provide a common frame structure for any bit rate below the backbone rate (150 Mb/s). This frame structure allows for bit rate flexible multiplexer design and also direct packet observability regardless of the previous multiplexing history.

We have also described a bit rate flexible multiplexing technique capable of multiplexing both statistical and deterministic (continuous) input tributaries through the same physical multiplexer. As a natural extension of the DTDM concept, for high speed transparent point-to-point transmission, we have developed an alternative TDM multiplexing technique which can provide all the benefits of synchronous multiplexing without requiring tight network synchronization.

One important conclusion of this paper is that it is possible to use the same transmission technique to carry both the continuous time-multiplexed signals and bursty traffic, thus eliminating the need to construct a mixed overlay network for the transitional purpose. Moreover, the multiplexing techniques described in this paper provide more functionality while at the same time no more design complexity than that of the conventional TDM multiplexer. Therefore, by adopting this flexible transport technique into the current network, DTDM will provide a practical migration strategy from the existing predominantly circuit switched network to the future broadband packet switched network.

ACKNOWLEDGEMENT

The author would like to thank H. J. Chao, C. Lo, R. P. Singh and L. T. Wu for valuable discussions and suggestions.

REFERENCES

[1] C. A. Cooper, "Synchronization for Telecommunications in a Switched Digital Network," *IEEE Transactions on Communications,* vol. COM-27, pp. 1028-1033, July 1979.

[2] "Digital Synchronization Network Plan," PUB 60110, AT & T., December 1983.

[3] G. R. Ritchie, "SYNTRAN - A New Direction for Digital Transmission Terminals," *IEEE Communications Magazine,* vol. 23, pp. 20-25, November 1985.

[4] R. J. Boehm et al., "Standardized Fiber Optic Transmission Systems- A Synchronous Optical Network View," *IEEE Journal on Selected Areas in Communications,* Vol. SAC-4, No.9, pp. 1424-1431, December 1986.

[5] A. Thomas, J. P. Coudreuse and M. Servel, "Asynchronous Time-division Techniques: An Experimental Packet Network Integrating Video Communication," Proceedings *International Switching Symposium,* Florence, Italy, May 1984.

[6] J. P. Coudreuse and M. Serval, "PRELUDE : An Asynchronous Time- division Switched Network," *International Conference on Communications* Seattle, Washington, June 1987.

[7] L. T. Wu, S. H. Lee and T. T. Lee, "Dynamic TDM - A Packet Approach to Broadband Networking," *International Conference on Communications,* Seattle, Washington, June 1987. *International Conference on Communications,* Seattle, Washington, June 1987.

[8] R. P. Singh and S. H. Lee, "Adaptive Clock Synchronization Schemes for Real-Time Traffic in Broadband Packet Networks," will appear *Proc. of the 8th European Conference on Electronics,* Stockholm, Sweden, June 1988.

[9] M. W. Beckner, T. T. Lee and S. E. Minzer, "A Protocol and Prototype for Broadband Subscriber Access to ISDNs," Proceedings *International Switching Symposium,* Phoenix, Arizona, March 1987.

[10] H. J. Chao, T. J. Robe and L. S. Smoot, "A CMOS VLSI Framer Chip for a Broadband ISDN Local Access System," *IEEE Digest of Technical Papers of 1987 Symposium on VLSI Circuits,* Karuizawa, Japan, May 1987.

[11] J. Elson and J. Brown, "Proposal for an Internally Synchronous SONET using Payload Pointers," Contribution to Standards Project - T1X1.4/86-022, February 1986.

[12] J. A. Bellisio, "Multilevel Multiplexing," U.S. Patent, Pending Application, No. 864037, May 1986.

[13] L. R. Linnell, "A Wide-band Local Access System using Emerging- technology Components," *IEEE Journal on Selected Areas in Communications,* Vol. SAC-4, No. 4, pp. 612-618, July 1986.

[14] M. Decina and U. de Julio, "International Activities on Network Synchronization for Digital Communications," *IEEE International Conference on Communications,* Boston, Massachusetts, June 1979.

[15] J. E. Abate et al., "The Switched Digital Network Plan," *Bell System Technical Journal,* pp. 1297-1320, September 1977.

[16] F. van den Dool, "Synchronization Aspects of ATD-IBC Network," Proceedings *International Switching Symposium,* Phoenix, Arizona, March 1987.

[17] L. T. Wu, E. Arthurs and W. D. Sincoskie, "A Packet Network for BISDN Applications," will appear *1988 International Zurich Seminar on Digital Communications,* Zurich, Switzerland, March 1988.

SONET: Now It's the Standard Optical Network

Ralph Ballart
Yau-Chau Ching

SONET (SYNCHRONOUS OPTICAL NETWORK) IS the name of a newly adopted standard, originally proposed by Bellcore (Bell Communications Research) for a family of interfaces for use in Operating Telephone Company (OTC) optical networks. With single-mode fiber becoming the medium of choice for high-speed digital transport, the lack of signal standards for optical networks inevitably led to a proliferation of proprietary interfaces. Thus, the fiber optics transmission systems of one manufacturer cannot optically interconnect with those of any other manufacturer, and the ability to mix and match different equipment is restricted. SONET defines standard optical signals, a synchronous frame structure for the multiplexed digital traffic, and operations procedures.

SONET defines standard optical signals, a synchronous frame structure for the multiplexed digital traffic, and operations procedures.

SONET standardization began during 1985 in the T1X1 subcommittee of the ANSI-accredited Committee T1 to standardize carrier-to-carrier (e.g., NYNEX-to-MCI) optical interfaces. Clearly, such a standard would also have an impact on intracarrier networks, and for that reason has been a subject of great interest for many carriers, manufacturers, and others. Initial T1 standards for SONET rates and formats and optical parameters have now been completed. The history and technical highlights of the SONET standard and its applications are the subject of this paper.

Since it began in the post-divestiture environment, SONET standardization can be thought of as a paradigm for the development of new transmission signal standards. Bellcore's original SONET proposal was not fully detailed because all the technical questions were not yet answered. However, some aspects of the proposal have been carried through the entire process and are now part of the final standards. These include:

- The need for a family of digital signal interfaces, since the march of technology is going to continually increase optical interface bit rates
- The use of a base rate SONET signal near 50 Mb/s to accommodate the DS3 electrical signal at 44.736 Mb/s
- The use of synchronous multiplexing to simplify multiplexing and demultiplexing of SONET component signals, to obtain easy access to SONET payloads and to exploit the increasing synchronization of the network
- Support for the transport of broadband (> 50 Mb/s) payloads
- Specification of enough overhead channels and functions to fully support facility maintenance

As standardization progressed, two key challenges emerged, the solution of which gave SONET universal application. The first was to make SONET work in a plesiochronous[1] environment and still retain its synchronous nature; the solution was the development of payload pointers to indicate the phase of SONET payloads with respect to the overall frame structure (see "SONET Signal Standard—Technical Highlights"). The second was to extend SONET to become an international transmission standard, and thereby begin to resolve the incompatibilities between the European signal hierarchy (based on 2.048 Mb/s) and the North American hierarchy (based on 1.544 Mb/s). Toward the latter goal, the International Telegraph and Telephone Consultative Committee (CCITT) standardization of SONET concepts began in 1986 and the first Recommendations (standards) were completed in June 1988.

This paper will not present a full technical picture of the national and international SONET standards. Instead, we will concentrate on those aspects of the standards and standardization process that are of particular interest. In the next section, a brief and instructional history of the SONET standard is presented. As philosopher George Santayana said, "Those who cannot remember the past are condemned to repeat it." We will then discuss key technical aspects of the SONET standard. Finally, an outline of future work is given in the final section.

[1] As defined in CCITT, corresponding signals are plesiochronous if their significant instants occur at nominally the same rate, any variation in rate being constrained within specified limits.

A History of SONET in T1 and CCITT

The standardization of SONET in T1 started in two different directions and in three areas. First, the Interexchange Carrier Compatibility Forum (ICCF), at the urging of MCI in 1984, requested T1 to work on standards that would allow the interconnection of multi-owner, multi-manufacturer fiber optic transmission terminals (also known as the mid-fiber meet capability). Of several ambitious tasks that ICCF wanted addressed to ensure a full mid-fiber meet capability, two were submitted to T1. A proposal on optical interface parameters (e.g., wavelength, optical power levels, etc.) was submitted to T1X1 in August 1984 and, after three and a half years of intensive work, resulted in a draft standard on single-mode optical interface specifications [1]. The ICCF proposal on long-term operations was submitted to T1M1 and resulted in a draft standard on fiber optic systems maintenance [2].

In February 1985, Bellcore proposed to T1X1 a network approach to fiber system standardization that would allow not only the interconnection of multi-owner, multi-manufacturer fiber optic transmission terminals, but also the interconnection of fiber optic network elements of varying functionalities. For example, the standard would allow the direct interconnection between several optical line terminating multiplexers, manufactured and owned by different entities, and a digital cross-connect system. In addition, the proposal suggested a hierarchical family of digital signals whose rates are integer multiples of a basic module signal, and suggested a simple synchronous bit-interleaving multiplexing technique that would allow economical implementations. Thus, the term Synchronous Optical NETwork (SONET) was coined. This proposal eventually led to a draft standard on optical rates and formats [3]. For the remainder of this paper, the focal points are the history and highlights of the rates and formats document. However, one should always be reminded that this document is only one part of the inseparable triplet: optical interface specifications, rates and formats specifications, and operations specifications.

As it turned out, the notions of a network approach and simple synchronous multiplexing had been independently investigated by many manufacturers. Some of them were already developing product plans, thus complicating the standards process. With the desire of the network providers (i.e., the OTCs) for expedited standards, SONET quickly gained support and momentum. By August 1985, T1X1 approved a project proposal based on the SONET principle. Because the issues on rates and formats were complex and required diligent but timely technical analyses, a steady stream of contributions poured into T1X1. Several ad hoc groups were formed and interim meetings were called to address them. The contributions came from over thirty entities, representing the manufacturers and the network providers alike.

In the early stage, the main topic of contention was the rate of the basic module. From two original proposals of 50.688 Mb/s (from Bellcore) and 146.432 Mb/s (from AT&T), a new rate of 49.920 Mb/s was derived and agreed on. In addition, the notion of a Virtual Tributary (VT) was introduced and accepted as the cornerstone for transporting DSl services. By the beginning of 1987, substantial details had been agreed upon and a draft document was almost ready for voting. Then came CCITT.

The SONET standards were first developed in T1X1 to serve the U.S. telecommunications networks. When CCITT first expressed its interest in SONET in the summer of 1986, major procedural difficulties appeared. According to the established protocol, only contributions that had consensus in T1X1 were forwarded, through U.S. Study Group C, to CCITT. As a result, some aspects of U.S. positions in CCITT appeared to lack flexibility without input from T1X1. Additionally, the views of other administrations in CCITT were not thoroughly understood in T1X1. There were also differences in schedule and perceived urgency. CCITT runs by a four-year plenary period and their meetings are six to nine months apart, while T1 approves standards whenever they are ready and its technical subcommittees meet at least four times a year. While T1X1 saw the SONET standard as a way to stop the proliferation of incompatible fiber optic transmission terminals, no such need was perceived by many other nations whose networks were still fully regulated and non-competitive.

TABLE I. CCITT Rec. G.702 Asynchronous Digital Hierarchies (in Mb/s)

Level	North America	Europe	Japan
1	1.544 (DS1)	2.048	1.544
2	6.312 (DS2)	8.448	6.312
3	44.736 (DS3)	34.368	32.064
4	—	139.264	97.728

The procedural difficulties were partially resolved when representatives from the Japanese and British delegations started to participate in T1X1 meetings in April 1987. These representatives not only gave to T1X1 the perspectives of two important supporters of an international SONET standard, they also served as a conduit between T1X1 and CEPT, the European telecommunications organization.

Separately, interests in an international SONET standard also gained support in the US. Spearheaded by Bellcore, informal discussions in search of an acceptable solution took place in a variety of forums, and contributions in support of this standard were submitted to both T1 and CCITT. Many of these informal discussions had the highest level of corporate support from several U.S. companies, including manufacturers and network providers.

In July 1986, CCITT Study Group XVIII began the study of a new synchronous signal hierarchy and its associated Network Node Interface (NNI). The NNI is a non-media-specific network interface and is distinct from the-user-network interface associated with Broadband ISDN. The interaction between T1X1 and CCITT on SONET and the new synchronous hierarchy was fascinating to the participants and will probably alter the way international standards are made in the future. The U.S. wanted an international standard, but not at the price of scrapping SONET or seriously delaying an American national standard upon which OTC networks were planned. The CCITT was not used to working so quickly on so complicated an issue, but was concerned about being supplanted by the T1 committee in the development of new standards.

The U.S. first formally proposed SONET to CCITT for use in the NNI at the February 1987 Brasilia meeting; this proposal had a base signal level (rate) near 50 Mb/s. Table I shows that the European signal hierarchy has no level near 50 Mb/s, and therefore CEPT wanted the new synchronous hierarchy to have a base signal near 150 Mb/s to transport their 139.264 Mb/s signal.

Thus, the informal European response was that the U.S. must change from bit interleave to byte interleave multiplexing to provide a byte organized frame structure at 150 Mb/s. However, there was still no indication from many administrations that an international standard was either desirable or achievable. It took T1X1 three months and three meetings to agree to byte-interleaving and the results were submitted to CCITT as a new T1X1 draft standard document. Thus, T1X1 never gave up the responsibility of developing a SONET standard for the U.S. and, while conceding changes to CCITT wishes, progress was made in other areas of the U.S. standard.

After CCITT met again in Hamburg in July 1987, a formal request was made to all administrations to consider two alternative proposals for an NNI specification near 150 Mb/s. The U.S. proposal was based on the SONET STS-3 frame structure; the STS-3 frame could be drawn as a rectangle with 13 rows and 180 columns of bytes. CEPT proposed, instead, a new STS-3 frame with 9 rows and 270 columns. (Commonly referred to as the 9-row/13-row debate, this prompted one amateur poet to chide that neither conforms to the correct SON(N)ET format of 14 lines.) An NNI near 150 Mb/s received unanimous support because it was assumed that future broadband payloads would be about that size. A North American basic module near 50 Mb/s could be easily derived in both proposals, with a frame structure of either 13 rows and 60 columns or 9 rows and 90 columns.

The Europeans wanted a 9-row frame structure to accommodate their 2.048 Mb/s primary rate signal. This signal has 32 bytes per 125 μs, but in the 13-row proposal could only be accommodated in the most straightforward way using three 13-byte columns, or 39 bytes. The Europeans decried this waste of bandwidth, and refused to consider any alternative (and more efficient) mapping of the 2.048 Mb/s signal into the 13-row structure. Their 9-row frame structure could carry the U.S. 1,544 Mb/s primary rate signal (requiring about 24 bytes/125 μs) in 3 columns of 9 bytes and the 2.048 Mb/s signal in 4 columns of 9 bytes.

The CEPT 9-row proposal called for changes in both the rate and format in the U.S., just as T1X1 was about to complete the SONET standard. However, the request also carried an attractive incentive from a CEPT subcommittee, who stated in a letter that these were the only changes necessary for an international agreement. In addition, the text of the CEPT proposal was based largely on the T1X1 draft document, so that it was complete. Therefore, after the Hamburg meeting, there was tremendous international pressure on the U.S. to accept the 9-row proposal. After some intense debates in T1X1, the U.S. agreed to change.

Unfortunately, the CEPT proposal did not have unanimous support from all CEPT administrations. While some administrations were anxious to get an international standard, a few became concerned that the 9-row proposal favored the U.S. DS3 signal over the CEPT 34.368 Mb/s signal. A CEPT contribution to the November 1987 CCITT meeting stated that it was too early to draft Recommendations on a new synchronous hierarchy. Little progress was made at that meeting and the international SONET standard was in serious jeopardy.

Many T1X1 participants were upset at the apparent change in CEPT's position. Since there were no alternative proposals from CCITT at its November meeting, T1X1 decided to approve the two SONET documents for T1 letter balloting. However, the balloting schedule was deliberately set such that it fell between the CCITT meeting at the beginning of February 1988 and the T1X1 meeting at the end of February 1988. This scheduling allowed a last ditch attempt for an international agreement. In CCITT, a mad rush to rescue the international standard also took place. In addition to a series of informal discussions, a pre-CCITT meeting was held in Tokyo to search for a compromise. Under the skillful helmsmanship of Mr. K. Okimi of Japan, the CCITT meeting in Seoul proposed one additional change to the U.S. draft standards. The new proposal called for a change in the order that 50 Mb/s tributaries are byte-multiplexed to higher SONET signal levels. It also put more emphasis on the NNI as a 150 Mb/s signal by including optional payload structures to better accommodate the European 34.368 Mb/s signal. The U.S. CCITT delegates eventually viewed this proposal as a minor change to the U.S. standard (minor to the extent that equipment under development would probably not require modification) and agreed to accept it. An extensive set of three CCITT Recommendations was drafted and approved by the working party plenary. The U.S. acceptance of these changes was predicated on the understanding that no additional changes of substance would be considered in approving the final versions of the Recommendations.

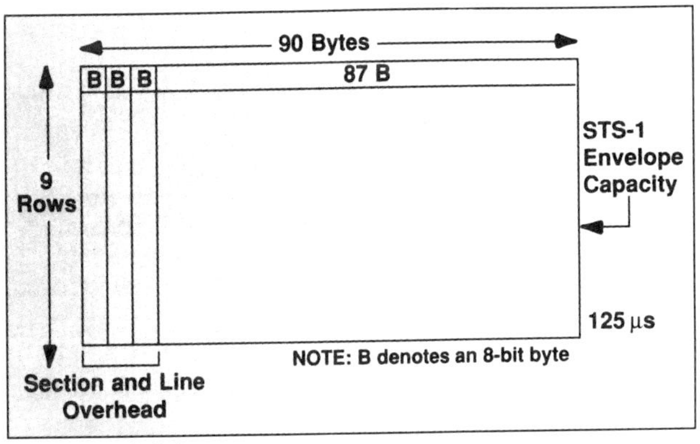

Fig. 1. STS-1 frame.

In February 1988, T1X1 accepted the new changes at its meeting in Phoenix. T1 default balloting based on the change was completed in May and the final passage of the American national standard is expected this summer. Editorial corrections to the CCITT Recommendations [4–6] were completed in June during the Study Group XVIII meeting and with their final approval later this year, the international SONET standard will be born!

SONET Signal Standard—Technical Highlights

In this section, we describe the technical highlights of the American national standards related to SONET. We use U.S. rather than CCITT terminology, although everything described is consistent with both the American national standards and the CCITT Recommendations.

SONET Signal Hierarchy

The basic building block and first level of the SONET signal hierarchy is called the Synchronous Transport Signal—Level 1 (STS-1). The STS-1 has a bit rate of 51.84 Mb/s and is assumed to be synchronous with an appropriate network synchronization source. The STS-1 frame structure can be drawn as 90 columns and 9-rows of 8 bit bytes (Figure 1). The order of transmission of the bytes is row by row, from left to right, with one entire frame being transmitted every 125 μs. (125 μs frame period supports digital voice signal transport, since these signals are encoded using 1 byte/125 μs = 64 kb/s.) The first three columns of the STS-1 contain section and line overhead bytes (see the following subsection). The remaining 87 columns and 9-rows are used to carry the STS-1 Synchronous Payload Envelope (SPE); the SPE is used to carry SONET payloads including 9 bytes of path overhead (see next section). The STS-1 can carry a clear channel DS3 signal (44.736 Mb/s) or, alternatively, a variety of lower-rate signals such as DS1, DS1C, and DS2.

No physical interface for the STS-1 signal has been defined as yet; the Optical Carrier—Level 1 (OC-1) is obtained from the STS-1 after scrambling (to avoid long strings of ones and zeros and allow clock recovery at receivers) and electrical-to-optical conversion. The OC-1 is the lowest-level optical signal to be used at SONET equipment and network interfaces.

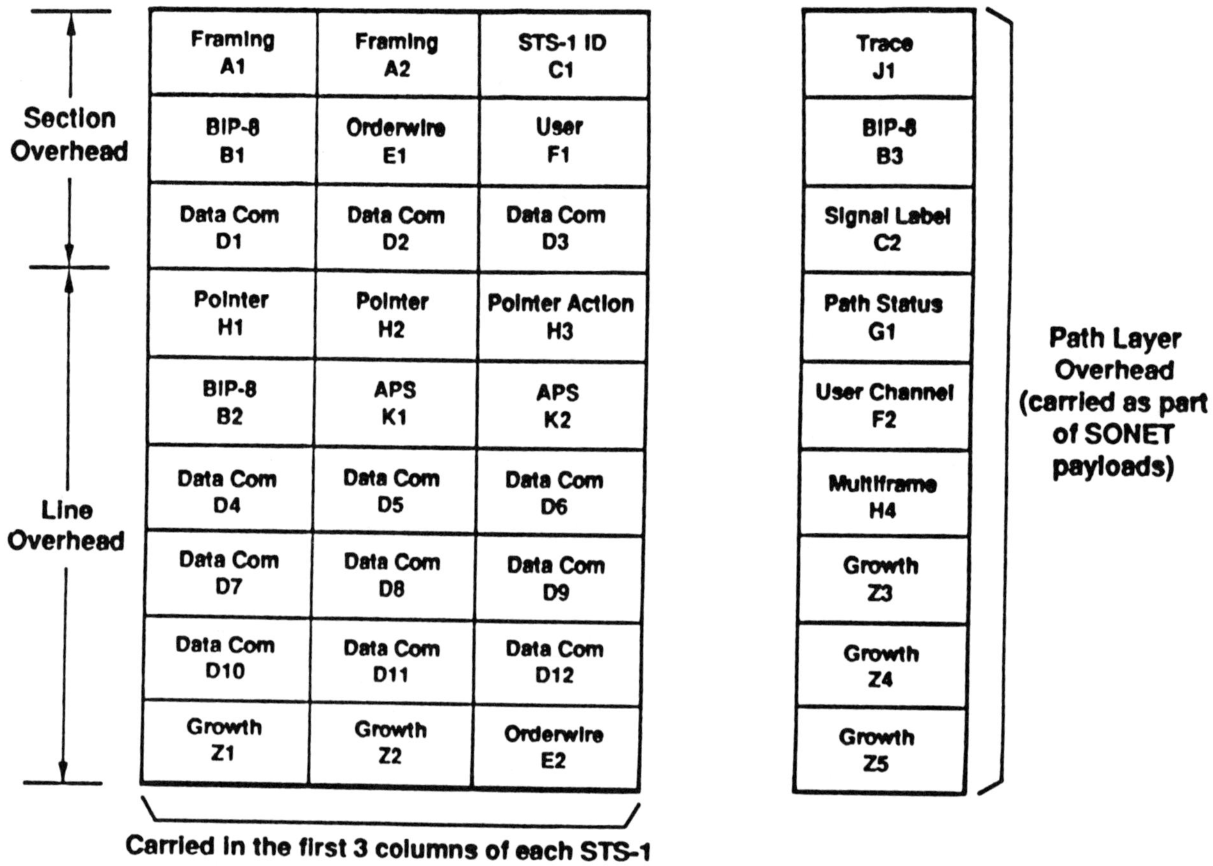

Fig. 2. SONET overhead bytes.

SONET Overhead Channels

The SONET overhead is divided into section, line, and path layers; Figure 2 shows the overhead bytes and their relative positions in the SONET frame structure. This division clearly reflects the segregation of processing functions in network elements (equipment) and promotes understanding of the overhead functions. The section layer contains those overhead channels that are processed by all SONET equipment including regenerators. The section overhead channels for an STS-1 include two framing bytes that show the start of each STS-1 frame, an STS-1 identification byte, an 8-bit Bit-Interleaved Parity (BIP-8) check for section error monitoring, an orderwire channel for craft (network maintenance personnel) communications, a channel for unspecified network user (operator) applications, and three bytes for a section level data communications channel to carry maintenance and provisioning information. When a SONET signal is scrambled, the only bytes left unscrambled are the section layer framing bytes and the STS-1 identification bytes. The second (link) layer of the section data communications channel protocol is LAPD while ISO 8473 is under study for the third (network) layer; higher layers of the protocol will be defined in future updates of the standard.

The line overhead is processed at all SONET equipment except regenerators. It includes the STS-1 pointer bytes (discussed below), an additional BIP-8 for line error monitoring, a two-byte Automatic Protection Switching (APS) message channel (both *1 + 1* and *1* by *N* protection are supported), a nine-byte line data communications channel, bytes reserved for future growth, and a line orderwire channel. The higher layers of the line data communications channel are not specified in the current version of the SONET standard.

The path overhead bytes are processed at SONET STS-1 payload terminating equipment; that is, the path overhead is part of the SONET STS-1 payload and travels with it. The path overhead includes a path BIP-8 for end-to-end payload error monitoring, a signal label byte to identify the type of payload being carried, a path status byte to carry maintenance signals, a multiframe alignment byte to show DSO signaling bit phase, and others.

Multiplexing

Higher rate SONET signals are obtained by first byte-interleaving *N* frame-aligned STS-1s to form an STS-*N* (Figure 3). Byte-interleaving and frame alignment are used primarily to obtain a byte-organized frame format at the 150 Mb/s level that is acceptable to the CCITT; as discussed below, frame alignment and byte-interleaving also help an STS-*N* to carry broadband payloads of about 150 or 600 Mb/s. All the section and line overhead channels in STS-1 #1 of an STS-*N* are used; however, many of the overhead channels in the remaining STS-1s are unused. (Only the section overhead framing, STS-1 ID, and BIP-8 channels and the line overhead pointer and BIP-8 channels are used in all STS-1s in an STS-*N*.) The STS-*N* is then scrambled and converted to an Optical Carrier—Level *N* (OC-*N*) signal. The OC-*N* will have a line rate exactly *N* times that of an OC-1. Table II shows the OC-*N* levels allowed by the American national standard.

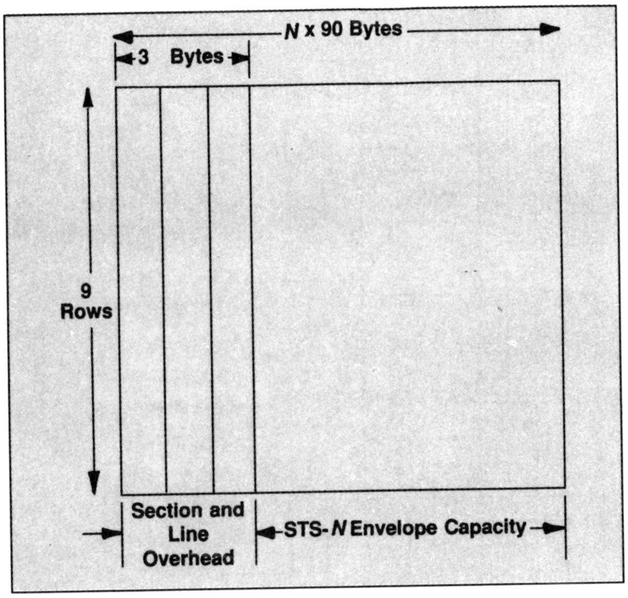

Fig. 3. STS-N frame

SONET STS-1 Payload Pointer

Each SONET STS-1 signal carries a payload pointer in its line overhead. The STS-1 payload pointer is a key innovation of SONET, and it is used for multiplexing synchronization in a plesiochronous environment and also to frame align STS-*N* signals.

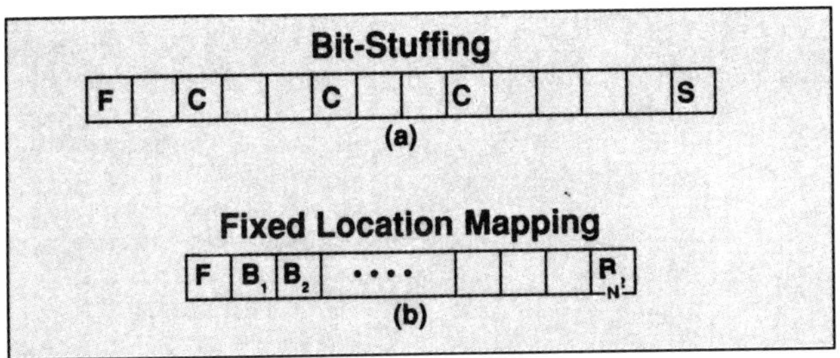

Fig. 4. Payload multiplexing methods.

Pointers and Multiplexing Synchronization

There are two conventional ways to multiplex payloads into higher-rate signals. The first is to use positive bit-stuffing to increase the bit rate of a tributary signal to match the available payload capacity in a higher-rate signal. As shown in Figure 4a, bit-stuffing indicators (labeled C) are located in fixed positions with respect to signal frame *F* and indicate whether the stuffing bit *S* carries real or dummy data in each higher-level signal frame. Examples of bit stuffing are the multiplexing of four DS1 signals into the DS2 signal and the multiplexing of seven DS2 signals into the asynchronous DS3 signal. Bit-stuffing can accommodate large (asynchronous) frequency variations of the multiplexed payloads. However, access to those payloads from the higher-level multiplexed signal is conceptually difficult, since the tributary signal must first be destuffed (real bits separated from the dummy bits) and then the framing pattern of the payload must be identified if complete payload access is required.

TABLE II. Levels of the SONET Signal Hierarchy

Level	Line Rate (Mb/s)
OC-1	51.84
OC-3	155.52
OC-9	466.56
OC-12	622.08
OC-18	933.12
OC-24	1244.16
OC-36	1866.24
OC-48	2488.32

The second conventional method is the use of fixed location mapping of tributaries into higher-rate signals. As network synchronization increases with the deployment of digital switches, it becomes possible to synchronize transmission signals to the overall network clock. Fixed location mapping is the use of specific bit positions in a higher-rate synchronous signal to carry lower-rate synchronous signals; for example, in Figure 4b, frame position B2 would always carry information from one specific tributary payload. This method allows easy access to the transported tributary payloads, since no destuffing is required. The SYNTRAN DS3 signal is an example of a synchronous signal that uses fixed location mapping of its tributary DS1 signals. However, there is no guarantee that the high-speed signal and its tributary will be phase-aligned with each other. Also, small frequency differences between the transport signal and its tributary signal may occur, due to synchronization network failures or at plesiochronous boundaries. Therefore, multiplexing equipment interfaces require 125-μs buffers to phase-align and slip (repeat or delete a frame of information to correct frequency differences) the tributary signal. These buffers are undesirable because of the signal delay that they impose and the signal impairment that slipping causes.

In SONET, payload pointers represent a novel technique that allows easy access to synchronous payloads while avoiding the need for 125-μs buffers and associated slips at multiplexing equipment interfaces. The payload pointer is a number carried in each STS-1 line overhead (bytes H1, H2 in Figure 2) that indicates the starting byte location of the STS-1 SPE payload within the STS-1 frame (Figure 5). Thus, the payload is not locked to the STS-1 frame structure as it would be if fixed location mapping was used but instead floats with respect to the STS-1 frame. (The STS-1 section and line overhead byte positions determine the STS-1 frame structure; note in Figure 5 that the 9-row-by-87-column SPE payload maps into an irregular shape across two 125-μs STS-1 frames.)

Any small frequency variations of the STS-1 payload can be accommodated by either increasing or decreasing the pointer value; however, the pointers cannot adjust to asynchronous frequency differences. For example, if the STS-1 payload data rate is high with respect to the STS-1 frame rate, the payload pointer is decremented by one and the H3 overhead byte is used to carry data for one frame (Figure 6). If the payload data rate is slow with respect to the STS-1 frame rate, the data byte immediately following the H3 byte is nulled for one frame and the pointer is incremented by one (Figure 7). Thus, slips and their associated data loss are avoided while the phase of the STS-1 synchronous payload is immediately known by simply reading the pointer value. Thus, SONET pointers combine the best features of the positive bit-stuffing and fixed location mapping methods. Of course, these advantages come at the cost of having to process the pointers; however, pointer processing appears readily implementable in today's Very Large Scale Integration (VLSI) technologies.

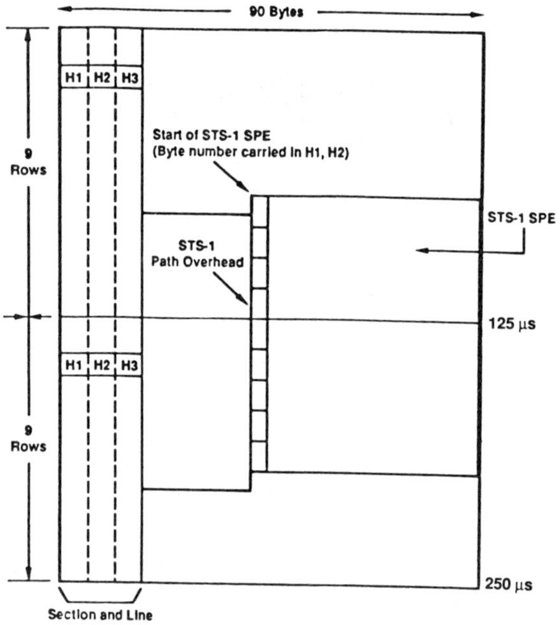

Fig. 5. STS-1 SPE in interior of STS-1 frame.

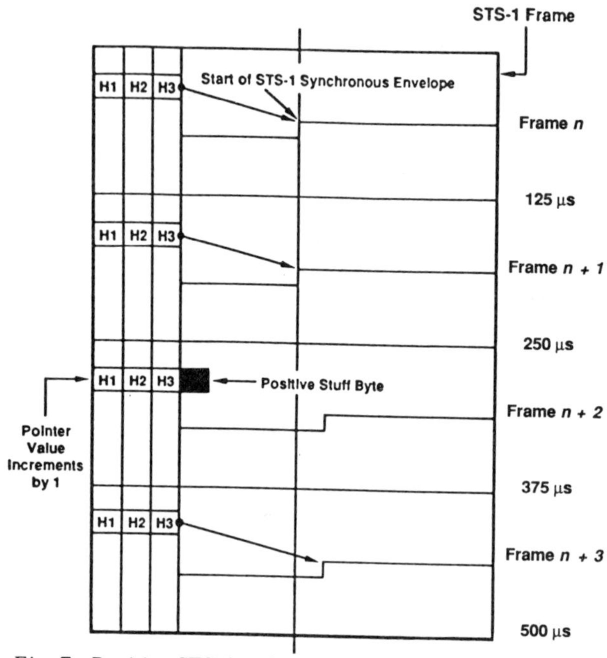

Fig. 7. Positive STS-1 pointer adjustment operation.

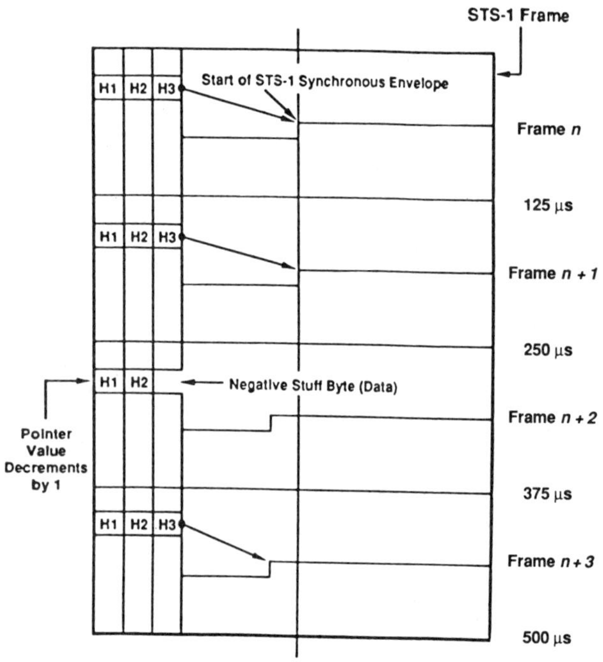

Fig. 6. Negative STS-1 pointer adjustment operation.

Broadband Payload Transport with Payload Pointers

As discussed above, STS-1 payload pointers can be used to adjust the frequencies of several STS-1 payloads in multiplexing to the STS-*N* signal level. As this is done, the various STS-1 section and line overhead bytes are frame-aligned. In Figure 8, two hypothetical and simplified SONET frames (A and B) are out of phase with respect to the arbitrary, outgoing (multiplexed) SONET signal phase. By recalculating the SONET pointer values and regenerating the SONET section and line overhead bytes, two phase-aligned signals (A and B) are formed. A and B can then be byte-interleaved to form a higher level STS-*N* signal. As shown, this can be done with minimum payload buffering and signal delay.

With frame alignment, the STS-1 pointers in an STS-*N* are grouped together for easy access at an OC-*N* receiver using a single STS-*N* framing circuit. If it is desired to carry a broadband payload requiring, for example, three STS-1 payloads, the phase and frequency of the three STS-1 payloads must be locked together as the broadband payload is transported through the network. This is easily done by using a "concatenation indication" in the second and third STS-1 pointers. The concatenation indication is a pointer value that indicates to an STS-1 pointer processor that this pointer should have the same value as the previous STS-1 pointer. Thus, by frame aligning STS-*N* signals and using pointer concatenation, multiple STS-1 payloads can be created. The STS-*N* signal that is locked together in this way is called an STS-*N*c, where the "c" stands for concatenated. Allowed values of STS-*N*c are STS-2c, STS-3c, STS-6c, STS-9c, etc. For broadband User Network Interfaces (UNI), STS-3c and STS-12c are of particular interest.

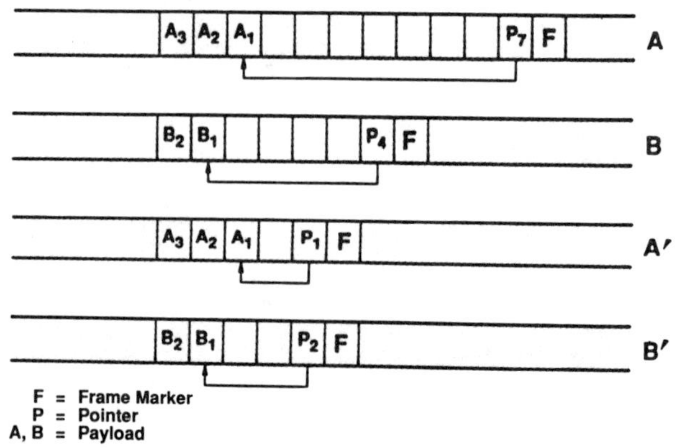

Fig. 8. Frame alignment using pointers.

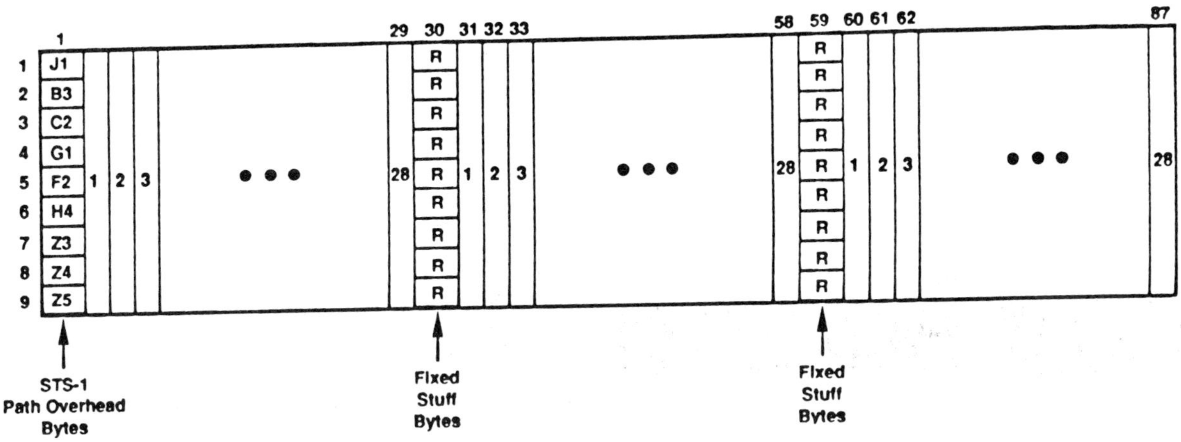

Fig. 9. VT-structured STS-1 SPE: all VT1.5.

As discussed in the section on the history of SONET standards, the Europeans had no interest in using the SONET STS-1 signal. Instead, they were interested in using a base signal of about 150 Mb/s to allow transport of their 139.264 Mb/s electrical signal and for possible Broadband ISDN applications. As the above discussion shows, the technical solution to this problem is the use of the STS-3c signal. In the U.S., we can continue to think of this signal as three concatenated STS-1 signals. In Europe and the CCITT, the STS-3c is considered as the basic building block of the new synchronous hierarchy and is referred to as the Synchronous Transport Module—Level 1 (STM-1).

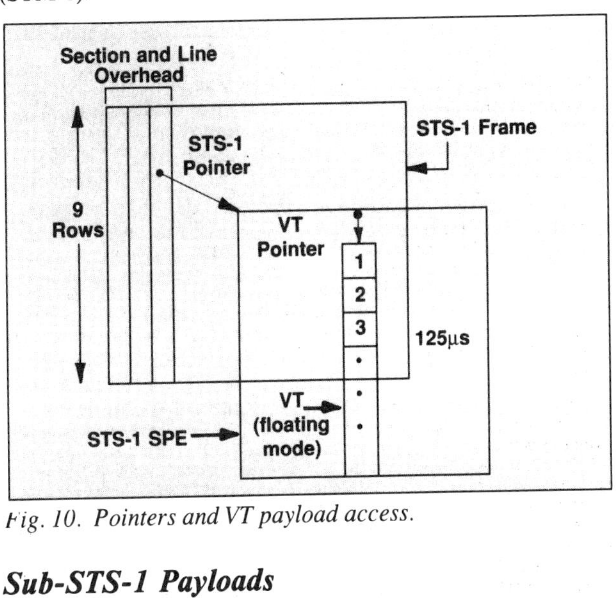

Fig. 10. Pointers and VT payload access.

Sub-STS-1 Payloads

To transport payloads requiring less than an STS-1 payload capacity, the STS-1 SPE is divided into payload structures called virtual tributaries (VTs). There are four sizes of VTs: VT1.5, VT2, VT3, and VT6, where each VT has enough bandwidth to carry a DS1, CEPT-1 (2.048 Mb/s), DS1C, and DS2 signal, respectively. Each VT occupies several 9-row columns within the SPE. The VT1.5 is carried in three columns (27 bytes), the VT2 in 4 columns (36 bytes), the VT3 in six columns (54 bytes), and the VT6 in twelve columns (108 bytes).

A VT group is defined to be a 9-row-by-12-column payload structure that can carry four VT1.5s, three VT2s, two VT3s, or one VT6. Seven VT groups (84 columns), one path overhead column, and two unused columns are byte-interleaved to fully occupy the STS-1 SPE. Figure 9 shows the STS-1 SPE configured to carry 28 VT1.5s. VT groups carrying different VT types can be mixed within one STS-1. As discussed in the section on history, the ability of the 9-row format structure to flexibly carry both the 1.544 and the 2.048 Mb/s signals was a necessary step in reaching an international agreement on SONET.

Two different modes have been adopted for transporting payloads within a VT. The VT operating in the "floating" mode improves the transport and cross-connection of VT payloads. A floating VT is so called because a VT pointer is used to show the starting byte position of the VT SPE within the VT payload structure. In this sense, the operation of the VT pointer is directly analogous to that of the STS-1 pointer, and has the same advantages of minimizing payload buffers and associated delay when mapping signals into the VT. Figure 10 shows conceptually how the STS-1 and VT pointers are used to locate a particular VT payload in an STS-1. The other VT mode is the "locked" mode. The locked VT does not use the VT pointer, but instead locks the VT payload structure directly to the STS-1 SPE. (Of course, the STS-1 SPE still floats with respect to the STS-1 frame.) The locked mode improves the transport and cross-connection of DS0 signals by maintaining the relative phase and frequency of DS0 signals carried in multiple locked VT's. When VT-organized, each STS-1 SPE carries either all floating or all locked VTs.

More than one specific payload mapping is possible with each of the VT modes described above. Asynchronous mappings are used for clear channel transport of nominally asynchronous signals using the floating mode of operation; conventional positive bit-stuffing is used to multiplex these signals into the VT SPE. "Byte synchronous" mappings have been defined in both the locked and floating modes for the efficient, synchronous transport of DS0 signals and their associated signaling; conventional fixed position mappings are used to carry the DS0's in the VT SPE (floating mode) or VT (locked mode). "Bit synchronous" mappings are used in both the locked and floating modes for the clear channel transport of unframed, synchronous signals. The VT mappings that have been defined in the current version of the American national standard are given in Table III.

TABLE III. Sub-STS-1 Mappings

Mappings	VT (Virtual Tributary) Modes	
	Floating	Locked
Asynchronous	DS1 CEPT-1 DS1C, DS2	—
Byte Synchronous	DS1 CEP-1	DS1 CEP-1 SYNTRAN
Bit Synchronous	DS1 CEP-1	DS1 CEP-1

Optical Parameters

The SONET optical interface parameters were developed in parallel with the SONET rates and formats. The optical parameters specified in the American national standard include spectral characteristics, line rate, power levels, and pulse shapes; jitter specifications will be developed in the next phase of the standard. The current optical specifications extend up to OC-48 (see Table II). It is expected that as more experience is gained with high data rate systems, the optical parameters associated with OC-18, OC-24, OC-36, and OC-48 will be updated.

The intent of this first optical interface standard is to provide specifications for "long reach" fiber transmission systems, i.e., systems using lasers. The second phase of SONET standardization in T1 will address "short reach" specifications for fiber transmission systems based on LEDs and low-power, loop lasers.

Conclusion and Future Work

The Synchronous Optical Network concept was developed to promulgate standard optical transmission signal interfaces to allow mid-section meets of fiber systems, easy access to tributary signals, and direct optical interfaces on terminals, and to provide new network features. The basic SONET signal format can transport all signals of the North American hierarchy up to and including DS3, and also future broadband signals. SONET will soon be an American national standard and a CCITT transmission signal hierarchy standard. The second phase of SONET T1 standardization will fully specify the data communications channel protocols, specify short-reach SONET optical interfaces for use in intra-office applications, and update SONET optical parameters for selected levels above OC-12.

SONET represents a successful test case for standards-making in the post-divestiture environment. Of course, the ultimate test for any standard is the development of products and services that are compliant with the new standard. For specific implementations, requirements beyond those contained in the standard are often needed. For example, Bellcore has issued a series of Technical Advisories giving additional requirements for SONET multiplexes, digital cross-connect systems, and digital switch interfaces. The first field trials of SONET equipment are expected in 1989.

References

[1] ANSI T1.106-1988, "American National Standard for Telecommunications—Digital Hierarchy Optical Interface Specifications, Single Mode," to be issued.

[2] T1M1.2/87-37R2 "Functional Requirements for Fiber Optic Terminating Equipment."

[3] ANSI T1.105-1988 "American National Standard for Telecommunication—Digital Hierarchy Optical Rates and Formats Specification," to be issued.

[4] CCITT Recommendation G.707, "Synchronous Digital Hierarchy Bit Rates," to be issued.

[5] CCITT Recommendation G.708, "Network Node Interface for the Synchronous Digital Hierarchy," to be issued.

[6] CCITT Recommendation G.709, "Synchronous Multiplexing Structure," to be issued.

Biography

Ralph Ballart is District Manager of Network Node Equipment Requirements at Bell Communications Research, Inc. (Bellcore), responsible for the development of requirements for SONET digital cross-connect systems and trunk side digital switch interfaces. Ralph is also Chairman of the T1X1.5 U.S. CCITT Drafting Sub-working Group that has responsibility for drafting contributions related to SONET to CCITT Study Group XVIII.

He received his B.S. degree in physics from the then Polytechnic Institute of Brooklyn in 1973 and his Ph.D. in physics from the University of Arizona in 1980. He joined Bell Telephone Laboratories in 1980 and worked on fundamental network planning tools and network applications for digital cross-connect systems. In 1984, he joined Bellcore and worked on requirements for SYNTRAN equipment. He was promoted to District Manager in 1985; his district participated in the technical development of SONET and its international standardization in CCITT. Dr. Ballart is a member of IEEE Communications Society.

Yau-Chau Ching is District Manager of SONET Interface Standards, responsible for strategic planning to support the SONET project in its initial applications as well as its later extension to Broadband ISDN. He is the Co-Chairman of T1X1.5 Working Group (Optical Hierarchy), where much of the SONET standardization work takes place. Before the inception of T1X1.5, he was the Co-Chairman of T1X1.4 Subworking Group on Optical Rates and Formats and an active member of T1X1.4 Subworking Group on Optical Interface. These latter groups were the predecessors of T1X1.5.

He received his B.E.E.E. degree from City College of New York in 1966 and his Ph.D. from New York University in 1969 under a National Science Foundation fellowship. From 1969 to 1984, he was with Bell Laboratories, where he did a series of exploratory development work in data compression, such as interframe video coding, digital speech interpolation, embedded ADPCM coding, and Fast Packet Network. From 1980 to 1983, he supervised a group responsible for the implementation of the Access Interface to the Fast Packet Network.

Since the divestiture of the Bell System, he has been with Bellcore, where his district was responsible for the generic requirements of fiber optic transmission systems. His district was also the originator of the SONET concept in Bellcore and its strong advocate in T1X1. Dr. Ching is a member of IEEE, Tau Beta Pi, and Eta Kappa Nu, and he holds 11 patents.

Design of an Integrated Services *Packet* Network

Jonathan S. Turner

Washington University
St. Louis, Missouri 63130
jst@wucs.UUCP

ABSTRACT

The Integrated Services Digital Network (ISDN) has been proposed as a way of providing integrated voice and data communications services on a universal or near-universal basis. In this paper, I argue that the evolutionary approach inherent in current ISDN proposals is unlikely to provide an effective long term solution and advocate a more revolutionary approach, based on the use of advanced packet switching technology. The bulk of this paper is devoted to a detailed description of an Integrated Services *Packet* Network (ISPN), which I offer as an alternative to current ISDN proposals.

Introduction

An integrated voice and data packet communications system has several advantages over existing methods.

- It uses a common set of switching and transmission facilities for both voice and data communication. This is less costly than current systems that use separate mechanisms.
- It allows voice communication to be done using less than 25 percent of the bandwidth currently needed, without sacrificing quality. This allows major savings in long distance transmission costs. It also allows customers to carry on two or three simultaneous voice conversations along with a substantial amount of data traffic over a standard copper loop.
- It provides much higher performance data communication and at lower cost than current systems. This is largely due to the integration of data communication with voice, which allows one to take advantage of the economies of scale possible in the large systems needed for a national telephone network.

The system is based on high performance packet switches which are large and fast enough to effectively support both voice and data communication on a large scale. Each packet switch can have a raw throughput of up to 1.5 Gbs, allowing it to support as many as 50,000 simultaneous voice conversations using a 32 Kbs voice encoding scheme. The one-way cross-network delay in a worst-case connection in a national network in the U.S. can be limited to about 150 ms. The key elements of the design are

- The use of high speed digital transmission facilities (1.5 Mbs) with excellent error performance.
- Simple link level protocols. In particular, there is no flow control or error correction done at the link level, which eliminates the need for state information in the link level protocol processors.
- A predominantly connection-oriented service. This allows the routing of most packets to be handled by a very simple method and facilitates bandwidth allocation and overload control. A connectionless (datagram) service can also be supported.
- Hardware implementation of basic switching and protocol functions. Switching is done using a large self-routing network containing roughly 1300 custom VLSI chips. All per packet protocol functions are handled by protocol processors (one for each link) consisting of one custom controller chip plus one large memory chip.
- Implementation of higher level protocol functions (including error correction and flow control) on an end-to-end and application-dependent basis.

The governing philosophy behind the design is that the communications network should provide transport of information at the highest possible level of performance, but nothing else. The network achieves generality, not by providing every service a user might conceivably require, but by providing only those services that every user requires. The system can be implemented with currently available technology at a cost that is comparable to that of conventional telephone switching systems.

The Trouble With ISDN

The Integrated Services Digital Network has been heralded as the mechanism that will usher in the Information Age. ISDN, it is said, will facilitate the development of new communications services, including a wide range of data services and maybe even video. It will allow such services to be implemented on a large scale at a cost most customers can afford, and spur the transformation to a 'post-industrial society.'

While I share the long range goals of ISDN proponents, I don't believe that current ISDN plans can take us very far. The fundamental problem is that current plans (see [1], for example) implicitly assume a network model based on circuit switched voice and a combination of circuit and packet switched data. This assumption is evident in current standardization efforts which focus on a transmission format for digital subscriber lines that provides two 64 Kbs circuit switched channels plus a 16 Kbs packet switched channel. While this plan can provide a limited data communications capability (it's certainly a vast improvement over the current situation), it is too inflexible to satisfy long term needs. What happens, for example to an application that requires a 75 Kbs channel? Do we implement it using the two 64 Kbs channels or one of them and the packet switched channel. Neither option is particularly attractive.

The source of the trouble is the reliance on circuit switching, which requires that the available bandwidth be divided up into fixed size channels. The channel size is a permanent feature which can't be changed easily, if at all. There's no reason to think that 64 Kbs is a particularly useful channel size. The only reason for picking it is that current telephone switching systems are based on that size. In fact, the driving force behind current ISDN plans is to preserve the investment in existing equipment while 'evolving' to a more flexible network.

Unfortunately, it won't work, because it's inherently a hybrid approach. The only thing integrated about ISDN is its name. Two (or more) switching networks are required to support ISDN as it's currently envisioned. Manufacturers may talk about their integrated architectures for ISDN, but what they mean is that they plan to put a packet switch and a circuit switch in the same box and call it an integrated system. This is not their fault—what else can they do? Packet switching and circuit switching are very different communications methods. They require different switching and transmission facilities and all the proposed schemes for combining them are little more than packaging.

What to do then? Is a hybrid network really necessary or is there an integrated solution that can satisfy the needs of both voice and data and remain flexible enough to meet satisfy new requirments as they arise? I claim there is such a solution, but it requires abandoning circuit switching and moving to new network designs based on packet switching.

Why Packet Switching?

Here are three reasons that make packet switching an attractive method for providing integrated voice and data services.

- *Adaptability to changing traffic.* Packet switching naturally provides the user with exactly the bandwidth required. As new services are developed with different bandwidth requirements, packet switching systems can adapt to the changing conditions easily. Circuit switching systems cannot.
- *Integrated internal architecture.* As outlined above, current ISDN plans require separate switching networks for different types of information. Packet switching can provide both an integrated customer interface and a single network solution for a wide range of communications needs, leading to substantial cost savings in switching systems and system administration.
- *Transmission efficiency.* Many data services are characterized by bursty communications patterns which make poor use of conventional circuit switched facilities. For example, interactive data users typically use only a few percent of the bandwidth available to them. Although it is less widely recognized, voice is also bursty. In the average telephone conversation, less than 40% of the available bandwidth is actually used. Packet switching can exploit this burstiness to double the number of conversations that can share a single transmission facility.

Given all these advantages, why has packet switching not been used extensively for voice? To answer this, we must take a closer look at the technical specifications and costs of conventional circuit switches and packet switches and see how they compare.

Circuit switching has been the technology of choice in the telephone network, since its origin about one hundred years ago. During that period, circuit switches have evolved from manually operated switchboards to small but automatic step-by-step switches, to larger panel switches to the computer-controlled electronic switching systems that currently dominate the scene. The current systems are very large—local switches can provide service to over 100,000 customers, large toll switches support as many as 50,000 simultaneous voice calls corresponding to a raw bandwidth of 6 gigabits per second (Gbs). The network as a whole is also very large. There are approximately 10^8 telephones in the United States at present and over 10,000 local switching offices. The performance of the switching systems is quite impressive—information passing through a modern digital switch is typically delayed no more than a few milliseconds, and new connections can be established in a fraction of a second. Equally impressive is the cost—while there is considerable variation, per line equipment costs for modern digital telephone systems is typically in the \$100–200 range.

Just as circuit switching has long dominated the telephone network, so has packet switching dominated the data communications scene, principally because of the advantages cited earlier. Packet switching is exemplified by the Arpanet, which was the first major example of a large scale data communications network. In the Arpanet, the endpoints of the communication are typically large time-shared computers, called hosts. Communication is provided by packet switches, each of which typically connects to a few hosts and several other packet switches. There are currently about 300 hosts in the Arpanet and 100 packet switches. The packet switches are implemented using general purpose computers (usually minicomputers), although more recent versions have been supplemented with front-end communications processors to reduce the load on the main processor. The transmission facilities used by the Arpanet include low speed modem connections (1–10 Kbs) and higher speed digital facilities (56 Kbs). The throughput of the packet switches is generally under 1 Mbs and delays can be substantial (50–100 ms per switch). The cost of packet switching as provided by the Arpanet is quite high, since each packet switch supports a small number of hosts. Commercial data networks do better, but there remains a large gap between per-host costs in data networks and per-line costs in the telephone network.

This comparison explains the conventional wisdom that packet switching is poorly suited to the needs of telephony and the resulting conclusion that an ISDN implementation must include a circuit switching component to support voice. The fallacy in the conventional wisdom is that the disadvantages attributed to packet switching by this comparison are not due to any inherent properties, but are side-effects of the conventional implementations. The requirements and design constraints that shaped the development of the Arpanet and commercial data networks were completely different from those that shaped the telephone network. The scale, the performance requirements and the cost sensitivities are all very different. It's because the *needs* differ that the resulting systems differ so in their technical specifications. In the remainder of this paper, I will attempt to correct the widely held, but erroneous view that packet switching is poorly suited to the needs of voice by describing a system capable of supporting voice and data communication on a large scale and at a cost that is competitive with conventional telephone systems.

ISPN Architecture

Let's begin by deciding what general properties a packet switching system must have if it is to be a suitable vehicle for providing voice and data communication on a large scale. First and most obviously, it must be big—comparable in raw bandwidth to conventional telephone systems. Second, it must be fast—long end-to-end delays are annoying in voice connections and hence unacceptable. While opinions differ on the exact amount of delay that can be tolerated, most experts would agree that 100 ms is acceptable, while more than 500 ms is not. Third, it must be inexpensive. Any system that seeks to replace circuit switching must be able to provide voice services at a competitive cost. It is not enough to offer a better product at a higher cost—for a system to succeed on a large scale, it cannot significantly increase costs for basic services.

How can we achieve the requisite scale, performance and economy in a packet switching network? Well, the telephone network is one place to look for the answer. One of the first things one notices is that the telephone network makes extensive use of high bandwidth digital transmission facilities. Modern digital switching systems are designed to interface directly to 1.5 Mbs transmission facilities carrying 24 voice channels in a 64 Kbs format. Interfacing costs are cheap, since they are shared by the 24 channels and the bit error rates are excellent. Conventional packet switching systems, on the other hand interface to much smaller channels, and must be prepared to cope with the much higher error rates present in modem connections. A second thing one notices about the telephone network is the amount of special purpose hardware used to provide the switching function. While telephone systems contain general purpose computers, their function is primarily connection establishment. The computers don't move the bits. That function is handled by large specialized networks. In contrast, most conventional packet switches include a general purpose computer that must do some processing on every packet and can quickly become a bottleneck.

These observations suggest that a high performance packet switch should (1) interface directly to high speed digital transmission facilities and (2) use special purpose hardware to perform all per packet processing. Now, those familiar with link level protocols used in conventional packet networks may recognize this as a challenging task. Typical protocols are quite complex, requiring extensive state information in the protocol processors and complicated error recovery procedures. This complexity almost demands a programmable processor (a hardware implementation would never be completely debugged), but a microprocessor-based implementation is unlikely to be fast enough to keep up with a 1.5 Mbs link, and also may be too expensive. Fortunately, there is a way out—simplify the protocol. Current packet switching protocols were designed for hostile environments—the low speeds and high error rates typical of modem connections. In a network composed entirely of high speed digital facilities, much of the complexity of typical protocols can be eliminated from the lower protocol layers. In particular, error correction and flow control can be taken out of the link layer and provided on an end-to-end basis rather than a link-by-link basis. This eliminates the need for extensive state information at the ends of each link and the synchronization and recovery procedures required to maintain it and in turn makes possible the construction of inexpensive, high performance protocol processors. In addition, if these functions are provided on an end-to-end basis, they can be pro-

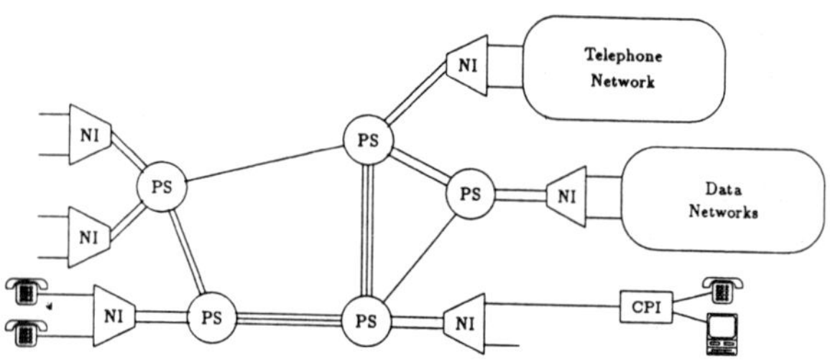

Figure 1: Network Architecture

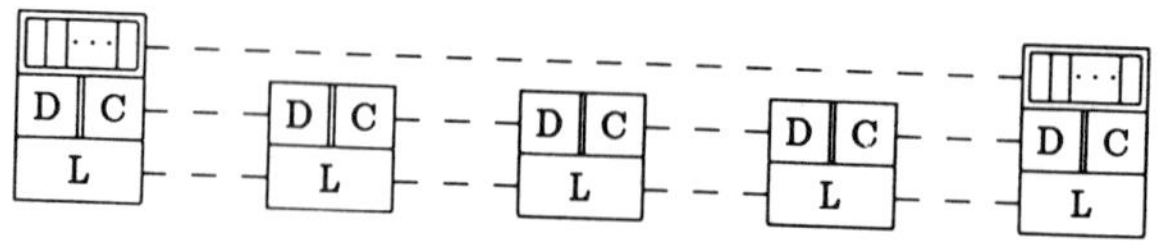

Figure 2: Protocol Structure

vided selectively. Hence, delay-sensitive applications like voice can avoid the performance penalites they can cause.

Returning to the question of scale, how large must our packet switches be? In order to compete with conventional telephone switches, they should be capable of supporting at least 50,000 simultaneous voice conversations. How many conversations can a 1.5 Mbs link carry? When operated in a circuit switching mode using the 64 Kbs digital encoding method currently employed, the answer is 24. When operated in a packet switching mode that number jumps to about 50. One can obtain another factor of two improvement by using newer voice encoding methods. 32 Kbs adaptive differential pulse code modulation (ADPCM) is a good choice, since it provides comparable quality to the method currently used, but requires only half the bandwidth. This leads to a figure of 100 voice channels per 1.5 Mbs link, implying that our packet switch must terminate 1000 full-duplex links if it is to support 50,000 simultaneous voice conversations. This in turn, means that our switch must include some mechanism capable of receiving over two million packets per second, and sending each one out on the appropriate outgoing link.

Network Overview

Based on the discussion in the previous section, we can begin to develop an architecture for the ISPN. The major components are shown in Figure 1. The Packet Switches (PS) each terminate up to 1000 High Speed Links (HSL). Using 32 Kbs ADPCM coding for voice, they can support over 50,000 simultaneous voice conversations. Residential customers connect to the network over Medium Speed Links (MSL), which operate at 100 Kbs. High speed access can be provided to business customers. The Customer Premises Interface (CPI) can take a variety of forms depending on the kind of service required. At the low end of the spectrum would be a simple controller providing service for a single telephone and implemented using a single chip eight bit microcomputer. Customers wanting several phones and data communication would require a more complex controller. Businesses would typically have an interface to a Private Branch Exchange (PBX) or Local Area Network (LAN). The Network Interfaces (NI) provide concentration, accounting data collection and network protection. A configuration designed for residential customers would support about 500 customers and would be connected to a local packet switch by four HSLs. Thus, one PS could have as many as 125 NIs and support over 60,000 customers.

NIs can also be designed to provide a conventional line interface as supported by current telephone systems. Even higher concentration ratios can be supported in this case—up to 2000 customers could be supported on a single NI connected to its host PS by four HSLs. While this configuration doesn't exploit the advanced capabilities of the system, it does provide a mechanism for easing the transition from a circuit-oriented network to a packet-oriented one. In a similar fashion, NIs can be designed to interface to the current telephone network. In this case, the NI would interface directly to up to 960 digital trunks.

The network provides two communications services.

- *Point-to-point channels.* These are two-way channels joining pairs of customers. The customer establishes a channel by sending a connection request message to the network specifying the destination and the average bandwidth needed. If the network accepts the connection, it in effect guarantees that the customer can expect to have the requested bandwidth available. The network may refuse to accept the connection if there are not adequate resources available. The connections do not provide perfectly reliable information transport. In particular, the network does not provide mechanisms for error correction and flow control. The network can provide connections at any speed up to 1.5 Mbs (although larger connections are more likely to be blocked). The bandwidth requirement may be asymmetric—that is, it may depend on the direction of transmission. No distinction is made between voice and data connections. A voice connection is simply one with an average bandwidth of about 12 Kbs.
- *Datagrams.* These are individually addressed packets, not associated with a pre-established connection. The network makes an effort to deliver them, but does not guarantee delivery.

The communications protocols are divided into three levels—the *link level*, the *network level* and the *customer level*. The interrelationships among the different levels are indicated in Figure 2. There is a single link level protocol, two network level protocols (one for connection-based communication and one for datagrams) and a variety of customer level protocols. The customer level protocols are not discussed in detail here, but the intention is that these protocols would be application-dependent and built on top

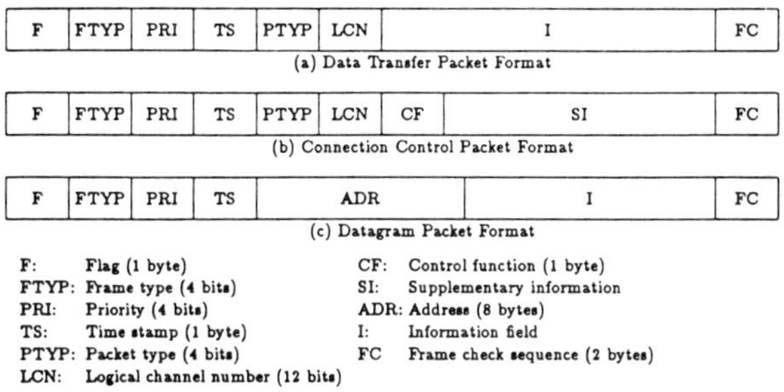

Figure 3: Packet Formats

of one of the two network level protocols. One of these would be a telephony protocol. Another might be an internet protocol such as the DARPA IP, to facilitate communication among different data networks. Still another might be a connection-oriented internet protocol.

There are essentially three packet formats of interest to the network—*connection control packets*, *data transfer packets* and *datagram packets*. These packet formats are shown in Figure 3.

Link Level Protocol. The link level protocol provides frame delimiting, link transparency, error detection, packet timing, and congestion control, but not error correction. There are four fields used by the link level protocol, the Frame Type field (FTYP), the Priority field (PRI), the Time Stamp field (TS) and the Frame Check field (FC). The FTYP field identifies the frame as a test frame, a datagram or a frame belonging to a connection. The Priority field (PRI) contains a customer-specified priority. The network preferentially discards low priority frames to alleviate short-term overload conditions in the network. The network uses the Time Stamp field (TS) to record the delay encountered by the packet as it crosses the network. More precisely, it records the number of milliseconds the packet spent in each PS and NI it passed through (see [11]). This is important for applications such as voice that are sensitive to delay variations and need a mechanism to remove the timing 'jitter' that packet networks can introduce. It can also be used in distributed programming applications for clock synchronization. The Frame Check field (FC) uses a sixteen bit cyclic redundancy code to detect errors in the frame. Frames with errors are simply discarded.

Network Level Protocols. There are two protocols at the network level, the Simple Datagram protocol (SDP) and the Simple Connection protocol (SCP). SCP packets contain a four bit Packet Type field (PTYP) and a twelve bit Logical Channel Number field (LCN). The Packet Type field (PTYP) identifies each packet as either a data packet or a control packet and contains a Congestion Control subfield, used to inform the NIs and customers of internal network congestion. SCP data packets also contain an Information field, which can have any length up to 144 bytes. SCP control packets are used to establish and control connections and contain a Control Function field (CF) and a Supplementary Information field (SI). SDP packets contain an eight byte address field. Two bytes are reserved as a socket number, to be interpreted by the customer protocols. The remaining six bytes define a hierarchical address space organized on a geographical basis, to permit routing based on local knowledge of network topology.

Network Performance

Given the network architecture described above, we can make some approximate calculations of network performance. To do this we hypothesize a worst-case reference connection in which two customers access the network using MSLs and communicate over a 4000 mile connection passing through two NIs and six PSs. We assume that the access links each carry two voice calls and that the seven internal HSLs are all operating at an occupancy of 85%.

The delay experienced by a packet passing across the reference connection contains several components.

- *Speed-of-light delay.* This is the constant delay caused by the finite speed of light. It is approximately 40 ms for a 4000 mile connection.
- *Packetizing delay.* This is the delay involved in producing enough data to put in a packet. Using 32 Kbs ADPCM voice encoding, information is generated at the source at the rate of 32 bits per millisecond. 512 bits are placed in each packet, giving a packetizing delay of 16 ms.
- *Queueing delay.* This is the delay encountered by the packet as it is waiting to be transmitted over each of the seven HSLs and two MSLs in the reference connection. We will analyze this component in detail below.
- *Switching delay.* This is the delay encountered by a packet as it passes through a PS or NI, excluding queueing delays. We will see in the next section, this can be kept below 0.5 ms per switch, hence we allow 4 ms of switching delay in the reference connection.

We use an M/M/1 queueing model to approximate the queueing delay on the seven HSLs in the reference connection. The formula for the average delay through n tandem queues is

$$q = \frac{bn}{s(1-\rho)}$$

where b is the packet length in bits, s is the bit rate of the transmission lines (1.544 Mbs) and ρ is the link occupancy (0.85). We use an average packet length of 650 bits to model a mixture of voice packets with 1000 bit data packets. For seven queues we calculate $q \approx 20$ ms. Of course, this is just the average delay. For voice, it is important to have an upper bound on the delay experienced by *most* packets. We can determine this by using an Erlang distribution with a rank of seven and a mean of 20 ms. With this technique we estimate that 99% of all packets in the

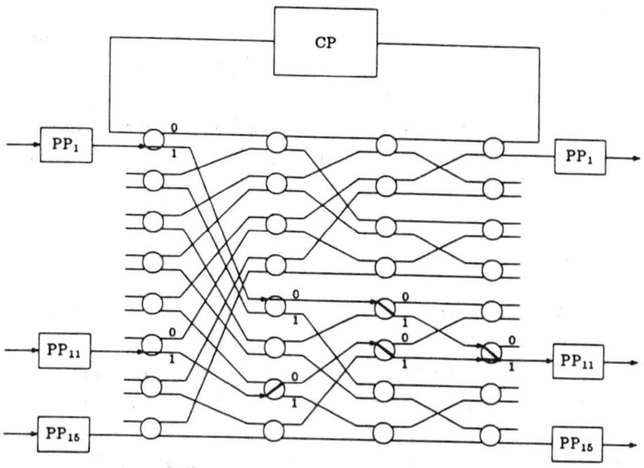

Figure 4: Packet Switch Structure

reference connection experience a delay of no more than 42 ms on the internal network links.

We still need to determine the queueing delays experienced on the access links. Here the M/M/1 model is not really applicable. On the access link going into the network, the delay is bounded by 12 ms, since we are assuming two simultaneous voice calls generating packets at 16 ms intervals. The worst thing that can happen is that packets are generated simultaneously from both voice encoders meaning that one packet must wait for the other to be transmitted. The delay of 12 ms results from the observation that voice packets are less than 600 bits long, and with an access line speed of 100 Kbs can be transmitted in 6 ms—hence it takes 12 ms to transmit a packet for each call. Packets experience a similar delay as they leave the network. Of course, the actual delays experienced at the access links is highly dependent on the nature of the services the customer is using at the time. One way to ensure good performance for voice at the access links is to use the packet priorities to give preference to voice packets. With this mechanism, we can add data to the traffic carried by the access link and be guaranteed a maximum delay of 22 ms on each end (assuming a non-preemptive priority mechanism). Based on these considerations, we assume 50 ms of access delay.

Summing the various delay components yields an estimate of about 150 ms for 99% of all packets.

Packet Switch Design

The network is built using large high performance packet switching systems, each terminating up to 1023 HSLs. The structure of such a packet switch is illustrated in Figure 4, which shows a small version with 15 HSLs.

The system is controlled by a Control Processor (CP) which performs all connection control functions, plus administrative and maintenance functions. The CP is a large general-purpose computer. Its role is analagous to that of the Control Processor in large telephone switching systems such as the No. 4 ESS [2].

Each HSL is terminated by a Packet Processor (PP), which performs the link level protocol for all packets and the network level protocol for data transfer packets. It also forwards connection control packets to the CP and datagram packets to the Datagram Routers (DR).

The heart of the switch is the Switch Fabric (SF) which consists of a large binary routing network. The important property of such networks is that the path each packet takes through the network is determined by successive bits of its destination address. The figure shows paths from two different SF input ports to output port 1011. Note that at the first stage the packets are routed out the lower port of the nodes (corresponding to the first '1' bit of the destination address), at the second stage they are routed out the upper port (corresponding to the '0' bit) and in the third and fourth stages they are routed out the lower ports. The self-routing property is shared by a variety of interconnection patterns, including the so-called delta, shuffle-exchange and banyan networks (see [5]). The PS uses a ten stage binary routing network with 1024 ports.

The Datagram Routers (not shown) are special-purpose devices used to route datagrams. The number of DRs can be engineered to suit the traffic. Each occupies a port on the SF, replacing one PP. Since most of the traffic is expected to be connection-oriented, the number of DRs required should be modest.

Duplication is required to provide the highly reliable operation expected in a large scale public communications network. The scheme used here (but not shown in the figure) is to duplicate the SF, but not the PPs. In normal operation, the duplicate SFs are operated in a load-sharing fashion, with each carrying half the traffic. If one side fails, all traffic is shifted to the other side.

Packet Processing

When a packet is received by a PP, it is placed in a buffer with several additional header fields added. The Destination field (D) identifies the destination port on the SF. The Source field (S) identifies the port where the packet arrived. The Length field (LNG) gives the packet length in bytes. The Switch Packet Type field (SPTYP) is used to identify various packet types within the PS. The Arrival Time field (AT) gives the time at which the packet arrived at the PS (this is used for processing the TS field).

For data transfer packets, the destination port is determined by the PP using the packet's LCN field and the PP's Logical Channel Translation Table (LCXT). Each entry in the LCXT contains an outgoing port number and a new LCN. The outgoing port is placed in the D field of the packet and the new LCN goes in the LCN field. The

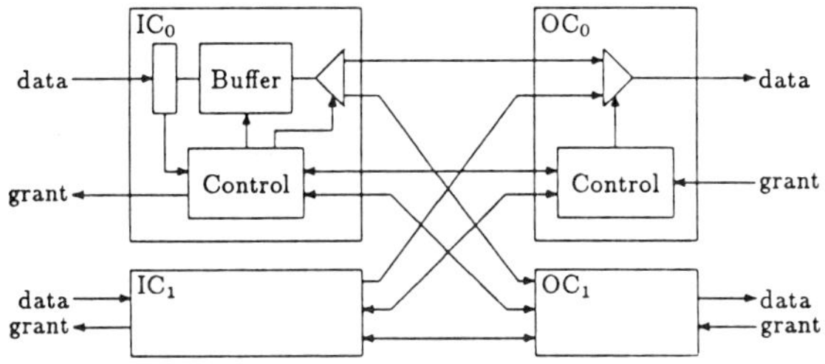

Figure 5: Switch Node

packet is then sent on to the switch fabric, which uses the D field to route the packet to the proper outgoing port as described earlier. When a packet arrives at the outgoing PP, it is buffered and then transmitted on the HSL with the extra header information stripped off.

The contents of the LCXTs is controlled by the CP, which can read and write the LCXT using special control packets sent to the PPs through the SF. Thus, the connection establishment process includes the sending of messages from the CP to the two PPs selected for the connection, updating their LCXTs appropriately.

Switch Fabric

The basic operation of the switch fabric has already been described. The SF's highly regular and parallel structure allows the construction of very large switches, without the bottlenecks that can arise in bus or ring based interconnection networks.

The nodes of the SF operate as miniature packet switches. Each node has a buffer at each input port capable of holding one maximum length packet. The data paths joining the nodes are bit serial and operate at 12 Mbs. This gives the SF an 8:1 speed advantage over the external HSLs. Thus, if all the external links are operated at an occupancy of 85%, the internal links will have an average occupancy of less than 11%. (This is assuming just one switch plane active. When both are active, the average occupancy is less than 6%.) There is also an upstream control lead joining each pair of adjacent nodes. This is used to implement a simple hardware flow control mechanism, which prevents buffer overflows within the SF.

A more detailed look at the switch node appears in Figure 5. It consists of two Input Controllers (IC) and two Output Controllers (OC). The ICs contain a buffer large enough to hold one packet and a controller implemented as a state machine. The OCs are simple state machines that arbitrate requests for their ports.

When a packet is received, the IC determines the proper outgoing port by examining the appropriate bit of the packet's destination field, then requests permission to use that port. If the desired port is immediately available, the packet is sent to it directly, bypassing the buffer.

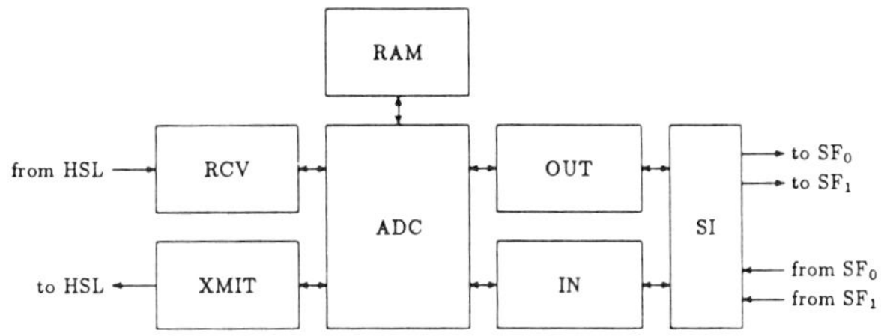

Figure 6: Packet Processor

Hence, a packet can pass through a switch node after experiencing a delay of just a few bit times. In fact, this is the normal case due to the relatively low occupancy of the internal data paths.

If the desired port is not available, the packet is shifted into the buffer. As soon as the desired port becomes available, the packet is sent out—even if not all of the packet has been received. If the port is still not available when the end of the packet is received, the IC holds its *grant* lead low to prevent the arrival of new packets. The grant lead is re-asserted as soon as the desired link becomes available, allowing a new packet to enter the buffer, while another is leaving.

Packet Processors

The structure of the Packet Processors (PP) is shown in Figure 6. It is organized around a 16 Kbyte RAM, which contains four packet buffers and the Logical Channel Translation Table (LCXT). The principal buffers are the Receive Buffer (RCB) used for packets received from the HSL on their way to the switch, and the Transmit Buffer (XMB) for packets going from the switch to the HSL. The Link Test Buffer (LTB) and the Switch Test Buffer (STB) are small buffers that provide loop-back paths for testing the HSL and switch respectively.

Access to the memory is provided through the Address Controller (ADC), which contains read and write pointers for the buffers and arbitrates memory access.

The Receive circuit (RCV) receives incoming packets from the HSL, removes the flag field, discards packets with errors, adds the extra header fields, initializes the length (LNG) and arrival time (AT) fields, converts from bit serial to eight bit parallel format and writes the packet to the RCB through the ADC.

The Output circuit (OUT) takes packets from the RCB, performs the logical channel translation described above and sends the packets onto the switch in bit serial format. The Input circuit (IN) takes packets from the switch and writes them to the XMB.

The Transmit circuit (XMIT) takes packets from the XMB, performs the time stamp calculation, strips the extra header information, adds the flag field and transmits the packets on the HSL.

The Switch Interface (SI) connects to the duplicated switch planes, normally routing packets to and receiving packets from the active plane. The SI can also send and receive packets from the standby plane—this is used for testing.

The PP can be implemented using two chips—one custom controller chip and one memory chip. The simplicity of the protocols makes this possible. There is no need to buffer unacknowledged frames that may need to be retransmitted as in conventional link level protocols that perform error correction and flow control. Similarly, there is no need for the recovery and synchronization procedures that are required by such protocols.

Performance of the Switch Fabric

Switching networks similar to the one considered here have been analyzed by several authors to determine their performance characteristics (see [4] and [8], for example). Unfortunately the earlier analyses don't quite suit the network considered here, since they assume packets are buffered at each stage. This is not the case for the network considered here, which permits packets to pass through a stage without being buffered at all. This buffering technique is called *virtual cut-through* by Kermani and Kleinrock [10], who analyze its performance in a somewhat different context. Their analysis does not apply to our situation either.

The analysis given by Jenq in [8] can be modified to suit our needs. Jenq provides an algorithm for calculating the delay through a binary routing network preceded by packet processors containing infinite buffers. The main difference between his model and the network described here is that it assumes that packets are buffered at each stage. This means that in his model, packets encounter a delay of at least nb/s, where n is the number of stages in the network, b is the packet length and s is the speed of the internal data paths. This is true even at very low occupancies. In our network, on the other hand, the delay experienced at low occupancies is $\approx 10n/s$, since each packet is delayed about 10 bit times per stage. Hence, at low occupancies the delay through the network can be obtained by subtracting $(b-10)n/s$ from the estimate obtained using Jenq's method. Since the network is normally operated at occupancies below 20%, this approach can be expected to yield satisfactory estimates.

Two other assumptions of Jenq's method need to be mentioned. First, it assumes that all packets have the same length—we will use it to estimate the delay under the assumption that all packets are of maximum length (b = 1264), which is a worst-case assumption for this network. Second, Jenq's analysis assumes that traffic is uniformly distributed throughout the network—more precisely, the destination field of each arriving packet is assumed to be selected at random from the integers 0–1023. This is called the *uniform traffic assumption* and while it appears reasonable, it ignores the effect of the network's connection-orientation on traffic patterns. This assumption is examined more closely in a later section and justified.

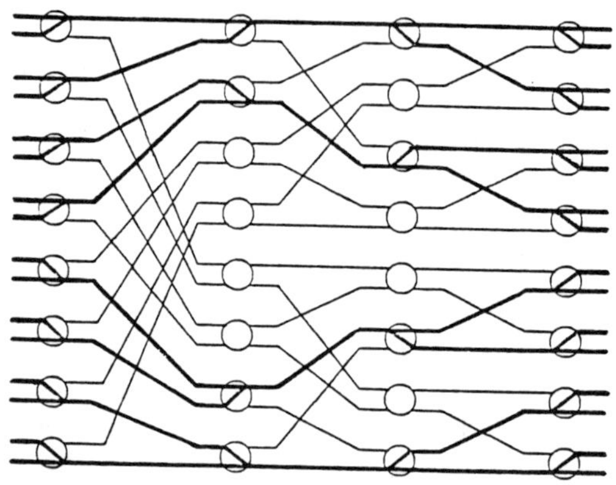

Figure 7: Congestion in Binary Routing Networks

Let $d_J(\rho, n)$ be the average delay through an n stage binary routing network for a given occupancy ρ, using the method described in section V of Jenq's paper [8]. We find, $d_J(.05, 10) = 1070\,\mu s$, $d_J(.1, 10) = 1090\,\mu s$, $d_J(.15, 10) = 1113\,\mu s$ and $d_J(.2, 10) = 1141\,\mu s$, Subtracting $1045 = (b - 10)n/s$, as described above gives corresponding estimates of 35, 45, 68 and 96μs for our network. When all the HSLs coming into the PS are operated at occupancies of 80%, the occupancy of the internal links is about 11%, leading to an average switch fabric delay of about 50 μs.

Switch Fabric Revisited

One problem with binary routing networks is that they can become congested in the presence of certain traffic patterns. This is illustrated in Figure 7, which shows a traffic pattern corresponding to several *communities of interest*. In this pattern, all traffic entering the first four inputs is destined for the first four outputs, all traffic entering the second group of four inputs is destined for the second group of four outputs, and so forth. Note that with this pattern, only one fourth of the links joining the second and third stages are carrying traffic. Thus, if the inputs are heavily loaded the internal links will be hopelessly overloaded and traffic will back up. In a 1024 × 1024 network, there are ten stages and the links between the fifth and sixth stages can in the worst-case be carrying all the traffic on just 32 of the 1024 links.

This problem can be solved by using two networks instead of one. One of these is called the Routing Network (RN) and is a standard binary routing network. The other network sits in front of the RN and is called the Distribution Network (DN). It has the same structure as the RN, but instead of routing packets based on their destination address, it attempts to distribute packets evenly across all its output ports. This is done by having each switch node route packets alternately out its two ports. The alternate-port strategy is modified if one or both ports is unavailable. In this case, the first port to become available is used. This approach breaks up any communities of interest and makes the combination of the DN and RN robust in the face of pathological traffic patterns.

Note, that if the DN used a purely random distribution algorithm, the uniform traffic assumption made in the performance analysis section would be validated. The distribution algorithm used by the DN is actually superior to a random one, since it distributes traffic more evenly and can route around congested areas. These observations justify the simple analysis based on the uniform traffic assumption.

One drawback of the DN is that it doubles the number of stages in the SF, thus roughly doubling the packet delay and the circuit complexity. This loss can be recovered by using larger switch nodes for the RN and DN. Instead of 2×2, we can use 4×4 nodes. With the larger nodes we can construct a network with half the number of stages required with 2×2 nodes. This also halves the delay and the circuit complexity.

Other Issues

The circuit complexity of the switch nodes is dominated by the buffer, which is basically a 1280 bit shift register. Thus a 4×4 node contains about 5000 bits of memory and will fit comfortably on a single chip with 20 pins. These chips are combined onto circuit boards, each board containing either 32 or 48 chips. A single RN will contain 32 of these boards.

The PPs can be implemented using one custom controller chip and one $16K \times 9$ memory chip. Sixteen PPs are packaged together on a single board, meaning that 64 of these boards are required for the entire PS. The Datagram Routers, which haven't been described in detail can be implemented in a similar fashion to the PPs—they're actually a bit simpler.

The packaging of the PS can be remarkably compact. A single equipment frame, 3 feet wide by 6 feet high by 1.5 feet deep is all that's required for the duplicated switch fabric and all the packet processors.

Summary

This paper has described the design of a high performance packet switching network capable of supporting both voice and data communication on a large scale and at a cost to the user, comparable to current telephone service. The key features of the design are the use of high speed digital transmission facilities and simple link level protocols, a predominantly connection-oriented service, hardware implementation of all per packet functions and implementation of higher level protocol functions on an end-to-end and application-dependent basis. I claim that the advanced packet switching technology described here is inherently better suited for the provision of advanced communications services than the circuit switching technology which underlies current ISDN proposals.

Many aspects of the packet switching system described here have been patented by AT&T Bell Laboratories. See references [14]–[20] for further details.

Acknowledgements. The work described here was carried out between 1981 and 1983 when I was a Member of Technical Staff at Bell Laboratories in Naperville, Illinois. Many individuals contributed ideas. I particularly want to thank Harold Andrews, Danny Creed, Bill Hoberecht, Warren Montgomery and Len Wyatt who helped turn the original idea into a workable design. I also want to thank Bob Cardwell and Eric Nussbaum for recognizing the potential and providing the backing to develop it. Special thanks go to Maurizio Dècina, whose refusal to accept half-baked ideas forced me to make sure that mine were always well-cooked.

REFERENCES

[1] Aoki, D. J., D. H. Florin, S. D. McKenna, G. R Welsh and P. E. White. "Digital Feature Evolution in the 5ESS^{TM} Digital Switch," *Proceedings of Globecom 83*, 11/83, 601–605.

[2] *Bell System Technical Journal*, vol. 56, no. 7, 9/77 (Special issue devoted to the No. 4 ESS).

[3] Dècina, Maurizio. "Performance Requirements for Integrated Voice/Data Networks," *IEEE Transactions on Communications*, vol. COM-30, no. 9, 9/82, 2117–2130.

[4] Dias, Daniel M. and J. Robert Jump. "Packet Switching Interconnection Networks for Modular Systems," *Computer*, vol. 14, no. 12, 12/83, 43–54.

[5] Feng, Tse-yun. "A Survey of Interconnection Networks," *Computer*, vol. 14, no. 12, 12/83, 12–30.

[6] Gruber, John G. "Performance Requirements for Integrated Voice/Data Networks," *IEEE Journal on Selected Areas in Communications*, vol. SAC-1, no. 6, 12/83, 981–1005.

[7] Hoberecht William L. "A Layered Network Protocol for Packet Voice and Data Integration," *IEEE Journal on Selected Areas in Communications*, vol. SAC-1, no. 6, 12/83, 1006–1013.

[8] Jenq, Yih-Chyun. "Performance Analysis of a Packet Switch Based on a Single-Buffered Banyan Network," *IEEE Journal on Selected Areas in Communications*, vol. SAC-1, no. 6, 12/83, 1014–1021.

[9] Kasahara, M., H. Shimizu, M. Imaizumi. "Basic Concept of the Digital Network for the Information Network System," *Proceedings of Globecom 83*, 11/83, 969–973.

[10] Kermani, Parviz and Leonard Kleinrock. "Virtual Cut-Through: A New Computer Communication Switching Technique," *Computer Networks*, vol. 3, 1979, 267–286.

[11] Montgomery, Warren A. "Techniques for Packet Voice Synchronization," *IEEE Journal on Selected Areas in Communications*, vol. SAC-1, no. 6, 12/83, 1022–1028.

[12] Rettberg, R., C. Wyman, D. Hunt, M. Hoffman, P. Carvey, B. Hyde, W. Clark and M. Kraley. "Development of a Voice Funnel System: Design Report," Bolt Beranek and Newman, Report No. 4098, 8/79.

[13] Turner, Jonathan S. and Leonard F. Wyatt. "A Packet Network Architecture for Integrated Services," *Proceedings of Globecom 83*, 11/83, 45–50.

[14] Turner, Jonathan S. "Packet Load Monitoring By Trunk Controllers," United States Patent #4,484,326, 11/20/84.

[15] Turner, Jonathan S. "Packet Switching Loop-Around Network and Facilities Testing," United States Patent #4,486,877, 12/4/84.

[16] Turner, Jonathan S. "End-to-End Information Memory Arrangement in a Line Controller," United States Patent #4,488,288, 12/11/84.

[17] Turner, Jonathan S. "Interface Facility for a Packet Switching System," United States Patent #4,488,289, 12/11/84.

[18] Turner, Jonathan S. "Packet Error Rate Measurements by Distributed Controllers," United States Patent #4,490,817, 12/25/84.

[19] Turner, Jonathan S. "Fast Packet Switch," United States Patent #4,491,945, 1/1/85.

[20] Turner, Jonathan S. "Fast Packet Switching System," United States Patent #4,494,230, 1/15/85.

[21] Turner, Jonathan S. "Duplicated Network Arrays," U.S. patent application filed.

[22] Turner, Jonathan S. "Integrated Self-Checking Packet Switch Node," U.S. patent application filed.

[23] Turner, Jonathan S. "Distributed Monitoring of Packet Transmission Delay," U.S. patent application filed.

[24] Turner, Jonathan S. and Leonard F. Wyatt. "Alternate Paths in a Self-Routing Packet Switching Network," U.S. patent application filed.

[25] Weinstein, Clifford J. and James W. Forgie. "Experience with Speech Communication in Packet Networks," *IEEE Journal on Selected Areas in Communications*, vol. SAC-1, no. 6, 12/83, 1022–1028.

Conceptual Issues for ATM

Jerry Gechter
Peter O'Reilly

Introduction

The Asynchronous Transfer Mode (ATM) has been strongly promoted, as the transport structure for future broadband telecommunication networks, within standards bodies CCITT Study Group XVIII (internationally) and T1S1 (North America) [1] [2]. ATM is a transport technique based on fast packet switching, where the information is packed into fixed-size cells of relatively short length, 32–120 octets, perhaps. (Although standards clearly do not exist at this time, there is a strong trend, particularly in Europe, towards very short cells.) These cells are identified and switched throughout the network by means of a label in the header—hence, the term "label multiplexing" is often used. This label contains the virtual connection identification and whatever other addressing is required, as well as perhaps priority and other control data. In OSI terminology, ATM has Layer 1 functionality but only some Layer 2 functions.

Other proposals for the transport mode are the Synchronous Transfer Mode (STM) and some hybrid of ATM and STM. STM basically implies an underlying multirate circuit-switched transport structure. STM is, to a certain extent, a natural extension of the current public network and of the narrowband ISDN interface standard. However, ATM appears to have succeeded over its rivals in becoming the target network for Broadband ISDN (BISDN) [3].

Whereas the authors of this paper do not question the technical feasibility of ATM at 150 Mb/s rates, the state of system definition is such that we cannot unequivocally support ATM as a concept. Furthermore, we feel that there is danger in premature standardization of system parameters while the system concept itself is still in flux. To be more specific, we feel that there are four major areas where the system concept needs to be clarified before standardization can proceed: grade of service, accounting, compatibility, and scaling. The authors are currently developing a circuit-switched class of service concept to deal with these problems; the approach is outlined in [4].

Grade of Service

One of the basic features of an ATM transport structure is that packets of real-time services will get dropped, even when the network is over-engineered to a significant extent; the discussion in the subsection below brings out this point clearly.

Whereas the traffic statistics of different service types are unknown today, it is likely that such statistics will become known as these services evolve. Thus, we will be able to dimension networks for different mixes of service types. However, the Grade Of Service (GOS) provided will only be as good as our estimates of the mix of traffic demand during peak periods, our models of the traffic statistics, and the capability of the network interface to monitor and control the packet/cell rate into the network. It is important to recognize that the efficiency of ATM as a multiplexing technique is still unknown, since the statistical character of the offered traffic is unclear.

If traffic statistics (e.g., variability of the bit rate of an application as a function of time) are truly unknown or are very dynamic, then there is only one way we can really engineer the network and protect network resources without running the risk of significant degradation of GOS. This is to use the peak rates for different service types to size the network by providing a loss probability of, say, 10^{-9} for each node, and to base the access control strategy on these peak rates. It should be realized that this is the only way in which GOS at the packet level can actually be guaranteed. For example, at call setup, an application could provide the network with information on its peak bit rate; acceptance or rejection of the call would then be based on the availability of the (peak) bandwidth requirements of the service.

However, with this strategy, the ATM network would end up providing what would essentially be an inefficient, expensive form of circuit-switching. The reason is that, even under extremely synchronous traffic conditions, the network will be unable to support the same traffic load as an equivalent circuit-switched STM network. Let's consider an example. A 100 Mb/s link can support five 20 Mb/s connections when STM is used. How many equivalent ATM connections can be supported when the peak rate is 20 Mb/s? Probably only four—this is because we generally do not allow 100% link utilization under ATM (we may allow, including the ATM overhead, say, 90% utilization). To support five connections, we need to reduce the peak bit rate using some compression technique; for this example, assuming a maximum permissible link utilization of 90% and an ATM overhead of 10%, the peak rate should be reduced to approximately 16 Mb/s. (Note that this 20% differential remains even when the number of connections is very large.) It should be recognized that the "fall-back" position is less desirable than conventional STM.

ATM only makes sense when the services being carried are known to be bursty in nature, when the nature of this burstiness is known at call setup (and is not violated during the call), and when the network has the capability to dynamically manage network resources. At this point in time, these are all open issues.

Sizing Buffers for Peak Rate Transmission

We wish to consider how to size buffers for the "fallback" position (namely, all network resources engineered to handle peak rate transmissions). Clearly, the worst case here, as far as losing packets is concerned, is when all sources are transmitting at their peak rate. If we know the exact mix of traffic when dimensioning the network, then we can use some form of *nD/D/1/K* model to dimension the buffers—this is some variation on *nD*, since not all *n* packet streams will have the same period. With a priori knowledge of the nuber and type of the sources and the period of the different streams, the buffers can be sized so that no packets are lost and the permitted bandwidth utilization level (including packet overhead, of course) can reach 100%. However, in an operational ATM network, unless we constrain the number of calls of each type to remain below the dimensioning level, the actual mix of traffic can easily be such that packets are dropped even though the utilization is less than the permissible level. In fact, with this dimensioning technique, there is no upper bound on the average loss probability without some control on the number of calls of each type allowed into the network.

Thus, we may need to consider a more robust model for sizing purposes. It is well known that an *M/D/1* model provides larger delays than an *nD/D/1* model, even though the latter's results approach those of the *M/D/1* as the number of sources *n* gets large. However, the *M/D/1* model is robust with respect to the traffic mix and, although in general it will provide over-dimensioned buffers, it will provide an upper bound on the average loss probability across all mixes of traffic. On the other hand, unlike the other model, there is no guarantee that packets will not be lost for any particular traffic mix. Using an *M/D/1* model, buffers can be sized so that the average loss probability is, say, 10^{-9}.

A simple example might be useful here, if only to put some numbers on these quantities. Consider a link carrying a throughput of 300,000 ATM cells per second where the ATM cell size is 60 octets and the link transmission rate is 150 Mb/s. With the assumption that the traffic flow consists of five synchronous streams, each with a period of 16.7 μs, then with storage for four cells, the average queueing delay is about 2.7 μs with no packet loss [6, Figure 5] (worst case delay is 12.8 μs). On the other hand, if the flow is Poisson (i.e., a mix of many traffic types), then the average delay is 38.4 μs and a buffer to hold 249 cells is required to maintain the average loss probability below 10^{-9}.

Accounting

While accounting questions usually tend to be considered after the fact, for ATM they are closely tied to fundamental system parameters and even to system feasibility. The basic question is how to charge for system usage on an ATM network. One either charges on a per-packet basis or does charging at a higher level process (e.g., per call setup). The disadvantage of per-packet charging is obvious: it increases the per-packet processing load on the system by complicating the header handling. This translates to fewer packets per second at an ATM interface, hence longer packets and/or lower speeds for a given processing technology. Qualitatively, there is an unsettling inelegance in wasting network resources by charging for every hundred bytes in a 100 Mb/s video or graphic bit stream. If the network will be supplying the greatest part of its bandwidth to high-speed users, then it would seem that there has to be a way to make charging activities commensurate with the real dimensions of usage.

However, once one gets away from per-packet charging, one quickly loses the advantages that ATM is intended to provide. Without per-packet charging, there are few usage parameters that can be negotiated with the network at call setup time. Enforcement of any usage measure will require counting packets, though peak bandwidth enforcement could probably be done without the full complexity of per-packet charging. Average bandwidth usage criteria would seem to be equivalent to per-packet charging.

Charging on the basis of peak bandwidth usage is telling the customer he will be charged at the same rate as a full-period circuit-switched connection, whether he chooses to fill his circuit or not. The rational customer response would be to find Customer Premises Equipment (CPE) based mechanisms to smooth traffic flow so as to reduce cost. Hence, the existence of a peak-based charging scheme would tend to reduce burstiness of offered load and thus reduce the advantages of ATM as compared to more conventional approaches.

Consequently, some notion of per-packet charging is probably necessary. From the above line of argument, charging requirements are so fundamental to the system definition that charging procedures must be established before other system parameters (such as cell length) can be set. As yet, this work has not been done.

Compatibility

It was recognized from the beginning that ATM networks will not necessarily be easy to fit with the current Public Switched Telephone Network (PSTN). The principal problem is the packetizing delay associated with any packet voice system. Unless packets are very short and internal network delays are kept at an insignificant level, some sort of echo control will be necessary. Even with echo control, interworking problems remain with echo control devices on existing long-haul circuits. While these problems are solvable by various means (including the partial filling of cells), it is an indication of the immaturity of the ATM concept that no specific solution concept or specific packetization delay objective has yet been chosen.

The basic question regards the extent to which interworking problems are to be solved within the ATM network. One commonly sees standards contributions directed toward choosing ATM system parameters to allow direct interworking with existing networks under current telephony norms. As noted above, this can affect cell size, internal network load levels, and even internal network configuration (so as to limit the number of potential delay points in an ATM voice connection).

The alternative approach is to decree that ATM system parameters should be chosen to optimize the system as a whole (including the charging considerations above), and that modifications will be made to the PSTN as necessary to support the introduction of ATM. This is only logical, but there is no evident consensus behind it.

The philosophical approach to compatibility could make ATM either a good or an extremely awkward system. Until the issue is resolved, support for ATM is signing a blank check.

Scaling

Whereas the current interest is in developing standards for a 150 Mb/s interface, it is expected that, in the relatively near future, standardization efforts will be directed towards a 600 Mb/s interface. With further advances in technology (e.g., coherent transmission systems), it is quite probable that standards will be required for even higher rates than this. At this point in time at least, the (switching) technology does not exist for ATM at these higher speeds. It may very well transpire that ATM is not scaleable, or that scaleability is not cost-effective.

In fact, it's not clear that technology will allow us to avoid a hybrid solution at higher rates. With today's technology, it is certainly true that space-division circuit switching can handle higher speeds than ATM implementations, and the same may well be true in the future. Therefore, with an all-ATM solution, we are betting that customers don't want the highest rates that technology can deliver.

If we are essentially going towards a hybrid solution, then the breakpoint between STM and ATM becomes a critical issue, with major impact on the parameters of the ATM subsystem. Furthermore, even if ATM is scaleable to significantly higher speeds, there is no available body of work to indicate that the system parameters (cell size, etc.) for a 150 Mb/s ATM interface are even relevant to higher speed interfaces.

Summary

The above set of arguments, when taken together, indicate a high-degree of uncertainty, not only in what an optimal ATM system should look like but also in the role of ATM in the future service environment. Consequently, while we recognize the business interest in an early BISDN standard, we feel there is a real danger that an early standard could have deficiencies serious enough to limit the market success of BISDN products.

Of the four areas of concern discussed in this paper, GOS is probably the most serious. This is because in any future broadband network, the underlying transport fabric should appear transparent to services and applications carried on the network. The real fear is that ATM transport will not be transparent and that upper layer protocols will need to be employed to take care of imperfections in the underlying transport structure. This, of course, violates the basic principle of protocol layering.

The authors' approach to these problems, including the creation of a circuit-switched service class and partial filling of cells, is outlined in [4]. Future publications [5] will provide detail on the service description and discuss the impact on ATM system parameters.

References

[1] S. E. Minzer, "Broadband User-Network Interfaces to ISDN," *Proc. ICC '87*, Seattle, pp. 364–369, June 1987.

[2] R. Handel, "Evolution of ISDN Towards Broadband ISDN," *Proc. INFOCOM '88*, New Orleans, pp. 434–441, Mar. 1988.

[3] CCITT Draft Recommendation I.121, "Broadband Aspects of ISDN," Seoul, Korea, Feb. 1988.

[4] J. Gechter and P. O'Reilly, "Circuit-Switched Service in Broadband ISDN," submitted to *ICC '89*.

[5] P. O'Reilly, J. Gechter, and S. S. Huang, "Specification of a Circuit-Switched Service for BISDN," GTE Labs working paper.

[6] A. E. Eckberg, "The Single Server Queue with Periodic Arrival Process and Deterministic Service Times," *IEEE Trans. on Commun.*, vol. COM-27, no. 3, pp. 556–562, Mar. 1979.

Jerry Gechter is Manager of the Intelligent Network Systems Department at GTE Laboratories. Since joining the Laboratories staff in 1983, he has worked in the areas of switching and network architectures; ISDN and Common Channel Signaling services; and network planning. During 1985, he headed a corporate workgroup on next-generation switching. His prior experience included three years in switching systems engineering at GTE ATEA in Belgium, and three years in 4 ESS network planning at Bell Laboratories. Dr. Gechter has a Ph.D. in mathematics from M.I.T. and taught for two years at the University of Washington in Seattle.

Peter O'Reilly is Manager of the Network Performance Analysis Department at GTE Laboratories. He has been with GTE Laboratories since 1983, where he has worked on the performance of integrated networks, burst switching, high-speed network architectures, and network management. Dr. O'Reilly has a Ph.D. in electrical engineering from the Georgia Institute of Technology, and is the coauthor of a textbook on the performance of local area networks. He is a Senior Member of the IEEE, and is an Adjunct Professor of electrical engineering at Worcester Polytechnic Institute.

SOME PERFORMANCE TRADE-OFFS ASSOCIATED WITH ATM FIXED-LENGTH VS. VARIABLE-LENGTH CELL FORMATS

Shyam Parekh and Khosrow Sohraby

AT&T Bell Laboratories, Holmdel, NJ 07733

ABSTRACT

This paper reports some comparison results of Asynchronous Transfer Mode (ATM) Fixed-Length Cell (FLC) versus Variable-Length Cell (VLC) formats based on memory requirements, delays and efficiency considerations, which are three of many issues involved in such a comparison. In particular, we develop models for the purpose of a comparison of these formats at an access multiplexer.

For the purpose of comparison, we maintain the same information flow rate under these formats. Our results show that at moderate outgoing link utilizations, the memory requirements for an FLC format are less than those for a VLC format. However, at higher outgoing link utilizations, the opposite is true, due to the inherent efficiency advantage of a VLC format. At moderate outgoing link utilizations, delays for short messages are found to be comparable. However, at higher outgoing link utilizations, a VLC format has an edge from the point of view of delays for short messages because of the efficiency advantage.

In summary, we conclude that the performance differences found here do not overwhelmingly favor either an FLC or a VLC format, and thus it is important to factor in other issues as well.

1. *INTRODUCTION AND SUMMARY OF CONCLUSIONS*

Recently, there has been much interest in Asynchronous Transfer Mode (ATM) as a transport mechanism for Broadband ISDN and other network protocols which support integration of a wide spectrum of end-user services [5],[9]. Historically, ATM has evolved from Asynchronous Time Division Multiplexing (ATDM) [1]. The basic idea behind ATDM is to allow statistical multiplexing among various sources without pre-assigning time slots in a synchronous fashion. CCITT SG XVIII, responsible for Broadband ISDN standards, introduced ATM as a generalization of ATDM and has been involved in the debate for standards for ATM. In the recent past, there have been proposals for ATM based on both Fixed-Length Cell (FLC) and Variable-Length Cell (VLC) formats [11],[12]. More recently, CCITT SG XVIII has finalized that ATM will be the transport mechanism for Broadband ISDN networks [13].

In both these formats, information is transported by basic entities called cells, where a cell consists of a header and an information field. In an FLC format, cells are of a fixed-size with a fixed-size header and a fixed-size information field. In an FLC format, a message arrives at an access multiplexer by a sequence of FLCs, where the last cell may only be partially filled in its information field. In contrast, in a VLC format a cell has a variable-size information field and a header which may also be of variable-size. In this study, we will allow only information fields to be of variable-size, for simplicity. In a VLC format, typically a message arrives at an access multiplexer either by a single VLC or by a sequence of maximum size VLCs followed by a VLC for the remaining portion of the message. The latter case occurs when an arriving message is longer than the longest allowable information field. We defer the study of the effects of such long messages to a future study and assume for this study that a longest message needs only a single VLC. Although CCITT SG XVIII has settled the issue in favor of an FLC format [11]-[13], a comparison of these formats may still be of interest for choosing one of the formats for the design of a private network. For such a comparison, one has to weigh a multitude of considerations including evolution, end point complexity, congestion control, efficiency, memory requirements, delays, integration of traffic, switch architecture, flexibility of protocols, billing procedures, etc. In this paper, we make a comparison at an access multiplexer considering only the issues of efficiency, memory requirements and delays. Our hope is that this study will contribute positively in our understanding of trade-offs involved in the overall comparison of these formats. Let us emphasize that this study reports a comparison only at an access multiplexer.

Clearly, a VLC format is more bandwidth-efficient. This is because a VLC format requires fewer headers to be transmitted overall and also because a packetizer in a VLC format is required to waste a fewer number of bits in the way of partially filled cells. This efficiency advantage of a VLC format may be diminished somewhat because of the fact that a VLC approach requires some technique for indicating the beginning and end of a cell, such as flags and bit-stuffing for transparency. A detailed study is required to evaluate this overhead for various techniques. In the present study, we will not take into consideration this issue of cell boundary demarcation.

Next, let us consider the issue of memory requirements. If messages arrive at an access multiplexer at a rate slower than that of the outgoing link then it is necessary to accumulate a complete cell before it is transmitted. Even for the case when the access and outgoing link rates are equal, almost all current architectures of access multiplexers require that cells should arrive completely before they are transmitted due to concerns regarding synchronization, error control, etc. Then, since in an FLC format a message is segmented into a number of cells and each individual cell can leave the access multiplexer independently, an FLC format, unlike a VLC format, permits "cut-through" on a cell basis for a message as it traverses the access multiplexer. This results in lower memory requirements for an FLC format as compared to those of a VLC format for a given outgoing link utilization.

Finally, let us consider the issue of delays. Observe that, if the duration when the access multiplexer is accumulating bits from a short message overlaps with those of longer messages, the FLCs of the short message may sneak in between the FLCs of the overlapping long messages. The situation here is quite analogous to a scheduling mechanism that allocates quanta of service in a round robin manner to a set of queues. In which case, a short job does sneak in between longer jobs. This suggests that an FLC format partially avoids the undesirable situations where a short message might get stuck behind a number of large messages in a multiplexer queue.

In this paper, we will examine the question whether the efficiency advantage in a VLC format outweighs other associated performance concerns regarding memory requirements and delays for short messages. To make the comparison fair, we maintain the rates at which information bits traverse the access multiplexer the same in the cases of FLC and VLC formats. Evidently, this requires a smaller number of bits to be transmitted overall for a VLC format, and thus a lower total outgoing link utilization.

Although, our comparison of FLC and VLC formats is made at an access multiplexer of an ATM based Broadband ISDN network, the models can be extended to the inside of a network with some minor modifications in the traffic arrival patterns. We will compare these formats for a hypothetical message size distribution. We feel that although the actual distribution might vary considerably from one network to another, the assumed distribution should enable us to bring about some of the trade-offs involved in FLC versus VLC formats.

From the simulation results, we conclude that an FLC format requires less memory at moderate outgoing link utilizations. However, at higher outgoing link utilizations, the opposite is true due to the inherent efficiency advantage of a VLC format. This shows, e.g., that a VLC format will avoid the need for congestion control up to a higher offered load. At moderate outgoing link utilizations, delays for short messages are found to be comparable. Here, the "sneaking in" advantage of an FLC format is seen to almost offset the efficiency advantage of a VLC format. However, at higher outgoing link utilizations, a VLC format incurs smaller delays for short messages due to the efficiency advantage.

The bottom line is that, the trade-offs between efficiency, memory requirements and delays of short messages, favor either an FLC or a VLC format depending on the outgoing link utilization and access rate. The magnitudes of differences found in our study do not seem to be high enough to to draw a firm conclusion regarding FLC versus VLC formats from this performance viewpoint alone.

2. MODELS AND ASSUMPTIONS

2.1 NO CUT-THROUGH ASSUMPTION

As discussed in § 1, in this study we assume that there is no "cut-through" as far as transmission of a cell is concerned in both formats. Our precise assumptions regarding memory occupation and passage of a cell through an access multiplexer are as follows.

(1) Memory space for a cell is reserved as soon as the first bit of the cell arrives and it continues to be reserved till the last bit is transmitted.

(2) A cell transmission on the outgoing link does not begin until the last bit arrives.

As we have pointed out before, when a cell arrives at the access multiplexer at a rate less than that of the outgoing link, it is not possible to start transmitting that cell before accumulating all the bits of the cell. Even in the case where the access and outgoing link rates are equal, a survey of existing access multiplexer designs has convinced us that no "cut-through" for cells is a reasonable assumption. Observe that it is precisely because of this implementation constraint that an FLC format allows partial "cut-through" for messages and a VLC format does not.

2.2 VARIABLE-LENGTH CELL (VLC) FORMAT

Here, we model the message arrivals at an access multiplexer by a Poisson process with the rate λ_{VLC}. This stream of arrivals models the superposition of message arrivals from various sources. We assume that each message is transported by a single VLC. Cells are read in starting at these Poisson arrival epochs. Let us denote by A, L, S and H the access rate (in Mbps), outgoing link rate (in Mbps), random message size (in bytes) and header size (in bytes), respectively. Clearly, an arriving VLC takes

$$\frac{8(S+H)}{A}\ \mu sec \tag{1}$$

to be read in. By the assumptions above, during this time the cell requires $(S+H)$ bytes of memory and the access multiplexer can not start transmission of the cell over the outgoing link. Notice that at an arbitrary time, there can be an arbitrarily large number of cells being read in simultaneously. Only after this read-in phase do cells queue up for transmission over the outgoing link. We model this read-in phase by an $M/G/\infty$ queue. Here, G denotes the time to read in a cell as given by Eqn. (1).

In tandem with the $M/G/\infty$ queue, we place a $./G^*/1-FCFS$ queue to model the multiplexer queue, where . denotes the output process from the $M/G/\infty$ queue and G^* denotes the transmission time of a cell. The multiplexer queue models the queue of the cells already read in and waiting to be transmitted on the outgoing link. Clearly, a cell takes

$$\frac{8(S+H)}{L}\ \mu sec \tag{2}$$

to be transmitted. Notice that the service times in the two queues are not independent. In fact, they are scalar multiples of each other. Fig. (1) shows this system of queues in tandem. Note that the outgoing link utilization (ρ_{VLC}) is given by

$$\frac{8\lambda_{VLC}(E\{S\}+H)}{L} \tag{3}$$

Here, as a measure of the memory requirements, we are interested in the stationary memory occupancy (in bytes) at an arbitrary time, which has the same distribution as the stationary memory occupancy as seen by a message arriving to the access multiplexer because of the Poisson assumption. The memory occupancy above refers to the sum of the occupied memory (in bytes) in the two phases. We are also interested in the the stationary delay (waiting time before the transmission begins) in the multiplexer queue of an arbitrary cell of the smallest size.

2.3 FIXED-LENGTH CELL (FLC) FORMAT

Here, we model the message arrivals at an access multiplexer by a ***Poisson wave process*** which is a special case of the branching Poisson process [2]. By a Poisson wave process, we mean that, at Poisson epochs, waves consisting of a random number of equally spaced arrivals commence. Notice that at an arbitrary time, there can be an arbitrarily large number of ongoing waves. Observe that a Poisson wave process is used to model the superposition of message arrivals from various sources in an FLC format.

As in the VLC case, we model an access multiplexer by two queues in tandem, the first being an ∞-server queue modeling the read-in phase and the second being a single server FCFS queue modeling the multiplexer queue. As stated above, we model the input to the system by a Poisson wave process with rate λ_{FLC}. The arrival epochs of successive cells in a wave are separated by

$$\frac{8(F+H)}{A} \ \mu sec, \tag{4}$$

where $(F+H)$ represents the size in bytes of an FLC with F bytes of information and H bytes of header. This also equals the time to read in a cell. Let C denote the random number of cells in a wave. Since we want to compare FLC and VLC formats, we assume that an arbitrary VLC carries the same amount of information as an arbitrary wave. Hence, the S random number of bytes is now divided among C random number of cells, where the last cell may be only partially filled. Notice that if S happens to be less than F, we will still need one cell in the corresponding wave. There are various factors which may necessitate a different header size for an FLC format from that for a VLC format. However, for the purpose of this study, we assume that the header sizes are same in both VLC and FLC formats, i.e., each FLC contains a header of H bytes. Fig. (2) depicts this model pictorially.

With the above model, observe that after a wave begins, cells are read-in at the access rate and then they are transferred to the multiplexer queue for transmission. In the first queue, the inter-arrival time of cells from the same wave equals their service time (read-in time). As in the case of a VLC format, the service time of a cell in the second queue is its service time in the first queue times the ratio of the outgoing link and the access rates. Observe that now the outgoing link utilization (ρ_{FLC}) is given by

$$\frac{8\lambda_{FLC}E\{C\}(F+H)}{L}. \tag{5}$$

Similar to the VLC case, here we are interested in the stationary memory occupancy as seen by an arriving wave and the delays of single cell waves in the multiplexer queue.

3. SIMULATION RESULTS

For the purpose of comparison, we consider S, the message size, with the following distribution:

probability	0.1	0.2	0.3	0.3	0.1
S (bytes)	16	64	256	512	2048

A header size (H) of 5 bytes in both FLC and VLC formats, and the outgoing link rate (L) of 150 Mbps are assumed. We also assume that an FLC has 64 bytes for information (F). As a result, in the FLC format, a message would occupy 1, 4, 8, or 32 cells (C) with probability 0.3, 0.3, 0.3, and 0.1, respectively. Note that with the above message size distribution, approximately 50 percent of the traffic is generated by the longest messages. This might be the situation when about 50 percent of the capacity is utilized by video. Let us point out that the above mentioned message size distribution is only a hypothetical one and that it might be quite different depending on the services offered by a network. For example, there might be a lot of signaling traffic in the network giving rise to a higher proportion for short messages. Nonetheless, we feel that the message size distribution assumed here should enable us to bring out the trade-offs involved in such a comparison.

Recall that we are assuming the same information flow rate in both cases. This amounts to keeping

$$\lambda_{FLC} = \lambda_{VLC}. \tag{6}$$

This, in turn, amounts to higher outgoing link utilization in the FLC format, because: (1) transmission of a message of 16 bytes requires a (whole) cell of 64 bytes to be transmitted, and (2) the FLC format has an additional overhead due to the breakage of messages into cells, each requiring a header of 5 bytes.

Tables (1) and (2) show the means and higher percentiles of the memory requirements for access rates (A) of 15 and 150 Mbps for both VLC and FLC formats. As expected, we have observed from our simulations that at moderate outgoing link utilizations and lower access rates, the memory requirements are dominated by the read-in phase in both formats. As seen from Tables (1) and (2), the memory requirements of the VLC format are higher than for the FLC format, except when the outgoing link utilization for the VLC format is quite high (and the outgoing link utilization of the FLC format is even higher). In such a situation, loading becomes the dominant factor. Observe from Table (2) that at $\rho_{VLC} = 0.90$ and $\rho_{FLC} = 0.97$, the FLC format requires more memory for 150 Mbps access rates. Also, observe from the tables that the memory requirements in the FLC case are not stochastically monotonic with the access rate; while they are for the VLC case. In the FLC case, this is because of the play-off between the burstiness in the arrival process and read-in time as the access rate varies. While in the VLC case, the memory requirements decrease as the access rate increases because of decrease in the read-in time.

Tables (3) and (4) show the means and higher percentiles of the delays of single cell waves in the FLC format and 16 bytes messages in the VLC format for different loadings and access rates. Observe from these tables that the delays are insensitive to the access rates in the VLC format. This is to be expected because the input process to the multiplexer queue is the same Poisson process irrespective of the access rate. However, in the FLC format, they are stochastically monotonic with the access rate which is consistent with the fact that the burstiness in the input process to multiplexer queue increases as the access rate increases. At $\rho_{VLC} = 0.75$

and ρ_{FLC} = 0.81, for the access rates of 15 and 150 Mbps, the delays are found to be comparable. However, at ρ_{VLC} = 0.90 and ρ_{FLC} = 0.97, the VLC format incurs smaller delays for short messages due to the efficiency advantage.

4. ANALYSIS

In this section we present some analysis for both VLC and FLC models developed in § 2. In both models, of interest are the memory requirements and the delays (waiting times) in the multiplexer queue.

As it is discussed below the most interesting part of the following analysis is the analysis of a queue fed by a Poisson wave arrival process (see § 2.3). A useful technique of obtaining approximate results for a queue which is fed by a non-standard point process is to match a few moments of the point process with those of a "friendly" process which is amenable to an exact analysis. For example, see [6] where the superposition of voice calls is modeled as a two state *Markov Modulated Poisson Process* (MMPP). For the problem at hand, a similar approach could be to use the known results for an m-state MMPP [6] by matching the necessary moments to find the unknown parameters, or to apply a "two moment approximation" to the Poisson wave arrival process [10]. Some of the past contributions of analytic approximations for queues with similar arrival processes include the approximations based on the peakedness characterization [3],[4], and those based on the Benés' formula [7]. Here we use a different approach by defining the effective "batchiness" for the Poisson wave arrival process.

We denote by n_1 and n_2 the number in the system in the infinite server read-in queue and the single server multiplexer queue, respectively. Also, let W denote the waiting time in the multiplexer queue and r denote the ratio of the outgoing link rate to the access rate. The memory requirements can be easily obtained from n_1 and n_2.

4.1 VLC MODEL

In this model, we begin by using the classical results for the $M/G/\infty$ queue. n_1 is Poisson with mean $r\rho_{VLC}$. Using the fact that the output process of this queue is Poisson [8] and ignoring the dependency between service times in the two queues, the multiplexer queue is simply an $M/G/1$ queue. The distribution of n_2 and W are readily available in standard queueing theory books. In comparison with simulation results this analytical approximation has shown to be very accurate.

4.2 FLC MODEL

As in the VLC model, we ignore the dependency of service times between the two queues, and we study an infinite server queue and a single server queue in isolation fed by a Poisson wave arrival process. As described in § 2.3, by a Poisson wave arrival process we mean that at Poisson epochs occurring with rate λ, waves which consist of a random number (C) of FLCs arrive. The arrival epochs of successive cells in a wave are separated by the interval equal to r times the transmission time of an FLC. Again of interest are the total number of cells in the queues and the waiting time distribution in the single server queue (multiplexer). We denote this two queues by $M^{[W]}/D/\infty$ and $M^{[W]}/D/1$, respectively. The mean number in the $M^{[W]}/D/\infty$ queue follows easily by the Little's law. We have

$$E\{n_1\} = \lambda E\{C\} b_1, \tag{7}$$

where b_1 denotes the read-in time of a cell. Higher moments of n_1 or its distribution is a more challenging problem which is currently under consideration.

The analysis of $M^{[W]}/D/1$ queue is non-standard because of the strong correlation between the successive arrivals in a wave. Before we provide the results for general r, we may consider two special cases.

In the limiting cases of $r \to \infty$ and $r \to 0$, we already know the exact results. It can be seen that as $r \to \infty$, the dynamics of the queue under consideration approach that of the queue with Poisson arrival with rate $\lambda E\{C\}$. Note that this limiting behavior is independent of the actual distribution of the wave duration and it only depends on its mean. The limiting case of $r \to 0$ becomes the standard Poisson batch arrival process, where the number of cells in a batch is distributed according to the number of cells in a wave. Considering the above extreme cases, it becomes natural to attempt to characterize a Poisson wave arrival process through a "batchiness" parameter. (See [3] for other ideas along this line.)

To motivate this idea, we note that as $r \to 0$, the mean waiting time in the multiplexer queue is given by

$$E\{W\} = \frac{1}{2}\frac{\rho b_2}{1-\rho} + \frac{b_2[B_0-1]}{2(1-\rho)} \tag{8}$$

where b_2 denotes the service (transmission) time in the multiplexer queue (see Eqn. (4)), ρ denotes ρ_{FLC}, for simplicity, which is equal to $\lambda E\{C\} b_2$ and $B_0 = E\{C^2\}/E\{C\}$. Let us hypothesize that for general r and ρ, $E\{W\}$ is given by the same formula as in Eqn. (8), except that B_0 is now replaced by a function $B(r,\rho)$. We define $B(r,\rho)$ as the batchiness of a Poisson wave arrival process. We note that for Poisson arrival process ($r \to \infty$) the batchiness is equal to one for all ρ, and for Poisson batch arrival process ($r \to 0$) the batchiness approaches to the value B_0.

Before we analyze the batchiness, we outline some of its obvious properties:

(1) $B(0,\rho) = B_0$ for all ρ

(2) $B(\infty, \rho) = 1$ for all ρ

(3) $B(r,\rho)$ is a decreasing function of r (for fixed ρ) and an increasing function of ρ (for fixed r).

The mean waiting time of $M^{[W]}/D/1$ queue follows easily from a result made use of in [4] when $r \leq 1$. Using the results there the batchiness is easily seen to satisfy

$$\frac{B(r,\rho)-1}{B_0-1} = 1 - r(1-\rho), \quad 0 \leq r \leq 1. \tag{9}$$

We may note that although the above result is rather restrictive for general r, it contains considerable information regarding the batchiness in light traffic, say $\rho < 0.3$ (see Fig. (3)). The case of general r seems to be difficult, but based on simulation results we have found a fit for the case where $r > 1$. We found

$$\frac{B(r,\rho)-1}{B_0-1} \approx \rho e^{-\frac{(1-\rho)(r-1)}{\rho}}, \quad r > 1. \tag{10}$$

The above approximation has been found to produce errors of less than 15% in the batchiness for all ρ's up to 0.9, and $1<r\leq 10$. The results for $0\leq r\leq 1$ are exact. The approximation tends to underestimate the batchiness for large r as the loading increases.

Fig. (3) depicts $\frac{B(r,\rho)-1}{B_0-1}$ vs. r. For small ρ, it quickly approaches to zero, so the exact result of Eqn. (9) provides most of the information regarding the batchiness. We may note that these curves together with Eqn. (8) can be used to obtain the mean waiting time in the $M^{[W]}/D/1$ queue.

ACKNOWLEDGEMENTS

The authors are thankful to A. E. Eckberg and J. M. Holtzman for stimulating discussions.

REFERENCES

[1] W. W. Chu, "A Study of Asynchronous Time Division Multiplexing for Time-Sharing Computer," AFIPS, Vol. 35, pp. 669-678, 1969.

[2] D. R. Cox and P. A. W. Lewis, "The Statistical Analysis of Series of Events," Methuen & Co. Ltd., London, 1968.

[3] A. E. Eckberg, "Generalized Peakedness of Teletraffic Processes," 10'th International Teletraffic Congress, Montreal, 1983.

[4] A. E. Eckberg, "Approximations For Bursty (and Smoothed) Arrival Queueing Delays Based on Generalized Peakedness," 11'th International Teletraffic Congress, Kyoto, 1985.

[5] D. G. Fisher, "Asynchronous Time Division Techniques in Broadband Communications," International Conference on Wideband Communications, London, October, 1986.

[6] H. Heffes and D. M. Lucantoni, "A Markov Modulated Characterization of Packetized Voice and Data Traffic and Related Statistical Multiplexer Performance," IEEE J. Select. Areas Comm., Vol. SAC-4, No. 6, September, 1986.

[7] M. Miyazawa, "Approximations for a Single Server Queue with a Branching Poisson Arrival Process," 11'th International Teletraffic Congress, Kyoto, 1985.

[8] L. Takacs, "Introduction to the Theory of Queues," Oxford Press, 1962.

[9] A. Thomas, J. P. Coudreuse and M. Serve, "Asynchronous Time-Division Techniques: An Experimental Packet Network Integrating Videocommunication," ISS 84, May, 1984.

[10] W. Whitt, "Approximating a point process by a renewal process: Two basic methods," Oper. Res., Vol. 30, No. 1, pp. 125-147, January-February, 1982.

[11] "Asynchronous Transfer Mode (ATM): Considerations About the Cell Size," RTT Belgium, CCITT SG XVIII, D. 1387, June, 1987.

[12] "ATM - Definition of the Cell Field Functionalities and Proposal for a Fixed-Length Cell Format (ATD)," France, CCITT SG XVIII, D.1236, June, 1987.

[13] "CCITT I.121 Recommendations on Broadband Aspects of ISDN," 1988.

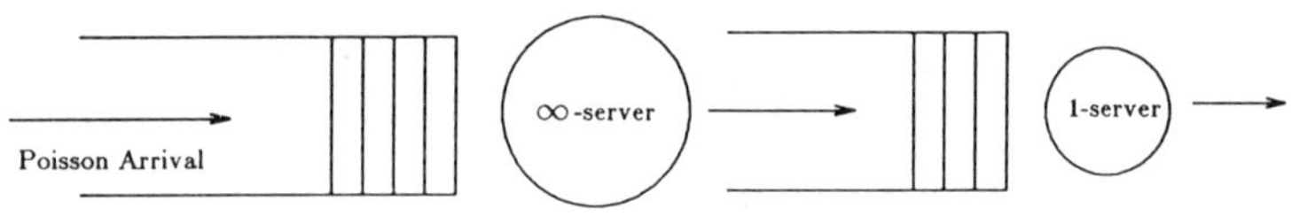

Model for VLC format

Fig. (1): $M/G/\infty+./G^{*}/1$ to model the read-in and the multiplexer phases

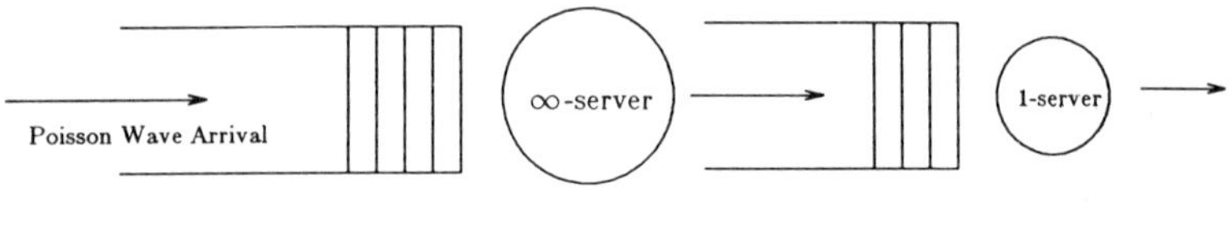

Model for FLC format

Fig. (2): Poisson wave arrival $/D/\infty+./D/1$ to model the read-in and the multiplexer phases

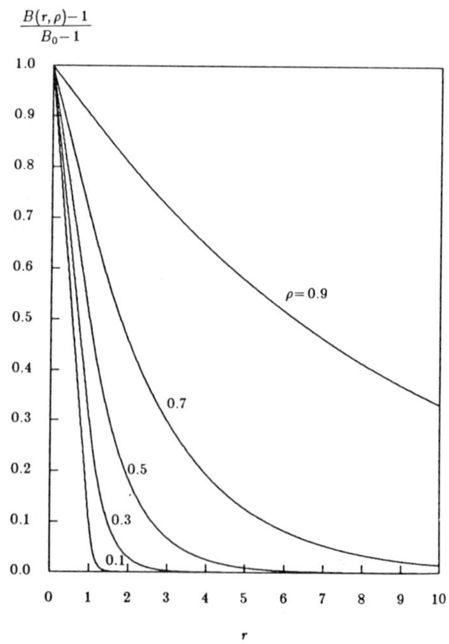

Fig. (3): Behavior of $\frac{B(r,\rho)-1}{B_0-1}$ vs. r for different loadings

Access Rate (Mbps)	FLC			VLC		
	mean (Kb)	90% (Kb)	99% (Kb)	mean (Kb)	90% (Kb)	99% (Kb)
15	1.7 ± 0.2	4.8 ± 0.6	11.1 ± 1.2	10.4 ± 1.0	16.7 ± 1.0	23.1 ± 1.0
150	1.4 ± 0.2	3.3 ± 0.7	9.2 ± 1.7	2.9 ± 1.0	6.8 ± 1.0	12.0 ± 1.0

Table (1): Memory requirements (in Kilobytes) (with 95% confidence interval);
$\rho_{FLC} = 0.81$ and $\rho_{VLC} = 0.75$

Access Rate (Mbps)	FLC			VLC		
	mean (Kb)	90% (Kb)	99% (Kb)	mean (Kb)	90% (Kb)	99% (Kb)
15	13.6 ± 2.4	32.4 ± 6.7	51.4 ± 8.0	15.5 ± 1.0	24.8 ± 1.0	37.0 ± 1.5
150	14.5 ± 2.5	34.4 ± 7.4	52.3 ± 7.8	6.4 ± 1.0	14.4 ± 1.0	26.5 ± 1.5

Table (2): Memory requirements (in Kilobytes) (with 95% confidence interval);
$\rho_{FLC} = 0.97$ and $\rho_{VLC} = 0.90$

Access Rate (Mbps)	FLC			VLC		
	mean (ms)	90% (ms)	99% (ms)	mean (ms)	90% (ms)	99% (ms)
15	0.05 ± .01	0.16 ± .04	0.51 ± .10	0.09 ± .01	0.25 ± .01	.52 ± .01
150	0.10 ± .01	0.30 ± .03	0.66 ± .07	0.09 ± .01	0.25 ± .01	.52 ± .01

Table (3): Multiplexer queue delays (with 95% confidence interval)
of single cell waves (FLC format) and 16 byte messages (VLC format);
$\rho_{FLC} = 0.81$ and $\rho_{VLC} = 0.75$

Access Rate (Mbps)	FLC			VLC		
	mean (ms)	90% (ms)	99% (ms)	mean (ms)	90% (ms)	99% (ms)
15	0.73 ± .13	1.79 ± .36	3.00 ± .66	0.27 ± .01	.68 ± .01	1.32 ± .02
150	0.81 ± .14	1.89 ± .37	3.08 ± .66	0.27 ± .01	.68 ± .01	1.32 ± .02

Table (4): Multiplexer queue delays (with 95% confidence interval)
of single cell waves (FLC format) and 16 byte messages (VLC format);
$\rho_{FLC} = 0.97$ and $\rho_{VLC} = 0.90$

Seoul, 25 Jan. - 5 Feb., 1988

Question: BBTG **D.1637**

Study Group 18 - Delayed Contribution

Source: Republic of Korea

Title: Proposal for Variable Cell Size in ATM

1. BACKGROUND

In the variable cell size strategy of ATM, the cell length is allowed to vary from a minimum length to a maximum length of cell. The bounds on the cell length would be dependent on the packet synchronizing aspects, queuing attributes in the switching and service aspects such as real time and data transfer services.

If the cell length is varied in N (0 to 8, integer) multiplicity of minimum cell length, some of the difficulties in a random cell size strategy can be improved such as,

- safe synchronizing process
- useful multiplexing applications,
  e.g. useful bandwidth allocation, more simple multiplexing of synchronous and asynchronous than in the variable undefined cell size strategy.
- more simple switching mechanism than in the variable undefined cell size strategy
- handling of hybrid frame structure in a simple way
- simple interworking between the ATM networks with different cell size,
  e.g. simple execution of cell segmentation and reassembly
- decreased processing and switching delay

2. VARIABLE CELL LENGTH

Some network performance parameters are influenced by the size of information field, independent of the header size. Therefore the categorization and identification of minimum and maximum value in information field is proposed as follows:

- minimum length: 32 bytes (256 bits)
- maximum length: 256 bytes (2048 bits)

And header size is fixed and proposed to 2 bytes, allowing 200 logical channels. The main functions of header information are proposed such as,

- virtual channel identification
- cell routing
- identification of the communication channel
- service identification
- routing priority
- header protection

Packet length indicator is proposed to 3 bits, and it is used to indicate the cell length in UNI or NNI.

The variable cell length has such mechanism as,

information field (32 bytes * N) + header (2 bytes),

when N = 1, 2, ...8

The optimal fixed cell length was proposed to (32 + 2) bytes by few countries [1][2][4]. Especially due to the allowable delay for voice communications, [4] proposed the range of 16/32 bytes for information field size. In the variable cell length with above structure, minimum cell length of (32 + 2) bytes may be observed that access delay and jitter are minimal through the adoption of an information field size (256 bytes). And when in the overload state, in particular longer information field size, the system becomes unstable, and the mean delivery time of voice information may increase sharply [7]. This undesirable effects can be alleviated by having variable voice information length, more shorter voice information length. However, the increased cost and complexity may be reduced by the structured variable cell strategy, which already proposed in this contribution.

3. Conclusion

It is possible to bound the cell size to a certain maximum value admissible for transparency of circuit switching, with the help of some buffering, and realize ATM with variable cell length. Such systems provide a saving of bandwidth, but at the cost of additional hardware and software complexity. However, if a slot unit of minimum cell size is applied to transmission or switching systems, their cost of hardware and software would be reduced slightly. And it is considered that extended cell length in constant proportion of N is not crucial in performance consideration of ATM, transfer delay, jitter, packetizing/depacketization delay and extra waiting time. Especially this structure of variable cell length will be very useful for internetworking between ATM networks with different variable or fixed cell length.

Therefore this contribution proposes the variable cell size strategy with structured rule in ATM to be added such as,

information field (32 bytes) * N (1 to 8, integer) + header (2 bytes)
(In UNI and NNI, 3 bits for cell length indication are added)

[REFERENCES]

[1] Belgium, RTT, "Asynchronous Transfer Mode (ATM): Considerations about Cell Size", Delayed Contribution D.1387, Hamburg Meeting, 1987

[2] France, "ATM - Advantages of a Fixed and Short Cell Length", Delayed Contribution D. 1237, Hamburg Meeting, 1987

[3] J. Neilson, Natesa Janakiraman and Bernard Pagurek, "Performance Analysis of Integrated Switch with Fixed or Variable Frame Rate and Movable Voice/Data Boundary", IEEE Transaction on Communications, Jan., 1984

[4] Netherlands, "Proposal for the Cell Format in ATD", Delayed Contribution D.1403, Hamburg Meeting, 1987

[5] M. Schwartz and B. Maglais, "Performance Evaluation of Variable Frame Multiplexer for Integrated Switched Network", IEEE Transaction on Communications, June, 1981

[6] U.K., "Virtual Cut - Through in Frame Relaying Network", Delayed Contribution D.1323, Hamburg Meeting, 1987

[7] K. Sriran and W. Whitt, "Characterizing Superposition Arrival Process in Packet Multiplexers for Voice and Data", IEEE Journal on Selected Areas in Communications, Sept., 1986

Bibliography

This bibliography is organized according to chapters and topics therein. Under each topic or chapter, references are alphabetically ordered.

1. LIGHTWAVE SYSTEMS AND SERVICE INTEGRATION

1.1 Lightwave Systems

J.J. Baldini, Guest Ed., "Photonic Switching," *IEEE Commun. Magazine,* Vol. 25, No. 5, May 1987. A special issue on photonic switching covering overview, devices, systems, and experiments.

E.E. Basch and T.G. Brown, "Introduction to Coherent Optical Fiber Transmission," *IEEE Commun. Magazine,* Vol. 23, No. 5, May 1985. A discussion of receiver sensitivity and configuration.

V.W.S. Cahn, K. Nakagawa, and D.W. Smith, Guest Ed., "Coherent Communications," *IEEE Journal of Lightwave Technology* jointly with *Journal on Selected Areas in Commun.,* Vol. LT-5, No. 4, April 1987. A special issue covering from basics of coherent optical fiber transmission to various systems and receivers.

P.S. Henry, "Introduction to Lightwave Transmission," *IEEE Commun. Magazine,* Vol. 23, No. 5, May 1985. An introductory discussion of the operating principles and limitations of lightwave systems.

P. Kaiser, J. Midwinter, and S. Shimada, "Status and Future Trends in Terrestrial Optical Fiber Systems in North America, Europe and Japan," *IEEE Commun. Magazine,* Vol. 25, No. 10, October 1987. A special issue on the future trends of terrestrial optical fiber systems, covering from devices to overall systems.

D.B. Keck, "Fundamentals of Optical Waveguide Fibers," *IEEE Commun. Magazine,* Vol. 23, No. 5, May 1985. A brief explanation of fiber types and signal transmission characteristics.

T. Li, "Advances in Optical Fiber Communications: An Historical Perspective," *IEEE Journal on Selected Areas in Commun.,* Vol. SAC-1, No. 3, April 1983. A survey of historical developments of fibers and devices. Contains a table of landmark developments of fiber technologies.

J.E. Midwinter and P.W. Smith, Guest Ed., "Photonic Switching," *IEEE Journal on Selected Areas in Commun.* jointly with *IEEE Journal of Lightwave Tech.,* Vol. LT-6, No. 7, August 1988. A special issue containing articles with reviews, descriptions on devices and systems.

S.D. Personick, "Review of Fundamentals of Optical Fiber Systems," *IEEE Journal on Selected Areas in Commun.,* Vol. SAC-1, No. 3, April 1983. A review article describing properties of optical fibers, transmitters, receivers, and couplers.

H.F. Taylor, ed., *Advances in Fiber Optics Communications,* Artech House, 1988. A comprehensive collection of articles on fiber optics, components, and systems.

1.2 Voice-Data Integration

S.R. Amstutz, "Burst Switching—An Introduction," *IEEE Commun. Magazine,* Vol. 21, No. 11, November 1983. Burst switching concept is described, showing overall architecture with link and hub switches.

A.G. Konheim and R.L. Pickholtz, "Analysis of Integrated Voice/Data Multiplexing," *IEEE Trans. on Commun.,* Vol. COM-32, No. 2, February 1984. A description and an analysis of the proportional boundary policy.

S. Li and J.W. Mark, "Performance of Voice/Data Integration on a TDM System," *IEEE Trans. on Commun.*, Vol. COM-33, No. 12, December 1985. A queuing analysis of an integrated voice-data multiplexer. A new analytic technique is introduced, which allows for evaluation of large multiplexer sizes of 24 or 48 slots in a frame.

H. Miyahar and T. Hasegawa, "Performance Evaluation of Modified Multiplexing Technique with Two Types of Packet for Circuit and Packet Switched Traffic," *Proc. ICC,* Boston, MA., 1979. The first proposal to allow for a variable frame structure with nominal frame boundaries.

K.S. Schneider, "Integrating Voice and Data on Circuit-Switched Networks," *IEEE Trans. on Aerospace and Electronic Systems,* Vol. AES-15, No. 4, July 1979. A variety of economic issues is explored for an integration of voice and data in a circuit switched network.

K. Sriram, P.K. Varshney, and J.G. Shanthikumar, "Discrete-Time Analysis of Integrated Voice/Data Multiplexers with and without Speech Activity Detectors," *IEEE Journal on Selected Areas in Commun.*, Vol. SAC-1, No. 6, December 1983. The first complete treatment of integrated voice-data multiplexer analysis. Numerical results are limited to small multiplexer sizes.

1.3 ISDN

M. Dècina, W.S. Gifford, R. Potter, and A.A. Robrock, Guest Ed., "Integrated Services Digital Network: Recommendations and Field Trials—I," *IEEE Journal on Selected Areas in Commun.*, Vol. SAC-4, No. 3, May 1986. A special issue including CCITT recommendations and field trials.

M. Dècina, W.S. Gifford, R. Potter, and A.A. Robrock, Guest Ed., "Integrated Services Digital Network: Technology and Implementation—II," *IEEE Journal on Selected Areas in Commun.*, Vol. SAC-4, No. 8, November 1986. A special issue with articles on ISDN terminals, implementations, and transmission technology.

N.Q. Duc and E.K. Chew, "ISDN Protocol Architecture," *IEEE Commun. Magazine,* Vol. 23, No. 3, March 1985. An explanation of ISDN protocol architecture, including diagrams of protocol layers at different channel arrangements.

J.J. Lawser and P.L. Oxley, "Common Channel Signaling Network Evolution," *AT&T Technical Journal,* Vol. 66, Issue 3, May-June 1987. Evolutionary developments of AT&T common channel signaling network are outlined from 1976 to the late 1980s.

W. Stallings, *ISDN: An Introduction,* Macmillan, 1989. Reviews of switching techniques, and an overview of ISDN services and architecture.

2. LOCAL AREA NETWORKS

2.1 Bus Networks

M. Gerla, L. Green, and R. Rutledge, Guest Ed., *IEEE Network,* Vol. 2, No. 1, January 1988. A special issue on LAN interconnections, including routing proposals for interconnected bridges and their comparison.

B.G. Kim and D.K. Choi, "The Expected (Not Worst-Case) Performance of Ethernet," to appear in *IEEE Trans. on Computers;* also *University of Lowell, Technical Report R-89-004.* An analytic derivation of the expected Ethernet performance.

R. Kohli, A.C. Mingo, and K.S. Vastola, "Throughput Analysis of Maxemchuk's Integrated Voice/Data Local Area Network Protocol," *Proc. Computer Networking Symposium,* Washington, DC, May 1988. Throughput of Maxemchuk's protocol is analyzed by considering conditional throughputs.

R.M. Metcalfe and D.R. Boggs, "Ethernet: Distributed Packet Switching for Local Computer Networks," *Commun. ACM,* Vol. 19, No. 7, July 1976. A classical article on Ethernet and its performance evaluation.

M.L. Molle, K. Sohraby, and A.N. Venetsanopoulos, "Space-Time Models of Asynchronous CSMA Protocols for Local Area Networks," *IEEE Journal on Selected Areas in Commun.*, Vol. SAC-5, No. 4, July 1987. A two-dimensional analysis of CSMA bus networks producing a more accurate performance than earlier time-dimensional analyses.

P.C. Ravasio, R. Marcogliese, and R. Novarese, "Voice Transmission over an Ethernet Backbone," *Local Computer Networks,* North-Holland, 1982. Experiments with packetized voice in an Ethernet. Delay and throughput are plotted with respect to the number of voice calls.

J.F. Shoch, Y.K. Dalal, D.D. Redel, and R.C. Crane, "Evolution of the Ethernet Local Computer Network," *IEEE Computer,* Vol. 15, No. 8, August 1982. Details of Ethernet specifications and design considerations.

2.2 Ring Networks

W.E. Burr, "The FDDI Optical Data Link," *IEEE Commun. Magazine,* Vol. 25, No. 5, pp. 18–23, May 1966. Physical layer of the FDDI protocol is illustrated with emphasis on transmitter and receiver devices and characteristics.

D. Dykeman and W. Bux, "Analysis and Tuning of the FDDI Media Access Control Protocol," *IEEE Journal on Selected Areas in Commun.,* Vol. SAC-6, No. 6, July 1988. An analysis of FDDI protocol with several priorities.

A. Hopper, S. Temple, and R. Williamson, *Local Area Network Design,* Addison-Wesley, 1986. Overview of local area networks and a detailed description of the slotted Cambridge ring network.

D. Karvelas and A. Leon-Garcia, "Performance Analysis of the Medium Access Control Protocol of the FDDI Token Ring Network," *Proc. GLOBECOM,* Hollywood, Florida, November 1988. An analysis of FDDI protocol with voice and data traffic.

F.E. Ross, "FDDI—A Tutorial," *IEEE Commun. Magazine,* Vol. 24, No. 5, May 1986. A description of the FDDI protocol.

K.C. Sevcik and M.J. Johnson, "Cycle Time Properties of the FDDI Token Ring Protocol," *IEEE Trans. on Software Eng.,* Vol. SE-13, No. 3, March 1987. A mathematical proof that token rotation time in the FDDI protocol is bounded.

3. METROPOLITAN AREA NETWORKS

D.P. Heyman, "Data-Transport Performance Analysis of Fasnet," *Bell System Technical Journal,* Vol. 62, No. 8, October 1983. An analysis of Fasnet performance with data traffic.

R.W. Klessig, "Overview of Metropolitan Area Networks," *IEEE Commun. Magazine,* Vol. 24, No. 1, January 1986. Overview of various topologies of MANs.

N.F. Maxemchuk, "Comparison of Deflection and Store-and-Forward Techniques in the Manhattan Street and Shuffle-Exchange Networks," *Proc. INFOCOM,* Toronto, April 1989. Delay characteristics are investigated for a deflection routing technique in two networks.

R.M. Newman, "Distributed Queuing: Performance Characterisation," *IEEE Computer Society 802.6 Documents,* March 1988. Performance of the DQDB protocol obtained by simulation.

C.S. Raghavendra and J.A. Silvester, "Double Loop Network Architectures—A Performance Study," *IEEE Trans. on Commun.,* Vol. COM-33, No. 2, February 1985. Performance analysis and examination of several routing algorithms in a double loop network.

4. BROADBAND ISDN

R.J. Boehm, "The SONET Interface and Network Applications," *Proc. GLOBECOM,* Hollywood, Florida, November 1988. Properties of the STS-1 (synchronous transport signal-level 1) frame format and examples of incorporating the SONET interface into the network hierarchy.

W.R. Byrne, T.A. Kilm, B.L. Nelson, and M.D. Soneru, "Broadband ISDN Technology and Architecture," *IEEE Network,* Vol. 3, No. 1, January 1989; see also *Proc. GLOBECOM,* Hollywood, Florida, November 1988. Descriptions of broadband interfaces and architectures of access and service nodes.

J.P. Coudreuse, W.D. Sincoskie, and J.S. Turner, "Broadband Packet Communications," *IEEE Journal on Selected Areas in Commun.,* Vol. 6, No. 9, December 1988. A special issue containing articles on various broadband switches, ATM architecture, and performance modeling. In particular, a nonblocking copy network is described. Layered ATM architectures are also discussed.

H. Ohnsorge, Guest Ed., "Broad-Band Communications Systems," *IEEE Journal on Selected Areas in Commun.,* Vol. SAC-4, No. 4, July 1986. A special issue outlining trends of various broadband communication components in individual countries.

P.E. White, J.Y. Hui, M. Decina, and R. Yatsuboshi, Guest Ed., "Switching Systems for Broadband Networks," *IEEE Journal on Selected Areas in Commun.,* Vol. SAC-5, No. 8, October 1987. A special issue containing articles on Starlite, Knockout, and other high-speed switches.